THE GENUS
LACHENALIA

Dedicated to my mother, Judy

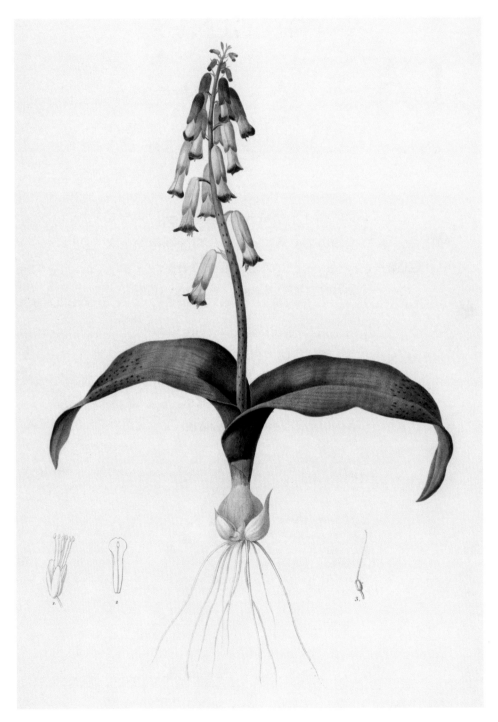

Coloured engraving of *Lachenalia aloides* by P. J. Redouté from *Les Liliacées* volume 1 (1802).

A BOTANICAL MAGAZINE MONOGRAPH

THE GENUS
LACHENALIA

Graham Duncan

WITH WATERCOLOURS BY
Winsome Barker, Fay Anderson, Ellaphie Ward-Hilhorst, Claire Linder Smith, Elbe Joubert, Vicki Thomas, Marieta Visagie and Rhona Collett

LINE DRAWINGS BY
Vicki Thomas

SCIENTIFIC EDITOR
Trevor Edwards

GENERAL EDITOR
Martyn Rix

For the right names of flowers are yet in heaven…
From Jubilate Agno by Christopher Smart (1722–1771)

Kew Publishing
Royal Botanic Gardens, Kew

©The Board of Trustees of the Royal Botanic Gardens, Kew 2012
Illustrations and photographs © the artists and photographers as stated in the captions

The author has asserted his right to be identified as the author of this work in accordance with the Copyright, Designs and Patents Act 1988.

All rights reserved. No part of this publication may be reproduced, stored in a retrieval system, or transmitted, in any form, or by any means, electronic, mechanical, photocopying, recording or otherwise, without written permission of the publisher unless in accordance with the provisions of the Copyright Designs and Patents Act 1988.

Great care has been taken to maintain the accuracy of the information contained in this work. However, the publisher, the editors and the author cannot be held responsible for any consequences arising from use of the information contained herein.

First published in 2012 by
Royal Botanic Gardens, Kew,
Richmond, Surrey, TW9 3AB, UK
www.kew.org

Distributed on behalf of the Royal Botanic Gardens, Kew in North America by the University of Chicago Press, 1427 East 60th Street, Chicago, IL 60637, USA

ISBN 978-1-84246-382-6

British Library Cataloguing in Publication Data
A catalogue record for this book is available from the British Library

Production editor: Sharon Whitehead
Cover design, typesetting and page layout: Christine Beard
Publishing, Design & Photography, Royal Botanic Gardens, Kew

Front cover: *Lachenalia sargeantii* painted by Elbe Joubert
Back cover: *Lachenalia violacea* painted by Claire Linder Smith

Printed in Spain by Grafos S.A.

For information or to purchase all Kew titles please visit
www.kewbooks.com or email publishing@kew.org

Kew's mission is to inspire and deliver science-based plant conservation worldwide, enhancing the quality of life.

Kew receives half of its running costs from Government through the Department for Environment, Food and Rural Affairs (Defra). All other funding needed to support Kew's vital work comes from members, foundations, donors and commercial activities including book sales.

CONTENTS

NEW TAXA PUBLISHED IN THIS WORK vi

LIST OF PAINTINGS vii

FOREWORD ix

ACKNOWLEDGEMENTS xi

PREFACE. xiii

1. HISTORY 1

2. CULTIVATION AND PROPAGATION
    Cultivation. 17
    Propagation 27
    Pests and diseases 29

3. LACHENALIAS AND THE ENVIRONMENT
    Conservation 33
    Phytogeography 35
    Ecology, habitat and adaptive strategies 41
    Phenology 50

4. LACHENALIA BIOLOGY
    Morphology. 57
    Pollination biology 79
    Seed dispersal 84
    Karyology 86

5. PHYLOGENY. 89

6. TAXONOMIC TREATMENT 95
    Insufficiently known names 444
    Excluded taxa 445

REFERENCES 447

GLOSSARY 461

APPENDIX. 464

GENERAL INDEX. 465

INDEX OF SCIENTIFIC NAMES 473

NEW TAXA PUBLISHED IN THIS WORK

Lachenalia argillicola G. D. Duncan, sp. nov.

Lachenalia bruynsii G. D. Duncan, sp. nov.

Lachenalia callista G. D. Duncan & T. J. Edwards, sp. nov.

Lachenalia canaliculata G. D. Duncan, sp. nov.

Lachenalia judithiae G. D. Duncan, sp. nov.

Lachenalia krugeri G. D. Duncan, sp. nov.

Lachenalia martleyi G. D. Duncan, sp. nov.

Lachenalia patentissima G. D. Duncan, sp. nov.

Lachenalia polypodantha Schltr. ex W. F. Barker subsp. *eburnea* G. D. Duncan, subsp. nov.

Lachenalia summerfieldii G. D. Duncan, sp. nov.

Lachenalia thunbergii G. D. Duncan & T. J. Edwards, sp. nov.

LIST OF PAINTINGS

Plate 1. *Lachenalia viridiflora* . . . . FAY ANDERSON (p. 111)

Plate 2. *Lachenalia orchioides* subsp. *glaucina* . . . . ELBE JOUBERT (p. 119)

Plate 3. *Lachenalia trichophylla* . . . . RHONA COLLETT (p. 122)

Plate 4. *Lachenalia vanzyliae* . . . . FAY ANDERSON (p. 132)

Plate 5. *Lachenalia aloides* . . . . WINSOME BARKER (p. 140)

Plate 6. *Lachenalia quadricolor* . . . . WINSOME BARKER (p. 145)

Plate 7. *Lachenalia flava* . . . . WINSOME BARKER (p. 147)

Plate 8. *Lachenalia callista* . . . . FAY ANDERSON (p. 154)

Plate 9. *Lachenalia punctata* . . . . WINSOME BARKER (p. 161)

Plate 10. *Lachenalia unifolia* . . . . WINSOME BARKER (p. 166)

Plate 11. *Lachenalia sargeantii* . . . . ELBE JOUBERT (p. 173)

Plate 12. *Lachenalia ventricosa* . . . . WINSOME BARKER (p.176)

Plate 13. *Lachenalia pusilla* . . . . WINSOME BARKER (p. 189)

Plate 14. *Lachenalia purpureo-caerulea* . . . . WINSOME BARKER (p. 202)

Plate 15. *Lachenalia mathewsii* . . . . ELLAPHIE WARD-HILHORST (p. 211)

Plate 16. *Lachenalia arbuthnotiae* . . . . ELLAPHIE WARD-HILHORST (p. 223)

Plate 17. *Lachenalia duncanii* . . . . CLAIRE LINDER SMITH (p. 230)

Plate 18. *Lachenalia latimeriae* . . . . WINSOME BARKER (p. 258)

Plate 19. *Lachenalia rosea* . . . . FAY ANDERSON (p. 265)

Plate 20. *Lachenalia martleyi* . . . . WINSOME BARKER (p. 271)

Plate 21. *Lachenalia salteri* . . . . FAY ANDERSON (p. 273)

Plate 22. *Lachenalia lutzeyeri* . . . . VICKI THOMAS (p. 276)

LIST OF PAINTINGS

Plate 23. *Lachenalia violacea* . . . . CLAIRE LINDER SMITH (p. 296)

Plate 24. *Lachenalia glauca* . . . . WINSOME BARKER (p. 300)

Plate 25. *Lachenalia zebrina* . . . . MARIETA VISAGIE (p. 305)

Plate 26. *Lachenalia juncifolia* . . . . WINSOME BARKER (p. 312)

Plate 27. *Lachenalia polyphylla* . . . . WINSOME BARKER (p. 318)

Plate 28. *Lachenalia convallarioides* . . . . FAY ANDERSON (p. 334)

Plate 29. *Lachenalia nervosa* . . . . CLAIRE LINDER SMITH (p. 347)

Plate 30. *Lachenalia macgregoriorum* . . . . ELLAPHIE WARD-HILHORST (p. 360)

Plate 31. *Lachenalia patula* . . . . WINSOME BARKER (p. 371)

Plate 32. *Lachenalia mutabilis* . . . . WINSOME BARKER (p. 380)

Plate 33. *Lachenalia verticillata* . . . . FAY ANDERSON (p. 386)

Plate 34. *Lachenalia peersii* . . . . VICKI THOMAS (p. 389)

Plate 35. *Lachenalia karoopoortensis* . . . . WINSOME BARKER (p. 395)

Plate 36. *Lachenalia sessiliflora* . . . . FAY ANDERSON (p. 404)

Plate 37. *Lachenalia congesta* . . . . MARIETA VISAGIE (p. 409)

Plate 38. *Lachenalia valeriae* . . . . FAY ANDERSON (p. 415)

Plate 39. *Lachenalia namaquensis* . . . . WINSOME BARKER (p. 423)

FOREWORD

As a botanist who has studied and cultivated bulbous plants for many years, I am always keen to delve into any recently published books on the subject to see if the authors have any interesting new angles, such as improved growing techniques. Usually these broad-spectrum books on bulbs follow a standard pattern with an A–Z of genera, species and cultivars and as such are informative and make useful additional reference works for the library. Much more exciting for me, however, are monographs in which one genus or group of plants has been selected for detailed study, resulting in a definitive account. Maybe monographs might be viewed as stuffy botanical works by those with a more general interest in the plant world, but it is these that supply the core of information on which the other wide-ranging books are based; and they certainly need not be stuffy!

Such is Graham Duncan's monograph of *Lachenalia*. There is certainly no-one better qualified to prepare this monograph. Graham has studied the genus in the wild and in cultivation for over 30 years and many will be familiar with his earlier publication, *The Lachenalia Handbook*, published in 1988. This undoubtedly gave the genus a boost of popularity and the present work will almost certainly take this to new heights. Lachenalias have for a long time been appreciated for their great range of form and colour as pot plants or for their garden use in favoured climates; hybridisation and selection continues in the quest to extend the range for horticulture and the cut-flower industry. Sadly, although some *Lachenalia* species have been grown for perhaps two centuries in Britain, they are just not hardy enough for outdoor cultivation and must be kept frost-free to survive. For gardeners in colder regions, however, there might be a breakthrough: the author identifies several species as being fully hardy.

The extraordinary degree of plant endemism and diversity in Southern Africa, particularly in the Cape region, is well known and many genera demonstrate this in the form of a proliferation of locally distributed species. *Lachenalia* is a prime example with 133 species defined in this new revision. *The Genus Lachenalia* is a work that botanists and bulb enthusiasts alike will appreciate: it has all the elements of a thorough, modern taxonomic monograph and takes into account a great fund of data, covering everything from the visually satisfying morphology to the more cryptic but fundamental cytological and molecular information. There is much more besides as this account is written by a specialist botanical researcher with a profound knowledge of the living plant and a great deal of hands-on experience at growing lachenalias and many other bulbs at Kirstenbosch.

In my experience, it is always worth taking note of the observations of people — be they professional or amateur, botanists or gardeners — who actually cultivate the plants they are studying, for they often notice subtleties that might well be missed by others. In the author's previous work on the genus, *The Lachenalia Handbook*, Miss Winsome F. Barker provided an introductory history of the genus and quoted an alleged saying by Jacquin that "a genus is not known until all its species are perfectly understood". In preparing this work, it would appear that Graham Duncan very much took this as his objective and the result is an exemplary account of this large and complex genus. That it is well researched and presented is no surprise as, in my years as editor of *Curtis's Botanical Magazine*,

Graham's papers always required little or no work to prepare them for publication. I am delighted to see that a recent issue of the journal (December 2011) is devoted to South African plants and that of the twelve papers presented there, nine are contributed by him.

One must of course also applaud the fact that the book is very well illustrated with photographs, for these convey so much about the plant's actual appearance and bring to life the written descriptions. The artwork adds another dimension, presenting the plants as seen through the eyes of a wealth of South African botanical illustrators past and present. It is appropriate that Kew should be publishing the new monograph. The first definitive account of the genus was by the Kew botanist John Gilbert Baker who published an extensive treatment in *Flora Capensis* in 1897.

Brian Mathew

ACKNOWLEDGEMENTS

Many people have assisted me in the completion of this work. I am particularly grateful to Dr Martyn Rix, Editor of *Curtis's Botanical Magazine*, for his initial support of my approach to Kew with a view to publication and for being overall editor. Dr Trevor Edwards, previously Curator of the Bews Herbarium at the University of KwaZulu-Natal in Pietermaritzburg, South Africa, now of the Botany Department at La Trobe University in Victoria, Australia, kindly performed the task of scientific editor. I also wish to record my sincere thanks to Dr Peter Bruyns, honorary researcher in the Botany Department at the University of Cape Town, for his advice on nomenclature and for compiling the Latin translations of the diagnoses. Mr Steve Bales, Group Art Custodian at FirstRand Bank, and Ms Mary Slack, provided some much-appreciated financial support.

I thank the Curators of herbaria (B, BM, BOL, C, GRA, K, NBG, PRE, SAM, TCD and WIND) who sent photographs, provided material on loan, or allowed me to study their collections on site. I am especially grateful to: Dr Koos Roux, Curator of the Compton Herbarium at Kirstenbosch, who assisted me in many ways; Terry Trinder-Smith, Curator of the Bolus Herbarium at the University of Cape Town; Dr Paul Wilkin, Curator of the monocot wing at the Kew Herbarium; Dr Robert Vogt, Senior Curator of the Botanical Garden and Botanical Museum of Berlin-Dahlem at the Free University of Berlin; Prof. John Parnell, Curator of the Trinity College Dublin Herbarium; Prof. Olof Ryding, Curator of the University of Copenhagen Herbarium; John Hunnex of the Natural History Museum, London; Bruno Erny, Curator of the Botanical Garden at the University of Basel, Switzerland; and Bruno Matter, formerly horticulturist at the latter institution.

I have been greatly assisted by a number of people who shared their discoveries, often taking me back to localities where they had observed *Lachenalia* plants to study, photograph and collect material. I am especially thankful to Gordon Summerfield, Cameron and Rhoda McMaster, Gerard Hansford, Ernst van Jaarsveld and Adam Harrower. Amongst those that have passed on, I acknowledge especially Margaret Thomas (1917–2006), a seasoned field collector and propagator of *Lachenalia*, and Prof. M. C. Botha (1922–2002), who in the 1970s maintained one of the most comprehensive living collections of *Lachenalia* species ever assembled.

I thank the following friends, colleagues and relatives who kindly assisted me in one way or another: Fay Anderson, Susyn Andrews, Liz Ashton, Jill Attwell, Fanie Avenant, Steve Bales, John Bell, Bertha Blackwood-Murray, Laurian Brown, Mike Clery, Lita Cole, Dr Jonathan Colville, Gillian Condy, Carly Cowell, Neil Cox, Charles Craib, Hildegard Crous, Dr John David, David Davidson, James Deacon, Phillip Desmet, Clifford Dorse, Mark and Carole Duckitt, the Duncan family, Prof. Niel du Plessis, Dr Connal Eardley, Dirk and Esna Ehlers, Mandy Fick, Patrick Fraser, Alan Garlick, Stephen Gibson, Gianpaolo Gilardi, Mary Gould, Thys Greeff, Dr Alberto Grossi, Nick Helme, Anthony Hitchcock, Mick Hoon, Prof. Steven Johnson, Elbe Joubert, Dr Peter and Barbara Knox-Shaw, Rupert Koopman, Barbara Kotze, Clement Kotze, Paul Kruger, Hubert Kurzweil, John Lavranos, Hugo Leggatt, Annelise le Roux, Caryl Logie, William Liltved, Heiner Lutzeyer, Christien Malan, Dr John Manning, Thomas Mihal, Linda Montgomery, Sandra

Muller, Alice Notten, Dr Ted Oliver, Tessa Oliver, Elizabeth Parker, Dr Mike Picker, Helene Preston, Dr John Rourke, Albertus and Christien Roux, Dr Brian Schrire, Anne Scott, Malcolm Shennan, Mary Slack, Michelle Smith, Dr Dee Snijman, Hannelie Snyman, Geert Sprangers, Hester Steyn, Nicolette Stoll, Vicki and Robbie Thomas, Dennis Tsang, Olivia Tyambetyu, James van Heerden, Ivan van Niekerk, Wouter and Judy van Warmelo, Pieter van Wyk, Cronjé and Antonette Visagie, Marieta Visagie, Caroline Voget, Leigh Voigt, Caitlin von Witt, Dr Piet Vorster and Odile Weber.

Finally, my grateful thanks to Dennis Tsang for his assistance with proof reading, and to Gina Fullerlove, Lloyd Kirton, John Harris, Lydia White, Sharon Whitehead and Christine Beard at Kew Publishing for their advice and support throughout the project.

PREFACE

More than a century has passed since the Kew botanist J. G. Baker's second and final monograph of *Lachenalia* appeared in 1897 in *Flora Capensis*. In that work, he recognised 42 species, eight of which (*L. bowieana* Baker, *L. latifolia* Tratt., *L. lilacina* Baker, *L. massonii* Baker, *L. pustulata* Jacq., *L. rhodantha* Baker, *L. succulenta* Masson ex Baker and *L. unicolor* Jacq.) are now considered synonymous with other species that he recognised at that time. *L. glaucina* Jacq. is now a subspecies of *L. orchioides* (L.) Ait. and *L. cooperi* Baker is presumed to be a hybrid. Nevertheless, the complement for the genus is now more than three times that published by Baker: 133 species, comprising 139 taxa.

The present monograph is based partly on an M.Sc. study that focused on morphological character variation within *Lachenalia* and a cladistic analysis of the genus, completed at the University of KwaZulu-Natal, Pietermaritzburg, South Africa, under the supervision of Prof. T. J. Edwards (now of the Botany Department, La Trobe University, Victoria, Australia) (Duncan, 2005; Duncan *et al.*, 2005a). It has also involved extensive field work, observation of the species in cultivation over a period of more than 30 years and a comprehensive study of preserved herbarium material. This thesis contains descriptions of, distribution maps for and illustrations of all *Lachenalia* taxa recorded in South Africa and Namibia.

A vast literature has been published on *Lachenalia* over a period of more than 320 years and, in accordance with the rules contained in the *International Code of Botanical Nomenclature*, a number of familiar names have given way to earlier ones, yet the changes instituted here are not excessive. In addition, a number of taxa have been relegated to synonymy whereas others have been upgraded to subspecific or specific level, the latter category including several first published in the late 18th century that are here returned to their original status.

From a horticultural view, it is natural to want to retain as many names as possible in order to easily distinguish the many different forms that exist within certain species. In this respect, I have adopted an approach that is as 'centrist' as possible, veering away from excessive fragmentation and lumping of species. *Lachenalia* has long been unfairly regarded as a genus of numerous similar-looking members that are difficult or impossible to identify; but the majority of species have clear-cut morphological differences that make their identification straightforward. Identification difficulties might arise amongst several species 'complexes', but the keys, illustrations and species descriptions included here will provide readers with the tools necessary to overcome these problems. It is essential to appreciate that similar-looking species are distinguished from one another not by a single morphological difference, but by a unique combination of morphological features that constitute a particular species concept.

In illustrating a genus as speciose as *Lachenalia*, it was not feasible to have all members painted by a single artist. I have been fortunate in having a number of competent and highly talented contemporary artists contribute paintings to add to the many executed by W. F. Barker during the early 20th century. The result is an eclectic mix of styles, complemented by photographs of *Lachenalia* plants taken in habitat and under cultivation.

I was most fortunate to have known Winsome F. Barker for 15 years, from 1979 until her passing in 1994, and she taught me the 'basics' of *Lachenalia*. She was unable to accompany me on field trips, but her regular visits to the *Lachenalia* collection in the Kirstenbosch Bulb Nursery, and the many fruitful discussions we had, enabled me to get to grips with an at times, bewildering genus. Winsome Barker had always intended to produce a monograph, but despite 59 years of research (her final paper appeared in 1989, when she was 82 years old), her advancing years and perhaps a lack of decisiveness led her to believe that she could not manage it. She did, however, lay part of the foundation for the present work in her extensive field collecting and meticulous preservation of pressed material, housed mainly in the Compton Herbarium (Kirstenbosch) and the Bolus Herbarium (University of Cape Town).

The context of some early species names can only be surmised, as is the case with William Aiton's *L. contaminata*, a name that could have referred to that plant's spicy scent, reddish tepal markings, or spotted scape and upper leaf surfaces; similarly, Thunberg's *L. reflexa* might refer to the leaves that sometimes become reflexed or to the scape that elongates in fruit and becomes reflexed. The descriptive Afrikaans colloquial name 'viooltjie' (little violin) accurately alludes to the squeaky sounds that emanate from *Lachenalia* flower stems when they rub together. By contrast, the English colloquial name for *Lachenalia* 'Cape cowslips' is misleading; none of the species remotely resemble cowslips (genus *Primula*). The registered trade name for *Lachenalia* hybrids developed in South Africa is 'Cape hyacinth'.

During many years of field work, I have experienced a number of highs and lows. Highlights include the first sight of thousands of the 'Critically Endangered' *L. mathewsii* in full flower near Vredenburg in 1985, the species having been considered extinct for almost 40 years; seeing the equally elusive *L. sargeantii* near Napier in 2004, flowering for the first time after a fire, 33 years after its last sighting in bloom; and in 2006, seeing *L. flava* at a new locality, a burnt hillside turned orange near Tulbagh. Searching for the inconspicuous *L. maximiliani* on the Pakhuis Pass was initially unsuccessful, but we later found that we had inadvertently parked on top of it. Having searched intensively for hours, an elusive plant would often be spotted immediately after we had given up all hope of finding it.

Disappointments have included recording the loss of numerous *Lachenalia* populations owing to agricultural expansion and urbanisation, including the type populations of *L. physocaulos* and *L. viridiflora* near Robertson and Vredenburg, respectively. Less traumatic has been the marking of promising specimens of *L. congesta* in bud near Sutherland and *L. stayneri* at Worcester for later photography, only to find them grazed to the ground by wild buck or hares, and attempting to photograph *L. polyphylla* at Gouda amid a blizzard of spring midges ('miggies' in Afrikaans). Certain species were only tracked down in the wild after numerous attempts. I searched for *L. glaucophylla* around Calvinia, *L. polypodantha* in Namaqualand and *L. xerophila* in Bushmanland for many years, but was unable to locate them partly because of insufficient rains and misleading locality records; all three were eventually located at new localities. Frequently, I was just too early or just too late to take photographs. Habitat photographs have been used in this book wherever possible, but where these were not available, photographs of cultivated plants have been substituted.

An understanding of a genus is greatly enhanced by observation of plants that have been in cultivation under uniform conditions for many years. Growing taxa from seed to flowering makes it possible to assess genotypic and phenotypic traits, and in this respect I have relied on the comprehensive living collection of *Lachenalia* species in the Kirstenbosch Bulb Nursery at the South

African National Biodiversity Institute. In the UK, the National Collection of *Lachenalia* was, for many years, ably maintained in Kent by Ben Clifton. Following his recent passing, this collection continues to be cherished in the hands of competent staff at Exbury Gardens in South Hampshire.

Mysteries persist. Winsome Barker discovered *L. macgregoriorum* in November 1962 just east of Nieuwoudtville, but the plant has never been seen in the wild again, despite exhaustive searches; fortunately, it survives in cultivation. I have been unable to trace, on any map, James Bowie's locality for his specimen of *L. nervosa* preserved in London's British Museum (Natural History). Bowie gave the locality as 'near Samson's River in the district of George', potentially the most easterly record for *L. nervosa*. Another great mystery remains unsolved: how do the bulbs of the three pyrophytic species *L. lutzeyeri*, *L. montana* and *L. sargeantii* remain completely dormant for up to 30 years or more and then flower within nine months of a bush fire?

Amongst gardeners and specialist bulb growers the world over, *Lachenalia* is more popular now than it has ever been. Members of the genus are ideal subjects for containers in temperate greenhouses or windowsills in cold climates, and for patios, rock gardens and flower beds in milder parts.

Ten new species and one new subspecies are described here; further discoveries and new distribution records will undoubtedly be made as remote areas become more widely explored. This monograph should not be regarded as the final answer, as much has yet to be achieved in obtaining a well-supported phylogeny for *Lachenalia*, but it represents a snapshot of the genus as we know it now that will hopefully assist readers in identifying and growing their plants.

Graham Duncan
Cape Town, May 2012.

1. HISTORY

The earliest record of a *Lachenalia* species is in a manuscript from the late 17th century, presently housed in the Trinity College Library in Dublin (TCD). Acquired from the library of Baron Hendrick Fagel, Secretary to the States General and Pensionary of Holland, in London in 1802, the manuscript was discovered in 1922 by Gilbert Waterhouse (1888–1977), Professor of German at the University of Dublin. It contains a watercolour painting, said to be by Heinrich Claudius, of the first species to be illustrated, now known as *L. hirta* (Thunb.) Thunb.. This painting had been included in the journal of Simon van der Stel, Dutch governor of the Cape of Good Hope, to illustrate his expedition to Namaqualand between 1685 and 1686. As described by Waterhouse (1932), the Trinity College manuscript is exactly the same section as that removed in about 1691–1692 from the Archives of the Dutch East India Company, and is widely considered to be the one Simon van der Stel used. The plant was found on 10th September 1685 and the painting can therefore be taken to be the earliest record, to which a definite date can be assigned, of the genus in colour (Figure 1).

Copies of the journal and its paintings were made, some of which were sent to Holland while others were kept at the Cape; one of two copies still in the Cape is the collection of 78 watercolours assembled in 1692, having been commissioned by Nicolaas Witsen, Mayor of Amsterdam, that became known as the *Codex Witsenii*. This collection is housed in the Library of the South African Museum in Cape Town. It includes a rather crude copy of the *L. hirta* painting described by Barnard (1947) and later reproduced in colour (Wilson *et al.*, 2002). A more pleasing rendition of the painting appears in another Codex housed in the South African Public Library collection in Cape Town (Kerkham, 1992a, 1992b). A further rendition of the *L. hirta* painting appeared in a manuscript by Jan Commelin, published in Cape Town and presently in the possession of the Staatsbibliothek Preussischer Kulturbesitz in Berlin. Unusually, the plants in this painting are set against a landscaped backdrop, with the bulb shown separately (Wijnands *et al.*, 1996).

Figure 1. Watercolour painting of *Lachenalia hirta*, probably by Heinrich Claudius, reproduced from Simon van der Stel's Journal of his Expedition to Namaqualand, 1685–6, courtesy of The Board of Trinity College, Dublin (TCD MS 984 fol. 124).

THE GENUS LACHENALIA
HISTORY

The first published figure of a *Lachenalia* plant appeared as a stylised copy of the familiar *L. hirta* painting in 1692, as figure 5 in Leonardi Plukenett's *Phytographia*. There, the taxon was described by the phrase name *Hyacinthus Africanus, Orchioides serpentarius, folio singulari, undato, pilisciliaribus, fimbriato, floribus ex aureo punicantibus*. Following his description, the words *Codicis Comptoniani* indicate that the painting had been taken from the Codex of the Right Reverend Bishop Henry Compton of London. Compton had visited Amsterdam in 1691 and had been given a personal copy of the *Codex Witsenii* containing 100 paintings, which were said to have been undertaken by Heinrich Claudius at the Cape, including the *L. hirta* figure (Gunn & du Plessis, 1978). Copies of the paintings in the Compton Codex were used over a period of two decades in the works of Plukenett and the London apothecary James Petiver. The latter published another version of the well-known *L. hirta* painting in his *Gazophylacii Naturae & Artis seu Herbarium Capense* (Petiver, 1709). The present whereabouts of the Compton Codex, if it still exists, are unfortunately unknown (Barker, 1988).

Another 17th century collection of over 300 Cape paintings, kept in South Africa and thought to be part of the *Codex Witsenii*, is the magnificent volume *Icones Plantarum et Animalium*, which is housed in the Africana Museum in Johannesburg. This collection was described in the *Journal of South African Botany* and contains three *Lachenalia* illustrations, the familiar copied figure of *L. hirta*, an excellent rendition of *L. bifolia* (= *L. bulbifera*) and *L. orchioides* subsp. *glaucina* (Macnae & Davidson, 1969).

Figure 2. Watercolour painting of *Lachenalia orchioides* subsp. *orchioides*, artist uncertain, possibly Heinrich Claudius, reproduced from *The Flora Capensis of Jakob and Johann Philipp Breyne*, courtesy of The Brenthurst Library, Johannesburg (ART.83/29).

In 1729, the German Johann Christian Buxbaum published his *Plantarum minus cognitarum Centuria III* with figures and phrase names for three lachenalias now known as *L. orchioides* subsp. *glaucina*, *L. bifolia* and *L. punctata*. He made no mention of the source of his material but provided long descriptions of the plants, and his figure 19 is the neotype for *L. bifolia*.

Amongst the watercolour paintings completed in about 1700 for the early 18th century Florilegium *The Flora Capensis of Jakob and Johann Philipp Breyne* were paintings of three members of the genus *Lachenalia*, now known as *L. orchioides* subsp. *orchioides* (Figure 2), *L. orchioides* subsp. *glaucina* and *L. contaminata*. The Florilegium was bound in 1724 and is currently housed in The Brenthurst Library in Johannesburg, having been published for the first time by The Brenthurst Press in 1978. Some of the paintings of the Florilegium could be those of the German artist Heinrich Claudius, who visited the Cape between 1681 and 1686 and became one of the best-known flower painters of his time, but none are signed by him and they cannot be attributed to him with certainty (Gunn & du Plessis, 1978). Three other collections of paintings include copies of these three lachenalias: one is housed in the William Sherard Collection in the Bodleian Library at Oxford University (Edwards, 1964); another is a volume in the Mary Gunn Library of the South African National Biodiversity Institute in Pretoria (Jessop, 1965) that includes a fourth species, *L. punctata*; and the third is a collection of unbound paintings in the Rijksherbarium at Leiden, The Netherlands. In 1739, the three familiar *Lachenalia* paintings appeared on the same engraved plate of the Breynes' subsequent work, *Prodromi Fasciculi Rariorum Plantarum*, in which figures 1, 2 and 3 represent *L. orchioides* subsp. *glaucina*, *L. orchioides* subsp. *orchioides* and *L. contaminata*, respectively (Figure 3). Introducing his binomial system of classification for plants and animals in 1753, Linnaeus cited figure 2 of this work as the iconotype of *Hyacinthus orchioides* L.. This figure matches the painting in *The Flora Capensis of Jakob and Johann Philipp Breyne*, and Linnaeus described the first taxon now included in the genus *Lachenalia* as follows:

Orchioides. 11. HYACINTHUS *corollis irregularibus sexpartitis.*
Hyacinthus orchioides africanus major bifolius maculatus, flore sulphureo obsoleto majore. Breyne. prodr. 3. p. 24. t. 11. f. 2. Habitat in Aethiopia.

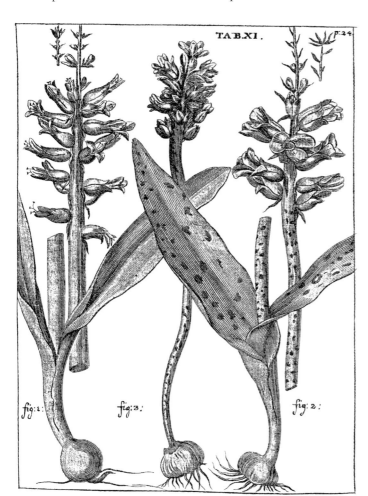

Figure 3. Engraved plate from Jakob and Johann Philipp Breyne's *Prodromi Fasciculi Rariorum Plantarum* (1739), illustrating the three lachenalias from the Codex *The Flora Capensis of Jakob and Johann Philipp Breyne* in The Brenthurst Library. Figure 1 represents *Lachenalia orchioides* subsp. *glaucina*, figure 2 represents *Lachenalia orchioides* subsp. *orchioides* [cited by Linnaeus as the iconotoype of *Hyacinthus orchioides* in his *Species Plantarum* (1753)], and figure 3 represents *Lachenalia contaminata*.

Nicolaus Joseph Jacquin (after 1774 known as Baron von Jacquin) was born to a French family at Leiden, Holland on 16th February 1727 (Figure 4). He studied medicine in Holland, France and Austria but his real interest lay in botany, which he pursued through frequent visits to the University Botanical Garden and the newly laid out Imperial Gardens at Schönbrunn, south-west of Vienna. It was while at Schönbrunn that Jacquin was introduced to Emperor Francis I who, on recognising his talent, requested that he catalogue the plant collections in the Imperial Gardens. Subsequently, the Emperor requested that Jacquin should lead a plant-collecting expedition to Central America and the West Indies, which took place from 1754 to 1759, to augment the collections at Schönbrunn. Following his return to Vienna, Jacquin published his first work, *Enumeratio stirpium in agro Vindobonensis*, which dealt with the native flora of Vienna and appeared in 1762. For the next five years, he occupied the Chair of Chemistry at Schemnitz, Hungary. Although they never met, Jacquin kept up a close correspondence with Linnaeus and became one of the staunchest supporters of the Linnaean binomial system. In 1768, he returned to Vienna, becoming Professor of Botany and Chemistry, and Director of the University Botanical Gardens. He held the latter position for 29 years until his retirement in 1797. Jacquin published several sumptuous volumes in his *Hortus Botanicus Vindobonensis* in the 1770s. These contained hundreds of plates that illustrated some of the rarer plants grown in the Botanical Garden (Garside, 1942).

Figure 4. Baron Nicolaus Joseph von Jacquin (1727–1817), reproduced from a painting in the University of Vienna, courtesy of Hunt Institute for Botanical Documentation, Carnegie Mellon University, Pittsburgh, PA.

Following the death of the Empress Maria Theresa, the new Emperor Joseph II, on the advice of Jacquin, sent the Austrian gardeners Francis Boos and George Scholl to collect plants for Schönbrunn on Mauritius and at the Cape. They arrived at the Cape in May 1786. Scholl sent many seeds and bulbs back to Schönbrunn via the Austrian Consul in Amsterdam. These specimens were used to make illustrations of many new lachenalias that were published in Jacquin's works from 1781 to 1795, mainly in volume 2 of his *Icones Plantarum Rariorum* (Schubert, 1945). Volume 2 contained 22 colour plates of *Lachenalia* including '*Lachenalia lanceaefolia*', now a synonym of *Ledebouria revoluta*, and *Polyanthes pygmaea*, which later became *Polyxena pygmaea* and is now *Lachenalia pygmaea*. The hand-coloured plates in this volume are unfortunately unsigned and the descriptions in Jacquin's *Collectanea* volumes, in which most of the descriptions were published, are without precise localities, the source given simply as the 'Cape of Good Hope'. Although Jacquin himself was an accomplished artist, very few of his own plates were ever published. He died in Vienna on 26th October 1817 at

Figure 5 (above). Lithograph of Baron Joseph Franz von Jacquin (1766–1839) by Josef Kriehuber, 1830.

Figure 6 (right). Hand-coloured lithograph of *Lachenalia tricolor* J. Jacq. (= *L. aloides*) by Joseph Hofbaur, reproduced from volume 1 of N. J. Jacquin's *Icones Plantarum Rariorum* (1781–1786).

the age of 90 (Garside, 1942). The Jacquin Herbarium that included the *Lachenalia* specimens used to illustrate volume 2 of N. J. Jacquin's *Icones Plantarum Rariorum* was housed in the Imperial Museum of Natural History in Vienna. During World War 2, the collection was divided into lots and stored at various less prominent locations in Vienna, but tragically all of the Liliaceae material, including the *Lachenalia* specimens, many of them holotypes, were destroyed by fire sometime during that period (Peter Bruyns, pers. comm.).

Jacquin's only son Joseph Franz, later Baron von Jacquin (Figure 5), was born at Schemnitz, Hungary on 17th February 1766 and qualified as a medical doctor at the Univeristy of Vienna in 1788. He assisted his father in his botanical pursuits from an early age, succeeding him as Director of the University Botanical Gardens, and died in Vienna in 1839 at the age of 74. Although his scientific publications were relatively few, it was he who, in 1780 at the age of just 14, coined the genus *Lachenalia* (perhaps on the advice of his father), naming it after Prof. Werner de Lachenal. The first illustration (by Joseph Hofbaur) of a plant under that genus was published on plate 61 in fascicle 1 of volume 1 of his father's *Icones Plantarum Rariorum*, in 1781, with 'Act. Helv. Vol. 9' written below it (Schubert, 1945) (Figure 6). Jacquin described the species as *L. tricolor* in respect of its three-coloured flowers and sent his manuscript to the editor of the Swiss journal *Acta Helvetica*, expecting it to be printed in volume nine of that year, but the volume was never published as the journal had ceased. The manuscript was actually published seven years later, accompanied by a figure on tab. II numbered a–d illustrating (a) an outer tepal, (b) an ovary and style, (c) an inner tepal and stamen and (d) a single flower, in the first volume of the newly styled *Nova Acta Helvetica* (Jacquin, 1787). In the meantime, J. A. Murray, evidently having seen Jacquin's manuscript and unwittingly anticipating Jacquin's publication of the genus *Lachenalia* and the species *tricolor* in 1780, included a brief description of *Lachenalia* in the 14th edition of *Linnaeus Systema Vegetabilium*, published in June 1784, citing the type species as *L. tricolor* (Murray, 1784). While Murray's publication of the genus was unintentional, he was the first to do so and the correct citation for it is therefore *Lachenalia* J. Jacq. ex Murray.

Werner (also known as Wernerus and Wernhardus) de Lachenal was a celebrated Swiss botanist, pharmacist and medic, born in Basel on 23rd October 1736. He studied anatomy and philosophy in Basel, and medicine in Montbéliard, France, becoming a Doctor of Medicine in 1763. In 1759, he published his first work *Specimen inaugurale Observationum botanicarum*, followed by *Observationes botanico-medicae* in 1776. De Lachenal was Professor of Anatomy and Botany at the University of Basel from 1776 to 1798 (Barnhart, 1965). He served as Dean of the medical faculty numerous times and maintained a voluminous correspondence with the Swiss anatomist and physiologist Professor Albrecht von Haller (1708–1777) of Berne. Von Haller was one of the intellectual giants of the 18th century, who proposed an alternative system of nomenclature to the binomial one of Linnaeus and whose magnificent *Flora of Switzerland* appeared in 1742. Werner de Lachenal died at Basel on 4th October 1800. A sandstone bust of him stands in the Botanical Garden at the University of Basel and his portrait hangs in the Naturhistoriches Museum there (Figure 7).

In July 1781, Carl Linnaeus junior published the second edition of his *Supplementum Plantarum et Specierum Plantarum*, including a description of *Phormium aloides* L.f., which later became *Lachenalia aloides* (L.f.) Engl. On page 336 of his 1784 publication, J. A. Murray included *Phormium aloides*, citing *Linnaeus's Supplementum Plantarum* (1781), evidently not realising that the names *P. aloides* and *L. tricolor* represented the same species.

In his *Dissertatio de Novis Generibus Plantarum*, C. P. Thunberg (1784) included four species under the genus *Phormium*: *P. tenax*, the 'New Zealand Flax'; *P. aloides*, including three varieties now recognised as good species; *P. orchioides*, including five varieties, three of which are good species; and *P. hirtum*, the plant illustrated in Simon van der Stel's journal in 1685 or 1686. A decade later, Thunberg revised his thoughts on the genus in his *Prodromus Plantarum Capensium* (1794), including these species under the generic name *Lachenalia* and elevating one of his varieties to species level as *L. reflexa*.

In 1787, William Curtis founded *The Botanical Magazine*, later to become *Curtis's*

Figure 7. Portrait of Professor Werner de Lachenal (1736–1800), courtesy of the Naturhistoriches Museum, Basel. Image: Bruno Matter.

Botanical Magazine, the longest continually published magazine of its kind and still appearing regularly after well over two centuries. Collectively, the many volumes of *Curtis's* contain the largest number of *Lachenalia* plates of any publication, the first of which appeared on plate 82 of volume 3 in 1789 as *L. tricolor*, not the plant described by J. Jacquin in 1780 but the species now known as *L. luteola*.

In 1799, a new edition of Linnaeus's *Species Plantarum* appeared and included more than 20 lachenalias. This was quickly followed in 1805 by Dr C. H. Persoon's *Synopsis Plantarum*, which included 25 species and the first (but invalid) attempt at combining *Phormium aloides* under *Lachenalia*. The year 1811 saw the publication of the enlarged second edition of volume two of W. T. Aiton's *Hortus Kewensis*, which included more than three times the number of recognised *Lachenalia* species contained in his father W. Aiton's volume one (1789).

The French biologist Jean-Baptiste Lamarck (1744–1829) wrote much of the botanical account for the *Encyclopédie Méthodique*, published between 1782 and 1832, including that of *Lachenalia* which included detailed descriptions of 22 species, each one with a French translation or description of the Latin specific epithet (Lamarck, 1813). In 1814, the second volume of *Archiv der Gewachskunde* by the Austrian Leopold Trattinnick (1764–1849) appeared, this volume included his new species *L. latifolia* (now included under *L. nervosa*) and was illustrated with 31 black-and-white copies of artworks taken from other publications.

In 1843, the German C. S. Kunth (1788–1850) published his *Enumeratio Plantarum*. He recognised 35 *Lachenalia* species, arranging them into two main groups, those with campanulate and those with tubular flowers, the former group subdivided into those with subsessile flowers and those with distinct pedicels. He treated the plant now known as *Lachenalia reflexa* under the separate genus *Coelanthus* as *C. complicatus*, as originally described by Willdenow. He also established the new genera *Periboea* and *Polyxena*, named after nymphs of Greek mythology (Periboea, the wife of Telamon, and Polyxena, the daughter of Priam) and both now included within *Lachenalia*.

The Frenchman C. A. Lemaire (1800–1871), a prominent botanist and editor of the monthly botanical journal *L'Illustration Horticole* published in Ghent, described and depicted remarkable and newly discovered plants with magnificent full-page chromolithographs and black-and-white woodblock illustrations. From 1855 to 1856, he transferred 20 *Lachenalia* species to the genus *Scillopsis* and a number of others to *Orchiastrum*, all of which were later reinstated under *Lachenalia* by J. G. Baker at Kew (1897a).

Richard A. Salisbury (1761–1829), the English botanist whose *The Genera of Plants* was published in 1866, recognised 27 species and divided *Lachenalia* into six genera (*Himas*, *Platyestes*, *Monoestes*, *Chloriza*, *Orchiops* and *Lachenalia*) under Section 1 of his Order Lachenaleae, three of which (*Lachenalia*, *Orchiops* and *Chloriza*) were recognised by Baker (1871) as subgenera in his first monograph.

John Gilbert Baker, who produced the only two monographs of *Lachenalia* published prior to the current work, was born in Cleveland, Yorkshire on 13th January 1834 (Figure 8). His interest in natural history, plants in particular, was evident from a very young age and he was appointed curator of the school herbarium in 1847. That same year, he left school to join his father in business at Thirsk in northern Yorkshire. He became a prolific author and by the age of 20, his knowledge of the plants of Yorkshire had grown to such an extent that he was able to collaborate with the moss expert J. Nowell in issuing a supplement to the *Flora of Yorkshire* (1854). In 1859, he became curator and secretary of the Botanical Exchange Club, which distributed plants around Britain; and in 1863, he published an authoritative work on the botany, geology, climate and physical geography of North Yorkshire. In 1864, Baker published a review of the British roses, and the following year, a monograph of British mints. Later that year, a fire at his business premises at Thirsk, Yorkshire completely destroyed his botanical library and herbarium, and the bulk of his work on North Yorkshire. Persuaded by friends to abandon his business interests in order to concentrate on scientific studies, Baker was fortunate in being invited by Dr Hooker (later Sir Joseph), Director of the Royal Botanic Gardens, Kew, to assist in completing the manuscript *Synopsis Filicum*, begun by Sir William Hooker, immediate past Director of Kew and father of Joseph Hooker. In January 1866, Baker took up duties as first assistant in the Kew Herbarium, a position he held until his appointment as Keeper of the Library and Herbarium in 1890. He served in this position until his retirement in 1899 at the age of 65, after which he continued with private studies at Kew. Baker completed *Synopsis Filicum* in 1868, and such was its popularity that it was re-issued in 1874 (Prain, 1921).

Baker produced several other important works on ferns and roses, and became an authority on genera and families of monocotyledonous plants, publishing numerous scientific papers in the *Journal of Botany* while at the same time contributing popular articles on the cultivation of monocotyledons to *The Gardener's Chronicle*. Between 1870 and 1880, he completed a monograph of the family Liliaceae *sensu lato* for the *Journal of the Linnean Society* (Botany). This included the genus *Lachenalia*, in which he recognised 29 species. He placed Willdenow's *Coelanthus complicatus* (*L. reflexa*) under *Lachenalia* in subgenus *Coelanthus*, grouping the rest of the species under the subgenera *Eulachenalia*, *Orchiops* and *Chloriza*, for which he provided a key. Numerous other major works followed, including his *Flora of Mauritius and the Seychelles* in 1877 and *Flora of the English Lake District* in 1885. He also produced monographs on the Hypoxidaceae and Iridaceae, and later collated many of the articles he had written for *The Gardener's Chronicle* and *Journal of Botany* into a series of valuable handbooks, the most well known of which was probably his *Handbook of the Amaryllideae*, published in 1888. Between 1877 and 1895, Baker also described over a thousand Madagascan plants.

Baker's second and final monograph of *Lachenalia* was published in 1897, in the 6th volume of *Flora Capensis*. There, he recognised 42 species in five subgenera *Eulachenalia*, *Coelanthus*, *Orchiops*, *Chloriza* and *Brachyscypha*. Three of his subgenera (*Brachyscypha*, *Coelanthus* and *Eulachenalia*) were distinct in floral morphology but *Chloriza* and *Orchiops* were difficult to separate and became merged as more species were discovered. Baker provided an indented key, with most of the species voucherised with collectors' details and herbarium citations. He recognised four species in his concept of *Polyxena* subgenus

Figure 8. Photograph of Dr John Gilbert Baker (1834–1920).

Eupolyxena, as well as two species of *Hyacinthus* from the south-western Cape, that are now included in *Lachenalia*.

Baker's work in classification and as a naturalist was well received and he was elected to honorary positions by numerous societies; he was Fellow of the Linnean Society and Honorary Life-fellow of the Royal Horticultural Society, and he received the Linnean Medal and the Victoria Medal of Honour in Horticulture. In 1919, the University of Leeds conferred an honorary doctorate on Baker, and he continued private studies at the Herbarium until his death at Kew on 16th August 1920 (Prain, 1921).

The German gardener, botanist, traveller and collector Rudolf Schlechter (1872–1925) undertook eight major collecting trips in southern Africa between 1891 and 1898 (Gunn & Codd, 1981; Jessop, 1964) (Figure 9). Schlechter's material included a large number of *Lachenalia* specimens, many of which were new to science, which were distributed to 15 local and foreign herbaria, some with his manuscript names appended. These were the subject of a paper by W. F. Barker (1983a), who provided a complete list of Schlechter's collections with identifications added. J. G. Baker (1904) described one of these as *L. schlechteri*; Barker (1978, 1979) described six new species from these collections; and a further three new taxa were described relatively recently (Duncan, 1996, 1997). A list of Schlechter's remaining manuscript names, with updated identifications added, is provided in the Appendix.

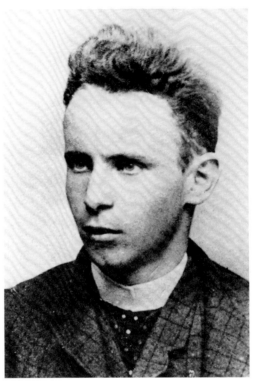

Figure 9. Photograph of Rudolf Schlechter (1872–1925), courtesy of the National Herbarium, South African National Biodiversity Institute, Pretoria.

The magazine *The Flowering Plants of South Africa* (now *Flowering Plants of Africa*) was launched in 1921 and has served as a medium for the publication of colour plates of many lachenalias, including a number of new species. Two other South African publications, *Bothalia*, the house journal of the South African National Biodiversity Institute (started in 1921), and the *Journal of South African Botany*, now the *South African Journal of Botany* (started in 1935), have served as the major vehicles for the publication of new species in *Lachenalia* during the 20th and 21st centuries.

Most of the subsequent taxonomic work on *Lachenalia* was done by Winsome Fanny (Buddy) Barker, first Curator of the Compton Herbarium at Kirstenbosch (Figure 10). Barker's interest in the genus began in the spring of 1929 during her first major botanical expedition as the Edward Muspratt Solly Scholar, under the Director of the National Botanical Gardens, Prof. R. H. Compton. The expedition took Barker, Louisa Bolus and a couple of colleagues to Nieuwoudtville, where a bowl of magnificent local lachenalias decorated their hotel table. The following day, Barker saw numerous *Lachenalia* species in flower in their natural habitat, and from then on she developed a deep interest in the genus. Under the guidance of Bolus, she began sketching and painting the *Lachenalia* species being cultivated in the Kirstenbosch Nursery, and later started to collect them in the wild.

On completion of her scholarship, Barker spent three years in the Bolus Herbarium, then located at Kirstenbosch, working for the Bentham Trustees, mounting herbarium specimens and doing illustrations to be sent to Kew. In 1933, she was appointed to the newly created post of Botanical Assistant on the staff of the Gardens, and continued her research into *Lachenalia* at the Bolus Herbarium.

In 1937, Barker spent a month studying *Lachenalia* at the Kew Herbarium and the British Museum (Natural History). At Kew, she was able to see important collections from Uppsala and Vienna that had been specially sent there on her account, including N. J. Jacquin's collection of specimens. Many of these were the types of Jacquin's new species described in *Collectanea*, which proved to be accurate matches of his illustrations in his *Icones Plantarum Rariorum*, but which perished in a fire in Vienna a few years later. In 1938, the Bolus Herbarium collection was transferred to a new building at the University of Cape Town, and the National Botanical Gardens began to build a herbarium collection of its own. In 1940, Barker was appointed first Curator of the Compton Herbarium, with a staff of two. Her

Figure 10. Miss Winsome Fanny Barker (1907–1994), photograph courtesy of the Compton Herbarium, South African National Biodiversity Institute, Kirstenbosch.

curatorial duties increasingly impinged on her *Lachenalia* research, but she officially visited Kew and the British Museum again in 1967, at which time she also visited several other European herbaria. Over a period of 43 years (1929–1972), Barker undertook extensive collecting trips and assembled the most comprehensive collection of *Lachenalia* herbarium specimens in the world. She illustrated many new species with her own line drawings and watercolour paintings, which often accompanied her taxonomic papers. On her retirement in 1972, and with time in hand, she continued work on *Lachenalia* in earnest as an honorary researcher at the Bolus Herbarium under Prof. Schelpe. Her first paper on *Lachenalia*, a synopsis of the genus, appeared in 1930 in the *Journal of the Botanical Society*. Her last, published in 1989 in the newly styled *South African Journal of Botany*, dealt with two new species, six new varieties and nomenclatural changes for a number of old species. One of Barker's greatest contributions to the knowledge of the genus was the use of seed morphology as a useful diagnostic character in delimiting many species. Her involvement with *Lachenalia* spanned 59 years, during which she published a total of 47 new species and 11 new varieties, including many collected by Rudolf Schlechter. She also published *Polyxena maughanii* (now *L. ensifolia* subsp. *maughanii*) and *Hyacinthus paucifolius* (now *L. paucifolia*), and *L. barkeriana* was named in her honour (Müller-Doblies *et al.*, 1987). Barker passed away in Wynberg, Cape Town on 27th December 1994 at the age of 87 (Rourke, 1995; Duncan, 1996).

In 1966, John Ingram of the Bailey Hortorium at Cornell University published a paper in *Baileya* on 13 species of *Lachenalia* widely cultivated in the USA. It included an indented key, brief descriptions of each species and several black-and-white images re-drawn from early published works. Twenty years later, Trevor Crosby, Superintendent of the Botanical Gardens of the Department of Plant Sciences at the University of Leeds, produced a more detailed work *The Genus Lachenalia* in *The Plantsman*. He provisionally subdivided the genus into five natural groups based on their probable relationships as deduced from a combination of chromosome studies, hybridisation experiments and flower shape. He provided a key, detailed descriptions of the most well-known species and cultivation notes, accompanied by a colour plate and his own line drawings (Crosby, 1986).

As a horticulturist at Kirstenbosch Botanical Garden, I had the good fortune to benefit from the vast wealth of knowledge accumulated on *Lachenalia* by W. F. Barker. In the mid 1980s, I began to work on a popular, illustrated guide to the genus, *The Lachenalia Handbook*, the main purpose of which was to provide a means of identifying the species with the aid of a simplified key, brief descriptions and colour photographs. A total of 88 species were included, arranged in two major groups according to stamen position, and subdivided into 10 smaller groups according to inflorescence type, with the species descriptions listed alphabetically within each subgroup (Duncan, 1988a).

In 1998, A. P. Dold and P. B. Phillipson provided a revision of the genus *Lachenalia* in the Eastern Cape that was based on field and herbarium studies. This included eight species, accompanied by an indented key, monochrome photographs and distribution maps.

Until the early 20th century, the classification of *Lachenalia* had been based on morphological evidence, but following advances in science and technology, additional methods of investigating the relationships between genera and species became available in the fields of cytology and cladistic analysis. Being cytologically variable, the genus *Lachenalia* has been investigated fairly extensively and numerous papers have appeared on chromosome number, the first of which was that of A. A. Moffett (1936). At present, the chromosome numbers of 94 species are known, with most of the recent research having been undertaken at the University of the Free State in South Africa (Spies *et al.*, 2008; Spies *et al.* 2009) and at Hiroshima University in Japan (Hamatani *et al.*, 2007).

In their partial revision of the tribe Massonieae, Müller-Doblies and Müller-Doblies (1997) considered *Lachenalia* to be distinct from *Polyxena* and placed these two genera in different subtribes. *Lachenalia* was placed in subtribe Lachenaliinae and *Polyxena* in subtribe Massoniinae, mainly on the basis of the authors' interpretation of the angle at which the perianth arises from the pedicel and the area of insertion of the stamens. In 2002, Dr Alison Summerfield (van der Merwe) completed a biosystematic study of the seven minor genera of the Hyacinthaceae including *Polyxena*, which was based on morphological, leaf anatomical, palynological, geographical and molecular data, to determine the phylogenetic relationships within the subtribe *Massoniinae*. Following a molecular phylogenetic analysis of the family Hyacinthaceae in sub-Saharan Africa, the genus *Polyxena* was included within the circumscription of *Lachenalia* (Manning *et al.*, 2004), and a further molecular study investigated phylogenetic relationships between the genus *Lachenalia* and other related liliaceous genera (Spies, 2004).

In 2005, a phylogenetic analysis based entirely on morphological data describing all 139 taxa recognised within *Lachenalia* and *Polyxena* at that time was completed (Duncan 2005; Duncan *et al.*, 2005). During the course of that study and subsequently, five new species were described in *Bothalia* (Duncan & Edwards 2002, 2006, 2007).

HISTORY OF HYBRIDISATION

Lachenalia has been in cultivation for well over two and a half centuries. Members of the genus have a long history of hybridisation in British, Irish and European horticulture, but in recent times, the development of the genus as a commercial ornamental crop has been centred in South Africa. *Lachenalia orchioides* was the first introduction into Britain when it arrived at the Chelsea Physic Garden in the early 1750s (Aiton, 1789). The Scottish plant collector Francis Masson added a further seven species to the collections at Kew in the 1770s, and the number of new introductions reached a climax during the period 1791–1830 (David, 2009).

The first recorded artificial hybrid flowered in 1880 in Britain, from seeds sown in 1877 by the Reverend John G. Nelson at Aldborough Rectory, Norwich. It was a cross between *L. luteola* (the female parent) and the plant now known as *L. flava* (previously *L. aurea*, amongst other names, the male parent) and was named *L.* 'Nelsonii'. This hybrid was subsequently exhibited at the Royal Horticultural Society, from which it received a First Class Certificate in 1881. Later, *L. luteola* and *L. aurea* came to be regarded merely as varieties of *L. aloides*, and hence *L.* 'Nelsonii' was not regarded as an interspecific cross and could not bear the botanical hybrid name *L.* × *nelsonii*. Now that both *L. luteola* and *L. flava* are restored to specific status, however, the botanical hybrid name *L.* × *nelsonii* is valid for that cross. The hybrid appears to have been of some interest even to the young Charles Darwin (1809–1882). Darwin, together with his sister Catherine, was drawn in chalk while holding a pot of a flowering *Lachenalia* hybrid, probably *L.* × *nelsonii*, by the artist Sharples (circa 1816 at Shrewsbury, Shropshire) (Figure 11). Remarkably, the hybrid is still in cultivation today in the UK and Europe, and stocks of bulbs received by Trevor Crosby from ten different sources in the 1970s appeared to match exactly the clone illustrated in colour in the German gardening magazine *Garten-Zeitung* in 1882 (Crosby, 1978).

Nelson is thought to have raised a second interspecific hybrid, between *L. reflexa* and *L. aurea*, which in 1885, some time afer Nelson's death, was exhibited to the Royal Horticultural Society by the nurserymen Barr & Son under the name 'Aldborough Beauty'. The same hybrid was repeated by Sprenger at the nursery of Messrs Dammann in Naples and sent to Kew, and was later described and illustrated as *L.* × *regeliana* (Sprenger, 1891). Material of the same cross was also received at Kew from F. W. Moore of Glasnevin (Baker, 1897a) and, according to Moore (1891, 1905), also became known as *L. aureo-reflexa* Baker. Preserved material of *L.* × *regeliana* and *L. aureo-reflexa* in the herbarium at Kew resembles similar crosses between *L. reflexa* and *L. aurea* made by Crosby. Sprenger (1891) also described and illustrated a cross between *L. reflexa* and *L. quadricolor* (previously *L. aloides* var. *quadricolor*) as *L.* × *comesii* Sprenger; the suberect perianth and very long perianth tube typical of *L. reflexa* is evident in both of these crosses.

The hybrid *L.* × *Cami* (named after Dr T. Cam of Hereford) was a cross between *L. aurea* and probably a maculate-leafed form of *L. luteola*, first described by Reuthe (1889) in *Gartenflora*. However, Baker (1897a) described this hybrid as being a cross between *L. aurea* and *L. pendula* (= *L. bulbifera*, now *L. bifolia*), a view not shared by Moore (1891) or Sprenger (1891). Neither of the pressed specimens of *L.* × *Cami* at Kew and the British Museum (Natural History) are reminiscent of *L. bifolia*. In fact, crosses claimed to be between *L. aloides* and *L. bifolia* made in the early days of hybridisation probably resulted in the pollination of individuals of *L. aloides* with one of its previously recognised varieties after cross-pollination with *L. bifolia* failed. Crosby's (1978) well-founded concern that, in accordance with the *International Code of Nomenclature of Cultivated Plants*, different crosses between varieties of the same species (i.e. *L. aloides*) do not warrant different botanical hybrid epithets is resolved in the present work as these 'varieties' are elevated to species status.

Figure 11. Chalk drawing by the artist Sharples of the six-year-old Charles Darwin (1809–1882) with his sister Catherine, c. 1816, holding a pot of *Lachenalia* × *nelsonii* at the family home in Shrewsbury, Shropshire, courtesy of English Heritage Photo Library.

Attempts made in the late 19th century to cross *L. bifolia* and *L. punctata* (= *L. rubida*) with other species proved unsuccessful, with either the seed failing to germinate or the 'hybrids' showing no sign of either of these parents in the progeny (Moore, 1905). Successful hybridisation between *L. bifolia* and *L. punctata*, and between a number of other species, was finally achieved many years later at the University of Leeds (Crosby, 1986) and at the Vegetable and Ornamental Plant Institute at Roodeplaat in South Africa (Hancke & Liebenberg, 1990).

A collection of 110 late 19th century watercolour paintings by Lydia Shackleton is housed in the National Botanical Garden at Glasnevin, Dublin and contains 28 paintings of the genus *Lachenalia*, including illustrations of the primary hybrids *L.* × *nelsonii*, *L.* × *regeliana* and *L.* × *Cami* and of a number of unnamed secondary hybrids. This collection gives a good representation of the species, cultivars and hybrids of *Lachenalia* in cultivation in about 1890, based on the collection at Glasnevin and the work of its Director F. W. Moore (1857–1949) (Moore, 1891). Unusually, most of these *Lachenalia* works have an herbarium specimen of the subject mounted adjacent to the painting (Morley, 1978). Tragically, the entire Moore collection of *Lachenalia* hybrids was lost in 1930 due to severe frost (Nelson, 2000).

Another plant of dubious hybrid parentage, known as *L.* × *boundii* and supposedly a cross between *L. bifolia* and *L. punctata*, was exhibited by Messrs Bound to the Royal Horticultural Society in 1925. The description in the Society's journal of 1929 is indicative of a form of *L. bifolia*. This view was reiterated by Grey (1938) and by Crosby (1978), and the appropriate name for it should therefore be *L. bifolia* 'Boundii'. Unsuccessful attempts were made in the early 20th century in the UK to cross *L. glaucina* (now *L. orchioides* subsp. *glaucina*) with forms of *L. aloides* (s.l.) (Jacob, 1919); this was achieved later in South Africa.

If the number of awards he received from the Royal Horticultural Society is any indication, Rev. Joseph Jacob of Whitewell, Shropshire achieved considerable success in hybridising forms of *L. aloides* (s.l.) in the mid to late 1920s. Unfortunately though, very few of the more than 80 cultivars and hybrids raised since *L.* × *nelsonii* first appeared remain in cultivation today. Understandably, little was achieved during the war years, then in the early1960s, breeding programmes were begun at Roodeplaat in South Africa and by Trevor Crosby at the University of Leeds.

The plant described in 1949 in *The New Zealand Gardener* as '*L.* Pearsonii', is said to have been raised in New Zealand from a cross between *L. bifolia* and *L.* × *nelsonii* in about 1924 by Aldridge, a previous Curator of Parks and Reserves in Auckland. It has bright orange-yellow, pendulous tubular flowers with red tips. As noted by Crosby (1978), a major discrepancy exists between the colour photograph of the plant on the cover of the journal and the rather poor description of it, in that the flowers lack the red borders to the tepals. Crosby thought it to be merely a form of *L. aloides* s.l., close to *L.* × *nelsonii*, whereas I have had similar individuals (possibly a cross with *L. flava*) showing no features reminiscent of *L. bifolia* appear amongst a batch of seedlings of otherwise pure *L. aloides* in the Kirstenbosch Nursery. '*Lachenalia pearsonii*', also sometimes named '*L. piersonii*' was widely available in the trade in South Africa in the 1960s and 1970s but is now much less frequently propagated, possibly because of loss of vigour through viral infection. The name '*L.* Pearsonii' as published in *The New Zealand Gardener* is in any event invalid, as it was not accompanied by a Latin description, and the epithet *pearsonii* had already been published for a wild species, *Scilla pearsonii* from southern Namibia (Glover, 1915), which was later transferred to *Lachenalia* as *L. pearsonii* (Barker, 1969).

In addition to producing fertile hybrids between *L. bifolia* and *L. punctata*, after much trial and error, Crosby succeeded in crossing both *L. flava* and *L. luteola* with *L. bifolia*, *L. orchioides* subsp. *glaucina*, *L. reflexa* and *L. viridiflora*. He also crossed both *L. bifolia* and *L. reflexa* with *L. orchioides* subsp. *glaucina* and *L. viridiflora* (Crosby, 1978). Some of the most desirable crosses are those of *L. flava* or *L. luteola* with *L. orchioides* subsp. *glaucina*, many of which inherit the strong sweet fragrance of the latter plant, although Crosby reported that the crosses between *L. luteola* and *L. orchioides* subsp. *glaucina* were achieved with great difficulty. Unfortunately, the Crosby *Lachenalia* collection no longer exists at Leeds, and it is not precisely known what became of the bulk of the collection that was taken over by a private nursery in the UK (David, 2009).

With most of the early hybrids lost and the few still in cultivation severely lacking in vigour, the opportunity to develop a range of new hybrids of brightly coloured, long-lasting flowering pot plants and bedding plants was taken at the Vegetable and Ornamental Plant Research Institute at Roodeplaat, South Africa in 1965. The progress of the breeding programme there has been categorised into three phases (Kleynhans, 2006). During the first phase (1965–1972) initial hybrids consisting of 39 interspecific hybrid combinations were made between 10 species (*L. aloides* s.l., *L. bifolia*, *L. contaminata*, *L. flava*, *L. mutabilis*, *L. orchioides*, *L. pallida*, *L. punctata*, *L. quadricolor* and *L. splendida*), of which only 18 combinations were successful and flowered (Lubbinge, 1980). Basic procedures for breeding were established, selections were made and material was tested by several South African growers who decided that these hybrids had commercial potential. During the second phase (1973–1982), many more hybrids were made and isolation barriers that prevent or limit the successful hybridisation of species with different flower shapes, such as polyploidy and pollen-tube length, were studied. Pollen-tube length was problematic because the pollen from small-flowered species that have relatively short pollen tubes, such as *L. contaminata*, is not suited to being a male parent. The pollen tube of such pollen is not adapted to move down the very long styles of the tubular-flowered species such as *L. bifolia*, but this problem is overcome when such species are used as the female parent (Lubbinge, 1980). Crossing species with the same basic chromosome number was fairly successful, but it was usually not possible to cross species that have different basic chromosome numbers.

One of the major problems experienced was the variable results obtained when forms of the same species (but with different ploidy levels, as in *L. bifolia*) were used in the same hybrid combinations with other species. Making intra-species crosses before combining these with other species resulted in improved hybridisation rates (Lubbinge, 1980; Kleynhans *et al.*, 2002). Another problem that had to be overcome was the susceptibility of the hybrids to the *Ornithogalum* mosaic virus (OMV), which causes severe streaking of the leaves and thus greatly reduces the market potential of the plants. This was overcome by reproducing virus-free material in tissue culture for supply to growers (Nel, 1983), and five cultivars were subsequently registered with Plant Breeders' Rights. Forcing techniques have been developed to treat the bulbs in order to produce optimum flower production year-round.

During the third phase of the breeding programme (1983–1992), it was realised that insufficient propagative material and information regarding the commercial production of the plants was available to local growers and those in The Netherlands. The compilation of production protocols was required before the product could be successfully commercialised (Kleynhans *et al.*, 2002). The fourth phase (1993–1996) saw the fine-tuning of commercial cultivation methods, such as multiplication and forcing procedures, and regular contact with growers was maintained to ensure that correct advice was received. In addition, many more crosses were made, geared more towards market requirements, such as suitability to pot cultivation and ornamental characteristics. The trade name 'Cape hyacinth' was registered for commercially produced *Lachenalia* cultivars (Niederwieser *et al.*, 1998). The fifth phase of the breeding programme (1997– present) has seen the first South African hybrids released to the market (1997–1998), and the start of large-scale production of stocks by commercial growers in order to commercialise the product successfully and comply with demand. The cultivars are sold both as dry bulbs and as flowering pot plants. Although it took many years, the project finally resulted in success and several cultivars are now easily obtainable both in South Africa and abroad.

2. CULTIVATION AND PROPAGATION

CULTIVATION

Amidst southern Africa's wealth of geophytic plants, the genus *Lachenalia* stands out as being generally easy to culture and possessing a wealth of horticulturally rewarding members. These neat and compact plants have long-lasting blooms (some remaining attractive for up to one month) that come in a range of shapes and colour combinations. Many also have invigorating scents and interesting leaf forms that further endear them to cultivation.

In maintaining a complete collection, it is theoretically possible to have one or more species in flower every month of the year. In the southern hemisphere, certain forms of the late-flowering *L. salteri* and *L. sessiliflora* have flowering periods that extend into late December, sometimes into early January. The flower buds of the summer-growing *L. pearsonii* usually emerge in mid-January, and can appear any time thereafter until early March. The flowering cycle continues with early forms of *L. corymbosa* in early March and *L. punctata* in late March, followed by an ever-increasing number of taxa until the peak flowering period is reached in September (see p. 51 (Table 2)).

The successful cultivation of lachenalias is dependent upon four main elements: a sufficiently free-draining, well-aerated growing medium; adequate light; correct watering and dormancy procedures; and protection from high humidity and sustained cold. All lachenalias can be successfully grown in containers but only certain species are suited to garden cultivation.

GROWING LACHENALIAS IN CONTAINERS

Containers supply the most convenient way in which to provide the most suitable growing environment. Pots can be moved to the most suitable positions, brought indoors where the plants' many attributes can be appreciated at close quarters, and safely stored away during summer dormancy (Duncan 1988a) (Figures 12–15).

Aspect

Almost all lachenalias require a sunny location, some preferring one with full morning sun and afternoon shade whereas others prefer to receive bright light for as much of the day as possible. *L. margaretiae* is the only species that requires light to moderate shade, but *L. capensis*, *L. fistulosa*, *L. leomontana*, *L. orchioides*, *L. thomasiae* and *L. vanzyliae* will flower fairly well in light shade. In conditions of insufficient light, the colour of the tepals becomes increasingly insipid, especially in *L. viridiflora*, and the prostrate leaf orientation of a number of species, such as *L. congesta* and *L. kliprandensis*, changes to spreading or suberect. The aspect must be well ventilated, with as little humidity as possible, and it is very important to position the containers so that the plants do not overheat on very hot days. For massed displays, containers can be arranged together in groups in sunny courtyards or patios, and apartment-dwellers can grow lachenalias in window boxes on sunny balconies. In temperate climates, where heavy winter rainfall is experienced, such as in the southern suburbs of the Cape Peninsula, the more delicate species and those from arid habitats must be grown under cover. Specialist *Lachenalia* growers in mild climates will be inclined to erect structures with raised benches, open sides and polycarbonate or glass-fibre roofs in which an ever-expanding collection can easily be maintained. Where temperatures regularly drop below freezing for extended periods in winter, lachenalias require the protection of a cool greenhouse, conservatory or sunny windowsill. Where insufficient light levels exist in winter, supplementary lighting is of great benefit.

Figure 12 (top left). *Lachenalia corymbosa*. Image: Graham Duncan.
Figure 13 (top right). *Lachenalia punctata*. Image: Graham Duncan.
Figure 14 (above). *Lachenalia mathewsii*. Image: Graham Duncan/SANBI.
Figure 15 (right). *Lachenalia liliiflora*. Image: Graham Duncan/SANBI.

Hardiness

Lachenalias require relatively low winter growing temperatures with a daytime optimum of 15–20°C (59–68°F) and 8–10°C (46–50°F) at night. Most species are half-hardy and withstand temperatures to 0°C (32°F) for short periods of a few days. To be safe, greenhouse thermostats can be set to 2°C during periods of severe frost. Nineteen species and particular genotypes of another six species can be regarded as frost hardy as they can withstand temperatures to -5°C (23°F) for short periods. Two species and genotypes of another three species from the Sutherland area can be termed fully hardy as they can withstand temperatures to -15°C (5°F) for short periods. The foliage of most lachenalias is damaged by severe frost, especially if exacerbated by wind chill.

Frost hardy species
(which can withstand temperatures down to -5°C (23°F) for short periods)

L. ameliae
L. attenuata (forms from Middelpos)
L. aurioliae
L. bowkeri (inland forms)
L. campanulata
L. canaliculata (forms from Victoria West)
L. comptonii
L. congesta
L. dasybotrya
L. doleritica
L. ensifolia (Karoo forms)
L. glaucophylla
L. inconspicua

L. isopetala
L. juncifolia (inland forms)
L. karooica
L. longituba
L. marlothii
L. multifolia
L. obscura
L. schelpei
L. summerfieldii
L. violacea (inland forms)
L. whitehillensis
L. zebrina

Fully hardy species
(which can withstand temperatures down to -15°C (5°F) for short periods)

L. attenuata (forms from Sutherland)
L. canaliculata (forms from Sutherland)
L. congesta
L. isopetala (forms from Sutherland)
L. longituba

Growing medium

Lachenalias grow naturally in a wide variety of soil types of greatly differing moisture regimes, ranging from nutrient-poor, dry sands to seasonally wet, mineral-rich clays. Despite these varying soil conditions, most lachenalias easily adapt to new soil types in cultivation provided certain guidelines are followed. All should be grown in well-drained, well-aerated sandy media with varying amounts of organic matter, depending on the species. Those from dry sands (such as *L. anguinea* and *L. punctata*) require minimal organic matter, and moisture applied at well-spaced intervals, whereas those from seasonally inundated clays (such as *L. bachmannii* and *L. orthopetala*) prefer an increased

Figure 16 (left). *Lachenalia orthopetala* needs regular heavy drenching. Image: Graham Duncan/SANBI.
Figure 17 (right). *Lachenalia pusilla* needs a sandy, well drained medium. Image: Graham Duncan/SANBI.

humus content and frequent heavy drenching. Each grower will discover his or her own preferred medium. For easily cultivated species such as *L. pallida* and *L. quadricolor*, a medium of equal parts well-decomposed, finely sifted compost or finely milled bark mixed with course river sand, grit or silica sand is recommended. For more sensitive species, such as *L. ameliae* and *L. physocaulos*, the sand content should be increased to three parts and the humus content reduced to one part. Highly sensitive species, such as *L. klinghardtiana* and *L. ventricosa*, should be grown in equal parts of coarse river sand and silica sand. A slightly acid to neutral pH is preferred by most species, except for *L. karooica*, which requires moderately alkaline media. For all species, a 2–3 cm layer of well-decomposed compost should be placed over the drainage chips at the bottom of the container.

Lachenalias are suited to a variety of container shapes and sizes, including conventional round pots in plastic or terracotta, square pots, troughs, window boxes and even hanging baskets. In the process of building-up a comprehensive collection, uniformity of pot shape and size saves space and square pots are highly recommended. Easily obtainable, ordinary 15 cm-diameter plastic pots are suited to dwarf species such as *L. angelica* and *L. corymbosa*; 20 cm-diameter pots are ideal for medium-sized plants such as *L. pallida* and *L. salteri*; and 25 cm-diameter pots are best for robust lachenalias.

Planting

Once night temperatures begin to fall markedly in late summer, the bulbs of *Lachenalia* become active, with new root growth appearing first, followed by leaf shoots. Bulbs are best planted from early to mid autumn, with late March to mid April being the most suitable time in the southern hemisphere. Species from sandy habitats that have relatively large bulbs, such as *L. anguinea* and *L. bifolia*, can be placed up to 4 cm deep, whereas those from heavy or stony soils that have very small bulbs, such as *L. namibiensis* and *L. patula*, should not be placed deeper than 1 cm; most species are planted 2–3 cm deep. The number of mature bulbs per container depends on the species, but all are gregarious and give a better display when massed together. Recommended planting densities are 6–8 bulbs per 25 cm-diameter container for larger lachenalias, and 12–15 bulbs per 20 cm-diameter container for dwarf species. Dwarf lachenalias such as *L. barkeriana* and *L. patula* perform extremely well in shallow containers, whereas most species require deep containers to grow and flower well. When touched, a crystal within the dry bulbs of *Lachenalia* causes an allergic reaction (contact dermatitis) in some people, resulting in mild to severe itching, swollen eyes and possible urticaria (hives).

Watering

A little procrastination is always a good thing and it's best to wait until cool weather has definitely set in before applying the first drench of the season some time from late March to mid April (late August in the northern hemisphere). The exception to this rule is the summer-growing *L. pearsonii*, which should receive its first drench in mid-summer. Several precocious species, including *L. bachmannii*, *L. corymbosa* and *L. quadricolor*, will often provide the cue to begin watering by sending up leaf shoots from bone-dry soil. Watering of the *Lachenalia* collection at Kirstenbosch usually commences around 15th April. For all species, an initial heavy drench is applied, and once leaf shoots begin to elongate, a heavy drench once weekly or every two weeks is suggested until early winter, depending on the species. Arid-habitat lachenalias, such as *L. buchubergensis*, *L. klinghardtiana* and *L. xerophila*, need only be watered heavily once every two weeks. Those from seasonally inundated habitats, such as *L. bachmannii*, *L. contaminata* and *L. orthopetala*, need a weekly drench and are best placed in deep saucers kept filled with water towards, and during, the flowering period.

Figure 18 (left). White form of *Lachenalia pallida*. Image: Graham Duncan.
Figure 19 (right). Blue form of *Lachenalia pallida*. Image: Graham Duncan.

With the exception of species from moist habitats, it is always preferable for the growing medium to be slightly dry rather than too wet. Watering is best done in the morning so that excess moisture evaporates speedily and to reduce the incidence of fungal attack at night.

Despite having perfectly healthy bulbs and sufficient moisture, a number of species, including *L. anguinea*, *L. flava*, *L. martiniae*, *L. nervosa*, *L. peersii* and *L. stayneri*, occasionally remain dormant throughout the growing season. As far back as 1872, it was stated in *The Gardener's Chronicle and Agricultural Gazette* that bulbs of *L. flava* remained completely dormant for up to two years (Anonymous, 1872). Should all of the bulbs within a particular container fail to sprout within four weeks of watering, it can be assumed that they will remain dormant for that season and that they can safely be allowed to dry off and stored until the following season. From early to late spring onwards, watering frequency in most species should be gradually reduced until the leaves have completely withered, following which a completely dry period should be maintained over the summer dormant period. Once dormant, the bulbs are best stored in dry soil at daytime room temperature (20–25°C/68–77°F) in a well-ventilated environment. Early-flowering species such as *L. barkeriana* and *L. punctata* enter dormancy in early spring whereas most lachenalias do so in late spring, and late-flowering ones such as *L. leomontana* and *L. salteri* do so from early to mid-summer. Exceptions are the summer-growing *L. pearsonii*, which becomes dormant in late winter, and *L. convallarioides* and *L. campanulata* which, although naturally deciduous and winter-growing, usually remain evergreen in cultivation in temperate climates, the latter species especially so. In acclimatising lachenalias sent from South Africa to the northern hemisphere, it is best to receive the bulbs after they have undergone a period of several months dormancy in South Africa, and then to pot them up as soon as they are received in a dry medium, maintaining them in this state until the northern hemisphere autumn begins in about September. Should any bulbs sprout during this period, they can be given some water and then placed in a cool spot until autumn when regular watering can commence.

Figure 20. Burning treatment of *Lachenalia montana* bulbs in terracotta pots stimulates flowering. Image: Graham Duncan/SANBI.

Burning pyrophytic lachenalias

The three pyrophytic species, *L. lutzeyeri*, *L. montana* and *L. sargeantii*, require acidic, quick-draining sandy media and intervention to stimulate leaf and/or flower development. *L. montana* and *L. sargeantii* regularly produce leaves in cultivation but *L. sargeantii* never flowers and *L. montana* very rarely does, whereas *L. lutzeyeri* rarely produces leaves and never flowers. The bulbs are best cultivated in deep terracotta pots and respond well to having a layer of approximately 4 cm of dry leaves, straw or twigs burnt over the soil surface for about 10 minutes in late summer (late January to late February in the southern hemisphere). Commence watering as usual in the autumn. Burning stimulates vigorous leaf growth in all three species, and this has resulted in the flowering of *L. montana* at Kirstenbosch (Figure 20).

Feeding

Many *Lachenalia* species occur in nutrient-deficient soils and many more are found in heavy, nutrient-rich clays. In cultivation, however, all species can be grown successfully without any supplementary feeding, although most respond very well to fertilisers that have a high potassium but low nitrogen content. Once the plants have begun active growth, monthly applications of an organic fertiliser that is high in macro- and micro-nutrients is all that is required. Fertilisers that are high in nitrogen cause excessive, soft leaf growth at the expense of flowers, sometimes resulting in the development of additional leaves and increasing susceptibility to fungal attack. Slow-release fertilisers can be incorporated into the upper part of the growing medium, or sprinkled onto the surface, and liquid fertilisers can be used at fortnightly intervals at a weaker rate than recommended. *Lachenalia* seedlings respond very well to liquid feeds of seaweed extract, applied every three or four weeks, either as a foliar spray or as a soil drench. It is important to note that a build-up of salts, especially in containers, can occur as a result of regular feeding and can damage the roots. It is best to flush out these excess salts once each month by watering with pure water, preferably rainwater, as water quality varies from time to time and can result in additional salt accumulation. Salt accumulation is seen as a white or brownish-yellow deposit on the soil or as a crust against the inside rim of containers.

A number of species with relatively long stems and lasting blooms make satisfactory cut flowers. They are best arranged in shallow vases containing not more than 2 cm of water, and it is advisable to cut the stems periodically to lengthen their vase-life.

Selection of species for cut flowers

L. aloides	*L. mutabilis* (certain forms)
L. arbuthnotiae	*L. pallida*
L. bifolia	*L. peersii*
L. callista	*L. purpureo-caerulea*
L. flava	*L. quadricolor*
L. glauca	*L. salteri*
L. luteola	*L. thomasiae*
L. moniliformis	*L. violacea*

Selection of species for moist growing conditions

L. arbuthnotiae
L. bachmannii
L. contaminata
L. corymbosa
L. longituba
L. mathewsii
L. orthopetala
L. polyphylla
L. reflexa
L. salteri
L. viridiflora
L. zeyheri

Selection of most attractive xerophytic species

L. angelica
L. anguinea
L. comptonii
L. duncanii
L. framesii
L. giessii
L. isopetala
L. kliprandensis
L. marlothii
L. minima
L. multifolia
L. namaquensis
L. namibiensis
L. pearsonii
L. patula
L. polypodantha
L. trichophylla
L. valeriae
L. verticillata
L. violacea
L. whitehillensis
L. zebrina

Selection of most strongly scented species

L. algoensis (sweet)
L. arbuthnotiae (sweet)
L. barkeriana (yeasty)
L. capensis (sweet)
L. comptonii (spicy)
L. congesta (sweet)
L. contaminata (spicy)
L. corymbosa (honey)
L. elegans (carnations)
L. ensifolia (spicy-sweet)
L. fistulosa (sweet)
L. glauca (coconut)
L. glaucophylla (spicy)
L. kliprandensis (spicy-sweet)
L. longituba (spicy-sweet)
L. lutea (sweet)
L. marginata (sweet)
L. marlothii (sweet)
L. multifolia (marzipan)
L. nervosa (cloves and carnations)
L. orchioides (sweet)
L. pallida (spicy)
L. paucifolia (honey)
L. peersii (carnations)
L. pusilla (yeasty)
L. pygmaea (almonds)
L. suaveolens (carnations)
L. violacea (spicy-sweet)

Selection of best species for pot cultivation

L. aloides	*L. margaretiae*
L. angelica	*L. mathewsii*
L. arbuthnotiae	*L. membranacea*
L. bolusii	*L. moniliformis*
L. callista	*L. mutabilis*
L. campanulata	*L. namaquensis*
L. comptonii	*L. namibiensis*
L. convallarioides	*L. nardousbergensis*
L. contaminata	*L. orchioides*
L. corymbosa	*L. orthopetala*
L. elegans	*L. pallida*
L. ensifolia	*L. patula*
L. fistulosa	*L. pygmaea*
L. flava	*L. quadricolor*
L. framesii	*L. salteri*
L. hirta	*L. splendida*
L. judithiae	*L. thomasiae*
L. juncifolia	*L. thunbergii*
L. kliprandensis	*L. trichophylla*
L. liliiflora	*L. vanzyliae*
L. lutea	*L. viridiflora*
L. luteola	*L. zeyheri*

LACHENALIAS FOR THE GARDEN

A number of lachenalias with large, brightly coloured, pendent tubular blooms make striking bedding or rock garden displays. The aloe-like blooms of *L. bifolia* always command attention (Figure 21), as do other lachenalias with dense bell- and urn-shaped flowers. However, very few species are able to withstand summer rainfall or indiscriminate garden irrigation during the summer dormant period, and most rapidly succumb to fungal disease if conditions are wet while they are dormant. Lachenalias are grown most successfully in permanent positions outdoors in sunny, dedicated rock garden pockets, kept absolutely dry in summer. Drainage in heavy soils can be improved by incorporating large quantities of well-decomposed organic matter and coarse river sand. Slightly sloping ground should be chosen for species that prefer the soil to dry out between waterings. To prevent disturbance by hill-forming and surface-foraging moles, pockets should be dug out to a depth of about 10 cm, lined with strong wire mesh and filled-in with a sandy medium containing plenty of finely sifted, well-decomposed organic matter. In very sandy soils, bulbs can be planted more deeply than in containers, and for best effect, bulbs should be set out close together (5–8 cm apart) in groups of the same species.

Figure 21. *Lachenalia bifolia* makes striking rock garden or bedding displays. Image: Graham Duncan.

Alternatively, bulbs can be grown in pots that are plunged into rock garden pockets or the front of borders once in active growth, then lifted after flowering and stored in dry conditions for the summer. Plunging pots of flowering specimens is not recommended as the fleshy stems and leaves are easily damaged in the process and flower heads tend to flop over. Garden-grown lachenalias can also be displayed to great advantage by interplanting them with either dwarf succulents, such as *Aloe brevifolia* and *Jordaaniella dubia*, or spring-flowering annuals, such as *Dorotheanthus bellidiformis*, *Felicia dubia* and *Steirodiscus tagetes*.

Species recommended for garden cultivation

L. aloides
L. arbuthnotiae
L. bifolia
L. contaminata
L. fistulosa
L. flava
L. juncifolia
L. lutea
L. luteola
L. mathewsii
L. membranacea
L. orchioides
L. orthopetala
L. pallida
L. peersii
L. purpureo-caerulea
L. quadricolor
L. rosea
L. salteri
L. splendida
L. thomasiae
L. thunbergii
L. vanzyliae

PROPAGATION

Maintaining genetic and morphological diversity within a species requires propagation by sexual means, that is, from seed, but the preservation of the identical morphological traits of a particular clone requires vegetative propagation from bulblets, bulbils, leaf cuttings or tissue culture (micropropagation).

SEED

To maintain species integrity, it is essential that none of the seeds have arisen from interspecific crosses. If harvested from the wild, species integrity is likely to be 100%, but if harvested from cultivated material where strict measures of isolation and hand-pollination are not enforced, species integrity will usually be severely compromised. If there is no intention to harvest seed from open-pollinated plants, it is best to remove old flower heads before capsules start developing, thereby preventing both energy from being wasted in seed formation and hybrid seeds from colonising adjacent containers (Duncan, 1978). Growing lachenalias from seed is a rewarding experience. Under ideal conditions, the seedlings of most species produce their first flowers during their third season (Duncan, 1988), although in exceptional circumstances *L. juncifolia* and *L. reflexa* sometimes flower in their second year (Duncan, 2010), whereas some species from arid habitats, such as *L. congesta*, can take 4–5 years.

Lachenalia seeds are best sown in late autumn (late April to May in the southern hemisphere) after cool weather has definitely set in. Sowing in deep seed trays, pots or seed beds is recommended as seedlings reach maturity faster in deeper media. In order to minimise the germination of weed seeds and infestation by damping-off fungi, a sterilised medium should be used. This might be composed of equal parts of finely sifted, well-decomposed compost and coarse river sand or silica (industrial) sand. Seed should be sown thinly to prevent overcrowding and to allow sufficient room for bulb development. The seeds of most species are sown 3 mm deep, whereas the minute seeds of *L. angelica* and *L. sessiliflora*, for example, can be sown a little shallower and the relatively large seeds of *L. giessii* and *L. isopetala*, for example, can be sown 4–5 mm deep. Containers and seed beds can be located in full sun if moisture is applied regularly, alternatively they could be in light shade. Seeds should be kept moist but not continually wet by watering every other day with a watering can fitted with a fine rose cap.

Day temperatures between 15–20°C (59–68°F) followed by night temperatures of 8–10°C (46–50°F) stimulate the germination of fresh seeds. Most species germinate within about 18 days, but the seeds of *L. patula* germinate within 11 days and those of *L. argillicola* take up to 58 days. It is worth noting that seeds of the late summer- and autumn-flowering species, including *L. argillicola*, *L. barkeriana*, *L. calcicola*, *L. ensifolia*, *L. paucifolia*, *L. pearsonii*, *L. pusilla* and *L. pygmaea*, usually fail to germinate if sown during the winter growing season in which they are harvested. Like those of the winter- and spring-flowering species, these seeds must first undergo a summer dormant phase before they can germinate in the following autumn. The exceptions to this rule are the seeds of the dwarf *L. corymbosa* and the high-altitude, early summer-flowering *L. campanulata*, the only two species known to germinate readily if sown directly after harvesting. Fresh seeds of the three pyrophytic species germinate readily, but seedlings of *L. lutzeyeri* remain dormant indefinitely from their second year onwards unless treated with fire.

Hand pollination

Lachenalia bulbs have a limited but variable life-span, depending on the species. Those species with small bulbs (in relation to plant size) that literally seed themselves to death, such as *L. mutabilis* and *L. ventricosa*, are relatively short-lived (with a lifespan of approximately 5 years) whereas larger species, such as *L. carnosa* and *L. kliprandensis*, can live for 20 years or more in ideal conditions. In order to maintain species in cultivation over the long term and to preserve genetic diversity within a collection, it is necessary to produce pure seed and to have seedlings at hand should adult bulbs be lost. This is achieved by isolation and hand-pollination. Pots should be isolated to exclude insect and bird pollinators, and a pollinating implement such as a fine water-paint brush, wax-coated match stick or ear bud can be used to dab the anthers and transfer pollen from the flowers of one individual to the stigmas of those on other plants of genetically different clones. Before commencing with pollination, the pollen should be dry, loose and easily removed, and the stigmas must be receptive and have a slightly sticky surface. Hand pollination is best performed from early to mid-morning, and each flower should be pollinated several times over a period of days to ensure that pollination has taken place. Seed capsules can take up to six weeks to mature and dehisce, and harvested seeds are best stored in paper packets to which an insecticidal powder such as carbaryl has been added to exclude caterpillars from seed stored at room temperature. Seeds stored this way remain viable for several years, but dry seed storage at low temperature, for example in the vegetable compartment of a fridge, extends seed viability by many more years.

BULBLETS, BULBILS AND DIVISION

Bulblets, also known as offsets or daughter bulbs, usually develop from the mother-bulb on the perimeter of the basal plate, on very short, stalk-like structures in the axils of the bulb scales, or at the tip of a long stolon. Many species produce bulblets on the basal plate, but only *L. bifolia*, *L. moniliformis* and *L. namaquensis* produce them at the tips of stolons. Bulblets also develop along the subterranean margins of the leaves of *L. bifolia* and *L. cernua*. Bulbils (aerial bulblets) form along the aerial leaf margins of *L. bifolia* and occasionally also develop either at the base of the floral bracts or at the rachis apex, especially if a flower head happens to break and touch the ground. In some species, the mother bulb divides into two or more large bulblets of similar size that will flower the following season. Bulblets and bulbils can be separated at the end of the growing season or in early autumn at planting time. Depending on size, they will need one or more seasons to reach flowering size. Large bulblets should be carefully tugged apart, not forcibly broken off, and injured surfaces should be allowed to dry for a few days and treated with a fungicide before planting.

LEAF CUTTINGS

Propagation by leaf cuttings was first reported by Cook (1931) and is an effective method of increasing species that have solitary bulbs or desirable clones that do not reproduce readily by bulblet formation. The leaves that are used as cutting material must be virus-free and in active growth, and should be harvested before flowering commences. A leaf can be removed by holding it firmly at the base and pulling it off with a swift tug, or by twisting it off at the base. Cuttings of about 3 cm in length are made with a sharp knife and placed in a slightly moist medium of equal parts of vermiculite and river sand or coarse grit, with about 1 cm of the base of the cutting inserted into the medium. The cuttings should be kept under semi-shaded, warm conditions, and great care should be taken to ensure that the medium never becomes too wet. The cutting taken from the proximal region of the leaf often produces the best results as the tissue in this portion is harder and less prone to rot. By the time the leaf section has withered, several bulblets will have formed at the base of the cutting and these are stored dry until planting time the following autumn (Du Plessis & Duncan, 1989). Leaf cuttings taken from species that have relatively broad leaves are naturally far more productive than those with very narrow, linear leaves. Under optimum conditions, bulblets that are produced from leaf cuttings can flower in one year.

MICROPROPAGATION

Propagating lachenalias by culturing minute leaf tissue portions in artificial media under sterile conditions is an effective means of producing large quantities of virus-free, desirable clones from limited material for the commercial market, within a short period. Protocols for this were first established by Nel (1983) who reported rapid propagation of *Lachenalia* hybrids *in vitro* that could produce 2,000 plants from a single leaf; typically, explants of approximately 1 cm are taken from the proximal region of the leaves. Adventitious buds can also be induced from *Lachenalia* bulb scales and scapes *in vitro* (Niederwieser & Ndou, 2002).

PESTS AND DISEASES

The following are suggested methods for the control of the more important pests and diseases of *Lachenalia*.

PESTS

Aphids

These small green or black sucking insects are sometimes found on developing flower buds. They should be controlled as soon as they are noticed because they cause flower deformation and premature yellowing of the leaves. Most importantly, they are transmitters of viral disease. Aphids can be controlled in an environmentally friendly way by making up a solution of 5 ml liquid soap in 1 litre of water and spraying them away. Alternatively, biological control by means of the parasitic wasps *Aphidius ervi* and *A. colemani*, which lay single eggs into juvenile aphids, is effective in sealed greenhouses. Ladybirds are natural predators of aphids.

Bulb mites

The translucent white bulb mite *Rhizoglyphus echinopus* causes necrotic lesions and craters to develop in *Lachenalia* bulbs (Journet, 2003) and is a transmitter of fungal disease such as that caused by *Fusarium oxysporum*. Bulbs are often attacked in storage, leading to plants that exhibit chlorotic, deformed leaves. The mites are resistant to many pesticides but can be controlled biologically by the predatory mite *Hypoaspis aculeifer*.

Caterpillars

The leaves and developing inflorescences of *Lachenalia* are sometimes subject to attack by caterpillars, causing massive, rapid damage. Caterpillars are active mainly at night, remaining well camouflaged by day. They can be picked off at night by hand, sprayed with a natural spray containing natural pyrethrins or treated with a carbaryl-based insecticide.

Mealy bugs

One of the most important pests affecting lachenalias, these oval-shaped sucking insects have a white waxy covering and are found in large numbers between the bulb scales and around the leaf bases. Their presence should immediately be suspected when oily white deposits are noted on the leaf bases, especially on the undersides. Mealy bugs weaken and stunt the plants, causing yellowing and distortion, and also excrete honeydew that coats the plant surface. They are often spread by ants that feed on the honeydew, and most importantly, are transmitters of viral disease. They are especially troublesome amongst container-grown plants in enclosed, greenhouse conditions. Preventative treatment with an annual drench of systemic insecticide, such as imidacloprid or thiacloprid, is recommended in autumn, once the plants are in active growth, and is effective for up to one year. Mealy bugs eventually become resistant to continued use of the same insecticide, but alternating products alleviates this tendency. Biological controls for mealy bugs such as the ladybird beetle *Cryptolaemus montrouzieri* and the parasitic wasp *Leptomastix dactylopii* are available, but these are only effective on aerial plant parts.

Mealy bugs can be removed by hand by squashing them against the bulb, or by dipping a cotton wool bud into a solution of equal parts of methylated spirits and water and wiping the insects away. Infested plants should be isolated and treated immediately; if viral symptoms appear, they should be disposed of by burning. Ideally, new acquisitions for a *Lachenalia* collection should be placed in quarantine until known to be free of infestation.

Slugs and snails

Slugs and snails feed on the leaves of broad-leaved lachenalias. They are important transmitters of viral disease and should be controlled as intensively as possible. Snails can be removed by hand and slugs picked up with tweezers, killed instantly in a container of boiling water and placed onto the compost heap. Alternatively, environmentally friendly natural snail bait containing the active ingredient iron phosphate can be used. This causes slugs and snails to cease feeding, become less active and die within a few days.

FUNGAL DISEASES

Fungal diseases are difficult to control and their prevention is certainly the best course of action to take. Fungicides are very toxic and should only be used as a last resort. Preventative measures include sterilising growing media, providing good ventilation around the plants, refraining from using overhead irrigation, removing and destroying damaged foliage, and keeping the area underneath potted plants free of dead material. Plants suspected of disease should be isolated and treated immediately, and if heavily infected, discarded by burning.

Damping-off

Lachenalia seedlings are susceptible to damping-off fungi (mainly by *Pythium* species) that attack the base of the leaf in autumn and winter, causing collapse and disintegration of the whole plant. These fungi are prevalent in insufficiently well-ventilated conditions, and where seeds have been sown too thickly or allowed to remain wet for too long. Jeyes Fluid can be used to drench the soil before sowing by mixing 75 ml in 10 l of water for each square metre of soil. Sprinkle the solution on the soil and cover with a plastic sheet for 10 days before sowing. Sowing media can also be sterilised with boiling water, or seeds can be dusted with a wettable powder such as captab before sowing. Seeds should always be sown sparsely so that the seedlings are well ventilated and the leaves can dry off quickly after watering.

Rotting

Fungal rotting of *Lachenalia* bulbs should be suspected when the leaves are slow to develop, turn yellow prematurely or become distorted, or if the growing shoot rots at the base or fails to appear at all. It is most often caused by the *Botrytis cinerea* and *Fusarium oxysporum* fungi, and is prevalent in conditions of poorly drained soil, excessive moisture and warm climate. *Botrytis* infection is seen as light brown areas that are wet or spongy. The rot usually starts at the base of the rootstock and spreads. *Fusarium* dry rot causes lesions to develop on bulbs during storage. Symptoms in actively growing bulbs are leaves that turn yellow and wilt. Dusting of stored rootstocks with 'Flowers of Sulphur' or captab wettable powder greatly reduces the incidence of fungal rotting, as does attention to keeping bulbs dry in summer and using sterilised growing media. During the growing period, rootstocks can sometimes be saved if they are treated in a timely manner. Immediately lift rootsocks that are suspected of having an infection, clean and cut the infected areas away, and dust with captab. Replant the rootstock in sterilised river sand and place in a cool shaded spot. If the plant recovers, it can be brought into bright light again.

Rust

The unsightly rust fungus *Uromyces lachenaliae* is prevalent in winter and is exacerbated under conditions of poor ventilation and insufficient light. It attacks the leaves, liberating clusters of bright orange spores that can rapidly destroy entire leaves. Especially susceptible species include *L. aurioliae*, *L. elegans*, *L. klinghardtiana*, *L. marlothii*, *L. minima*, *L. punctata*, *L. schelpei*, *L. undulata* and *L. xerophila*. Improved ventilation and more intense light assists in reducing the occurrence of this fungus. In highly susceptible areas, preventative spraying with a triforine-based fungicide might be required.

VIRAL DISEASE

Lachenalias are particularly prone to the *Ornithogalum* mosaic virus (OMV), which can produce symptoms within 6 weeks of infection and for which there is no cure. The virus is transmitted through the sap of an infected plant to a healthy plant by transmitting agents such as aphids, mealy bugs, slugs and snails, and on cutting instruments such as secateurs. The virus causes a variety of symptoms: deformed or stunted leaves and inflorescences; discolouration patterns on the leaves and flowers in the form of mosaics, mottling or streaks of light yellow or dull red; and necrotic spots. Valuable plant material that displays symptoms should immediately be drenched and sprayed against transmitting agents, and isolated in insect-proof enclosures. The seeds can be harvested as viral infections of *Lachenalia* are not transferred this way, and the infected bulbs then destroyed speedily, preferably by burning.

3. LACHENALIAS AND THE ENVIRONMENT

CONSERVATION

In common with numerous geophytes that are endemic to South Africa's winter rainfall zone, increasing numbers of *Lachenalia* species are under threat or facing extinction in the wild (Duncan, 2003f). Currently, 51 taxa fall within the threat categories of 'Critically Endangered', 'Endangered' and 'Vulnerable'; six are 'Critically Endangered' (*L. arbuthnotiae, L. convallarioides, L. mathewsii, L. moniliformis, L. purpureo-caerulea* and *L. viridiflora* (Figure 22)), 14 are 'Endangered' (*L. aloides, L. bachmannii, L. elegans, L. lactosa, L. liliiflora, L. mediana* subsp. *rogersii, L. neilii, L. nervosa, L. orchioides* subsp. *glaucina, L. paucifolia, L. physocaulos, L. polyphylla, L. salteri* (Figure 23) and *L. stayneri*) and 31 are 'Vulnerable' (*L. alba, L. angelica, L. capensis, L. corymbosa, L. dasybotrya, L. dehoopensis, L. doleritica, L. duncanii, L. fistulosa, L. haarlemensis, L. krugeri, L. leipoldtii, L. macgregoriorum, L. margaretiae, L. marginata* subsp. *neglecta, L. martiniae, L. martleyi, L. mediana* subsp. *mediana, L. minima, L. orchioides* subsp. *parviflora, L. orthopetala, L. peersii, L. quadricolor, L. reflexa, L. sargeantii* (Figure 24), *L. sessiliflora, L. schelpei, L. summerfieldii, L. thomasiae, L. ventricosa* and *L. youngii*). In emphasising the most recent conservation status of each species and the causal threats in the species treatments, I hope that an increased appreciation of the plight facing many lachenalias will result.

Figure 22. The Critically Endangered *Lachenalia viridiflora* is threatened by coastal housing development on the Cape west coast. Image: Graham Duncan.

Figure 23 (left). The Endangered *Lachenalia salteri* is threatened by coastal housing development in the southern Cape. Image: Graham Duncan.

Figure 24 (right). The Vulnerable *Lachenalia sargeantii* is potentially threatened by alien plant infestation. Image: Graham Duncan.

The primary cause of loss of diversity in the genus has been the destruction, degradation or alteration of wild habitats by human interference. Amongst these, agricultural expansion for winter cereal crops, housing and industrial development account for the greatest loss, followed by road construction, overgrazing and trampling by livestock, vineyard expansion, alien plant infestation, mining, road widening and eutrophication. Latterly, a disturbing new cause of loss has been large-scale ploughing of virgin bush for the booming rooibos tea industry in the Western Cape.

One of the major obstacles impeding the protection of threatened lachenalias is that, in certain instances, the entire gene pool of a species falls outside the boundaries of formal nature reserves or protected areas. Notable examples include *L. mathewsii*, *L. moniliformis* and *L. viridiflora*, which exist exclusively on privately owned land. Of key importance is the role that the farming community can play in setting aside tracts of land on which threatened species occur and in forming partnerships with nature conservation agencies. As a measure towards promoting the importance of plant conservation, the South African National Biodiversity Institute (SANBI) began an awareness program in 2003: The Custodians for Rare and Endangered Wildflowers (CREW). The main objective of this program is to coach local volunteers in assessing and monitoring threatened species that occur in tracts of surviving vegetation. The program links volunteers with local conservation agencies and assists in promoting the conservation of key sites. In addition, SANBI forms part of the Kew Millenium Seed

Bank (MSB) partnership, the largest *ex situ* conservation project in the world, in which wild-collected seeds of mainly threatened and useful plants are placed in long-term cold storage in the seed bank at Wakehurst Place, Sussex, thus preventing their extinction. As of May 2012, wild-collected seed stocks of 48 lachenalias (including 21 threatened species) have been lodged within the MSB. Ultimately, stored seeds can be used for restoring species to their natural habitats where feasible. The restoration of original habitat that has been excessively degraded or completely destroyed is, however, not always possible. A real contribution to conservation can also be made by home gardeners and specialist growers of *Lachenalia* in cultivating, propagating and distributing pure stocks of these species.

PHYTOGEOGRAPHY

DISTRIBUTION

The distribution data are based on preserved herbarium material in B, BLFU, BM, BOL, C, GRA, K, LINN, M, NBG, PRE, S, SAM, TCD, UPS, WIND and Z, and recorded in full degree and quarter degree grid square increments according to the reference system of Edwards & Leistner (1971).

Lachenalia is endemic to South Africa and Namibia; it extends from the western and southern parts of Namibia southwards into South Africa, occurring in the Richtersveld, western Bushmanland, throughout Namaqualand, the north-western, western, south-western and southern parts of the Western Cape, the Great and Little Karoos, and the western, southern, south-eastern and north-eastern parts of the Eastern Cape (Duncan 1988a, 2005). The distribution extends as far inland as the central and north-eastern parts of the Northern Cape, and the south-western part of the Free State (Duncan, 1996).

The most northerly record for the genus is *L. giessii* from Namibia, recorded from the farm Zaris in the Maltahöhe grid (2416 CD) in the Tsarisberge of western Namibia (*Müller* 1349, in WIND). The most northerly records in South Africa are in the northern Richtersveld in the Oranjemund grid (2816 BD) where four species occur (*L. bolusii*, *L. buchubergensis*, *L. klinghardtiana* and *L. nordenstamii*), and in the north-eastern Northern Cape in the Kimberley grid (2824 CA and DA) where *L. karooica* occurs. The most easterly record is that of *L. campanulata* from Ongeluks Nek close to the border with south-eastern Lesotho, in the Matatiele grid (3028 AD) (Duncan, 2005, 2006). The most southerly records are at Cape Agulhas in the Bredasdorp grid (3420 CC) where *L. bifolia*, *L. rosea* and *L. sessiliflora* occur. The main centre of diversity, which supports 38 species, is divided between the Fynbos and Succulent Karoo biomes in the 3319 (Worcester) grid, in the valleys and flats of the interior mountains of the winter rainfall zone of the south-western Cape (Map 1). This grid contains a relatively high number of species that are restricted to just one known locality, or with extremely limited ranges. Within this grid, the largest number of species per quarter degree square is in the 3319 AC grid, located in the Tulbagh Valley, where 16 species are found (Table 1). Overall, the quarter degree square with the most species is shared by 3218 BB (Clanwilliam) in the Western Cape and 3119 AC (Calvinia) in the Northern Cape, both of which have 18 species (Table 1). There is a marked decrease in species diversity towards the summer-rainfall areas of the Eastern Cape, the Great Karoo (particularly in the vast southern part known as 'Die Vlakte'), Bushmanland, and the south-eastern parts of Namibia. There are no records of *Lachenalia* from the 3122 and 3223 degree squares of the Great Karoo, but this is probably due to undercollecting. *L. karooica* occurs in several disjunct localities, mainly in the summer-rainfall areas of the north-western Eastern Cape, south-western Free State and north-eastern Northern Cape, with an

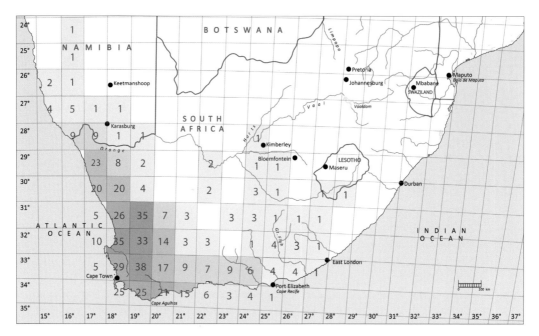

Map 1. Species richness in *Lachenalia* in geographical degree squares.

outlier at Prieska in central Northern Cape. Similarly, the almost exclusively Eastern Cape species *L. bowkeri* has a potentially wider distribution, with outliers west of Colesberg and east of Murraysburg in the north-eastern and eastern Great Karoo, respectively. The deciduous nature of the genus, inhospitable terrain and the fact that growth and flowering are dependent on sufficient seasonal rainfall, are contributory factors to the poor distribution records for these areas (Duncan, 1988a).

Table 1

Ten most speciose quarter degree squares for *Lachenalia*.

Grid square	Region	Number of species
3218 BB	Clanwilliam	18
3119 AC	Calvinia	18
3319 AC	Tulbagh	16
3318 CD	northern Cape Peninsula	15
3318 DC	Bellville	14
3219 AA	Pakhuis	14
3318 DA	Philadelphia	13
3319 BC	Hex River	13
3418 AB	southern Cape Peninsula	13
2917 DB	Springbok	13

Factors promoting speciation and distribution

The two most important modes of speciation in *Lachenalia* appear to have been allopatry and ploidy changes. Under circumstances of environmental change, such as increasing aridity, evolution may be due to the isolation of marginal populations by physical barriers that disrupt gene flow, such as mountain ranges. This results in populations becoming reproductively isolated and diverging morphologically, leading to allopatric speciation (Axelrod, 1972; Stebbins, 1952). Speciation can also occur through the establishment of founder populations as a result of extraordinarily long-distance dispersal of seed beyond the normal gene pool of the parent population. If the parent and founder populations experience different environmental changes in conjunction with random genetic drift, morphological divergence and eventual speciation might result.

Owing to their high biodiversity, Africa's arid regions are thought to be of considerable age (Jürgens, 1997). Many geophytic genera have speciated extensively in the arid southern and western parts of the subcontinent. Axelrod and Raven (1978) note that geographic speciation in this area is likely to have been rapid, and that much of the richness at the species level is of recent origin, leading to high species:genus ratios (Linder, 2006). In *Lachenalia*, the arid regions of the Worcester–Robertson Karoo, Little Karoo, Ceres Karoo, Bokkeveld Plateau, Knersvlakte and central Namaqualand, as well as the semi-arid Olifants River and Tulbagh Valleys and the coastal parts of the south-western and western Cape, have high levels of diversity. Most of the *Lachenalia* species of these regions are endemic, implying local radiation. Considering the large numbers of closely related *Lachenalia* species typical of these highly diversified lineages, much of the speciation is probably recent.

Disjunct populations often represent the initial stages of allopatric speciation. The large topographical diversity present in southern Africa has given rise to highly variable climatic conditions, superimposed upon very different soil types, which generates a wide range of spatially disjunct niches. The exceptional species richness in the geophyte flora of the region is a product of these factors (Cowling *et al.*, 1989). A number of *Lachenalia* species with winter growth cycles that occur in hot, arid areas such as the Nama Karoo and Grassland biomes in the interior of South Africa, such as *L. bowkeri* (Figure 25), *L. canaliculata*, *L. karooica* and *L. zebrina*, occur within bands of transitional rainfall between the extremes. These species may be able to accumulate water opportunistically, allowing them to maintain their winter growth cycle.

Several widely distributed lachenalias occur across markedly different soil types. *L. unifolia* occurs in deep, acid sandy granitic or sandstone-derived soils over much of the western parts of its range but is also found in alkaline, stony clays in the valleys of the Little Karoo. Similarly, the widely distributed *L. bifolia* occurs on deep alkaline sand dunes and flats (Figure 26), as well as in acidic, humus-rich loam on granite outcrops of the Cape west coast (Figure 27). *L. punctata* occurs on alkaline sand just above sea level on the Cape west

Figure 25. A form of the winter-growing *Lachenalia bowkeri* from Colesberg in the eastern Karoo, a predominantly summer rainfall area. Image: Graham Duncan.

Figure 26 (left). *Lachenalia bifolia* in deep alkaline sand dunes at Bloubergstrand, northern Cape Peninsula. Image: Graham Duncan.

Figure 27 (right). *Lachenalia bifolia* in a shallow, acidic, humus-rich depression of a granite outcrop near Vredenburg. Image: Graham Duncan.

coast and on acid sandstone outcrops on the southern Cape Peninsula. The widespread *L. ensifolia* subsp. *ensifolia* occurs in stony, poorly drained soils in the western parts of its range, and in granitic and calcareous, well-drained soils in the southern parts. Even species with relatively narrow distributions can occur in markedly different soil types; for example, *L. valeriae* is usually encountered in deep red sandy plains in western Namaqualand but also occurs in crevices of shales in central Namaqualand.

A number of lachenalias have only been found on single substrates and most have highly restricted distributions; examples include *L. salteri*, which only occurs in seasonal bogs in acid sandstone of the southern Cape (Figure 180); *L. doleritica* (Figure 28) and *L. neilii*, which only occur in the heavy, fine-grained doleritic clay soils of the Bokkeveld Plateau; and *L. alba*, a Bokkeveld endemic, which only occurs in Dwyka Tillite, a hard, stony clay soil (Figure 159). Similarly, *L. karooica* is restricted to dolomitic limestone outcrops in the Northern Cape, Free State and Eastern Cape; *L. lutzeyeri*, *L. montana* and *L. sargeantii* (Figure 29) are confined to rocky, sandstone mountain slopes at the southern Cape; and *L. sessiliflora* is restricted to limestone flats and ridges at the southern Cape (Figure 264). Examples of parapatric species are *L. mathewsii* and *L. punctata*. The distributions of the two species overlap near Vredenburg, but they are always found on different soil types: *L. mathewsii* occurs on seasonally wet, granite-derived gravelly flats, whereas *L. punctata* grows in relatively dry, deep alkaline sand. A further example of parapatric species is *L. anguinea* and *L. framesii* in western Namaqualand, the two species sometimes occur within centimetres of each other, but *L. anguinea* is always in deep red sand (Figure 30), whereas *L. framesii* is always in quartz patches (Figure 31).

Figure 28 (left). *Lachenalia doleritica* occurs on doleritic clay flats around Calvinia. Image: Graham Duncan.

Figure 29 (right). *Lachenalia sargeantii* is confined to rocky sandstone mountain slopes of the southern Cape. Image: Graham Duncan.

Figure 30 (left). *Lachenalia anguinea* in a deep red sand dune in western Namaqualand. Image: Graham Duncan.

Figure 31 (right). *Lachenalia framesii* on quartz flats in western Namaqualand. Image: Graham Duncan.

Figure 32. *Lachenalia longituba*, endemic to seasonally inundated clay sites of the Roggeveld Plateau. Image: Graham Duncan.

Figure 33 (left). *Lachenalia moniliformis*, endemic to a single locality in the Worcester Valley in the south-western Cape. Image: Graham Duncan.

Figure 34 (right). *Lachenalia lutzeyeri*, only known from its type locality near Gansbaai in the southern Cape, with *Watsonia stenosiphon*. Image: Graham Duncan.

High levels of endemism are evident throughout the distribution range of *Lachenalia*, although the highest concentration occurs in the western and south-western Cape; examples include *L. margaretiae* and *L. thomasiae* (Cederberg), *L. mathewsii* and *L. viridiflora* (west coast), *L. flava* (Tulbagh Valley and Bainskloof), and *L. moniliformis* (Figure 33) and *L. stayneri* (Worcester Valley). Smaller concentrations of narrow endemic species occur in the southern Cape (e.g. *L. lutzeyeri* and *L. sargeantii*), the Bokkeveld Plateau at Nieuwoudtville (e.g. *L. macgregoriorum* and *L. neilii*), the Hantam Plateau at Calvinia (*L. schelpei*), the Roggeveld (*L. congesta* and *L. longituba* (Figure 32)), and western Namaqualand (*L. angelica*). Eleven species are currently known only from their type localities (*L. angelica*, *L. duncanii*, *L. leomontana*, *L. lutzeyeri* (Figure 34), *L. macgregoriorum*, *L. moniliformis*, *L. patentissima*, *L. pearsonii*, *L. schelpei*, *L. summerfieldii* and *L. thunbergii*).

The south-western and western parts of the distribution range of *Lachenalia* are the most speciose. The 3319 (Worcester) grid has the greatest diversity, with 38 species in the flats and valleys created between major mountain ranges such as the Hex River Mountains and the Skurweberge, with their accompanying edaphic and microclimatic variation (Map 1). This habitat heterogeneity is also evident in the two second-most speciose areas for the genus, the 3218 (Clanwilliam) grid (35 species) and the 3119 (Calvinia) grid (35 species), where disruption of gene flow created by major mountain ranges such as the Cederberg, Piketberg, Koue Bokkeveld, Olifantsrivierberge and Bokkeveld mountains provide topographic variation.

ECOLOGY, HABITAT AND ADAPTIVE STRATEGIES

ECOLOGY AND HABITAT

Lachenalia occurs in six of the nine biomes recognised in South Africa (Mucina & Rutherford, 2006): Fynbos (90 species, 57 endemics), Succulent Karoo (62 species, 26 endemics), Nama Karoo (5 species, 0 endemics), Albany Thicket (5 species, 0 endemics), Grassland (5 species, 0 endemics) and Desert (4 species, 2 endemics). The genus occurs in at least 162 of the vegetation types recognised by Mucina & Rutherford (2006), of which Cederberg Sandstone Fynbos supports the largest number of species (21). Other species-rich vegetation types include Namaqualand Klipkoppe Shrubland (20 species), Hantam Karoo (19 species), Breede Shale Renosterveld (16 species), Swartland Shale Renosterveld (14 species), Swartland Granite Renosterveld (14 species) and Leipoldtville Sand Fynbos (13 species). A number of species traverse more than one biome: for example, *L. trichophylla* is found in Fynbos and Succulent Karoo, *L. inconspicua* in Succulent- and Nama Karoo, *L. bolusii* in Desert, Succulent Karoo and Fynbos, and *L. karooica* in Succulent- and Nama Karoo and in Grassland.

The genus occurs in all three principal floristic regions (Cape Floristic, Succulent Karoo and Maputaland–Pondoland Regions) and in nine of the 18 centres of endemism recognised in southern Africa (Kamiesberg, Gariep, Knersvlakte, Little Karoo, Worcester–Robertson Karoo, Hantam–Roggeveld, Albany, Drakensberg Alpine and Griqualand West) (Van Wyk & Smith, 2001).

The altitudinal range of *Lachenalia* species varies significantly from just above sea-level (*L. bifolia*, *L. punctata* (Figure 35) and *L. pallida* (Figure 36)) on the littoral dunes and granite outcrops of the Cape west coast, to well over 2,000 m in the Sneeuberge Mountains west of Cradock and the Witteberge Mountains east of Lady Grey in the Eastern Cape interior (*L. campanulata* (Figure 37)) (Duncan, 1998b). *Lachenalia* occurs across a very wide range of habitats, from arid conditions in the

Figure 35. *Lachenalia punctata* in deep alkaline sand just above sea level near Saldanha. Image: Graham Duncan.

far north-western parts of its range in south-western Namibia and the Richtersveld, to areas of high winter rainfall in the south-western parts, such as the southern Cape Peninsula. Most *Lachenalia* species occur in open habitats in full sun, often amongst low succulent scrub in rocky areas. Varying degrees of shade are tolerated within *Lachenalia* populations that occur in rocky habitats; some protection results in larger, more floriferous specimens. *L. margaretiae* is the only species that always occurs in full shade whereas six other species (*L. capensis*, *L. fistulosa*, *L. latimeriae*, *L. orchioides*, *L. rosea* and *L. thomasiae*) sometimes occur in semi-shade. *L. arbuthnotiae* and *L. bachmannii* (Figure 38) occur along the margins of seasonal pools or in standing water in acid sand or clay, respectively. *L. contaminata*, *L. mathewsii*, *L. mediana* subsp. *rogersii*, *L. orthopetala*, *L. polyphylla*, *L. reflexa*, *L. salteri* and *L. zeyheri* (Figure 39) are always associated with seasonally wet sites. *L. salteri* is always associated with seepage in nutrient-deficient, acid sandy soil (Duncan, 2003d). *L. quadricolor* (Figure 40) and *L. viridiflora* (Figure 22) inhabit humus-rich crevices and shallow depressions of granite outcrops that become waterlogged for short periods in winter. Along the western margin of Namibia and South Africa, *L. nutans* and *L. anguinea* are restricted to arid deep red sand, and *L. zebrina* occurs in arid conditions on stony shale flats (Figure 41).

Most lachenalias are gregarious and occur in small to medium-sized groups, almost always in association with other vegetation such as low succulent scrub, geophytes, annuals, grasses, restios and low shrubs. Populations can number hundreds or even thousands of individuals for species such as *L. bachmannii* from the fertile clay ground of the Swartland, *L. splendida* from the Knersvlakte (Figure 42) and *L. pallida* (Figure 43), which is widespread in the south-western Cape. The bulbs of most lachenalias produce offsets, but certain species, such as *L. congesta* from the Roggeveld Plateau and *L. duncanii* from south-western Bushmanland, are always solitary. Many species occur sympatrically; for example, *L. mathewsii* and a mauve form of *L. pallida* occur sympatrically on seasonally moist gravel flats near Vredenburg on the Cape west coast, and 25 km to the south, *L. paucifolia* and *L. quadricolor*

Figure 36 (top left). *Lachenalia pallida* on a granite outcrop on the Cape west coast. Image: Graham Duncan.

Figure 37 (top right). *Lachenalia campanulata* grows in grass tussocks at well over 2,000 m on the interior mountains of the Eastern Cape. Image: Hubert Kurzweil.

Figure 38 (bottom left). *Lachenalia bachmannii* grows in large colonies associated with seasonal pools in clay soil. Image: Graham Duncan.

Figure 39 (bottom right). The dwarf *Lachenalia zeyheri* occurs in winter-inundated sites, flowering as temperatures rise in late spring. Image: Graham Duncan.

Figure 40. *Lachenalia quadricolor* inhabits crevices and depressions of granite outcrops. Image: Graham Duncan.

Figure 41. The xerophyte *Lachenalia zebrina* occurs in arid conditions on stony shale flats. Image: Graham Duncan.

Figure 42. A large colony of *Lachenalia splendida* on clay flats near Vredendal. Image: Graham Duncan.

occur on a granite outcrop at Langebaan within a few metres of each other. Similarly, *L. montana* and *L. peersii* grow side by side on south-facing lower mountain slopes at Hermanus on the south coast. Granite outcrops are especially favoured by lachenlias: four taxa (*L. bifolia*, *L. longibracteata*, *L. pallida* and *L. wrightii*) occur sympatrically at a locality near Vredenburg that used to support a fifth taxon, *L. viridiflora*. Similarly, five taxa (*L. comptonii*, *L. juncifolia*, *L. multifolia*, *L. suaveolens* and *L. violacea*) occur in close association at a locality in the southern Tanqa Karoo; at a locality south-west of Tulbagh, *L. flava* and *L. longibracteata* occur in mixed populations (Figure 44); and at another locality near Tulbagh, five taxa (*L. contaminata*, *L. longibracteata*, *L. pallida*, *L. polyphylla* and *L. unifolia*) occur sympatrically. The marked absence of natural hybrids throughout the genus can be ascribed mainly to the high level of chromosomal variation that results in barriers that prevent crossing of species even though their flowering periods might overlap.

The lowest temperature experienced across the range of habitats in which lachenalias grow is -15°C during the winter months at Sutherland on the Roggeveld Plateau, where *L. canaliculata*, *L. congesta*, *L. isopetala* and *L. longituba* occur. Sub-zero winter temperatures and snowfalls also occur regularly where *L. campanulata* grows on the interior mountains of the Eastern Cape. *L. karooica* in the vicinity of Fauresmith in the south-western Free State and *L. canaliculata* at Victoria West in the central Great Karoo are often subjected to severe frost. Mid-summer temperatures often rise above 40°C in the Richtersveld, Namaqualand, and the Tanqua-, Little- and Great Karoos. Precipitation patterns vary markedly, from an annual average of 43.4 mm in the Richtersveld in the far north-western Northern Cape (and probably much lower in south-western Namibia), to 1,500 mm at Newlands on the eastern slopes of Table Mountain in the Cape Peninsula.

THE GENUS LACHENALIA
LACHENALIAS AND THE ENVIRONMENT

Figure 43. Yellow forms of *Lachenalia pallida* around Hopefield and Malmesbury often occur in dense colonies numbering thousands of individuals. Image: Graham Duncan.

Figure 44. *Lachenalia flava* (golden yellow) and *L. longibracteata* (yellow) in mixed populations near Tulbagh. Image: Graham Duncan.

ADAPTIVE STRATEGIES
Dealing with aridity

Adaptation within *Lachenalia* has given rise to xeromorphic features that reduce water loss during the growing period and protect the bulb during the dormant period. The leaves and roots of all members of the genus are deciduous *in situ*, and the bulbs of all species except *L. pearsonii* are adapted to survive the hot, dry summer months in a dormant state. In arid habitats, such as the Richtersveld, Namaqualand and Bushmanland, lachenalias are frequently subjected to prolonged droughts and erratic rainfall patterns during the winter growing period, but the bulbs of species in these regions are adapted to remain dormant for many years until favourable conditions return. Lachenalias from arid areas are often small; for example, leaf and bulb reduction is clearly evident in *L. patula*, which is endemic to the barren Knersvlakte in the north-western Western Cape. The extremely succulent, subterete leaves of this species are able to store large quantities of water and it has the smallest bulb in the genus, which is surrounded by extremely tough, cartilaginous outer scales (Figure 45). Its remarkably shallow-seated bulbs and shallow, horizontally spreading root system allow it to take advantage of heavy dew falls and light rain showers, as do the marginally larger, shallow-seated bulbs of *L. namibiensis* from south-western Namibia, which have cartilaginous outer scales similar to those of *L. patula*. Protective outer scales have evolved in numerous other lachenalias, including *L. bowkeri*, *L. maximiliani* and *L. undulata*, and are especially well developed in *L. isopetala* (Figure 46). Many

Figure 45 (left). The extremely succulent leaves of *Lachenalia patula*. Image: Graham Duncan.

Figure 46 (right). The bulbs of *Lachenalia isopetala* are adapted to arid conditions in having protective hard outer scales. Image: Graham Duncan.

Figure 47. The bulbs of *Lachenalia glaucophylla* lack protective hard outer scales but survive extended periods of drought in being deep-seated. Image: Graham Duncan.

Figure 48. The small, shallow bulbs of *Lachenalia angelica* benefit from drenching sea mists in coastal western Namaqualand. Image: Graham Duncan.

Lachenalia species from arid habitats have solitary bulbs and reproduce by abundant seed formation instead of by expending energy in producing offsets.

Several lachenalias from arid areas lack protective outer scales and these have deep-seated bulbs that occur in deep sandy soils. Such species include *L. nutans* from open flats in deep red coastal sands in south-western Namibia, forms of *L. anguinea* from western Namaqualand, and a form of *L. glaucophylla* (Figure 47) from red dunes south-west of Middelpos. All are capable of remaining dormant for many years during seasons of insufficient rainfall. Several lachenalias from the arid coastal parts of Namaqualand, the Richtersveld and south-western Namibia benefit from drenching sea mists in winter. These include *L. buchubergensis* from Namibia, *L. klinghardtiana* from Namibia and the northern Richtersveld, and *L. angelica* (Figure 48) from western Namaqualand.

Defence

The upper leaf surfaces of many lachenalias, including *L. arbuthnotiae*, *L. punctata* and forms of *L. luteola*, are lightly to heavily marked with deep purple, brown or green spots or blotches. The leaf bases and lower leaf surfaces of several other species are heavily marked with transverse bands ranging from magenta at, and just below, soil level and to deep purple just above soil level. The central and uppermost portions of the leaves of these species either shade to green (as in *L. anguinea*, *L. hirta*), or have deep purple or maroon spots or blotches (as in *L. verticillata* and *L. macgregoriorum*, respectively). Leaf base and leaf upper surface markings are thought to provide herbivore avoidance by confusing search images through camouflage (Duncan 2005) (Figure 49). The extremely succulent, short and erect, subterete leaves of *L. patula* turn deep maroon in bright sunlight and are perfectly camouflaged against the black or white quartz pebbles and dolerite stones amongst which the species grows. The leaf margins of *L. duncanii* and *L. nervosa* are sclerotic, and this may deter edge-feeding invertebrate predators. Long, simple trichomes on the upper or lower leaf surfaces of several species, such as *L. comptonii* and *L. hirta*, and long and short stellate trichomes on the upper surfaces of certain forms of *L. trichophylla* (Figure 50), may also discourage grazing by herbivores. Trichomes are likely to have evolved under a whole regime of selective pressures because they form the interface between the plant and the environment. Clearly, they are involved in mitigating physical stresses such as cold, light and heat loading by insulating the mesophyll. They also reduce the flow of air across stomata and thus reduce transpiration rates (Esau, 1977). A number of researchers have also shown a link between indumentum and

Figure 49. Leaf upper surface markings, as in *Lachenalia luteola*, are thought to provide herbivore avoidance by confusing search images through camouflage. Image: Graham Duncan.

protection against insect predation, oviposition and nutrition of larvae (Levin, 1973).

Many lachenalias have dull-coloured flowers in shades of green, brown or grey, which could be a defensive mechanism that minimises attention from predators. Cowling and Pierce (1999) hypothesise that the long subterranean, horizontal stolons produced from the base of the bulb of *L. namaquensis*, which terminate in bulblets that are produced just below ground level, are a defensive strategy that allows these plants to escape the attention of molerats.

Fire

The three pyrophytic species *L. lutzeyeri*, *L. montana* and *L. sargeantii* occur in the southern Cape and have deep-seated bulbs. The bulbs of *L. lutzeyeri* can be buried up to 100 mm deep, allowing them to survive very hot fires and interfire periods when no leaves are produced. These species flower at around the same time of year, and produce relatively few fibrous roots but numerous contractile roots that pull the bulbs deep into the soil.

Figure 50. Long stellate trichomes on the upper surfaces of certain forms of *Lachenalia trichophylla* may discourage grazing by herbivores. Image: Graham Duncan.

PHENOLOGY

Collectively, the genus *Lachenalia* has a very long flowering period of approximately twelve months *in situ* (depending on rainfall and fire patterns). The earliest species (*L. pearsonii*) begins to flower from mid-January, the latest (*L. lutzeyeri*, *L. salteri* and *L. sessiliflora*) can flower until late December and sometimes extend their flowering to early January (*L. salteri*) and rarely to late January (*L. lutzeyeri*). September is the month in which the most taxa (94) flower (Table 2).

The collective growing period for the genus is of slightly shorter duration, extending from mid January until early November. With the exception of the mainly summer-growing *L. pearsonii*, *Lachenalia* species' growth coincides with autumn and winter rainfall, both in winter and predominantly summer rainfall areas. Most species are synanthous, the growth cycle being characterised by initial rapid vegetative growth following the onset of cool, wet weather in the autumn and early winter. With the onset of warmer weather in spring, leaf and root senescence marks the beginning of a dormant period that lasts throughout the summer (Duncan & Anderson, 1999). Under cultivation, two Eastern Cape endemics, *L. campanulata* and *L. convallarioides*, as well as *L. longituba* from the Sutherland district, remain evergreen unless forced into dormancy by sustained drought.

The leaves of two of the earliest-flowering species (*L. punctata* and certain forms of *L. corymbosa*) are almost always partially hysteranthous. The leaves of the summer-growing *L. pearsonii* enter dormancy in late winter, and those of *L. punctata* in early spring, much earlier than the winter-,

Table 2

Flowering phenology in *Lachenalia*.

	Jan	Feb	Mar	Apr	May	Jun	Jul	Aug	Sep	Oct	Nov	Dec
L. alba									●	●		
L. algoensis								●	●			
L. aloides						●	●	●				
L. ameliae									●			
L. angelica									●	●		
L. anguinea							●	●	●			
L. arbuthnotiae								●	●			
L. argillicola						●	●					
L. attenuata								●	●			
L. aurioliae						●	●	●				
L. bachmannii								●	●			
L. barkeriana				●	●							
L. bifolia				●	●	●						
L. bolusii								●	●			
L. bowkeri								●	●			
L. bruynsii								●	●			
L. buchubergensis					●	●						
L. calcicola					●							
L. callista								●	●			
L. campanulata										●	●	●
L. canaliculata								●	●			
L. capensis								●	●			
L. carnosa								●	●			
L. cernua									●	●		
L. comptonii								●	●			
L. concordiana							●	●				
L. congesta					●	●	●					
L. contaminata								●	●	●		
L. convallarioides							●	●	●	●		
L. corymbosa			●	●	●							
L. dasybotrya								●	●			
L. dehoopensis								●				

THE GENUS LACHENALIA
LACHENALIAS AND THE ENVIRONMENT

	Jan	Feb	Mar	Apr	May	Jun	Jul	Aug	Sep	Oct	Nov	Dec
L. doleritica									▓	▓		
L. duncanii								▓	▓			
L. elegans										▓	▓	
L. ensifolia subsp. ensifolia			▓	▓	▓	▓						
L. ensifolia subsp. maughanii					▓	▓						
L. fistulosa									▓	▓		
L. flava							▓	▓	▓	▓		
L. framesii							▓					
L. giessii												
L. glauca									▓	▓		
L. glaucophylla									▓	▓		
L. haarlemensis									▓			
L. hirta								▓				
L. inconspicua							▓	▓				
L. isopetala										▓	▓	
L. judithiae												
L. juncifolia											▓	▓
L. karooica						▓	▓					
L. karoopoortensis												
L. klinghardtiana						▓	▓					
L. kliprandensis												
L. krugeri							▓					
L. lactosa									▓			
L. latimeriae								▓				
L. leipoldtii												
L. leomontana										▓		
L. liliiflora									▓			
L. longibracteata							▓	▓				
L. longituba				▓	▓	▓						
L. lutea												
L. luteola												
L. lutzeyeri	▓										▓	▓
L. macgregoriorum												
L. magentea									▓	▓		
L. margaretiae										▓	▓	
L. marginata subsp. marginata						▓	▓	▓				

THE GENUS LACHENALIA
LACHENALIAS AND THE ENVIRONMENT

	Jan	Feb	Mar	Apr	May	Jun	Jul	Aug	Sep	Oct	Nov	Dec
L. marginata subsp. *neglecta*							■	■				
L. marlothii							■	■				
L. martiniae								■	■			
L. martleyi									■	■		
L. mathewsii									■			
L. maximiliani							■	■				
L. mediana subsp. *mediana*								■	■			
L. mediana subsp. *rogersii*								■	■			
L. membranacea								■	■			
L. minima						■						
L. moniliformis								■	■	■		
L. montana									■	■	■	
L. multifolia									■			
L. mutabilis							■	■	■			
L. namaquensis								■	■			
L. namibiensis								■	■			
L. nardousbergensis									■			
L. neilii									■			
L. nervosa									■	■	■	
L. nordenstamii					■	■						
L. nutans							■	■				
L. obscura						■	■	■	■	■		
L. orchioides subsp. *orchioides*						■	■					
L. orchioides subsp. *glaucina*									■	■		
L. orchioides subsp. *parviflora*									■	■		
L. orthopetala									■	■		
L. pallida									■	■		
L. patentissima									■	■		
L. patula									■			
L. paucifolia			■	■	■							
L. pearsonii	■	■	■									
L. peersii										■		
L. perryae								■	■			
L. physocaulos									■	■		
L. polyphylla									■	■		

THE GENUS LACHENALIA
LACHENALIAS AND THE ENVIRONMENT

	Jan	Feb	Mar	Apr	May	Jun	Jul	Aug	Sep	Oct	Nov	Dec
L. polypodantha subsp. polypodantha								■	■			
L. polypodantha subsp eburnea									■			
L. punctata			■	■	■	■	■					
L. purpureo-caerulea									■	■	■	
L. pusilla				■	■	■						
L. pygmaea				■								
L. quadricolor						■	■	■				
L. reflexa						■	■	■				
L. rosea									■	■	■	■
L. salteri	■									■	■	■
L. sargeantii										■	■	
L. schelpei						■	■					
L. schlechteri												
L. sessiliflora	■											
L. splendida							■	■				
L. stayneri												
L. suaveolens												
L. summerfieldii												
L. thomasiae											■	
L. thunbergii												
L. trichophylla							■	■				
L. undulata					■	■						
L. unifolia												
L. valeriae							■					
L. vanzyliae								■	■	■		
L. variegata							■	■	■			
L. ventricosa							■	■				
L. verticillata												
L. violacea												
L. viridiflora					■	■						
L. whitehillensis										■		
L. wrightii								■	■			
L. xerophila												
L. youngii									■	■		
L. zebrina								■	■			
L. zeyheri								■	■			

spring- and summer-flowering species. A group of ten species (*L. angelica*, *L. comptonii*, *L. isopetala*, *L. liliiflora*, *L. macgregoriorum*, *L. nervosa*, *L. orthopetala*, *L. peersii*, *L. purpureo-caerulea* and *L. sessiliflora*) are often proteranthous *in situ*, as their flowers appear at the end of the growing season in late spring or early summer, usually after the leaves have withered (Duncan, 1987; Duncan & Linder Smith, 1999b). Only *L. pearsonii* from inland southern Namibia follows a summer-rainfall growth cycle in the wild. This species normally flowers in January and February at the hottest time of year, following late summer rains, but it is opportunistic in areas of extremely erratic rainfall and in cultivation it readily adapts to a winter growth cycle.

Three pyrophytic species from the mountains of the southern Cape, *L. lutzeyeri*, *L. montana* and *L. sargeantii*, require bush fires in the preceding summer and autumn seasons for flowering to occur (Duncan, 1998b). An example of erratic flowering in one of these was observed recently in *L. sargeantii*. The species was discovered in full bloom in November 1970 after a summer burn near Bredasdorp. Only a few individuals flowered there the following year, and the species was next seen in flower 33 years later, in a new locality in the southern Cape (Duncan & Edwards, 2005; Duncan *et al.*, 2005) (Figure 51). The typical phenological pattern displayed by these fire-dependent lachenalias is lush vegetative growth produced in the winter season immediately following summer or autumn fires, followed by prolific flowering in early summer, and dormancy from midsummer to mid-autumn. During the second winter season following a burn, most of the bulbs remain completely dormant, with an extremely small number of individuals producing leaves in winter and flowers in early summer; in subsequent winter seasons in the absence of fires, no leaf growth or flowers are produced.

The flowering performance of many species, including *L. fistulosa*, *L. flava*, *L. orchioides*, *L. peersii*, *L. rosea* and *L. salteri*, is greatly enhanced by fire but not dependent upon it (Duncan, 2003c, d; Duncan & Anderson, 1997). Usually, the leafing period is directly correlated to the main rainy season, but several species that occur mainly in summer-rainfall regions follow a winter-rainfall growth cycle. Thus, *L. karooica* from south-western Free State, the eastern Northern Cape and the north-western Eastern Cape grows from April to September and flowers from June to September; *L. convallarioides* from the south-eastern part of the Eastern Cape grows from April to November and flowers from late September to early November; and the high altitude *L. campanulata*, which occurs mainly in the mountains of the Eastern Cape interior, grows from February to December and flowers from October to December (Duncan, 1998b).

Figure 51. The pyrophytic *Lachenalia sargeantii* requires bush fires in the preceding summer or autumn seasons for flowering to occur. Image: Graham Duncan.

Flowering phenology in wild populations of *L. congesta* is unusual within the genus. For almost all *Lachenalia* species, the flowering of all of the adult individuals within a population commences more or less simultaneously and continues for a two- to three-week period. *L. congesta* is the exception: the flowering period within a population of this species is staggered, often extending over about two months, with certain individuals commencing flowering in late June, followed by a continuous and increasing number of individuals until the peak flowering period is reached in late July; sporadic individuals come into flower up until the end of August. A proportion of adult *L. congesta* individuals do not flower in a given season, despite having healthy leaves and seemingly ideal conditions. The habitat of *L. congesta* has highly erratic rainfall and droughts can last for extended periods; non-flowering individuals ensure that the bulbs of at least some individuals have sufficient reserves to flower and produce seed during the next season in which sufficient rains fall, while those that flowered in the previous season have time to build up their resources (Duncan, 2009).

4. LACHENALIA BIOLOGY

MORPHOLOGY

Lachenalia is variable in its morphology. Species delimitation is usually unambiguous, but certain species display extensive variation that has led to considerable taxonomic confusion, with minor morphological differences being overemphasised (Duncan, 1992a).

HABIT

Lachenalia is a deciduous, tunicate bulbous geophyte (Duncan, 1988a; du Plessis & Duncan, 1989) (Figure 52). All species growing in the winter- and summer-rainfall zones follow a winter-growing cycle, with the exception of one species from the summer rainfall area (*L. pearsonii* from inland southern Namibia) that has a summer-growing cycle (Duncan & Anderson, 1999). Plant size varies from 5 mm high with rosulate leaves appressed to the ground and a subcapitate inflorescence produced at ground level, as in *L. pusilla*, to 600 mm tall for *L. anguinea*, with a many-flowered, racemose inflorescence produced on a robust, inflated scape and a single long, canaliculate, flaccid leaf.

The arid habitats where many *Lachenalia* species occur such as the Richtersveld, Namaqualand, Bushmanland and Karoo, are frequently subject to prolonged drought and erratic rainfall patterns during the normal winter-growing period, but the bulbs of species in these areas are adapted to remain dormant for many years until favourable conditions return (Duncan, 2003b).

A number of *Lachenalia* species have solitary bulbs and never reproduce vegetatively, examples include *L. kliprandensis* and *L. verticillata*. Most species, however, reproduce by the formation of bulblets (offsets or daughter bulbs). Some produce several bulblets in a ring around the base of the bulb, as in *L. contaminata*, *L. orthopetala* and *L. viridiflora*, but most produce only one or two bulblets each growing season, as in *L. corymbosa* and *L. multifolia*. Bulblets form mainly in the axils of the oldest leaf bases but also occur in the axils of the younger leaf bases. *L. bifolia* and certain forms of *L. cernua* sometimes produce bulblets on the subterranean margins of their leaves, and *L. bifolia* also sometimes produces bulbils on the aerial margins. In addition, seven species are soboliferous (*L. aurioliae* (certain forms), *L. bifolia*, *L. juncifolia* (certain forms), *L. maximiliani*, *L. montana*, *L. moniliformis*, *L. namaquensis* and *L. sargeantii*), increasing in numbers by the formation of bulblets that are produced at the tips of horizontal or erect, subterranean, short or long stolons up to 50 mm long, sometimes resulting in large colonies.

Bulblets and bulbils begin to form towards the end of the winter-growing period, from early to late spring, and undergo dormancy in the same way that mature bulbs do from early summer to early autumn. Towards the end of the growing season, several species, including *L. montana* and *L. neilii*, produce bulblets on very short, erect, stalk-like structures of up to 4 mm in length, arising from the base of the bulb in the axils of the bulb scales. In cultivation, the bulbs of a number of species, notably *L. martiniae*, *L. nervosa*, *L. peersii*, *L. polypodantha*, *L. xerophila* and *L. zebrina*, periodically remain completely dormant during the normal winter-growing season for one or more consecutive growing seasons, irrespective of available moisture (Duncan & Linder Smith, 1999b; Duncan & Visagie, 2001). This sporadic phenomenon has been noted over many years in the collection at Kirstenbosch Botanical Garden in which one or more bulbs from a particular container remain dormant while others from the same container continue to grow normally.

ROOTS

Two types of adventitious roots are produced, fibrous (non-contractile) roots and thicker contractile roots, both are annual and desiccate completely during the hot, dry dormant period. Root development begins around the periphery of the basal plate as soon as night temperatures begin to drop towards the end of summer, from mid- to late February in the southern hemisphere, and is followed by the production of leaf shoots in early autumn. Fibrous roots are produced in abundance in all species and are brittle, branched or unbranched, with a diameter ranging from 0.5 mm to 1.0 mm, and a length of up to 45 mm. Contractile roots are most developed in the species from fire-prone habitats, including *L. lutzeyeri*, *L. montana* and *L. sargeantii*, and in those that occur in deep littoral sand dunes (e.g. *L. bifolia* and *L. punctata*). The contractile roots of *L. lutzeyeri*, *L. montana* and *L. sargeantii* pull the bulbs deeply into the soil, ensuring that they are not damaged by excessive heat during bush fires. Contractile roots are harder, thicker (1.5–2.5 mm) and longer (up to 60 mm) than fibrous roots and, following contraction, have transverse wrinkles.

BULB

The storage and regenerative organ in *Lachenalia* is a well-developed, hypogeal bulb. Its development is sympodial as it is constructed by a series of two (inner and outer) consecutive units or modules attached to a dome-shaped, compressed stem, the basal plate. The new (inner) module is situated inside the old (outer) module (Du Toit *et al.*, 2001). At the beginning of the growing period, the old module consists of a dry, papery or cartilaginous tunic (formed from the second cataphyll in the previous growth cycle) and, depending on the species, one, two or more thickened leaf bases from the discarded laminas of the previous growth cycle.

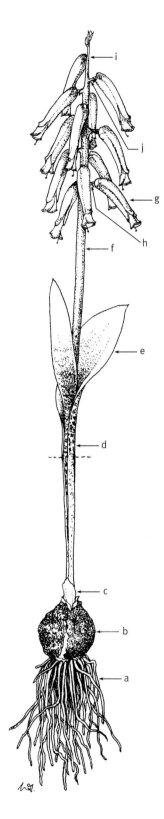

Figure 52. General morphology of a mature *Lachenalia* plant (*L. punctata*) × $^2/_3$. **a** = contractile roots; **b** = globose, tunicate bulb; **c** = cataphyll; **d** = leaf bases; **e** = leaf; **f** = scape; **g** = perianth; **h** = pedicel; **i** = rachis; **j** = bract. Drawn by Vicki Thomas.

The new module usually comprises two cataphylls, and two or more scale leaves that enclose the leaf primordia and the inflorescence. The first (outermost) cataphyll of the outer module disintegrates at the end of the flowering period (Roodbol & Niederwieser, 1998).

Bulb size and depth is extremely variable. The species with the smallest bulb is *L. patula* whose adult bulb reaches a diameter of only 5–8 mm, whereas *L. bifolia* has the largest bulbs, with a diameter of up to 35 mm. Species that occur in deep white or red sand on littoral or inland dunes, such as *L. punctata* and *L. anguinea*, respectively, have bulbs buried up to 170 mm deep. Similarly, the three species that are stimulated into flowering by the effect of natural bush fires (*L. lutzeyeri*, *L. montana* and *L. sargeantii*), occur in deep acid sandstone and have bulbs buried up to 100 mm deep. Most species, however, have bulbs that grow between 5–15 mm below the soil surface. This is often associated with substrate conditions such as shale-derived soils (for *L. obscura* and *L. stayneri*) or shallow depressions in granite rock sheets (for *L. pallida* and *L. paucifolia*) or quartzitic outcrops (for *L. framesii* or *L. patula*).

Most species have globose bulbs but subglobose and ovoid shapes also occur (Figure 53). Bulb scales are usually white but in 10 species, including *L. kliprandensis* and *L. splendida*, they are light to dark yellow and in *L. carnosa* scale colour is polymorphic within populations. When *Lachenalia* bulbs are exposed to the sun, their surfaces usually turn deep purple; *L. contaminata* and *L. orthopetala* provide particularly good examples of this. The tunics adhere to the bulbs and have one of two forms: cartilaginous, light to dark brown or black outer tunics that overlay membranous inner layers, or light to dark brown or greyish-brown spongy outer tunics that overlay membranous inner layers, the latter condition being the most common. Cartilaginous tunics form a protective layer and usually occur in species from arid or semi-arid habitats, such as *L. isopetala* and *L. schelpei*. The thickness of the layer varies in widely distributed species such as *L. undulata*, depending on the substrate in which the population is growing; populations in hard, rocky ground have thicker, harder outer tunics than those in softer, loamy or sandy soils. The bulb apex is distinctly extended in a number of species from arid habitats and may be surrounded by spongy (*L. congesta* and *L. nordenstamii*) or cartilaginous tunics (*L. giessii* and *L. minima*). The uppermost portion of the outer tunics of certain lachenalias from arid and semi-arid areas, such as *L. isopetala* and *L. namibiensis*, is fasciculate, producing a cluster of short to long, more or less flat papery bristles or segments from the splitting of the upper part of the outermost tunics. Gamophylly (fusion) of the thickened leaf bases of the previous growth cycle within the bulb only occurs in two geoflorous species, *L. barkeriana* and *L. pusilla* (Müller-Doblies et al., 1987).

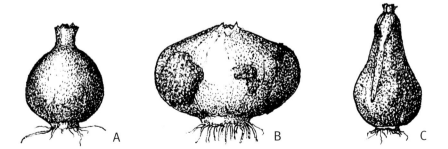

Figure 53. Variation in bulb shape in *Lachenalia*. (A) The globose bulb of *L. vanzyliae*; (B) the subglobose bulb of *L. bifolia*; and (C) the ovoid bulb of *L. schelpei*. All life size. Drawn by Vicki Thomas.

CATAPHYLLS

Tubular, sheathing membranous cataphylls occur in all species of *Lachenalia*. Müller-Doblies *et al.*, (1987) reported three cataphylls for *L. pusilla*, but normally only two cataphylls are present in mature bulbs. The first (outermost) cataphyll of the new bulb module usually remains below soil level and disintegrates during the senescence of the aerial parts of the plant (Du Toit *et al.*, 2001). The second (inner) cataphyll of the new bulb module extends conspicuously above the bulb, and in most species, the distal portion is partially exposed early in the growing season. It can be translucent, green or dull red, but in most species it contains little chlorophyll. The length of the subterranean and aerial portions of the inner cataphyll are highly variable, even within populations; for example, in a population of *L. punctata* growing at Bloubergstrand on the northern Cape Peninsula, it varies from 15 mm to 50 mm. The inner cataphyll usually has distinct, depressed longitudinal veins and the aerial portion is either plain or maculate, but this varies, even within populations. Inner cataphyll apex shape is either obtuse or acute, and in some species it is polymorphic. In most lachenalias, the inner cataphyll adheres tightly to the leaf bases, but in a few species, it loosely surrounds the upper part of the leaf bases. Both loose and tightly adherent cataphylls occur within single populations of *L. bifolia*.

LEAVES

Leaf morphology in *Lachenalia* is remarkably diverse. Some lachenalias have solitary leaves whereas others have many leaves arranged in rosettes, and leaf orientation ranges from prostrate to erect. Most species produce one or two spreading or suberect, lanceolate leaves. The upper leaf surfaces are usually smooth and immaculate, but they can be densely maculate, pustulate, or covered with simple or stellate trichomes. The lower leaf surfaces are usually smooth and immaculate, but they can be densely maculate or covered with simple trichomes (Duncan, 1988a). The colour of the upper and lower leaf surfaces varies mainly through shades of light to dark green, and a number of species have greyish-green or intensely glaucous leaves; the latter condition is seen, for example, in *L. glaucophylla* and *L. isopetala*. In some species, leaf colour is polymorphic and the colours of the upper and lower surfaces vary within populations from uniformly dark green to partially dark green and dark magenta, or to uniformly dark magenta. The extremely short, erect leaves of *L. patula* turn bright maroon in strong sunlight and are perfectly camouflaged against the quartz pebbles amongst which they grow. Several lachenalias that have prostrate leaves (*L. angelica*, *L. congesta* and certain forms of *L. trichophylla*) have lower leaf surfaces that are light to dark magenta. This colouration is due to the presence of anthocyanins in the vacuoles of the subepidermal cell layer (Duncan, 2005).

Leaf length and width are highly variable. *L. angelica* has the shortest leaves (20–35 mm long) and *L. anguinea* has the longest (up to 450 mm). For certain widely distributed species like *L. bifolia*, leaf length varies greatly across the distribution range, and there is frequently considerable intrapopulational variation in the leaf length of species such as *L. mutabilis*. *L. polyphylla* has the narrowest leaves (0.8–1.0 mm wide) and *L. bifolia* has the broadest leaves (up to 90 mm wide). Depressed longitudinal grooves occur between the major veins on the upper leaf surfaces, to a greater or lesser extent; those of *L. nervosa* are particularly deep and conspicuous, hence the specific epithet. The leaves of certain forms of the early-flowering *L. corymbosa* and *L. punctata* are often partially hysteranthous, not being fully developed at flowering time, whereas those of a number of late-flowering species, including *L. macgregoriorum* and *L. sessiliflora*, are often proteranthous in the wild, withering just before the inflorescences emerge; in most species, the leaves are synanthous.

The leaf bases of most lachenalias are immaculate, but in some species they are prominently banded or spotted. The occurrence of banding on the leaf bases and lower region of the lower leaf surface is particularly evident in (but not restricted to) species that have single linear leaves, such as *L. hirta* and *L. unifolia*. In *L. zebrina*, the lower leaf surface and aerial portion of the leaf base is marked with broad, dark purple transverse bands, shading to dark magenta at ground level and to bright magenta on the subterranean portion of the clasping base, whereas the upper surface is always immaculate. In *L. campanulata* the leaf base is marked with minute spots of 1–2 mm in diameter, whereas in *L. klinghardtiana*, the spots can be large and up to 6 mm in diameter.

There are five leaf number classes in *Lachenalia*: always two (e.g. *L. congesta*), always solitary (e.g. *L. hirta*), one or two (e.g. *L. pallida*), usually two but sometimes more (e.g. *L. corymbosa*) and three or more (e.g. *L. orthopetala*). In approximately 35% of species, single- and two-leafed individuals occur within the same population. With the exception of *L. pusilla*, that has lanceolate leaves, the multi-leafed species produce linear leaves. The highest leaf number counted for an adult bulb is 18 leaves for *L. multifolia* (Duncan, 2005).

Leaf orientation can be prostrate, spreading to suberect or erect, with most species having spreading to suberect leaves. Leaf orientation is phenotypically plastic as the prostrate condition frequently becomes spreading or suberect at lower light intensities in cultivation. Prostrate leaves frequently occur in members of the South African Amaryllidaceae and Asparagaceae, and in *Lachenalia* they are evident in seven species: *L. angelica*, *L. congesta*, *L. kliprandensis*, *L. nervosa*, *L. pusilla*, *L. stayneri* and *L. trichophylla*. The precise significance of this leaf trait is a matter of debate, but prostrate leaves clearly provide maximum exposure to sunlight in areas of low vegetation cover. Prostrate leaves might also be effective in avoiding herbivory, in reducing competition from neighbours, in creating a CO_2-rich environment underneath the leaves or in regulating evapotranspiration (Esler *et al*., 1999).

Most lachenalias have smooth upper leaf surfaces, but a small number of species are always pustulate (e.g. *L. purpureo-caerulea*) and in several species this condition is polymorphic within populations (e.g. *L. nervosa* and *L. pallida*). Pustules are usually dome-shaped and their size and density is highly variable, ranging from a sparse to moderate or dense covering of minute pustules of 0.5–1.0 mm in diameter (e.g. *L. valeriae*), to a moderate or dense covering of medium-sized pustules (e.g. *L. liliiflora*), or a sporadic, moderate or dense covering of large pustules of 2–4 mm in diameter (e.g. *L. stayneri*). Pustules only occur on upper leaf surfaces and vary in colour through shades of green, brown or deep purple. The anatomy of pustules is consistent within the genus and comprises the elongation of palisade chlorenchyma cells (Duncan, 2005). *L. moniliformis* produces unique circular, raised fleshy bands along the upper two-thirds of its leaves, giving them the appearance of a string of beads, hence its specific epithet.

Immaculate upper leaf surfaces occur in most species, with smaller groups having surfaces that are always maculate and other species showing polymorphism within populations; the latter condition is evident in *L. viridiflora* (Barker 1972, 1979b). Maculation is produced within a single cell layer through the irregular occurrence of anthocyanins in the vacuoles of subepidermal cells, and occurs as spots or blotches, or a combination of these. Spot colour varies through shades of light to dark purple, magenta, brown or green; frequently all four colours are present in different parts of the same leaf. Spot colour is highly variable in species such as *L. bifolia* and *L. punctata*, where individuals within the same population can have leaves with only dark green spots or with only dark purple spots. In *L. arbuthnotiae*, the upper surface is usually heavily spotted with deep purple, whereas the lower surface has only sporadic spots, but in species such as *L. glaucophylla* and *L. verticillata*, the lower surface is usually heavily spotted with only sporadic spotting of the upper surface.

Three basic leaf shapes dominate within the genus: ovate, lanceolate and linear leaves. Most lachenalias have lanceolate leaves, followed by those with linear and ovate leaves. A high level of variation occurs within each of these leaf shapes; for example, the leaves might be narrowly (*L. congesta*) to broadly (*L. kliprandensis*) ovate, narrowly (*L. haarlemensis*) to broadly (*L. vanzyliae*) lanceolate, or narrowly (*L. moniliformis*) to broadly (*L. hirta*) linear. Leaf shape is stable within species, has diagnostic potential and does not appear to correlate to habitat as both narrow- and broad-leafed species occur in arid environments. Leaf margins vary from flat to undulate or crisped, with most lachenalias having flat margins. Margin shape is phenotypically plastic as the crisped margins are often much reduced or frequently lost in cultivated specimens such as *L. bolusii* and *L. verticillata*.

Juvenile leaf macromorphology and ontogeny

As monocotyledon families are clearly characterised by their seedling type, the differentiation of these types must have happened early in monocot evolution. The developmental stage at which selective pressures would be expected to be maximal is that of early seedling development (Stebbins, 1974). Some large families, including Asparagaceae, have conservation of seedling types even though they have undergone considerable adaptive radiation under a range of widely different ecological conditions. Thus, seedlings provide an important character set for differentiating plants at family level. In many species, germination only occurs under favourable conditions, and therefore divergent radiation of vegetative characters is not strongly selected. A consequence of this is that seedling structure frequently presents conservative character states that are useful in defining phylogenetic relationships (Tillich, 2000).

Seed germination in all *Lachenalia* species is hypogeal (Boyd, 1932). The lower portion of the cotyledon consists of a short, tubular sheath, the hypophyll, and the upper portion of a very short, cylindrical stalk, the hyperphyll, which connects the cotyledon to the seed. In *Lachenalia*, a hypocotyl is not distinguishable on the cotyledon but instead is included in the short cotyledonary sheath, so that the primary root appears to be attached directly to the collar, at the base of the cotyledon. A ring of strong root hairs develops in the upper part of the primary root, just below the collar.

The leaf surfaces of first-year seedlings are smooth in all *Lachenalia* species; pustules and trichomes only develop during the second season of growth. The seeds take a single season (roughly six months) from germination to bulb formation. Three forms of primary seedling leaf orientation are present in the genus: prostrate, suberect or erect. Most species have erect seedling leaves, with prostrate leaves being more common than suberect leaves. The three basic types do not appear to correlate to particular habitat types but progressively flatter leaves do correlate with a progressive increase in leaf width. Prostrate primary seedling leaves frequently contain high concentrations of magenta anthocyanins in the upper mesophyll, resulting in brownish colouring of the upper surface. The orientation of the primary seedling leaf is stable within taxa, but sometimes deviates slightly or markedly from that of the adult leaves. Primary seedling leaf orientation is phenotypically plastic as the prostrate leaf condition becomes spreading or suberect where light intensity is insufficient for plants in cultivation.

Most *Lachenalia* species have terete seedling leaves, with the minority having flat seedling leaves. Terete seedling leaves have erect or suberect orientation whereas flat seedling leaves are always prostrate.

Leaf anatomy and micromorphology

Considerable primary, secondary and tertiary microsculpturing of the epidermis exists in *Lachenalia*, and epidermal cells occur as a single layer in all species. Vesture of the upper and lower leaf surfaces is diagnostic for several species; for example, *L. ameliae* and *L. comptonii* have soft, simple, multicellular trichomes restricted to their upper surface leaf surfaces, whereas *L. hirta* has stiff, simple multicellular

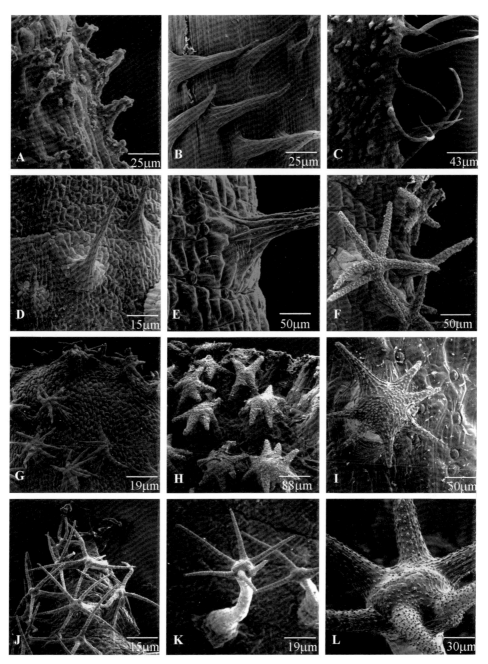

Figure 54. Scanning electron micrographs of variation in leaf trichome morphology in *Lachenalia*: (A) upper surface simple trichomes of *L. ameliae*; (B) lower surface simple trichomes of *L. hirta*; (C) lower surface and marginal simple trichomes of *L. hirta*; (D) upper surface simple trichomes of *L. comptonii*; (E) trichome stalk of upper surface simple trichome of *L. comptonii*; (F) upper surface and marginal stellate trichomes of *L. angelica*; (G) upper surface apical stellate trichomes of *L. angelica*; (H) upper surface short stellate trichomes of *L. trichophylla*; (I) upper surface single, short stellate trichome of *L. trichophylla*; (J) upper surface apical long stellate trichomes of *L. trichophylla*; (K) upper surface long stellate trichomes of *L. trichophylla*; and (L) close-up view of trichome branches of *L. trichophylla*. Image: Graham Duncan.

trichomes that are restricted to the lower leaf surface, and *L. angelica*, *L. polypodantha* and *L. trichophylla* have stellate trichomes that are restricted to the upper leaf surface (Figure 54). The density and length of simple trichomes varies considerably both within and between individuals of the same population. In *L. comptonii*, trichomes range from 3–12 mm in length, in *L. ameliae* they are 1–5 mm long and in *L. hirta* they vary from 1.5 mm to 10.0 mm in length. In certain forms of *L. ameliae*, trichome occurrence varies within populations, some individuals having both long and short trichomes, others possessing only long trichomes. The stellate trichomes of *L. angelica*, *L. polypodantha* and *L. trichophylla* vary considerably in length. The trichome stalks of the very short, stellate trichomes of *L. angelica* and *L. polypodantha* vary from 0.1 mm to 0.3 mm in length, whereas those of *L. trichophylla* vary from 0.3 mm to 15.0 mm in length, depending on provenance. The stalks of the stellate trichomes of certain 'southern' populations of *L. trichophylla* are uniformly short, whereas those of 'northern' populations vary from short to very long on the same leaf (Barker, 1980a). Trichome characters are stable, expressing consistently *ex situ*.

The upper and lower surfaces of all *Lachenalia* species are covered with a dense covering of epicuticular wax platelets, which have irregular margins and are nearly perpendicular to the cell surfaces. The leaves are amphistomatic and stoma shape varies from almost round as in *L. dehoopensis* to narrowly elliptical as in *L. ventricosa* or elliptical as in *L. peersii*. Four xerophytic species (*L. buchubergensis*, *L. multifolia*, *L. pearsonii* and *L. polypodantha*) have conspicuously few stomata on both the upper and lower leaf surfaces. The stomata of most species occur in shallow crypts, whereas those of three xerophytes (*L. isopetala*, *L. nordenstamii* and *L. zebrina*) are positioned in deep crypts, but generally stomatal position is not correlated to any particular habitat or leaf shape.

Two main types of epidermal cells occur, elongate and fusiform cells (as seen in surface view); they are regular or irregular in shape and axially elongated, with the vast majority of lachenalias having elongate cells. Elongate epidermal cells vary in shape from relatively short and wide to long and narrow, and are mostly four-sided, with more or less straight walls. Those species with prostrate leaves, such as *L. angelica* and *L. stayneri*, have the shortest epidermal cells in their upper leaf surface but leaf shape does not correlate to cell shape.

The secondary microsculpturing of the cell walls is variable, from almost flat (as in *L. capensis*) to strongly convex (as in *L. giessii*), with sculpturing varying from slightly to medium convex. Six types of tertiary upper leaf surface sculpturing occur, namely finely striate, laevigate, median longitudinally ridged (the most commonly encountered condition across the genus), median longitudinally ridged with sides transversely ridged, median finely striate with sides transversely ridged, and transversely ridged (Figure 55). Four types of tertiary sculpturing occur on the lower leaf surfaces: finely striate, laevigate, median longitudinally ridged, and transversely ridged.

Mesophyll

The differences in the transverse anatomy of *Lachenalia* leaves are mainly quantitative and of limited taxonomic use. Leaf anatomy does not appear to correlate to leaf morphology, habitat or phytogeography. Mesophyll anatomy in *Lachenalia* can be placed into three broad types, but is unstable in some species in which two of the three types occur, and in others where cell layers are intermediate and grade into one another. Commonly, two to five layers of adaxial chlorenchyma, three to five layers of central spongy chlorenchyma and two to three layers of abaxial chlorenchyma occur (type one). The second most frequent type comprises two to three layers of adaxial palisade chlorenchyma, two to three layers of central spongy chlorenchyma, and two to three layers of abaxial chlorenchyma (type two). The least common type consists of one layer of adaxial chlorenchyma, two

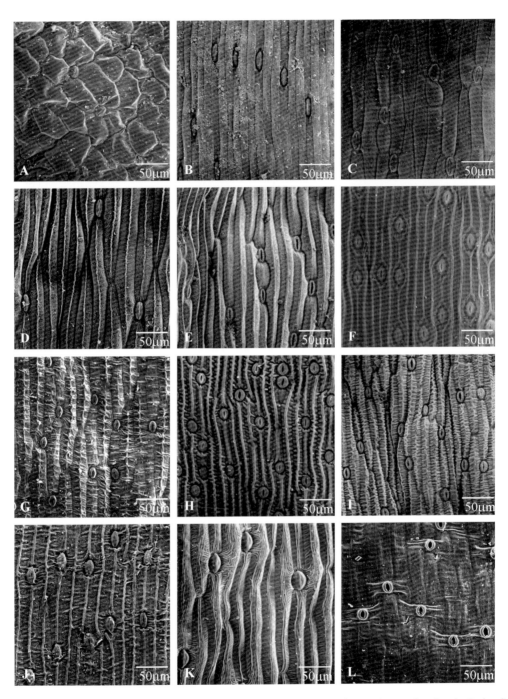

Figure 55. Scanning electron micrographs of variation in upper leaf surface tertiary sculpturing in *Lachenalia*. Laevigate cells of (A) *L. stayneri* and (B) *L. mathewsii*; finely striate cells of (C) *L. sargeantii*; cells with a longitudinally ridged median of (D) *L. margaretiae*, (E) *L. physocaulos* and (F) *L. orchioides* subsp. *orchioides*; cells with a longitudinally ridged median and transversely ridged sides of (G) *L. pusilla*, (H) *L. lutea*, (I) *L. leipoldtii*, (J) *L. elegans* and (K) *L. bifolia*; and (L) transversely ridged cells of *L. dehoopensis*. Image: Graham Duncan.

to three layers of adaxial palisade chlorenchyma, two to three layers of central spongy parenchyma and two to three layers of abaxial chlorenchyma (type three).

Vascular bundles are arranged in a single row in transverse sections, the xylem portion uppermost and the phloem beneath. Midvein bundle size is usually the largest, with the remaining bundles alternating with two to three small bundles between single large bundles for species with canaliculate or flat leaves (such as *L. ensifolia* and *L. macgregoriorum*). Alternatively, in those species with terete or subterete linear leaves (such as *L. multifolia* and certain forms of *L. corymbosa*), the bundles are more or less the same size. The highest number of vascular bundles counted in a leaf section is 42 bundles for *L. macgregoriorum*, and the lowest number is seven bundles for *L. multifolia* (Duncan, 2005).

Leaf margin anatomy

Leaf margin anatomy in *Lachenalia* comprises five distinct types and is diagnostic in some species (Figure 56). Flabellate marginal anatomy occurs most frequently and consists of four subtypes: vestigial flabellate margins in which the marginal cells are up to three times larger than the epidermal cells in transverse section, and are arranged in a narrow fan around large central cells; ordinary flabellate margins, in which similar marginal cells are arranged in a distinct fan in transverse section, as in *L. marginata* subsp. *marginata*; elongate flabellate margins, in which the central marginal cells are much longer than the peripheral cells in transverse section; and tapered flabellate margins, in which all cells are very narrow in transverse section and taper to a fine point, as in *L. salteri*.

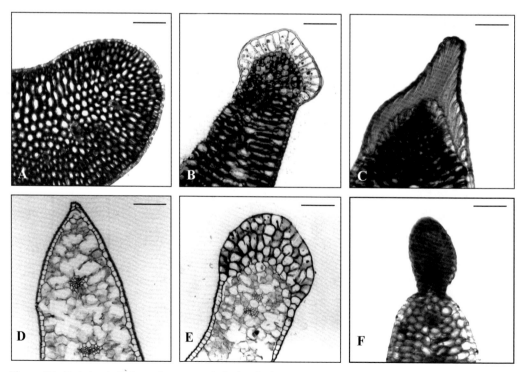

Figure 56. Variation in leaf margin anatomy in *Lachenalia*. Transverse sections of (A) simple margin of *L. montana*; (B) ordinary flabellate margin of *L. marginata* subsp. *marginata*; (C) tapered flabellate margin of *L. salteri*; (D) 'Polyxena' margin of *L. ensifolia*; (E) multi-epidermal margin of *L. verticillata*; and (F) sclerotic margin of *L. duncanii*. Scale bars = 150 μm. Image: Graham Duncan.

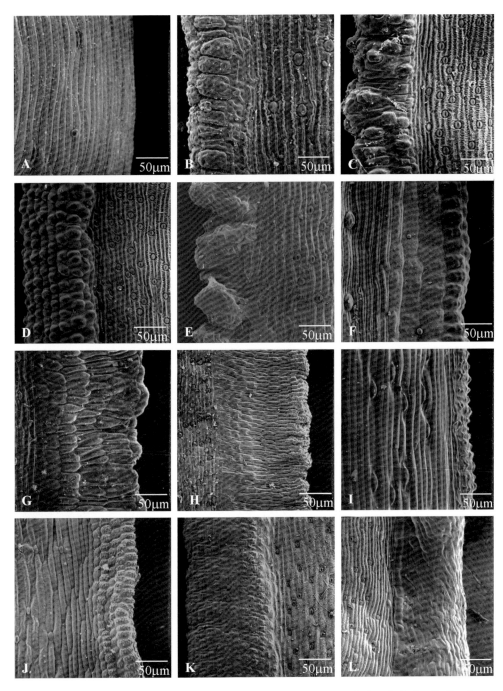

Figure 57. Scanning electron micrographs of variation in leaf margin micromorphology in *Lachenalia*. Simple margin of (A) *L. moniliformis*; vestigial flabellate margins of (B) *L. arbuthnotiae* and (C) *L. leipoldtii*; ordinary flabellate margin of (D) *L. lutea*; elongate flabellate margins of (E) *L. obscura*, (F) *L. leomontana* and (G) *L. rosea*; tapered flabellate margin of (H) *L. salteri*; 'Polyxena' margin of (I) *L. campanulata*; multi-epidermal margin of (J) *L. verticillata*, and sclerotic margins of (K) *L. nervosa* and (L) *L. duncanii*. Image: Graham Duncan.

The second most frequent margin-type comprises simple margins, in which the marginal cells are similar in size to the lamina epidermal cells in transverse section, as in *L. montana*. 'Polyxena' margins, in which a single additional marginal cell occurs in transverse section, are present in very few species such as *L. ensifolia*. Multi-epidermal margins with three to four layers of marginal cells are present only in *L. stayneri* and *L. verticillata*. In sclerotic margins, three to five layers of marginal cells with hard, lignified cell walls occur in transverse section; these are found only in *L. duncanii* and *L. nervosa*.

Leaf margin micromorphology corresponds to the five types of leaf margin anatomy (Figure 57). In simple margins there is no thickening of the epidermal cell wall, as in *L. moniliformis*. In flabellate margins, epidermal cell wall thickening occurs, and the more or less fan-like arrangement of marginal cells results in axially elongate cells, as can be seen in the vestigial flabellate margins of *L. arbuthnotiae* and *L. leipoldtii*, the ordinary flabellate margins of *L. lutea*, the elongate flabellate margins of *L. obscura*, *L. leomontana* and *L. rosea*, and the tapered flabellate margins of *L. salteri*. In 'Polyxena' margins there is no thickening of the epidermal cell wall and the single additional epidermal cell results in dome-shaped marginal cells, as in *L. campanulata*. In multi-epidermal margins, cell wall thickening occurs and the three to four layers of epidermal cells are isodiametric, as in *L. verticillata*. In sclerotic margins, the heavily lignified cell walls result in axially elongate cells, as in *L. duncanii* and *L. nervosa*.

Three types of leaf marginal hairs have been recorded in *Lachenalia* but glabrous margins are the most common. *L. argillicola*, *L. ensifolia*, *L. giessii*, *L. longituba*, *L. namibiensis*, *L. pearsonii* and *L. polyphylla* have ciliolate margins, *L. ameliae*, *L. comptonii* and *L. hirta* have ciliate margins, and stellate trichomes occur on the margins of *L. angelica*, *L. polypodantha* and *L. trichophylla*.

INFLORESCENCE

The inflorescence in *Lachenalia* is unbranched and axillary. In habitat, a single inflorescence is usually produced annually per bulb, but in cultivation, *L. longituba* sometimes produces a second inflorescence in early summer (Duncan, 2003a). The orientation of inflorescences in all species is erect to suberect, depending on the degree of exposure to light. Most *Lachenalia* inflorescences support at least 15 flowers, but certain species, such as *L. angelica*, *L. longituba* and *L. nordenstamii*, are usually few-flowered with fewer than 10 flowers. Flower number is unstable and strongly influenced by seasonal fluctuations in rainfall. Periods of heavy winter rainfall often result in robust inflorescences with unusually large flower numbers. The flowers are arranged spirally in most species, but sometimes occur in discrete three- or rarely four-flowered whorls.

The inflorescences of *Lachenalia* fall into two main types, racemose and spicate (Figure 58). Most species have racemose inflorescences with pedicel lengths varying from 0.5 mm to 20 mm. The inflorescences of two geoflorous, dwarf species (*L. barkeriana* and *L. pusilla*) are condensed into subcapitate racemes. *L. calcicola*, *L. ensifolia*, *L. longituba* and *L. pygmaea* have corymbose racemes, whereas *L. argillicola* and *L. paucifolia* have subcorymbose racemes, and the various forms of *L. corymbosa* can have corymbose, subcorymbose or ordinary racemes. In *L. sargeantii*, the inflorescence is usually an ordinary raceme but occasionally a subcorymbose raceme. In less than a quarter of

Figure 58 (right). Variation in inflorescence type in *Lachenalia*. (A) The corymbose raceme of *L. corymbosa*; (B) the spike of *L. orchioides* subsp. *orchioides*; (C) the long-pedicelled raceme of *L. flava*; (D) the whorled spike of *L. verticillata*; (E) the short-pedicelled raceme of *L. cernua*; and (F) the geoflorous, subcapitate raceme of *L. pusilla*. All × $^2/_3$. Drawn by Vicki Thomas.

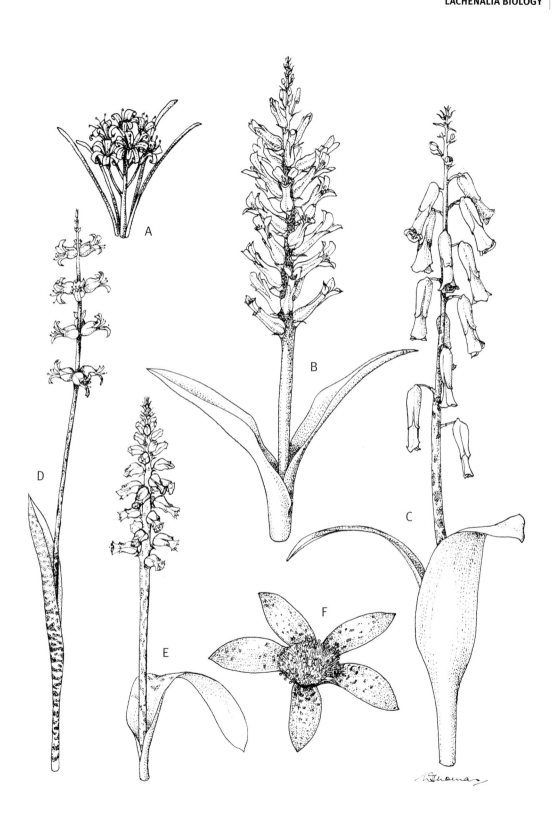

Lachenalia species, the flowers are sessile, forming a spike. Both spikes and racemes occur in different populations of *L. bowkeri*, *L. capensis*, *L. karooica*, *L. longibracteata*, *L. marlothii* and *L. mutabilis*. In a few species (such as *L. trichophylla*), spikes and racemes sometimes occur within single populations.

Four pedicel orientations occur in racemose species – erect, suberect, spreading and curved downwards – with most species having suberect pedicels. Most species have white pedicels but in some species the pedicels are light to dark green, yellow, orange, red, blue, turquoise, mauve or magenta.

SCAPE

The scapes of all lachenalias are smooth and vary in length from 5–10 mm (*L. barkeriana*) or up to 300 mm (*L. anguinea*), with those of most species being 100–150 mm long. Scapes are terete in all species but their diameters and colours are highly variable. Scape diameter is usually more or less constant along the entire length of the scape, but sometimes increases slightly to markedly towards the apex. Scape diameter varies from 2 mm (*L. polyphylla*) up to 15 mm (*L. anguinea*), but in most lachenalias it varies between 3 mm and 5 mm. Scape colour usually ranges from light to dark green, and is white in those species with subterranean scapes. Several species have uniformly deep brownish-magenta (such as *L. angelica*, *L. orthopetala*) or bright magenta scapes (*L. patula*). Phenotypic variation in scape colouration is marked; for example, the scapes of individual *L. montana* plants can be uniformly green or brownish purple within the same population. The scapes are covered with a delicate powdery bloom and many are variously blotched or spotted for their entire length. Blotch size and colouring is highly variable, ranging from minute (up to 0.5 mm in diameter) and dull magenta (as in *L. polyphylla*) to medium-sized (up to 2 mm in diameter) and bright magenta, dark brown or grey (as in *L. whitehillensis*, *L. lactosa* and *L. haarlemensis*, respectively), to very large (up to 4 mm in diameter) and brown to deep purple (as in *L. zebrina* and certain forms of *L. violacea*). Most lachenalias have aerial scapes but in eight species they are subterranean (*L. barkerianana*, *L. calcicola*, *L. congesta*, *L. ensifolia*, *L. kliprandensis*, *L. longituba*, *L. pusilla* and *L. pygmaea*). Scape elevation is adaptable in *L. congesta* and *L. kliprandensis* as the subterranean condition is often lost in cultivation when plants are grown in insufficient light.

In most lachenalias, the scape is not inflated, thus the diameter of the uppermost portion is the same as, or up to 2 mm broader than, that of the lower portion. In a small group (including *L. physocaulos* and *L. xerophila*), however, the scapes are distinctly inflated in the uppermost portion, being more than twice the diameter of the lower portion. In *L. mutabilis*, this condition is polymorphic and both non-inflated and inflated scapes occur within certain populations. Scape inflation is plastic in *L. anguinea*, *L. klinghardtiana* and *L. violacea* as the inflated condition is often much reduced or lost in cultivated specimens.

RACHIS

Rachis shape varies from slightly to distinctly angular. Rachis colour is similar to that of the scape in most species, but in some it becomes lighter towards the apex. In others, it becomes more intensely coloured towards the apex; for example, certain forms of *L. mutabilis* become light to bright electric blue. In a number of species, colour changes between the scape and the rachis are abrupt, probably because they are governed by different genes (Vogelpoel, 1986); for example, *L. angelica* has a deep maroon scape but a white rachis.

Apical anthocyanin expression in the rachis is a very common condition in *Lachenalia*. Attractive carotenoids are present in less than 10% of species in both the inner and outer mesophyll (as in *L. callista* and *L. flava*) or in only the inner mesophyll (as in *L. aloides* and *L. luteola*). Chlorophylls are present at the apex of the rachis in only a few species, including *L. bowkeri* and *L. undulata*, and aerenchymatous structural white tissue occurs in the inner and outer mesophyll of *L. angelica, L. barkeriana, L. ensifolia, L. longituba, L. patula* and *L. pusilla*.

FLOWERS

Bracts

Floral bracts with basal cupules occur in all *Lachenalia* species, and various bract arrangements occur. The function of the bracts is probably protection of the inflorescence while it is being pushed through other tissues or the substrate, and they might also assume secondary functions in pollinator attraction. Bract colour varies from translucent white to green, brown, orange, red or purple, and there are two main bract shapes, triangular and obtuse. The triangular bracts have fleshy bases and translucent, membranous upper parts, whereas the obtuse bracts are fleshy throughout. *L. isopetala* is unique in being the only species with papery bracts. Most lachenalias have triangular bracts, and within this group there are two subtypes. In the larger subtype, the bracts in the lower portion of the inflorescence are triangular and ovate to broadly ovate, becoming progressively longer and narrower towards the middle and upper part of the inflorescence, with narrowly lanceolate bracts at the apex. The smaller subtype has triangular, ovate bracts throughout the inflorescence, which become progressively smaller towards the apex. Obtuse, cup-shaped bracts are the second major type. These occur throughout inflorescences but are relatively large at the base and become progressively smaller towards the top.

Bract size is highly variable: *L. polypodantha* and *L. zebrina* have minute, inconspicuous basal bracts (1 × 1 mm), whereas *L. bifolia* has relatively large, basal bracts (6 × 4 mm). The uppermost bracts reach 16 mm long in *L. longibracteata*. Most species have ovate lower bracts with the minority having cup-shaped lower bracts. Lanceolate upper bracts predominate over ovate and cup-shaped upper bracts.

Perianth

Perianth shape is the product of tepal posture, tepal shape and floral symmetry, and is extremely variable in *Lachenalia*. The tepals of all lachenalias are biseriate, arranged in outer and inner whorls of three. Divergence in perianth shape appears to be driven by a number of pollination vectors, including sunbirds and insects. Floral symmetry is slightly to strongly zygomorphic in all members of subgenus *Lachenalia*, and actinomorphic in all members of subgenus *Polyxena*. Five major perianth shapes occur at anthesis: tubular, oblong-campanulate, narrowly campanulate, widely campanulate and urceolate (Figure 59). In 34 *Lachenalia* species, the perianth is tubular and the outer and inner tepals either radiate up to 10 degrees from the longitudinal axis (as in *L. callista*) or converge (as in *L. reflexa*). In seven species (*L. argillicola, L. calcicola, L. corymbosa, L. ensifolia, L. longituba, L. paucifolia* and *L. pygmaea*), tepal orientation is widely or narrowly spreading, but their perianth shape is nevertheless regarded as tubular by virtue of the overall tubular shape of the perianths. In 58 species, the perianths are oblong-campanulate, with the outer tepals straight or weakly convex and the inner tepals radiating at an angle of 15–20 degrees from the longitudinal axis (e.g. *L. doleritica*). In 9 species, including *L. orthopetala*, perianth shape is narrowly campanulate, and the outer and inner tepals radiate at an angle of 45–50 degrees from the longitudinal axis. The perianth is widely campanulate in 13 species, including *L. comptonii*, in which the outer and inner tepals radiate at an angle of 60–85 degrees from the longitudinal axis. In 19 species, the perianth is urceolate, and the outer and inner tepals converge; in these flowers (e.g. *L. membranacea*), the tepals are contracted at the mouth, then radiate at an angle of 25–40 degrees from the longitudinal axis, with the inner tepal apices being slightly to distinctly recurved.

Overall flower morphology varies from the minute, widely campanulate flowers of *L. glaucophylla* (5 mm long) to the comparatively long, tubular flowers of *L. callista* (up to 40 mm long). Intraspecifically, perianth length varies considerably within variable species such as *L. bifolia* and *L. orchioides*.

The apical region of the outer tepals usually have a swelling known as a gibbosity. The size of this structure varies from small and inconspicuous (1 × 1 mm, as in *L. fistulosa* and *L. minima*) to relatively large and prominent (2–3 × 2–3 mm, as in *L. marginata* and *L. sessiliflora*). Gibbosity colour varies from light to dark green in most species, through light magenta to dark pink or maroon, and many shades of brown. The median keels of the inner tepals usually match the gibbosities in colour. Within species, gibbosity size is stable but gibbosity colour is often variable.

All *Lachenalia* species are gamotepalous, the tepals are fused at their bases to form a cup or flat disc, or a short or long tube. The base of the perianth arises obliquely from the scape or pedicel in all members of subgenus *Lachenalia*, and radially in all members of subgenus *Polyxena*. The open flowers of most species are borne in a suberect, spreading or drooping position, with smaller groups bearing erect and pendulous flowers. The terminal flowers in all lachenalias are much smaller and sterile; they are frequently more intensely coloured than those on the rest of the inflorescence (as in *L. callista* and *L. mutabilis*), and probably assist in pollinator attraction. In certain species that usually have sessile flowers, such as *L. mutabilis*, the terminal, sterile flowers often become distinctly pedicellate. The flowers of all lachenalias are acropetalous.

Certain widespread species vary considerably with respect to perianth length, flower colour and flowering period. The perianth in forms of *L. bifolia* in the southern and eastern parts of its distribution range are considerably shorter than those from the west. Perianth tubes may be long (6–27 mm) or short (up to 3 mm), most species have short tubes. Three perianth-tube shapes occur, tubular, cup-shaped and disc-shaped, with cup-shaped tubes predominating.

Outer tepals are usually ovate but oblong and lanceolate shapes also occur. Outer tepal length falls into two classes: most species have short (up to 9 mm) tepals, the remainder have long tepals

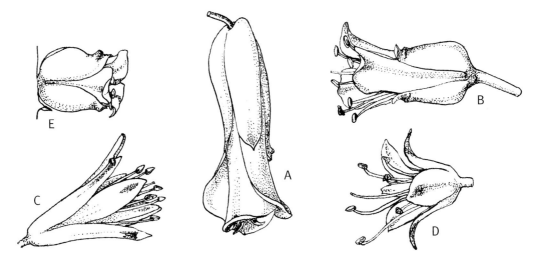

Figure 59. Variation in perianth shape in *Lachenalia*. (A) The tubular perianth of *L. callista* (× 1½); (B) the oblong-campanulate perianth of *L. doleritica* (× 2⅓); (C) the narrowly campanulate perianth of *L. orthopetala* (× 2⅓); (D) the widely campanulate perianth of *L. comptonii* (× 2⅓); and (E) the urceolate perianth of *L. membranacea* (× 3). Drawn by Vicki Thomas.

(12 mm or longer). The apices of the outer tepals are flat in most species, or recurved. Three distinct inner tepal shapes occur, linear-oblong, obovate and oblong-obovate, with most lachenalias having obovate inner tepals. Inner tepal length falls into two discrete classes: short (up to 13 mm) and long (16 mm or longer). The arrangement of the outer and inner tepals also has two classes: either subequal or with the inner tepals distinctly longer than the outer tepals, the latter condition being the most frequent. In most species, the inner tepals are of equal length, but in some, such as *L. capensis* and *L. martiniae*, the ventral inner tepal is distinctly shorter than the upper inner tepals or, as in *L. framesii* and *L. verticillata*, distinctly longer than the upper inner tepals.

Flower pigments

Plastid and/or anthocyanin pigments are present in various regions of the outer and inner tepals of all species of *Lachenalia* and combine to form a wide range of colours. White flowers lack pigments, and the reflected light is a product of aerenchymatous tissues (Fox, 1979).

Plastid pigments

In *Lachenalia*, plastid pigments occur in chloroplasts and chromoplasts in the mesophyll tissue and are always absent from the epidermis. They occur throughout the mesophyll of both the outer and inner tepals. Plastids and vacuoles that contain anthocyanin pigments can occur together in the same mesophyll cells of some species, in other species they occur in separate cells. Xanthophylls occur in the relatively few species that have pale yellow flowers. Carotenoids occur in flowers that are bright yellowish-orange or orange; they are found in chromoplasts throughout the mesophyll, but are concentrated in the outermost three or four layers of the tepals and are overlain by a colourless epidermis. Chlorophyll occurs mainly in the gibbosities of the outer tepals, and in the keels of the outer and inner tepals.

Anthocyanin pigments

Water-soluble anthocyanin pigments occur within the vacuoles of epidermal cells in most angiosperm families (Kay *et al.*, 1981; Vogelpoel, 1995), but in most lachenalias, they occur in vacuoles in the tepal mesophyll. Anthocyanin pigments do occur in the epidermis of three species: in *L. aloides*, *L. quadricolor* and certain forms of *L. luteola*, they occur at the apices of the inner tepals. These pigments also occur in the turquoise outer and inner tepals of *L. viridiflora*. In most species anthocyanins are restricted to the outer mesophyll, but in some such as *L. patula* and *L. youngii*, they are also present in the innermost cell layer. In *L. aloides*, *L. quadricolor* and certain forms of *L. luteola*, anthocyanins are restricted to the outermost mesophyll layer in the red portion of the outer tepals, where the epidermal layer contains magenta anthocyanin. The anthocyanin-containing mesophyll layer overlies several layers of mesophyll that contain carotenoid pigment within chromoplasts. Where outer and/or inner tepals are suffused with different colours close together, like the blue and pink in certain forms of *L. salteri*, the differently coloured pigments occur in discrete areas. The anthocyanin pigments in *Lachenalia* are predominantly magenta but blue, pink and purple also occur.

Stamens

The stamens of *Lachenalia* are biseriate to a greater or lesser degree. The filaments are variously inserted from the base to the mouth of the perianth tube and the distance between the two levels of insertion is greater in species that have long tubular flowers; the outer stamen series is inserted at a lower level than the inner series. The anthers of the outer series dehisce about 24 hours before those of the inner series. Degree of stamen exsertion varies considerably within some species, such as *L.*

contaminata and *L. mutabilis*; in these species, most forms have included or slightly exserted stamens but both species include forms that have well exserted stamens. Filaments are terete and slender in all species, with the width being more or less stable, except in *L. barkeriana* where the upper portion is swollen, and in *L. nordenstamii*, where the filaments are relatively stout. Filament length varies from 1 mm (*L. paucifolia*) to 39 mm (*L. bifolia*) and is variable within species. Filaments usually exceed the style in length, but in some species they are shorter than, or equal in length to, the style.

The anthers of most lachenalias are dorsifixed, but in long-tubed members of subgenus *Lachenalia*, they are basifixed. The anthers are oblong and approximately 1.0 × 0.5 mm, but those lachenalias with large tubular flowers, such as *L. aloides* and *L. bifolia*, have bigger anthers of 2.5 × 1.0 mm. The anthers are versatile, and dehisce longitudinally. Anther colour is variously dark green, blue, purple, brown, maroon or dull red prior to anthesis, and the pollen is light to bright yellow in most species, rarely cream or brown. Following anthesis, the pollen remains yellow in most species, but the pollen of several species, such as *L. convallarioides* and *L. juncifolia*, turns black after anthesis. Stamens are either included within the perianth (shorter than the perianth), or shortly exserted up to 2 mm, or well exserted (more than 2 mm) beyond the perianth. The species with the most-exserted stamens are *L. barkeriana* and *L. pusilla*, and these stamens protrude up to 10 mm.

Four filament orientation classes occur at anthesis: straight, narrowly spreading, recurved and declinate (Figure 60). Straight filaments occur in more than half of all *Lachenlia* species, including *L. isopetala*. Smaller groups have narrowly spreading filaments that radiate 15–18 degrees from the longitudinal axis and are not recurved in the upper half (as in *L. orthopetala*), or filaments that are recurved in the upper half and radiate 22–25 degrees from the longitudinal axis (as in *L. neilii*). Filament orientation is further modified to bend downwards 30–40 degrees from the longitudinal axis (termed declinate) in *L. mathewsii*. Filament orientation is probably pollinator-driven.

White filaments occur in more than 90% of *Lachenalia* species, and bicoloured filaments that are white in the lower half and purple or magenta in the upper half occur in the remainder. Two classes of filament length occur: short filaments of up to 17 mm and long filaments of 20 mm or longer. Outer and inner filaments of equal length are restricted to certain species that have narrowly campanulate and widely campanulate flowers, but all others have subequal filaments. The outer filaments are inserted in the middle of the perianth tube, except in *L. reflexa*,

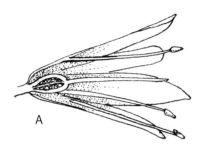

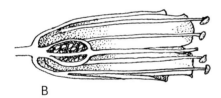

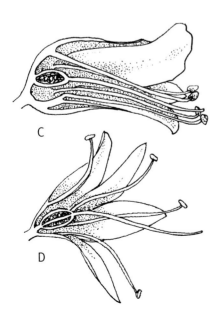

Figure 60 (left). Variation in filament orientation in *Lachenalia*. (A) Narrowly spreading filaments of *L. orthopetala* (× 2¹/3); (B) straight filaments of *L. isopetala* (× 3); (C) declinate filaments of *L. mathewsii* (× 3); and (D) recurved filaments of *L. neilii* (× 4). Drawn by Vicki Thomas.

L. viridiflora and all members of subgenus *Polyxena* (where they are inserted at the mouth of the tube) and in all long-tubed members of subgenus *Lachenalia* except *L. algoensis* (where they are inserted at the base of the perianth tube). Inner filaments are inserted at the mouth of the tube, except in long-tubed members of subgenus *Lachenalia* (except *L. algoensis*) where insertion is in the middle of the tube.

Lachenalia pollen is consistently monosulcate. The exine is reticulate or rarely rugose. There is slight variation in the shape, degree of reticulation and size of the pollen grains (21–30 × 12–15 μm).

Gynoecium

The gynoecium in *Lachenalia* has a simple, slender style consisting of three stylar channels, a capitate stigma, and a trilocular, obovoid or ellipsoid ovary.

Style and stigma

Styles are either straight (radiating up to 10 degrees from the longitudinal axis) or declinate. Straight styles occur in species that have straight, narrowly spreading and recurved filaments, whereas declinate styles usually correspond to declinate filaments. Styles are white in most species but where the distal portion of the filaments is purple or magenta, the distal portion of the style is similarly coloured.

L. paucifolia (1–2 mm), *L. karoopoortensis* (2 mm), *L. congesta* (2–3 mm) and *L. sessiliflora* (2–3 mm) have the shortest styles. The longest styles occur in *L. bifolia* (22–42 mm) and *L. punctata* (25–30 mm) but for most lachenalias, style length falls between 7 mm and 12 mm. Stigmas become receptive two to seven days after anthesis. During the fruiting stage, the styles of many species protrude conspicuously beyond the stamens and the perianth apex as the ovary enlarges.

Ovary

The tricapellary ovary in *Lachenalia* is syncarpic with axile placentation. Two ovary shapes occur, obovoid and ellipsoid, but most species have ellipsoid ovaries. Simple septal nectaries occur in the upper half of the ovary dome in all lachenalias but nectar quantities vary greatly. Those species with long tubular flowers that are bird-pollinated, such as *L. bifolia* and *L. punctata*, contain much greater quantities of nectar than mellitophilous species that have oblong-campanulate, narrowly campanulate, widely campanulate or urceolate flowers.

Ovules are produced in two rows and ovule production per locule is highly variable, ranging from four (*L. congesta*) to 26 (*L. angelica*), with most species having 10–15 ovules per locule, although this varies each season.

FRUIT

The ripe fruits of *Lachenalia* are dry, membranous or papyraceous, loculicidal capsules that dehisce longitudinally (Figure 61). Ellipsoid capsules are common (e.g. in *L. bifolia*) but obovoid (e.g. in *L. anguinea*), ovoid (e.g. in *L. longituba*) and obcordate (e.g. in *L. zebrina*) fruits also occur. Capsule shape is stable within species and, in most species, conforms to that of ovary shape. *L. buchubergensis*, *L. nordenstamii* and *L. zebrina* are exceptional in having ellipsoid ovaries but obcordate, broadly winged, aerodynamic capsules. *L. ensifolia* and *L. longituba* are also unusual in having ellipsoid ovaries that develop into ovoid capsules, and *L. paucifolia* has ellipsoid ovaries that develop into obovoid capsules. Capsule size is highly variable across the genus, ranging from the small, obovoid capsules of *L. campanulata* (up to 4 mm long) to the long, ellipsoid capsules of *L. reflexa* (up to 15 mm long).

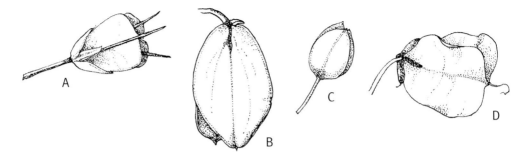

Figure 61. Variation in capsule shape in *Lachenalia*. (A) obovoid capsule of *L. anguinea* (× 2); (B) ellipsoid capsule of *L. bifolia* (× 2); (C) ovoid capsule of *L. longituba* (× 2); and (D) obcordate capsule of *L. zebrina* (× 2). Drawn by Vicki Thomas.

In most lachenalias, the capsules retain the same posture as the flowers, but in some (e.g. *L. mediana* and *L. rosea*), the pedicels bend upwards in fruit to achieve a suberect or erect posture. In *L. multifolia*, the pedicels bend downwards into a hanging position.

In nine species with subterranean or abbreviated scapes (*L. argillicola*, *L. barkeriana*, *L. calcicola*, *L. corymbosa*, *L. ensifolia*, *L. longituba*, *L. paucifolia*, *L. pusilla* and *L. reflexa*) the scape elongates and falls into a horizontal position immediately prior to capsule dehiscence, dispersing most of the seeds a short distance from the mother plant. In *L. barkeriana*, *L. ensifolia*, *L. pusilla* and *L. pygmaea*, the scape breaks off shortly after capsule dehiscence, allowing the whole infructescence to be carried away by wind, effectively dispersing the remaining seeds over a wide area.

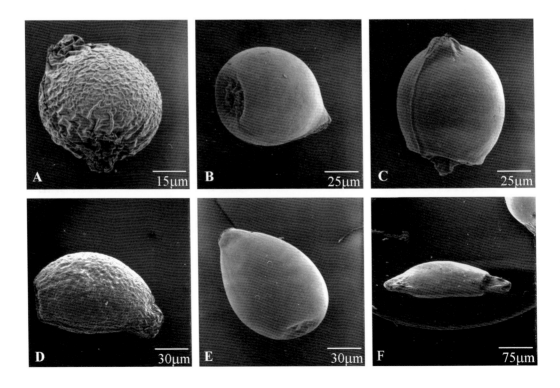

SEED

Seed characters have played an important role in delimiting genera of the tribe Massonieae (Jessop, 1975; Wetschnig *et al.*, 2002; Pfosser *et al.*, 2003) and were first used as additional diagnostic features in *Lachenalia* by Barker (1978). The diagnostic characters of *Lachenalia* seeds include shape and size, strophiole length, inflation and topography, and primary and secondary sculpturing of the testa.

Seed morphology varies across the genus but is stable within species (Figure 62). There are three basic seed shapes: most species have globose seeds (e.g. *L. trichophylla* and *L. comptonii*), whereas others have ovoid (e.g. *L. corymbosa* and *L. ensifolia*) or oblong (e.g. *L. nordenstamii*) seeds. Seed size varies from minute to relatively large, with *L. angelica* having the smallest seeds (0.4 × 0.4 mm) and *L. isopetala* the largest (1.8–2.0 × 1.8–1.9 mm). The longest seeds are those of *L. buchubergensis* (3.0–3.5 mm), but most lachenalias have medium seed lengths of 0.9–1.4 mm.

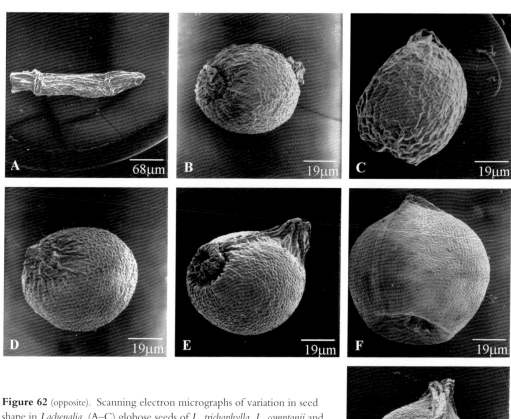

Figure 62 (opposite). Scanning electron micrographs of variation in seed shape in *Lachenalia*. (A–C) globose seeds of *L. trichophylla*, *L. comptonii* and *L. juncifolia*; (D–E) ovoid seeds of *L. corymbosa* and *L. ensifolia*; (F) oblong seed of *L. nordenstamii*. Image: Graham Duncan.

Figure 63 (above). Scanning electron micrographs of variation in testa sculpturing. (A–C) rugose primary sculpturing of *L. buchubergensis*, *L. patula* and *L. polyphylla*; (D–F) reticulate secondary sculpturing of *L. angelica*, *L. framesii* and *L. hirta*; (G) colliculate secondary sculpturing of *L. montana*. Image: Graham Duncan.

Testa surfaces are usually glossy or sometimes matte black, with smooth primary sculpturing and no secondary sculpturing. A few lachenalias, including *L. buchubergensis*, *L. patula* and *L. polyphylla*, have rugose primary sculpturing, and 11 species, including *L. angelica*, *L. framesii* and *L. hirta*, have reticulate secondary sculpturing. Unusual colliculate secondary sculpturing occurs in two species, *L. montana* and *L. sargeantii* (Figure 63).

Lachenlia seeds have an appendage known as a strophiole that occurs in the micropylar region (Figures 64, 65). Strophiole length is extremely variable across the genus but is stable within species and falls into one of three distinct groups: rudimentary strophioles of up to 0.3 mm in length, medium strophioles of 0.5–0.8 mm in length, and long strophioles of 1.1 mm in length or longer.

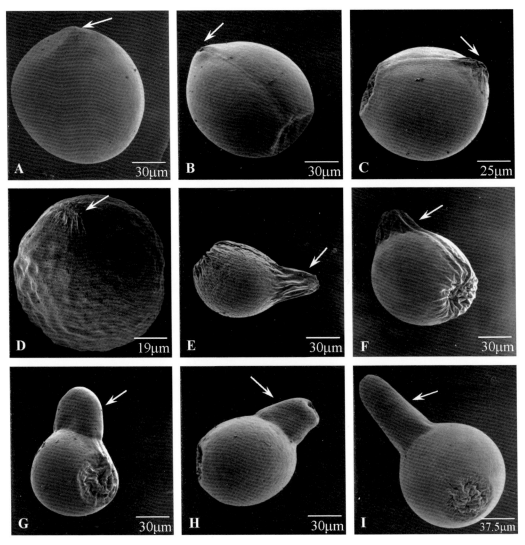

Figure 64. Scanning electron micrographs of variation in strophiole length in *Lachenalia*. (A–D) rudimentary strophioles of *L. isopetala*, *L. namibiensis*, *L. pallida* and *L. paucifolia*; (E–H) medium strophioles of *L. latimeriae*, *L. dasybotrya*, *L. ventricosa* and *L. viridiflora*; (I) long strophiole of *L. vanzyliae*. Image: Graham Duncan.

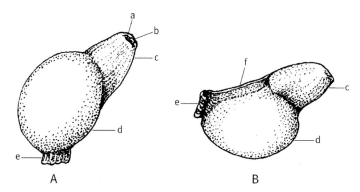

Figure 65. Seed and strophiole morphology in *Lachenalia*. A & B, lateral views of globose seeds of *Lachenalia ventricosa* (× 50). a = micropyle; b = hilum; c = strophiole; d = seed; e = chalazal collar; f = raphe. Drawn by Vicki Thomas.

More than half of *Lachenalia* species have medium strophioles, as in *L. latimeriae* and *L. dasybotrya*, for example. The shortest strophioles, including those of *L. isopetala* and *L. namibiensis*, are 0.1 mm long; the longest are those of *L. bifolia* and *L. vanzyliae* (respectively 1.4–1.5 mm and 1.4–1.6 mm long). Strophioles are ridged and non-inflated in most species, rarely smooth and inflated (Figure 65). The inflated strophiole of *L. bifolia* seeds is noteworthy in being translucent white. A longitudinal ridge known as a raphe, forms below the hilum; it is distinctly inflated in *L. bifolia*, *L. punctata* and *L. xerophila* and joins with the inflated strophiole in these three species. A collar forms in the chalazal region of the seed in some species and can be short (up to 0.4 mm long) or relatively long (0.8 mm long or longer), but this collar is absent in most species.

POLLINATION BIOLOGY

The flowers of most lachenalias are self-incompatible. However, two unusual species with subcapitate racemes (*L. barkeriana* and *L. pusilla*) and another with corymbose racemes (*L. longituba*) have shown a breakdown of self-incompatibility in the Kirstenbosch Bulb Nursery and set copious seeds regardless of animal visits when grown under windless, isolated conditions. Seven other species (*L. contaminata* (certain forms), *L. ensifolia*, *L. judithiae* (certain forms), *L. juncifolia* (certain forms), *L. lactosa*, *L. nervosa* and *L. reflexa*) regularly produce a heavy seedset under cultivation at Kirstenbosch without any obvious evidence of cross-pollination.

The differences in floral structure within *Lachenalia* probably relate to a variety of pollination and dispersal strategies. Pollinators are attracted to the flower primarily by colour, fragrance and shape, and are offered rewards of nectar and pollen. Two obvious pollination syndromes are encountered in the genus: mellitophily and ornithophily. In addition, generalist flowers are visited by monkey beetles (*Pachycnema crassipes* Fabricius (Figure 66) and *Peritrichia cinerea* Olivier (Hopliini: Scarabaeidae)) and the exotic beewolf (*Philanthus triangulum* Fabricius) (Hymenoptera: Crabronidae) that consume pollen, and by six nectar-feeding insects upon which pollen is deposited: blister beetles (*Lytta nitidula* Fabricius) (Coleoptera: Meloidae) (Figure 67), hoverflies (Diptera: Syrphidae) including the exotic drone fly (*Eristalis tenax* L.) (Figure 68), march flies (Diptera: Bibionidae) (Figure 69), three butterfly species, the Painted Lady (*Vanessa cardui* L.) (Lepidoptera: Nymphalidae)(Figure 70), the African Monarch (*Danaus chrysippus aegyptius* L.) (Lepidoptera: Nymphalidae) and the exotic Cabbage White (*Pieris brassicae* L.) (Lepidoptera: Pieridae) and a moth, the day-flying Heady Maiden (*Amata cerbera* L.) (Lepidoptera: Amatidae) (Figure 71).

THE GENUS LACHENALIA
LACHENALIA BIOLOGY

Figure 66 (left). A monkey beetle (*Pachycnema crassipes*) consuming pollen of *Lachenalia bachmannii*. Image: Graham Duncan.

Figure 67 (right). A blister beetle (*Lytta nitidula*) feeding on nectar of *Lachenalia glaucophylla*. Image: Graham Duncan.

Figure 68 (above). A drone fly (*Eristalis tenax*) feeding on nectar of *Lachenalia corymbosa*. Image: Graham Duncan.

Figure 69 (right). A march fly (Diptera: Bibionidae) after feeding on nectar of *Lachenalia liliiflora*. Image: Adam Harrower.

Figure 70 (left). The Painted Lady butterfly *Vanessa cardui* feeding on nectar of *Lachenalia corymbosa*. Image: Graham Duncan.

Figure 71 (right). The Heady Maiden day-flying moth *Amata cerbera* feeding on nectar of *Lachenalia contaminata*. Image: Graham Duncan.

Rodents are a possible, but as yet unconfirmed, pollination vector for two yeast-scented species, *L. barkeriana* and *L. pusilla*. The flowers of several lachenalias, including *L. punctata* and *L. capensis*, are visited by ants that consume nectar but play no part in pollination (see Figures 112, 125 respectively).

MELLITOPHILY

Cape honey bees (*Apis mellifera* subsp. *capenis* Eschscholtz) (Figure 72) and African honey bees (*Apis mellifera* subsp. *scutellata* Lepeletier) (Hymenoptera: Apidae) visit the full range of differently shaped flowers in *Lachenalia*, including the long-tubed, heavily scented, geoflorous flowers of *L. longituba*. Certain non-scented species with aerial long-tubed flowers, such as *L. aloides*, *L. punctata* and *L. quadricolor*, are also visited by honey bees although they are adapted primarily to sunbird pollination. Specialisation for bee pollination in *Lachenalia* includes mechanically strong flowers with adequate landing facilities and a tepal surface that provides an adequate foothold (Faegri & van der Pijl, 1979). The anthers of bee-pollinated species are prominent and usually bright yellow. The bees are attracted to colour and also to scent that is emitted to a greater or lesser degree by all bee-pollinated species. Certain species are heavily scented, such as *L. fistulosa* and *L. arbuthnotiae* (sweetly scented), *L. glauca* (coconut-scented) (Duncan, 1988a), *L. nervosa* (scented like a mixture of cloves and carnations) (Duncan & Linder Smith, 1999b) and *L. peersii* (scented like carnations) (Duncan, 2003c). Honey-scented species include *L. corymbosa* and *L. paucifolia*, whereas *L. ensifolia* and *L. longituba* are spicy sweet-scented and *L. pygmaea* is almond-scented. Bees alight on the lower inner tepal, enter the flowers and drink nectar, then either collect pollen or leave the flower. Pollen deposition is nototribic, and in leaving the flower, the bee usually brushes against the stigma. Vogel (1954) pointed to the specialised floral ecology of one species, the geoflorous *L. pusilla*, which has a mellitophilous brush mechanism (also found in *L. barkeriana* and *L. pygmaea*) in which the exserted, erect stamen filaments act as the main visual attractant.

Figure 72 (top left). The Cape honey bee *Apis mellifera* subsp. *capensis* visiting the flowers of *Lachenalia pallida*. Image: Graham Duncan.

Figure 73 (top right). The solitary bee *Anthophora diversipes* feeding on nectar of *Lachenalia pallida*. Image: Dennis Tsang.

Figure 74 (bottom left). The solitary bee *Amegilla nivea* feeding on nectar of *Lachenalia isopetala*. Image: Graham Duncan.

Figure 75 (bottom right). A male Southern Double-collared Sunbird *Cinnyris chalybeus* drinking nectar from *Lachenalia viridiflora* at St. Helena Bay. Image: Graham Duncan.

Two solitary bee species (Hymenoptera: Apidae) have been seen to feed on the nectar of *Lachenalia* species. *Anthophora diversipes* Friese visits *L. membranacea*, *L. obscura*, *L. orchioides* subsp. *orchioides* and *L. pallida* (Figure 73), whereas *Amegilla nivea* Lepeletier has been observed pollinating *L. isopetala* (Figure 74). Honey bees visit the flowers of *Lachenalia* from mid-morning (10 a.m.) until mid-afternoon (about 3 p.m.), and solitary bees are most active from early to mid-morning between 8 a.m. and 11a.m.

ORNITHOPHILY

Lachenalia species with unscented, brightly coloured, tubular flowers with long perianth tubes and an abundance of nectar (*L. aloides*, *L. bifolia*, *L. callista*, *L. flava*, *L. luteola*, *L. patentissima*, *L. punctata*, *L. quadricolor*, *L. reflexa*, *L. thunbergii*, *L. vanzyliae* and *L. viridiflora*) conform to the bird pollination syndrome (Faegri & van der Pijl, 1979; Proctor & Yeo, 1972). These species all appear to be specialised for pollination by sunbirds (Nectariniidae), notably Southern Double-collared Sunbirds (*Cinnyris chalybeus* L.), Malachite Sunbirds (*Nectarinia famosa* L.) and Orange-breasted Sunbirds (*Anthobaphes violacea* L.), which occur throughout their distribution range. The flowers of these *Lachenalia* species are mainly shades of yellow, orange, red and pink: *L. aloides* (bright yellow and red), *L. bifolia* (bright red or orange-red), *L. callista* (bright yellow, green and orange), *L. flava* (bright golden yellow), *L. luteola* (yellowish green or a combination of bright yellow, reddish-orange and magenta), *L. patentissima* (bright yellow), *L. punctata* (bright pink), *L. quadricolor* (a combination of yellow, green, reddish-orange and magenta), *L. reflexa* (bright greenish yellow), *L. thunbergii* (bright yellowish green), *L. viridiflora* (bright turquoise) and *L. vanzyliae* (a combination of green, white and turquoise). Actual pollination sightings include Southern Double-collared Sunbirds on *L. aloides*, *L.*

Figure 76 (left). A male Malachite Sunbird visiting the flowers of *Lachenalia flava* near Tulbagh. Image: Neil Cox.

Figure 77 (right). Foraging damage by sunbirds, resulting in detached flowers at the base of a *Lachenalia luteola* plant at Kommetjie, southern Cape Peninsula. Image: Graham Duncan.

bifolia, *L. luteola*, *L. punctata*, *L. quadricolor* and *L. viridiflora* (Figure 75), Malachite Sunbirds on *L. flava* (Figure 76), and Orange-breasted Sunbirds on *L. luteola* and *L. vanzyliae*.

Most of the sunbird-pollinated species have scapes that are strong enough to allow visiting sunbirds to cling to them while probing the perianth tubes for nectar. *L. reflexa* and *L. viridiflora* are anomalous in having rather short, weak scapes, and their sunbird pollinators have to probe the flowers directly from the ground or from surrounding vegetation. Illustrating the convergent adaptation of flowers to similar pollinators, the long perianth tube and tubular perianth of the putatively sunbird-pollinated *L. reflexa* are similar to those of other sunbird pollinated species, but the erect or suberect flower orientation in this species and in *L. viridiflora*, and the morphology of the tubular perianth in *L. reflexa*, differs from those of other sunbird-pollinated lachenalias in that it narrows distinctly towards the mouth; all other sunbird-pollinated species have pendulous or cernuous flowers and an absence of narrowing tepals.

When feeding on species with sturdy scapes, the sunbird clings to the base of the scape and inserts its curved beak into the tubular perianth, and in so doing, it lifts the perianth into a horizontal or suberect position. Considerable damage is often caused during foraging, with detached flowers and tepals often seen lying at the base of the plants (Figure 77). Sunbirds have been seen to visit the flowers of *L. bifolia*, *L. punctata*, *L. quadricolor* and *L. viridiflora* from early morning (around 8 a.m.) until late afternoon (around 5 p.m.), of *L. flava* from noon to 2 p.m. and of *L. vanzyliae* from 11 a.m. to 4 p.m.

POLLINATION BY RODENTS

The geoflorous inflorescences of *Massonia depressa* (the outgroup species used in a morphological cladistic analysis of *Lachenalia* (Duncan *et al.*, 2005)) emit a strong yeasty odour and have been shown to be pollinated at night by at least four rodent species, including two gerbil species, in the Succulent Karoo of the north-western Cape (Johnson *et al.*, 2001). The possibility exists that the ground-level flowers of *L. barkeriana* and *L. pusilla* might also be pollinated by rodents as their flowers emit an extremely strong yeasty odour by day and by night, much stronger than that of *M. depressa*, and *L. barkeriana* and *M. depressa* occur in a Succulent Karoo habitat. Preliminary investigations of scent chemistry indicate, however, that the flowers of these lachenalias may be bee-pollinated as the compounds isolated are not typical of rodent-pollinated flowers (S. Johnson, pers. comm.). *L. barkeriana* and *L. pusilla* have the shortest flowering period (no more than two weeks) of any *Lachenalia* species.

SEED DISPERSAL

LOCAL DISPERSAL

The dispersal of most *Lachenalia* seeds is local. The scape remains attached to the bulb for several weeks or months following the drying of the infructescence and the seeds drop to the ground as they become dislodged from the open capsules by strong winds. *Lachenalia* seeds are non-buoyant, but local dispersal by raindrops does play a small role in the distribution of seeds of ten early-flowering dwarf species, *L. argillicola*, *L. barkeriana*, *L. calcicola*, *L. corymbosa*, *L. ensifolia*, *L. longituba*, *L. paucifolia*, *L. pusilla*, *L. pygmaea* and *L. reflexa*. Their ripe seeds are exposed to rainshowers from late winter to early spring and raindrops dislodge the seeds from the dry capsules and project them for distances measured at up to 140 mm.

ANEMOCHORY

In other species, the base of the scape rapidly detaches from the bulb and the dry infructescence is blown away, scattering the seeds from the open capsules. This method of dispersal is encountered mainly in species from arid habitats, such as *L. patula* from the Knersvlakte and *L. xerophila* from the Richtersveld, Namaqualand and Bushmanland. Broadly winged capsules are limited to three species, *L. buchubergensis* and *L. nordenstamii* from the arid Richtersveld and south-western corner of Namibia and *L. zebrina*, which is widespread in the Great Karoo. In these species, the capsules contain relatively few, large seeds (one to six per capsule). The pedicels fracture in strong gusts of wind and the detached capsules are dispersed over a wide area.

MYRMECOCHORY

A number of species flower close to the ground (*L. argillicola*, *L. barkeriana*, *L. calcicola*, *L. corymbosa*, *L. ensifolia*, *L. longituba*, *L. paucifolia*, *L. pusilla*, *L. pygmaea* and *L. reflexa*) and their scape elongates and becomes decumbent in late winter and early spring. The glossy black seeds are spread from this position. Barker (1930b) reported that *Lachenalia* seeds had been said to be dispersed by ants, "for each seed is furnished with an oily appendage, known botanically as an elaiosome, and the ants in their endeavour to secure this desirable nutriment have perforce to take the whole seed along. I have not been able to determine what exactly happens, since the ants I observed at Kirstenbosch were the small Argentine ones, and although they appeared to be trying to carry off the seed, this was not actually effected". Müller-Doblies *et al.* (1987) reported that the prominent strophiole of *L. barkeriana* and *L. pusilla* acts as an elaiosome, allowing the seeds to be dispersed by ants (myrmecochory) (Figures 78, 79). Following initial dispersal by ants, the infructescences are blown away and the remaining seeds are dispersed. This method of dispersal has been observed in *L. barkeriana* in the wild by Nordenstam (Müller-Doblies *et al.*, 1987). In *Lachenalia*, the strophiole is not a true elaiosome as it does not contain any oil, but it may mimic true elaiosomes. This method of dispersal is probably also used by other species that have prominent strophioles, such as *L. punctata*, whose relatively long scapes bend downwards to rest at, or just above, ground level under the weight of the ripening capsules (the only member

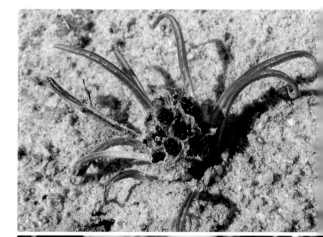

Figure 78 (top). The ripe infructescence of *Lachenalia barkeriana*, whose seeds are initially dispersed by ants. Image: Carly Cowell.

Figure 79 (bottom). The ripe infructescence of *Lachenalia pusilla*, whose seeds are initially dispersed by ants. Image: Cameron McMaster.

of the genus in which the scapes do this), eventually releasing the seeds onto the ground. The seeds of other species with long strophioles, such as *L. aloides* and *L. reflexa*, are initially dispersed by wind as they are dislodged from the widely flared, open capsules; subsequently, they are probably carried away by ants.

KARYOLOGY

The karyology of *Lachenalia* is extremely variable, including differences in basic chromosome numbers, differing ploidy levels and the presence of B-chromosomes (supernumerary chromosomes) within some species.

Chromosome counts are available for 94 species (Moffett, 1936; de Wet, 1957; Fernandes & Neves, 1962; Riley, 1962; Gouws, 1964; Zakharyeva & Makushenko, 1969; Mogford, 1978; Ornduff & Watters, 1978; Nordenstam, 1982; Crosby, 1986; Hancke & Liebenberg, 1990; Duncan, 1993, 1996; Johnson & Brandham, 1997; Hamatani et al., 1998; Kleynhans & Spies, 1999; Hamatani et al., 2004, 2007; Sâto, 1942; Spies et al., 2000, 2002, 2008, 2009; van Rooyen et al., 2002). The basic chromosome numbers $x = 5, 6, 7, 8, 9, 10, 11, 12, 13$ and 15 have been recorded, with most species having $x = 7$, and $x = 8$ being the next most common. Ploidy levels vary from 2x to 8x and are predominantly diploid, but tetraploid, hexaploid and octoploid counts have been made. Intraspecific polyploidy has been reported in *L. orchioides*, in which diploids and tetraploids occur (Ornduff & Watters, 1978; Johnson & Brandham, 1997); tetraploids and octoploids occur in *L. membranacea* (Ornduff & Watters, 1978; Johnson & Brandham, 1997); and *L. bifolia* (=*L. bulbifera*) forms a polyploid complex (Kleynhans & Spies, 1999). Polyploidy is fairly common in the genus, being most prevalent in species that have a basic chromosome number of $x = 7$ (Spies, 2004). The high level of 'deviating' counts (recorded in 16 species) in the literature could be due to the mis-identification of B-chromosomes and the incorrect identification of specimens (Spies *et al.*, 2002). Spies *et al.* (2002) also postulate that the different basic chromosome numbers for *Lachenalia* indicate that aneuploidy (having a diploid chromosome number that is not an exact multiple of the haploid number because one chromosome set is incomplete or fused) played a major role in the evolution of the genus, but that it is doubtful that as many as 16 species could contain aneuploids.

Lachenalia mutabilis is currently regarded as a single morphologically variable species complex, for which different basic chromosome numbers ($x = 5, 6$ and 7) have been reported. Ornduff and Watters (1978) reported $x = 5$ for one collection, and Johnson and Brandham (1997) reported $2n = 10$ for six collections and $2n = 14$ for two collections. Spies *et al.* (2000) reported $x = 6$ and $x = 7$ for 35 specimens representing 16 populations, and found that different specimens collected within the same population showed no variation in chromosome number, but that there was variation from one population to another. They also suggested that *L. mutabilis* might represent an aneuploid series, but thorough meiotic studies, including cross fertilisations between *L. mutabilis* specimens, will be needed to test this hypothesis. The differences in basic chromosome number within this species is not correlated to geographical position, as certain populations that occur relatively close to each other have different basic chromosome numbers. Spies *et al.* (2000) also reported that polyploidy is scarce in *L. mutabilis*, and found that only 5% of 35 specimens representing 16 populations exhibited this phenomenon. Scape inflation is stable in certain populations of *L. mutabilis* but unstable in others; although polymorphism in scape inflation is acceptable at the species level, speciation is almost certainly taking place in the *L. mutabilis* complex because morphological divergence is being entrenched by breeding barriers at the chromosome level.

The robust and widespread *L. bifolia* is an example of a polyploid complex, and has a basic chromosome number of x = 7, with ploidy levels ranging from diploid to octoploid, and chromosome numbers of 2n = 14, 28, 42, 49 and 56 (Kleynhans & Spies, 1999). Ploidy levels for this species are constant within populations. The differences in ploidy level are probably the product of a founder effect; for example, the octoploid populations identified by Kleynhans and Spies (1999) in the southern Cape are probably derived from tetraploid populations in this area (Duncan, 2005). The *L. bifolia* complex has undergone chromosomal divergence generating a number of species, but these are difficult to distinguish on morphological grounds (Duncan, 2005). Morphological variation in *L. bifolia* is thus better correlated with geographical distribution than with ploidy level (Kleynhans & Spies, 1999).

In a further study of genetic variation in *L. bifolia*, Kleynhans and Spies (2000) tested the feasibility of detecting intra-specific variation using randomly amplified polymorphic DNA (RAPD). The genetic-distance values obtained from 21 accessions of this species complex revealed a high level of variation, and showed that different hexaploid and tetraploid accessions were grouped together in different RAPD clusters, suggesting that accessions with the same chromosome number are not necessarily closely related.

Speciation as a result of ploidy changes occurs when gene flow is disrupted by chromosomal incompatibility, such as that resulting from polyploidy and aneuploidy. Here, the species may occur sympatrically but do not share gene pools. The differentiation of numerous morphological and cytological forms within species such as *L. bifolia* and *L. mutabilis* appear to point to a state of rapid evolution among certain groups.

Johnson and Brandham (1997) reported that all the basic chromosome numbers x = 7–13 and 15 produce structural diploids with twice the basic chromosome number (i.e. a number of 2n = 20 represents a diploid based on x = 10 rather than a tetraploid based on x = 5). They speculated that diploids with 2n = 30 (x = 15) could actually be allotetraploids derived from taxa with x = 7 and x = 8.

B-chromosomes have been reported for 16 species in *Lachenalia* (Spies et al., 2002). They differ morphologically from the chromosomes of the normal complement in being smaller, unpaired and heterochromatic, but these characteristics are variable (Jones & Rees, 1982). In a cytogenetic study, Hancke and Liebenberg (1990) found B-chromosomes in somatic and meiotic material of some species and in F_1 hybrids. In addition to the normal complement, one to three B-chromosomes were observed in some of the cells of several species and in F_1 hybrids. These chromosomes were similar in size to the smallest chromosome of the normal complement. This irregular occurrence of B-chromosomes accounts for variation in chromosome counts between vegetatively similar plants of some species, such as *L. orchioides* (de Wet, 1957; Moffett, 1936; Riley, 1962; Johnson & Brandham, 1997).

Differences in basic chromosome number of *Lachenalia* result in breeding barriers between sympatric species whose flowering periods overlap. This explains the exceptionally low incidence of natural hybrids between some sympatric species. As an example, in a large population of *L. mathewsii* (x = 7) and *L. pallida* (x = 8) growing sympatrically near Vredenburg on the Cape west coast, the flowering time overlaps for a period of at least two weeks but no natural hybrids have been observed over a ten-year period, despite intense pollination activity by honey-bees visiting both species. Similarly, in another population near Vredenburg where the flowering periods of *L. pallida* (x = 8) and *L. unifolia* (x = 11) overlap, no hybrids are known; and at Wilgerbosdrift Farm north of Piketberg, the flowering periods of *L. membranacea* (x = 7) and *L. pallida* (x = 8) overlap, with no known hybrids. Where taxa are sympatric and have the same basic chromosome number, flowering phenology is important in maintaining species barriers; for example, in sympatric populations of *L. liliiflora* ($x = 8$) and *L. orthopetala* ($x = 8$) near Durbanville, the flowering period of *L. orthopetala* ends just before that of *L. liliiflora* begins.

The only known record of a natural hybrid in *Lachenalia* is between two long-tubed species (formerly regarded as colour varieties of *L. aloides*), the bright-orange-yellow-flowered *L. callista* and the unusually coloured turquoise-and-green-flowered *L. vanzyliae*, both of which occur on Piketberg Mountain in the Western Cape. In this instance, the single individual (*Thomas s.n.*, in NBG) is clearly a hybrid between the two species, having flowers of intermediate colour and length (Barker, 1984). As no hybrid swarms occur between the two species (which occur in close proximity) and both species almost certainly have the same sunbird pollinator, recognition of the two taxa as separate species is warranted.

Artificial hybridisation between species that share the same basic chromosome number is fairly successful, but between species with different basic numbers its success rate is low (Kleynhans *et al.*, 2009).

5. PHYLOGENY

The unusually high level of morphological variation within certain species of *Lachenalia* has led to considerable taxonomic confusion (Duncan, 1992a). Furthermore, molecular and morphological analyses of *Lachenalia* have not yet produced well-resolved phylogenies. Molecular analyses have probably been limited by the sequencing of insufficiently informative gene regions, whereas in morphological analyses, the number of informative characters available has been insufficient to determine the relatedness of all of the analysed taxa; not surprisingly the molecular results do not agree with the morphological results. *The Lachenalia Handbook* (Duncan, 1988a) was intended as "A guide to the genus, with introductory notes on history, identification and cultivation, with descriptions of the species and colour illustrations"; a number of authors (Spies *et al.*, 2002; Spies, 2004; Hamatani *et al.*, 2008) have repeatedly misinterpreted the species arrangement in that book as indicating relatedness, whereas it was intended entirely for identification purposes. Similarly, several authors have erroneously concluded that the extreme variability within the genus makes all of its members difficult to identify; the converse is true and most species are distinct and easily identified.

Numerous studies (Pfosser & Speta, 1999; van der Merwe, 2002; Wetschnig *et al.*, 2002; Pfosser *et al.*, 2003; Manning *et al.*, 2004; Spies, 2004; Duncan *et al.*, 2005; Hamatani *et al.*, 2008) have shown that *Polyxena* is embedded within *Lachenalia* and that the recognition of *Polyxena* renders *Lachenalia* paraphyletic. With the inclusion of *Polyxena*, *Lachenalia* becomes a well-defined genus that is easily distinguished from other genera in the Asparagaceae by floral, vegetative and seed characters (Duncan 2005, Duncan *et al.*, 2005).

MOLECULAR ANALYSES

A molecular study of the transfer-RNA intergenic spacer gene region *trn*L-F (Spies, 2004) set out to determine, first, the phylogenetic relationships between the genus *Lachenalia* and related taxa; second, whether *Lachenalia* could be subdivided into subgenera; and, third, whether the different basic chromosome numbers corresponded with monophyletic groups. Only 45 of the 122 taxa analysed were resolved in a strict consensus cladogram, but an Adams consensus cladogram gave an improved resolution and indicated four main groups within the genus, with some of the clades correlating with the basic chromosome numbers. At specific level, however, the resolution of these analyses was poor and could not resolve the phylogeny within the groups. Basic chromosome numbers were superimposed onto the four main groups obtained in the Adams consensus cladogram: most of the taxa in subgroups within one of the main groups had basic chromosome numbers $x = 7$ or 8; whereas taxa with basic chromosome numbers of $x = 6, 7, 8, 9, 10, 11$ and 13 were spread between the $x = 7$ and $x = 8$ groups within the other main group. It was concluded that at least one additional gene region should be sequenced to indicate clearly the relationships within the genus.

In another recent molecular study (Hamatani *et al*., 2008) using internal transcribed spacer (ITS) sequences in 34 *Lachenalia* taxa (including three '*Polyxena*' taxa), correlation was shown between the ITS sequence data and karyotypes, except for taxa with the basic chromosome numbers $x = 7$ and $x = 8$. It was concluded that the latter taxa might share a common ancestry, and that taxa with a basic chromosome number other than $x = 7$ or 8 originate from another ancestor.

A degree of correlation has also been found between groups identified in Spies (2004) from *trn*L-F sequences, and groups identified in a study of inter-species crosses: a shared basic chromosome number indicated species within the phylogenetic groups that had a tendency to cross successfully (Kleynhans *et al*., 2009).

MORPHOLOGICAL ANALYSES

In order to examine the potential of morphological characters to produce a natural classification of *Lachenalia*, and to determine its relationship with the small endemic, sympatric genus *Polyxena*, I undertook a cladistic study of all taxa recognised at the time (Duncan 2005, Duncan *et al*., 2005a, Duncan, 2006).

Variation within a species can occur in several macro-morphological characters, such as overall plant size, leaf number, pedicel length, degree of stamen exsertion, flower size, colour and orientation, and flowering period. Variable species often display population stability in features such as bulb and perianth shape, filament orientation and seed morphology. Certain other species, such as *L. bifolia, L. contaminata, L. juncifolia, L. longibracteata, L. mutabilis, L. orchioides, L. pallida* and *L. violacea*, are exceedingly variable.

The study was based on extensive field research, and study of the living collection at Kirstenbosch National Botanical Garden and the comprehensive collection of preserved material housed in the Compton Herbarium (NBG) at Kirstenbosch. Seventy-three characters, comprising 57 qualitative and 16 quantitative characters, were used in a morphological cladistic analysis of all of the taxa recognised in the genera *Lachenalia* and *Polyxena* in 2004. Qualitative characters, such as bulb and seed shape, were used where these formed clear evolutionary states, and quantitative characters, such as tepal and filament length, were used only when dimensional data formed clear, discontinuous units. The characters comprised 38 binary and 35 multistate characters and were polarised using outgroup comparison to postulate the direction of evolutionary change: numerical values were assigned to each hypothesised character state change, with zero indicating the outgroup state and values above zero indicating the derived state(s); these values were coded in a data matrix. Character states were unordered, autapomorphies were excluded (because they provide no grouping information) and unknown character states were indicated by a question mark (?). The endemic South African geophyte *Massonia depressa* was chosen as the outgroup as it forms part of the tribe Massonieae recognised by Pfosser and Speta (1999) and by Pfosser *et al*. (2003), and because, according to the molecular phylogenies of Wetschnig *et al*. (2002), it shares the most recent common generic ancestor with *Lachenalia* and *Polyxena*.

The computer programme PAUP★ (Version 4.0b10) was used for the cladistic analyses (Swofford, 1999). Owing to the large dataset (139 taxa comprising the ingroup, one taxon comprising the outgroup) and the relatively few (73) morphological characters studied, the heuristic method was used to search for the most parsimonious trees. Using the data matrix, an initial parsimony analysis was performed in which all 73 characters were unordered and unweighted, with the trees rooted using the outgroup comparison method. During stepwise addition, starting trees were obtained using random addition sequences. One thousand replicates were performed, with five trees held at each step. The delayed transformation (DELTRAN) method was used for character-state optimisation.

Parsimony analysis generated 71,153 equally parsimonious trees, each with a shortest possible length of 1,054 steps. The strict consensus cladogram suggested that the evolutionary development of *Polyxena* was paraphyletic with that of *Lachenalia*, with *Polyxena* forming the most basal clades. The uppermost major clade containing most *Lachenalia* taxa was polychotomous, including numerous terminal species pairs, unresolved taxa, several larger clades and four synapomorphies; for example, a terminal clade including the sister species *L. buchubergensis* and *L. nordenstamii* was supported by the common occurrence of seeds with long chalazal collars; a terminal clade including the sister species *L. barkeriana* and *L. pusilla* was supported by concrescent leaf bases (a specialisation of the bulb in which the bases of the foliage leaves of the previous growing season remain intact); a clade including all the taxa previously regarded as varieties of *L. aloides* was supported by shared medial inner filament insertion; and five of the '*L. aloides*' taxa, *L. bifolia* and *L. punctata* were linked by long filaments.

The strict consensus tree showed a low level of resolution and indicated a high level of homoplasy across the tree, as evidenced by numerous parallelisms and reversals. As a measure of confidence to test the stability of resolved nodes within the unweighted tree, a Bremer support analysis (also known as a 'decay index') (Bremer, 1988, 1994) was performed, for which the 'subparsimonious search procedure' was employed. Under this method, support for each of the clades resolved by the data set is quantified as the difference in length between the most parsimonious trees and the shortest tree in which the clade is not resolved. The analysis indicated a general trend of low support for most major nodes and highest support for nodes closer to the individual taxa.

Owing to the high level of homoplasy present in the unweighted trees in the first analysis, a weighted analysis using the successive approximations character weighting scheme (SACW) was performed in an attempt to resolve some of the homoplasy, using the same methodology as that used for the first analysis. The scheme assigns relative weights to characters such that those with more informative features are valued higher than others, thereby reducing the influence of the most homosplasious characters. The strict consensus tree showed a greatly improved level of resolution compared with that of the unweighted consensus tree in the first analysis. The major polychotomy in the unweighted analysis was resolved in the weighted analysis, but homoplasy remained at a high level. *Polyxena* again formed the most basal clades and the reduction in value placed on the most homoplasious characters had the effect of repositioning the clades, resulting in a significantly altered tree topology. When the outgroup was disregarded, and the ingroup was treated as unrooted in both analyses, the relationships within the ingroup were apparent. In both analyses, similar well-defined groups occurred, with the taxa placed next to or adjacent to one another, indicating no major conflict in the topologies of the two trees. The weighted tree comprised six synapomorphies. In addition to the four synapomorphies obtained in the unweighted tree, 52 taxa in the uppermost part of the tree were supported by seeds with inflated strophioles, and a clade near the base of the tree comprising 20 taxa was supported by cup-shaped lower bracts.

The major polychotomy in the uppermost part of the unweighted consensus tree could have been the result of insufficient data: the number of available characters was small in relation to the number of taxa analysed. In many instances, the consensus trees of both analyses corroborate intuitive classifications in terms of the relationships between species, and in certain instances, the trees accord with the alliances speculated in original species diagnoses. Almost all infraspecific taxa (including the 'varieties' of *L. aloides*) were grouped together or adjacent to one another.

Bulb shape is an example of a highly homoplasious vegetative character. Subglobose and ovoid bulbs have evolved from globose bulbs on numerous occasions across the genus, in geographically widely separated areas. The evolution of subglobose bulbs from globose bulbs has occurred mainly in deep sandy soils and in higher rainfall areas, whereas ovoid bulb shape has evolved mainly in clay-based soils in arid habitats. Similarly, inner and outer tepal arrangement is a highly homoplasious reproductive character. Inner tepals that are significantly longer than the outer tepals have evolved from subequal inner and outer tepals on numerous occasions, with numerous reversals. The adaptive value of the different inner and outer tepal arrangements is not immediately apparent but might be linked to pollinator attraction.

The most significant of the synapomorphies obtained in the unweighted analysis were medial inner filament insertion and long filaments. These synapomorphies are probably linked to the specialised sunbird pollination syndrome. In the weighted analysis, the significance of the additional synapomorphy of inflated strophioles is unclear, but may be linked to the myrmecochory observed in *L. barkeriana* by Nordenstam (Müller-Doblies *et al.*, 1987). With the exception of *L. polyphylla*, which has an oblong-campanulate perianth, all of the taxa sharing synapomorphic cup-shaped lower bracts have either narrowly campanulate or widely campanulate perianths, and in most of these taxa, the perianths are white. Cup-shaped lower bracts in combination with narrowly campanulate or widely campanulate white perianths could form part of a pollinator syndrome.

Most character states appear to have evolved more than once across *Lachenalia*; this can be ascribed to their high fitness value in providing expedient solutions to evolutionary problems. There is no doubt that the levels of homoplasy are not artificial constructs but rather represent convergent responses in a rapidly radiating genus. This situation is expected because the common phyletic base poses restrictions on the possible morphological answers to environmental challenges.

With regard to vegetative characters, a clear pattern that emerges is the convergent reversal of certain character states in response to aridity. Certain species from arid or semi-arid regions that have solitary bulbs and cartilaginous outer bulb tunics have evolved in tandem, examples being *L. dasybotrya* from the western edge of the Great Karoo, *L. martiniae* from the Olifants River Mountains and two species from southern Namibia, *L. giessii* and *L. pearsonii*. Similarly, fasciculate apices of the outer bulb tunics have evolved in tandem with solitary bulbs and cartilaginous outer bulb tunics in all the above-mentioned species. *L. namibiensis* from southern Namibia and *L. patula* from the Knersvlakte also display this character trait. Extended bulb apices are partly the product of deep-seated bulbs, another adaptation to aridity. These have evolved in tandem with solitary bulbs in *L. congesta* from the Roggeveld Plateau, *L. ameliae* from the Tanqua and Little Karoos, *L. zebrina* from the western and central Great Karoo, and *L. nordenstamii* from the Richtersveld and south-western Namibia.

With regard to reproductive characters, sunbird pollination appears to have driven convergence: basifixed outer filament insertion occurs in *L. aloides*, *L. bifolia*, *L. callista*, *L. flava*, *L. luteola*, *L. punctata*, *L. quadricolor*, *L. thunbergii* and *L. vanzyliae*, long outer tepals (12 mm or longer) (also in *L. algoensis*) and long (16 mm or longer), oblong-obovate inner tepals also occur in all of these species. Other characters that are common to sunbird-pollinated species include tubular perianths with long, cylindrical tubes, and oblong outer tepals.

Of the ornithophilous species, *L. bifolia* and *L. punctata* have relatively wide distribution ranges extending along the coast of the north-western, western, south-western and southern parts of the Western Cape and the western Northern Cape. The collective distribution of the remaining sunbird-pollinated species (*L. aloides*, *L. callista*, *L. flava*, *L. luteola*, *L. patentissima*, *L. quadricolor*, *L. thunbergii* and *L. vanzyliae*) is similar but extends further inland in the Western Cape. *L. viridiflora* has a highly restricted distribution on the west coast of the Western Cape, and the variable *L. reflexa* occurs in the western and south-western parts of the Western Cape, whereas the putatively sunbird-pollinated *L. algoensis* extends along the south coast of the Western Cape into the Eastern Cape.

Most *Lachenalia* species are bee-pollinated and share suites of character states, including oblong-campanulate, narrowly campanulate, widely campanulate or urceolate perianth shapes, short, ovate outer tepals, and short, linear-oblong or obovate inner tepals.

Reproductive characters are often tightly constrained by the morphology of pollinators, and may be fairly constant. By contrast, phenotypic plasticity in vegetative characters has high fitness value because these characters interface with unpredictable climatic factors and may vary with distribution or across seasons. The intrinsic plasticity of vegetative characters, especially of quantitative characters, makes the recognition of discrete evolutionary units difficult. As a consequence, character selection in this morphological analysis was unavoidably skewed in favour of reproductive characters (46 characters) over vegetative characters (27 characters).

Although the same character states have evolved in widely different areas and in different habitats, their distributions do not necessarily provide unambiguous clues regarding the evolutionary history of the genus because present distributions could be very different from those of the past.

Speta (1998) and Pfosser and Speta (1999) have shown that it is difficult to find morphological characters that yield synapomorphies for groups of taxa within the Hyacinthaceae (now a part of the Asparagaceae). In *Lachenalia*, morphological characters alone are insufficient to trace the evolutionary history of the genus, and other approaches such as molecular and cytological methods will be needed in conjunction with morphology to generate a robustly supported phylogeny of the genus (Duncan *et al.*, 2005a; Duncan, 2006). Recent molecular studies on Cape elements of *Oxalis* (Oberlander *et al.*, 2004), which have undergone similar rapid massive radiations, have produced molecular results that are similarly plagued with homoplasy, and expanded sequencing might be required to resolve questions relating to rapid speciation in the Cape Floristic Region.

6. TAXONOMIC TREATMENT

The hyacinth family (Hyacinthaceae) is no longer considered distinct from the broader Asparagaceae (APG III, 2009; Chase *et al.*, 2009). The current classification of *Lachenalia* follows:

Order Asparagales

Family Asparagaceae

Subfamily Scilloideae (= family Hyacinthaceae)

Tribe Hyacintheae (= subfamily Hyacinthoideae)

Subtribe Massoniinae (= tribe Massonieae)

Genus *Lachenalia*

Lachenalia forms a monophyletic group and its closest relative is *Massonia*, followed by *Veltheimia* (Wetschnig *et al.*, 2002; Pfosser *et al.*, 2003). *Massonia* differs from *Lachenalia* mainly in its very large, leafy basal bracts and its relatively thick filaments that are fused together at the base. *Veltheimia* is set apart by its relatively large, globose or ovoid bulb with perennial fleshy roots, by the presence of bracteoles and almost completely fused tepals, and in having large, broadly winged, inflated papery capsules (Duncan & Visagie, 2009).

SPECIES CONCEPTS

The species concept adopted in this work is a combination of the morphological and the biological species concepts, adapted from the phylogenetic species concept of Nixon and Wheeler (1990). A species is defined as the smallest aggregation of freely interbreeding populations or lineages that is reproductively isolated from other such populations, diagnosable by a unique combination of character states in comparable individuals.

Morphological characters in *Lachenalia* vary in importance when delimiting a species. Relatively uninformative characters such as leaf and bulb size, leaf maculation colour, flower number, pedicel length and overall plant size can vary tremendously depending on environmental conditions. Stable characters such as bulb, leaf and perianth shape, filament orientation, capsule and seed shape, and strophiole length are not greatly affected by environmental conditions, and are therefore of much greater value in setting species limits. The extremely low incidence of natural hybrids in *Lachenalia* is an indication of the importance of breeding barriers resulting from differences in chromosome number, pollinators and habitat preference. Phenological, ecological and geographic differences also provide useful features in circumscribing species. The continuous course of speciation is clearly illustrated in cryptic species such as those that exist within the *L. bifolia* and *L. mutabilis* complexes; these are at a point of divergence beyond which they will gradually attain distinction through genetic drift and the present work is a snapshot in geological time of these processes.

PHENOTYPIC AND GENOTYPIC VARIATION

The genus *Lachenalia* is phenotypically and genotypically diverse (Duncan, 1992a, 2005). Phenotypic plasticity, the extent to which a particular genotype can give rise to different phenotypes, is marked in certain species such as *L. bifolia* and *L. carnosa*. Extremely variable species, such as *L. juncifolia*, *L. trichophylla* and *L. undulata*, cannot be taxonomically subdivided because of the numerous intermediate forms that exist between their typical and extreme forms. Nevertheless, certain characters are under tight genetic control. Cultivated plants often retain characters (such as leaf maculation and pustule density) that were previously thought to be plastic (Duncan, 2005). When studied in conjunction, field studies and cultivation records provide a better understanding of phenotypic plasticity. Phenotypically stable characters in *Lachenalia* include bulb shape, bulb outer tunic texture, leaf pustules and trichomes, leaf shape, leaf margin anatomy, primary seedling leaf shape, lower and upper bract shape, perianth shape, perianth tube length, flower arrangement, tepal shape, floral symmetry, tepal arrangement, filament orientation, capsule and seed shape, testa primary and secondary sculpturing, and strophiole length and inflation. Examples of plastic characters include leaf orientation, leaf margin shape, scape elevation at anthesis, scape inflation and inflorescence length.

In addition to phenotypic plasticity, polymorphism (i.e. two or more distinct morphological expressions within a single population) is also encountered in *Lachenalia*. For example, in certain populations of *L. mutabilis* both inflated and non-inflated scapes occur; in *L. pallida* smooth and densely pustulate upper leaf surfaces occur in single populations, and in *L. viridiflora*, leaf upper surfaces can be spotted or unspotted within the same population.

The subgeneric classification followed here is based on floral symmetry groups identified in the morphological cladistic analyses (Duncan et al., 2005a). Taxa that have zygomorphic perianths represent all taxa traditionally placed under the genus *Lachenalia* (subgenus *Lachenalia*), and taxa with actinomorphic perianths represent all those previously placed under the genus *Polyxena* (subgenus *Polyxena*). Within subgenus *Lachenalia*, the five major perianth shapes identified in the cladistic analyses were considered the most important and practical features for the purpose of sectional subdivision. The species arrangement within the sections follows the cladistic synthesis of Duncan et al. (2005a).

Lachenalia J. Jacq. ex Murray, *Linnaeus Systema Vegetabilium*, ed. 14: 314 (1784); Baker, *Journal of the Linnean Society* (Botany) 11: 401 (1871); Baker, *Flora Capensis* 6: 421 (1897). Type species: *Lachenalia tricolor* J. Jacq. (= *Lachenalia aloides* (L.f.) Engl.).

Synonymy

Coelanthus Willd. ex J. A. Schultes & J. H. Schultes, in Roemer & Schultes, *Systema Vegetabilium* 7(2): xlvi (1830); subgenus *Coelanthus* (Willd. ex J. A. Schultes & J. H. Schultes) Baker, *Journal of the Linnean Society* (Botany) 11: 402 (1871); Baker, *Flora Capensis* 6: 422 (1897). Type: *Coelanthus complicatus* Willd. ex J. A. Schultes & J. H. Schultes (= *Lachenalia reflexa* Thunb.).

Periboea Kunth, *Enumeratio Plantarum* 4: 293 (1843). Type: *Periboea corymbosa* (L.) Kunth; *Hyacinthus corymbosus* L. (= *Lachenalia corymbosa* (L.) J. C. Manning & Goldblatt).

Polyxena Kunth, *Enumeratio Plantarum* 4: 294 (1843). Type: *Polyxena pygmaea* (Jacq.) Kunth; *Polyanthes pygmaea* Jacq. (= *Lachenalia pygmaea* (Jacq.) G. D. Duncan).

Himas Salisb., *The Genera of Plants*: 21 (1866). Type (designated here): *Lachenalia angustifolia* Jacq. (= *Lachenalia contaminata* Ait.).

Platyestes Salisb., *The Genera of Plants*: 21 (1866). Type (designated here): *Lachenalia purpureo-caerulea* Jacq.

Monoestes Salisb., *The Genera of Plants*: 21 (1866). Type (designated here): *Lachenalia unifolia* Jacq.

Chloriza Salisb., *The Genera of Plants*: 21 (1866); subgenus *Chloriza* (Salisb.) Baker, *Journal of the Linnean Society* (Botany) 11: 403 (1871); Baker, *Flora Capensis* 6: 422 (1897). Type (designated here): *Lachenalia mediana* Jacq.

Orchiops Salisb., *The Genera of Plants*: 21 (1866); subgenus *Orchiops* (Salisb.) Baker, *Journal of the Linnean Society* (Botany) 11: 402 (1871); Baker, *Flora Capensis* 6: 422 (1897). Type (designated here): *Lachenalia orchioides* (L.) Ait.

Brachyscypha Baker, *Journal of the Linnean Society* (Botany) 11: 394 (1871); subgenus *Brachyscypha* (Baker) Baker, *Flora Capensis* 6: 423 (1897). Type: *Brachyscypha undulata* (Thunb.) Baker; *Massonia undulata* Thunb. (= *Lachenalia pusilla* Jacq.).

Deciduous, perennial, winter- or summer-growing *geophytes* 10–600 mm high. *Bulb* globose, subglobose or ovoid, subterranean, sometimes produced into an extended, fasciculate neck, clumping or solitary; tissue usually white, rarely yellow; tunic multilayered, outer layers light to dark brown or black, spongy, papery or cartilaginous, inner tunics light brown, membranous; shallow- to deep-seated, roots fibrous or contractile. *Cataphylls* 2, rarely 3, membranous, tubular, translucent white, longitudinal veins brown, green or red, subterranean to partially aerial, adhering to or loosely surrounding leaf bases, apex obtuse or acute. *Leaves* 1–18, usually 2, synanthous, proteranthous or partially hysteranthous, linear, lanceolate or ovate, flat to canaliculate, subterete or terete, prostrate to spreading, suberect or erect, fleshy to coriaceous, upper surface smooth to densely pustulate, rarely covered with simple or stellate trichomes, upper surface plain or lightly to heavily blotched or spotted, lower surface plain or with sporadic or dense blotches or spots, or heavily marked with horizontal bands, rarely covered with simple trichomes; margins flat, undulate and/or crisped, soft to sclerotic; leaf bases clasping to loose, subterranean to partially aerial, white, green or maroon, plain, lightly to heavily banded, spotted or blotched; primary seedling leaves flat or terete, prostrate to erect. *Inflorescence* sparse to dense, spicate or racemose, or forming corymbose, subcorymbose or subcapitate racemes, congested at or just above ground level to elongate, sterile apex short to long; scape erect to suberect, solid, subterranean to aerial, cylindrical, uninflated to strongly inflated above, plain to heavily spotted or blotched, covered with a powdery bloom; bracts small, ovate to lanceolate or cup-shaped, membranous or rarely papery, bracteoles absent; pedicels absent to well developed, patent, suberect, erect to curved downwards, white, green, brown, yellow, orange or red. *Perianth* weakly to strongly zygomorphic, rarely actinomorphic, tubular, oblong-campanulate, narrowly campanulate, widely campanulate or urceolate, unscented or weakly to strongly fragrant, spreading, suberect or erect, cernuous or pendent, plain or multicoloured in shades of white, yellow, brown, blue, turquoise, grey, pink, reddish orange or red; tepals fused basally into a flat disc, oblique or radial cup, or short to long tube, outer and inner tepals subequal or inner tepals protruding slightly to well beyond outer tepals; outer tepals ovate, lanceolate or oblong, erect to recurved, apex usually with prominent gibbosity; inner tepals linear-oblong, lanceolate, obovate or oblong-obovate, erect or weakly to strongly recurved, median keel darker, lower inner tepal sometimes shorter or distinctly longer than upper inner tepals. *Stamens* more or less straight, narrowly spreading, recurved or declinate, included to well-exserted, biseriate, filaments terete, slender, rarely clavate, plain white, rarely purple to magenta above, variously inserted from base to mouth of perianth tube, outer stamen series inserted slightly to well below inner series; anthers oblong, basifixed or dorsifixed, dehiscing longitudinally, dull maroon to purple; pollen monosulcate, yellow, rarely ageing to black, exine reticulate or rugose. *Ovary* ellipsoid or obovoid, sessile or rarely shortly stipitate; ovules biseriate, few to numerous per locule; style included to well exserted beyond perianth, straight or declinate, terete, white, rarely purple or magenta above; stigma capitate, apical. *Capsule* ellipsoid, ovoid, obovoid or obcordate, erect to patent or pendulous, dry, membranous or papery, loculicidal, dehiscing longitudinally. *Seeds* globose, ovoid to oblong, testa black, matte or glossy, smooth, rugose, reticulate or colliculate, strongly adhering; strophiole rudimentary to long, smooth, ridged or inflated. *Chromosome number:* x = 7–13, 15.

Lachenalia includes 133 species (139 taxa), confined mainly to the winter rainfall zone of southwestern Namibia and Northern, Western and Eastern Cape, South Africa, extending inland to summer rainfall parts of southern Namibia, Northern Cape, Free State and Eastern Cape, South Africa.

Notes. The keys below apply to species in habitat; under cultivation, plants may become much larger, additional leaves may develop, leaf margins may lose their crisping/undulation, leaf orientation may change, species with inflated scapes may lose this trait and subterranean scapes may become aerial. Refer to morphology text (pp. 57–79, especially figures 58–60).

Key to the subgenera

1a Perianth slightly to strongly zygomorphic, arising obliquely from the pedicel or rachis. . . . . **1. subgenus *Lachenalia*** (p. 108)
1b Perianth actinomorphic, arising radially from the pedicel . . . . . **2. subgenus *Polyxena*** (p. 426)

Key to the sections of subgenus *Lachenalia*

1a Outer tepals oblong; perianth tubular . . . . **1.1 section *Lachenalia*** (p. 108)
1b Outer tepals ovate; perianth urceolate or campanulate 2

2a Inner tepals distinctly contracted at mouth; perianth urceolate . . **1.5 section *Urceolatae*** (p. 378)
2b Inner tepals radiating at mouth; perianth campanulate 3

3a Inner tepals radiating 15–20 degrees from longitudinal axis; perianth oblong-campanulate . . . . **1.2 section *Oblongae*** (p. 191)
3b Inner tepals radiating 45 degrees or more from longitudinal axis 4

4a Outer tepal apices straight; inner tepals straight to slightly recurved, radiating 45–50 degrees from longitudinal axis; perianth narrowly campanulate. . . . . **1.3 section *Angustae*** (p. 326)
4b Outer tepal apices slightly to strongly recurved; inner tepals strongly recurved, radiating 60–85 degrees from longitudinal axis; perianth widely campanulate . . . . . **1.4 section *Latae*** (p. 346)

Key to the species of section *Lachenalia*

1a Perianth tube 6 mm long or longer. 2
1b Perianth tube 3 mm long or shorter 20

2a Leaves ovate 3
2b Leaves linear, or narrowly to broadly lanceolate. 4

3a Upper leaf surface covered with minute to long hairs; margin not thickened, leaf prostrate . . . . **4. *L. trichophylla***
3b Upper leaf surface smooth; margin thickened, leaf spreading to suberect . . . . **5. *L. marginata* subsp. *marginata***

4a Leaves linear 5
4b Leaves narrowly to broadly lanceolate. 7

5a	Lower leaf surface hairy	**19. *L. bruynsii***
5b	Lower leaf surface smooth	6
6a	Pedicels 1–2 mm long, perianth suberect	**18. *L. schlechteri***
6b	Pedicels 4–15 mm long, perianth spreading to cernuous	**17. *L. unifolia***
7a	Perianth base ventricose; lower inner tepal shorter than upper inner tepals	8
7b	Perianth base not ventricose; lower inner tepal equal to or longer than upper inner tepals	9
8a	Inner tepal apices strongly recurved; perianth light yellow to greenish yellow; scape prominent	**6. *L. algoensis***
8b	Inner tepal apices flat; perianth bright yellow; scape hidden between leaf bases	**1. *L. reflexa***
9a	Perianth suberect	10
9b	Perianth pendulous, cernuous or widely spreading	11
10a	Perianth sessile, light to bright yellow or greenish yellow, blue or violet, strongly sweet-scented; lower inner tepal slightly to much longer than upper inner tepals, inner tepal apices spreading	**3. *L. orchioides***
10b	Perianth pedicellate, turquoise, unscented; lower inner tepal equal to upper inner tepals, inner tepal apices more or less straight	**2. *L. viridiflora***
11a	Inner tepals 14–20 mm long	12
11b	Inner tepals 23 mm long or longer	14
12a	Leaf solitary; perianth widely spreading, outer tepals bright yellow	**13. *L. patentissima***
12b	Leaves 2; perianth pendulous, outer tepals translucent white with turquoise keels, or greenish yellow, yellowish green, mustard, bright yellow, or reddish orange above	13
13a	Inner tepals translucent white; median keels bright green, broad above	**7. *L. vanzyliae***
13b	Inner tepals greenish yellow to bright yellow; median keels light green, narrow throughout	**9. *L. luteola***
14a	Outer tepals bright red or orange-red	**15. *L. bifolia***
14b	Outer tepals not bright red or orange-red	15
15a	Inner tepals bright yellow, greenish yellow, golden yellow or orange	16
15b	Inner tepals light to bright pink	**16. *L. punctata***
16a	Scape purple, inner tepals golden yellow or orange	**12. *L. flava***
16b	Scape green, plain or with brownish purple blotches, inner tepals bright yellow or greenish yellow	17
17a	Inner tepals with broad red, purplish red or purplish magenta apices	18
17b	Inner tepals with greenish yellow apices	19
18a	Outer tepals 15–17 mm long, apices red or purplish red	**10. *L. aloides***
18b	Outer tepals 10–11 mm long, apices purplish magenta	**11. *L. quadricolor***
19a	Inner tepals 32–33 mm long, apices widely spreading	**14. *L. callista***
19b	Inner tepals 23–24 mm long, apices slightly spreading	**8. *L. thunbergii***

20a	Scape subterranean; inflorescence subcapitate	21
20b	Scape aerial; inflorescence spicate or racemose	22
21a	Leaves linear; scape clavate; filaments swollen above	**26. *L. barkeriana***
21b	Leaves lanceolate; scape terete; filaments not swollen above	**27. *L. pusilla***
22a	Inflorescence spicate	23
22b	Inflorescence racemose	25
23a	Filaments well exserted; scape inflated above	**21. *L. ventricosa***
23b	Filaments included to shortly exserted; scape non-inflated	24
24a	Lower inner tepal shorter than upper inner tepals	**25. *L. capensis***
24b	Lower inner tepal longer than upper inner tepals	**5. *L. marginata* subsp. *neglecta***
25a	Leaves linear, perianth pendulous, pedicels bright magenta	**20. *L. sargeantii***
25b	Leaves lanceolate, perianth suberect, spreading or cernuous, pedicels not bright magenta	26
26a	Inner tepals 3 mm longer or more than outer tepals, flowers sweet-scented, perianth ageing to dull red	**25. *L. capensis***
26b	Inner and outer tepals subequal, flowers not sweet-scented, perianth not ageing to dull red	27
27a	Perianth grey with magenta tips; pedicels 0.5 mm long	**22. *L. buchubergensis***
27b	Perianth yellowish brown, maroon or greenish; pedicels 1.5–3.0 mm long	28
28a	Leaves intensely glaucous, deeply canaliculate, upper surface immaculate; perianth yellowish brown to deep maroon	**23. *L. isopetala***
28b	Leaves green, weakly canaliculate, upper surface heavily spotted; perianth greenish	**24. *L. schelpei***

Key to the species of section *Oblongae*

1a	Inflorescence racemose	14
1b	Inflorescence spicate	2
2a	Outer tepals lilac, grey or light purple	3
2b	Outer tepals white, cream, yellow, green or blue	5
3a	Filaments lilac, outer tepals lilac	**68. *L. splendida***
3b	Filaments white, outer tepals grey or light purple	4
4a	Inner tepals strongly recurved; filaments recurved	**53. *L. karooica***
4b	Inner tepals straight to slightly recurved; filaments declinate	**48. *L. bowkeri***
5a	Flowers arranged spirally	10
5b	Flowers arranged in 3- or rarely 4-flowered whorls	6
6a	Leaf bifacial, narrowly lanceolate, midrib prominent on lower surface	**54. *L. concordiana***
6b	Leaf unifacial, lanceolate or broadly lanceolate, midrib absent	7
7a	Lower inner tepal strongly canaliculate, protruding 2 mm beyond upper inner tepals	8
7b	Lower inner tepal weakly canaliculate, equal to upper inner tepals, or protruding up to 1 mm	9

8a	Filaments straight	**52.** *L. inconspicua*
8b	Filaments declinate	**51.** *L. canaliculata*
9a	Inner tepals with magenta margins, flowers unscented	**50.** *L. obscura*
9b	Inner tepals with white, greenish yellow or brownish margins, flowers strongly scented	**36.** *L. marlothii*
10a	Perianth greenish white, leaves broadly lanceolate	**41.** *L. undulata*
10b	Perianth blue or yellow, leaves lanceolate	11
11a	Upper bracts 8–16 mm long	**37.** *L. longibracteata*
11b	Upper bracts 0.5–6.0 mm long	12
12a	Plants dwarf, 20–100 mm high; upper leaf surface minutely pustulate; early winter-flowering	**40.** *L. minima*
12b	Plants not dwarf, 160–400 mm high; upper leaf surface non-pustulate; late spring-flowering	13
13a	Perianth slightly curved; inner tepals slightly spreading, protruding 1 mm beyond outer tepals, apices obtuse	**39.** *L. arbuthnotiae*
13b	Perianth straight; inner tepals strongly spreading, protruding 3 mm beyond outer tepals, apices subacute	**38.** *L. lutea*
14a	Leaves always 2	15
14b	Leaves 1 or more	28
15a	Leaves lanceolate	16
15b	Leaves linear	24
16a	Leaves prostrate	17
16b	Leaves spreading to suberect	18
17a	Upper leaf surface olive green, pustules purplish, depressed longitudinal grooves prominent, leaf margins white; perianth magenta	**31.** *L. nardousbergensis*
17b	Upper leaf surface bright green, pustules green, depressed longitudinal grooves indistinct, leaf margins maroon; perianth dull whitish blue	**79.** *L. stayneri*
18a	Perianth purplish-blue; upper leaf surface densely pustulate	**30.** *L. purpureo-caerulea*
18b	Perianth cream or white; upper leaf surface smooth	19
19a	Leaf margins sclerotic; filaments mauve above	**42.** *L. duncanii*
19b	Leaf margins soft; filaments white	20
20a	Lower leaf surface heavily spotted, leaves narrowly lanceolate	**43.** *L. summerfieldii*
20b	Lower leaf surface plain, leaves broadly lanceolate	21
21a	Filaments declinate; leaves spreading	22
21b	Filaments recurved; leaves suberect	23
22a	Leaf bases aerial; perianth greenish yellow, outer tepals 8–9 mm long	**44.** *L. doleritica*
22b	Leaf bases subterranean; perianth white, outer tepals 5–7 mm long	**45.** *L. dasybotrya*
23a	Perianth suberect, white; pedicels 0.5–2 mm long	**46.** *L. alba*
23b	Perianth spreading to slightly cernuous, greenish white; pedicels 3–5 mm long	**47.** *L. neilii*

24a	Filaments well exserted (protruding more than 2 mm)	**80. *L. juncifolia***
24b	Filaments included to shortly exserted (protruding up to 2 mm)	25
25a	Inner tepals 3 mm longer than outer; leaf bases banded	**49. *L. dehoopensis***
25b	Inner and outer tepals subequal; leaf bases plain	26
26a	Filaments declinate; perianth bright yellow	**34. *L. mathewsii***
26b	Filaments straight; perianth cream or pink	27
27a	Pedicels arcuate; scape rigid; perianth cream	**67. *L. montana***
27b	Pedicels suberect; scape flexible; perianth pink	**66. *L. youngii***
28a	Leaves almost always solitary, rarely 2	29
28b	Leaves 1 or 2, or alternatively leaves 3 or more	41
29a	Leaves linear	30
29b	Leaves lanceolate to broadly lanceolate	32
30a	Lower leaf surface and margins hairy	**83. *L. hirta***
30b	Lower leaf surface and margins glabrous	31
31a	Pedicels 1–4 mm long	**84. *L. attenuata***
31b	Pedicels 9–14 mm long	**85. *L. wrightii***
32a	Scape distinctly inflated above	33
32b	Scape never distinctly inflated above	34
33a	Perianth strongly cernuous, outer and inner tepals subequal; filaments 6–10 mm long; capsule obovoid	**69. *L. nutans***
33b	Perianth slightly cernuous, inner tepals 2–3 mm longer than outer tepals; filaments 10–11 mm long; capsule ellipsoid	**70. *L. klinghardtiana***
34a	Filaments well exserted (protruding more than 2 mm)	35
34b	Filaments included to shortly exserted (protruding up to 2 mm)	36
35a	Perianth cream to brown; filaments white; leaf broadly lanceolate, conduplicate	**77. *L. zebrina***
35b	Perianth greenish-grey; filaments bright mauve; leaf narrowly lanceolate, canaliculate	**58. *L. haarlemensis***
36a	Leaf leathery; outer tepals with variegated spots	37
36b	Leaf soft; outer tepals without variegated spots	38
37a	Leaf margin thickened; filaments declinate	**35. *L. variegata***
37b	Leaf margin not thickened; filaments straight	**36. *L. marlothii***
38a	Perianth pure white; pedicels 4–7 mm long; leaf horizontal, upper leaf surface heavily spotted	**65. *L. leomontana***
38b	Perianth whitish-grey or bluish-grey; pedicels 1–3 mm long; leaf slightly spreading to suberect, upper leaf surface plain	39
39a	Leaf margins undulate and crisped; inner tepals 10–11 mm long	**78. *L. martiniae***
39b	Leaf margins flat; inner tepals 7–9 mm long	40

40a	Perianth suberect, inner tepal apices ageing to pinkish magenta	**55. *L. maximiliani***
40b	Perianth spreading to cernuous, inner tepal apices not ageing to pinkish magenta	**56. *L. perryae***
41a	Leaves lanceolate to broadly lanceolate; leaf bases not swollen	42
41b	Leaves linear; leaf bases distinctly swollen	61
42a	Inner tepals protruding up to 1 mm beyond outer tepals	43
42b	Inner tepals protruding 2 mm or more beyond outer tepals	49
43a	Outer tepals white or cream	44
43b	Outer tepals rose-pink, bluish-grey, light green to bluish-green, greyish-mauve or light blue	46
44a	Scape heavily maculate; perianth tube 0.5–1.0 mm long	**59. *L. lactosa***
44b	Scape immaculate; perianth tube 2–3 mm long	45
45a	Pedicels 3–4 mm long; filaments straight, shortly exserted; plants 35–100 mm high	**33. *L. margaretiae***
45b	Pedicels 5–15 mm long; filaments declinate, well exserted; plants 120–450 mm high	**32. *L. thomasiae***
46a	Filaments included or shortly exserted up to 2 mm; perianth suberect	**60. *L. rosea***
46b	Filaments well exserted; perianth cernuous	47
47a	Inner tepals purple or greyish-mauve above, scape sturdy, inflated above	48
47b	Inner tepals white above, scape slender, not inflated above	**76. *L. whitehillensis***
48a	Filaments purple in upper two-thirds, outer tepals light green to bluish-green, flowers spicy sweet-scented	**74. *L. violacea***
48b	Filaments white or light mauve in upper two-thirds, outer tepals greyish-mauve, flowers heavily coconut-scented	**75. *L. glauca***
49a	Pedicels and capsules remain in same position in fruiting stage as in flowering stage	50
49b	Pedicels and capsules re-orientate to an erect or suberect position in fruiting stage	56
50a	Scape inflated above	51
50b	Scape not inflated above	53
51a	Leaf bases subterranean; inner tepals pinkish-magenta	**71. *L. physocaulos***
51b	Leaf bases aerial; inner tepals white	52
52a	Inner tepals 6–7 mm long, outer tepal gibbosities small, greenish or pinkish brown; seed globose	**72. *L. leipoldtii***
52b	Inner tepals 7–9 mm long, outer tepal gibbosities large, dark brown; seed oblong	**73. *L. xerophila***
53a	Filaments recurved	**53. *L. karooica***
53b	Filaments straight or declinate	54
54a	Upper bracts 8–16 mm long	**37. *L. longibracteata***
54b	Upper bracts 2–3 mm long	55
55a	Outer tepals grey, rarely light purple, speckled with dull blue at base	**48. *L. bowkeri***
55b	Outer tepals light pink or lilac, unspeckled at base	**57. *L. latimeriae***

56a	Filaments more or less straight	63. *L. salteri*
56b	Filaments declinate	57
57a	Scape rigid; leaves deeply canaliculate; plants pyrophytic	64. *L. lutzeyeri*
57b	Scape flexible; leaves flat to weakly canaliculate; plants non-pyrophytic	58
58a	Leaves leathery, firm, margins cartilaginous; scape heavily purple- or dark green-blotched	59
58b	Leaves succulent, flexible, margins soft; scape plain	60
59a	Lower inner tepal diverging markedly from upper inner tepals, inner tepal apices strongly recurved	61. *L. judithiae*
59b	Lower inner tepal not diverging markedly from upper inner tepals, inner tepal apices flat	62. *L. martleyi*
60a	Outer tepals grey, light blue or rarely light pink; upper leaf surface always smooth; strophiole well developed, inflated, smooth, 1.1–1.2 mm long	28. *L. mediana*
60b	Outer tepals white, cream, yellow, dark blue, mauve, magenta or light to deep purple; upper leaf surface smooth or pustulate, strophiole rudimentary, ridged, 0.1–0.2 mm long	29. *L. pallida*
61a	Leaves covered with minute papillae in lower third, smooth above; filaments 7–9 mm long	82. *L. polyphylla*
61b	Leaves without papillae but with raised fleshy bands along entire length; filaments 9–11 mm long	81. *L. moniliformis*

Key to the species of section *Angustae*

1a	Leaves prostrate; scape subterranean, rarely slightly aerial	94. *L. kliprandensis*
1b	Leaves spreading to erect; scape strongly aerial	2
2a	Pedicels 10–25 mm long, scape inflated above	86. *L. anguinea*
2b	Pedicels 1–9 mm long, scape never inflated above	3
3a	Inner tepals 11–15 mm long	4
3b	Inner tepals 4–9 mm long	5
4a	Leaves lanceolate; inner tepal apices obtuse	87. *L. liliiflora*
4b	Leaves linear; inner tepal apices acute	88. *L. orthopetala*
5a	Perianth cernuous	6
5b	Perianth spreading to suberect	7
6a	Filaments 4 mm long, leaves lanceolate, attenuate, uppermost portion canaliculate	89. *L. convallarioides*
6b	Filaments 8–13 mm long, leaves linear, uppermost portion subterete	93. *L. magentea*
7a	Inner tepals 5 mm long; pedicels 3–5 mm long	90. *L. zeyheri*
7b	Inner tepals 7–8 mm long; pedicels 1–2 mm long	8
8a	Leaves 3–11, leaf bases subterranean	91. *L. contaminata*
8b	Leaves always 2, leaf bases aerial	92. *L. bachmannii*

Key to the species of section *Latae*

- **1a** Leaves ovate, upper surface with prominent depressed longitudinal grooves between veins; flowers heavily scented . . . . **95. *L. nervosa***
- **1b** Leaves lanceolate or linear, depressed longitudinal grooves indistinct; flowers unscented or slightly scented 2
- **2a** Scape rigid, 1 mm wide, leaf prostrate, becoming suberect at anthesis . . . . **96. *L. angelica***
- **2b** Scape flexible, 2 mm wide or more, leaves spreading to suberect 3
- **3a** Filaments included to shortly exserted (protruding up to 2 mm) 4
- **3b** Filaments well exserted (protruding more than 2 mm) 8
- **4a** Filaments narrowly spreading 5
- **4b** Filaments declinate or recurved 7
- **5a** Leaves bifacial, narrowly lanceolate, midrib prominent on lower surface 6
- **5b** Leaves flat to canaliculate, linear to broadly lanceolate, midrib absent on lower surface . . . . **101. *L. giessii***
- **6a** Perianth tube 3 mm long; inner tepals 7–9 mm long; filaments 6–7 mm long . . . . **102. *L. namibiensis***
- **6b** Perianth tube 1 mm long; inner tepals 4.5 mm long; filaments 4 mm long . . **103. *L. pearsonii***
- **7a** Leaves very succulent, up to 60 mm long, apices apiculate; pedicels 5–7 mm long; filaments declinate . . . . **105. *L. patula***
- **7b** Leaves not very succulent, up to 180 mm long, apices canaliculate or subterete; pedicels 2–4 mm long; filaments recurved . . . . **97. *L. campanulata***
- **8a** Leaf solitary 9
- **8b** Leaves 2 or more 11
- **9a** Filaments stout, upper half maroon; perianth brownish maroon . . . . **107. *L. nordenstamii***
- **9b** Filaments slender, upper half white or light to dark violet; perianth not brownish maroon . . 10
- **10a** Perianth ivory, bluish white or light to dark violet; filaments declinate; leaf flat, dark green, upper surface covered with minute to long stellate hairs, lower surface unspotted . . . . **106. *L. polypodantha***
- **10b** Perianth cream; filaments recurved; leaf conduplicate, markedly glaucous, upper surface smooth, lower surface spotted . . . . **99. *L. glaucophylla***
- **11a** Leaves 5–18, linear, terete, spirally twisted, leaf bases swollen; filaments narrowly spreading . . . . **104. *L. multifolia***
- **11b** Leaves 2, lanceolate, leaf bases flat; filaments recurved 12
- **12a** Filaments purple above; upper leaf surface hairy, perianth white with green median keels . . . . **98. *L. comptonii***
- **12b** Filaments magenta; upper leaf surface smooth, perianth maroonish magenta with darker median keels . . . . **100. *L. macgregoriorum***

Key to the species of section *Urceolatae*

1a	Leaves spreading to suberect; scape strongly aerial	2
1b	Leaves prostrate; scape subterranean, rarely slightly aerial	**120.** *L. congesta*
2a	Inflorescence spicate	7
2b	Inflorescence racemose	3
3a	Pedicels 1–2 mm long	4
3b	Pedicels 4–8 mm long	5
4a	Outer tepals cream to light yellow	**112.** *L. cernua*
4b	Outer tepals light blue to grey	**109.** *L. aurioliae*
5a	Filaments straight	6
5b	Filaments declinate	**108.** *L. mutabilis*
6a	Leaf bases aerial, tightly clasping, banded with maroon or magenta; flowers light blue	**126.** *L. bolusii*
6b	Leaf bases subterranean, loose, unbanded; flowers white	**111.** *L. peersii*
7a	Flowers arranged in 3-flowered whorls	8
7b	Flowers arranged spirally	10
8a	Lower inner tepal equal to upper inner tepals; leaf margins not thickened	**109.** *L. aurioliae*
8b	Lower inner tepal 1–2 mm longer than upper inner tepals; leaf margins thickened	9
9a	Leaf greenish-grey, margins strongly undulate and crisped; inner tepals 7–8 mm long; filaments 7–9 mm long	**123.** *L. krugeri*
9b	Leaf markedly glaucous, margins flat or slightly undulate; inner tepals 8–10 mm long; filaments 5–6 mm long	**110.** *L. verticillata*
10a	Inner tepals 5–10 mm long	11
10b	Inner tepals 13–15 mm long	**118.** *L. sessiliflora*
11a	Inner tepals protruding up to 1 mm, subterminal tepal markings maroon	**114.** *L. karoopoortensis*
11b	Inner tepals protruding 2 mm or more, subterminal tepal markings not maroon	12
12a	Outer tepals 8–9 mm long; style 7–8 mm long; leaves broadly ovate	**119.** *L. carnosa*
12b	Outer tepals 4–8 mm long; style 2–7 mm long; leaves lanceolate, linear or ovate	13
13a	Perianth slightly to strongly suberect	16
13b	Perianth spreading to cernuous	14
14a	Inner tepals 7–9 mm long; flowers unscented; scape often inflated	**108.** *L. mutabilis*
14b	Inner tepals 6–7 mm long; flowers sweet-scented; scape rarely inflated	15
15a	Inner tepals with broad, translucent membranous apices; outer tepals light yellow to greenish yellow, rarely light blue	**115.** *L. membranacea*
15b	Inner tepals with narrow, solid bright white apices; outer tepals pink, maroon or deep purple	**116.** *L. suaveolens*

16a	Filaments straight, flowers heavily sweet-scented	17
16b	Filaments declinate, flowers unscented	18
17a	Inner tepals 6–7 mm long, outer tepals bright blue	**113. *L. elegans***
17b	Inner tepals 8–9 mm long, outer tepals white, cream, yellow, light blue, lilac or violet	**117. *L. fistulosa***
18a	Leaf upper surface and/or margins covered with short to long simple trichomes	**124. *L. ameliae***
18b	Leaf upper surface and margins without simple trichomes	19
19a	Leaves linear, perianth pinkish magenta	**125. *L. namaquensis***
19b	Leaves lanceolate, perianth greenish yellow or white with magenta markings	20
20a	Lower and upper inner tepal apices all bright magenta, perianth greenish yellow; upper leaf surface always smooth	**121. *L. framesii***
20b	Lower inner tepal apex bright magenta, upper inner tepal apices white or greenish yellow; upper leaf surface densely pustulate, rarely smooth	**122. *L. valeriae***

Key to the species of subgenus *Polyxena*

1a	Perianth tube 12–27 mm long	2
1b	Perianth tube 3–8 mm long	4
2a	Style 24–36 mm long; filaments well exserted; flowers almond-scented; seeds globose	**127. *L. pygmaea***
2b	Style 11–20 mm long; filaments included to shortly exserted; flowers spicy sweet-scented; seeds ovoid	3
3a	Perianth tube subterranean; outer tepals 4–5 mm wide	**128. *L. longituba***
3b	Perianth tube aerial; outer tepals 2–3 mm wide	**129. *L. ensifolia***
4a	Tepals 3–5 mm long, white with white gibbosities, flowers clustered deeply between leaf bases	**130. *L. calcicola***
4b	Tepals 7–10 mm long, lilac with lilac gibbosities, or white with mauve or purple gibbosities, flowers appearing above leaf bases	5
5a	Filaments 1 mm long, tepals bright lilac, gibbosities lilac	**131. *L. paucifolia***
5b	Filaments 3–7 mm long, tepals mauve or white, gibbosities mauve or purple	6
6a	Leaves linear; tepal keels dark mauve to purple, capsule obovoid	**132. *L. corymbosa***
6b	Leaves lanceolate; tepal keels white to light lilac, capsule ellipsoid	**133. *L. argillicola***

1. Subgenus LACHENALIA

Perianth slightly to strongly zygomorphic, arising obliquely from the pedicel or rachis.

126 species (131 taxa)

1.1. Section LACHENALIA

Perianth tubular, tube cylindrical, 6–12 mm long, or cup-shaped and up to 3 mm long; outer tepals oblong, inner tepals linear oblong or oblong-obovate, outer and inner tepals radiating up to 10 degrees from the longitudinal axis or converging; filaments straight, included, rarely shortly to well exserted.

27 species (30 taxa).

1. LACHENALIA REFLEXA

Lachenalia reflexa Thunb., *Prodromus Plantarum Capensium*: 64 (1794).
TYPE: South Africa, Cape, precise locality unknown, *C. P. Thunberg* 8558 (UPS!, holo., microfiche).
SYNONYMY: *Coelanthus complicatus* Willd. ex Roem. & Schult., *Systema Vegetabilium* 7 (2): xlvi (1830). Type: South Africa, Cape, precise locality unknown, *C. L. Willdenow* 6565 (B!, holo.).

ETYMOLOGY. *reflexa*: downwardly bent (reflexed) leaves.
DESCRIPTION. *Dwarf geophyte*, 20–160 mm high. *Bulb* subglobose, 10–15 mm in diameter, offset-forming; tunic multilayered, outer layers dark brown, spongy; inner cataphyll translucent white, upper half heavily tinged with brownish-magenta, adhering to leaf bases, apex obtuse or acute. *Leaves* 1–2, lanceolate, 50–180 × 10–25 mm, suberect to spreading or reflexed, yellowish-green to glaucous, deeply canaliculate, upper surface plain to heavily spotted with light to dark brown; leaf margins entire or slightly to heavily undulate; leaf bases clasping, 10–45 mm long, lower portion white, upper portion light yellowish-green, occasionally maroon-spotted; primary seedling leaf terete, erect. *Inflorescence* racemose, 1–many-flowered, sterile apex short; scape erect, 10–40 mm long, yellowish-green, hidden between leaf bases; rachis yellowish-green, flexuose; pedicels suberect to erect, 1–2 mm long, yellowish-white; lower bracts ovate, becoming lanceolate above, 3–5 × 2–4 mm, white. *Perianth* zygomorphic, tubular, suberect, straight to distinctly curved above, bright yellow to greenish-yellow, maturing to dull red; tube cylindrical, 7–10 mm long, base strongly oblique, ventricose; outer tepals oblong, 7–9 × 4–6 mm, apical gibbosities green; inner tepals narrowly obovate, 12–15 × 3–5 mm, all overlapping, narrowed towards mouth, keels light green, lower inner tepal 2–3 mm shorter than upper inner tepals. *Stamens* included, straight; filaments white, 8–11 mm long. *Ovary* ellipsoid, 4–5 × 3–4 mm, light green, lower half with minute blue speckles; style included, straight, 13–15 mm long, white. *Capsule* ellipsoid, 12–15 × 4–5 mm. *Seed* globose, 0.9–1.1 × 1.0–1.1 mm, glossy, black; strophiole inflated, 0.6–0.7 mm long, smooth. *Chromosome number* $2n = 14$ (Crosby, 1986; Johnson & Brandham, 1997; Hamatani *et al.*, 1998); $2n = 16$ (de Wet, 1957). Figure 80.
FLOWERING PERIOD. June to August, with a peak in July.

HISTORY. The Swedish naturalist Carl Peter Thunberg (1743–1828) first described this species as an unnamed variety (α) of *Phormium orchioides* in volume 15 of his *Nova Genera Plantarum* (Thunberg, 1784), and ten years later placed it in *Lachenalia* in his *Prodromus Plantarum Capensium*, as *L. reflexa* (Thunberg, 1794). The German botanist and pharmacist C. L. Willdenow appended the name *Coelanthus complicatus* to a specimen in his herbarium (*Willdenow* 6565) collected at an unknown locality at the Cape, and it was published as *C. complicatus* Willd. ex Roem. & Schult. in a note in volume 7 of *Systema Vegetabilium* (Schultes & Schultes, 1830). The name was upheld in *Enumeratio Plantarum* (Kunth, 1843), with *Phormium orchioides* Thunb. var. α and *Lachenalia reflexa* Thunb. relegated to synonymy under it, but the species was returned to *L. reflexa* by Baker (1871) and upheld in his second monograph (Baker, 1897a).

DISTINGUISHING CHARACTERS AND AFFINITIES. Phylogenetic analysis of morphological data placed *L. reflexa* as sister to *L. viridiflora* (Duncan *et al.*, 2005a). The latter resembles *L. reflexa* in its short racemes of usually suberect flowers with ventricose perianth bases, but differs in its turquoise outer tepals and translucent white, equal inner tepals. *L. viridiflora* has longer scapes (up to 70 mm long) and its leaves are flat to slightly canaliculate with flat margins. Its seeds are slightly narrower (0.9 mm) than those of *L. reflexa* and have slightly longer strophioles (0.7–0.8 mm).

DISTRIBUTION, HABITAT AND CONSERVATION STATUS. *Lachenalia reflexa* is confined to a small area of the south-western Cape extending from Darling in the north, south to the southern Cape Peninsula and inland to Drakenstein and Franschhoek (Map 2). It occurs in Cape Flats Sand Fynbos, Swartland Granite Renosterveld and Cape Winelands Shale Fynbos vegetation (Mucina & Rutherford, 2006).

Figure 80. *Lachenalia reflexa* in habitat near Darling. Image: Graham Duncan.

The plants grow in small to large colonies in seasonally inundated, flat sandy or loamy ground, in full sun or light shade. The species was very common on the Cape Flats but its habitat has been greatly reduced by industrial and housing development, as well as by agricultural expansion, invasive alien plants and trampling by livestock in the remainder of its range. Despite the devastating effects of urbanisation on the Cape Peninsula, *L. reflexa* survives in isolated locations including Rondebosch Common, Kenilworth Racecourse, Maynardville Park in Wynberg, the Edith Stephens Wetland Park in Wetton, and within the grounds of St Saviours Church in Claremont. It has a conservation status of Vulnerable (Helme & Raimondo, 2009g).

Map 2. Known distribution of *Lachenalia reflexa*.

NOTES. The tubular bright yellow, long-tubed, unscented flowers of *Lachenalia reflexa* produce an abundance of nectar and conform to the bird pollination syndrome. It appears to be specialised for pollination by sunbirds, although actual sightings have not yet been made. A particularly striking, long-flowered form that occurs in the Rondeberg Private Nature Reserve near Darling has strongly curved flowers. The species is at least partially self-fertile, regularly setting abundant seed under isolated conditions in cultivation. It flowers prolifically in the winters immediately following summer bush fires.

Lachenalia reflexa shares an unusual seed dispersal strategy with nine other members of the genus (*L. argillicola*, *L. barkeriana*, *L. calcicola*, *L. corymbosa*, *L. ensifolia*, *L. longituba*, *L. paucifolia*, *L. pusilla* and *L. pygmaea*) in that immediately prior to capsule dehiscence, the subterranean scape elongates, pushing the infructescence upwards before falling into a horizontal position. Most of the seeds fall out of the capsules close to the mother plant. The scapes break off within a few weeks and are blown away, scattering the remaining seeds further afield (Duncan, 2005).

Known as 'yellow soldier' in Australia, the species has become naturalised in the western part around Perth, where it is becoming invasive in sandy or calcareous soils in low open woodland; it has been placed on the Australian *Alert List for Environmental Weeds*.

2. LACHENALIA VIRIDIFLORA

Lachenalia viridiflora W. F. Barker, *Journal of South African Botany* 38 (3): 179–183 (1972).
TYPE. South Africa, Western Cape, Vredenburg district, Witklip Farm, in rock pans and crevices on huge granite boulders, *W. F. Barker* 10171 (NBG!, holo.; PRE!, BOL!, K!, iso.).

ETYMOLOGY. *viridiflora*: turquoise or greenish-turquoise flowers.
DESCRIPTION. *Dwarf geophyte* 40–150 mm high. *Bulb* subglobose, 7–20 mm in diameter, offset-forming; tunic multilayered, outer layers light brown, spongy; inner cataphyll translucent white, adhering to leaf bases, apex obtuse or acute. *Leaves* 2, lanceolate to broadly lanceolate, 50–100 × 15–30 mm,

PLATE 1. Watercolour painting of *Lachenalia viridiflora* from St. Helena Bay (*Duncan* 410, in NBG) courtesy of Fay Anderson. Artist: Fay Anderson.

2. LACHENALIA VIRIDIFLORA

Figure 81. *Lachenalia viridiflora* on a granite outcrop, St. Helena Bay. Image: Graham Duncan.

spreading to suberect, flat to slightly canaliculate, bright green, upper surface with depressed longitudinal grooves, plain to densely spotted with green or light to dark brown, lower surface plain to occasionally spotted with dark magenta; margins coriaceous, light to dark maroon; bases clasping, 5–40 mm long, white; primary seedling leaf terete, erect. *Inflorescence* racemose, few- to many-flowered, sparse to fairly dense, sterile apex short; scape erect, 15–70 mm long, light green or light to dark magenta, plain or with brown to dull magenta speckles or blotches; rachis light to dark magenta or light to bright turquoise; pedicels suberect, 1–2 mm long, white; lower bracts ovate, becoming lanceolate above, 2–6 × 1–3 mm, translucent white. *Perianth* zygomorphic, tubular, spreading to suberect; tube cylindrical, 7–10 mm long, greenish-turquoise to bright turquoise, base strongly oblique, ventricose; outer tepals oblong, 7–9 × 5 mm, greenish-turquoise to bright turquoise, plain or with darker blotches, margins translucent white, apical gibbosities brown to deep turquoise, keels deep turquoise; inner tepals oblong obovate, 10–13 × 4–5 mm, translucent white, lower inner tepal canaliculate, narrower than upper inner tepals, keels deep turquoise, margins translucent white, apices slightly recurved. *Stamens* included, straight; filaments 10–11 mm long, white. *Ovary* ellipsoid, 5 × 2 mm, bright green; style included, straight, 14–15 mm long, white. *Capsule* ellipsoid, 12–13 × 4–5 mm. *Seed* globose, 0.9–1.0 × 0.9 mm, glossy, black; strophiole inflated, 0.7–0.8 mm long, smooth. *Chromosome number* 2n = 14 (Crosby, 1986, Nordenstam, 1982; Hancke & Liebenberg, 1990; Johnson & Brandham, 1997; Hamatani *et al.*, 2007). Plate 1, Figures 22, 81.

FLOWERING PERIOD. May to July, with a peak from early to mid-June.

HISTORY. *Lachenalia viridiflora* was discovered in fruit in August 1953 by a former Kirstenbosch horticulturist, Harry Hall (1906–1986). These plants were growing at Steenberg's Cove, St Helena Bay, and in June 1955, Hall found flowering specimens further north at Stompneus. W. F. Barker collected the type material (*Barker* 10171) at Witklip Farm, Vredenburg in June 1964, and the following year, Dr Ted Oliver made another collection at the same locality. The species was described in the *Journal of South African Botany* (Barker, 1972) and a watercolour painting by Fay Anderson was published in *Flowering Plants of Africa* (Barker, 1979b). More recently, a fairly large population has been recorded at St Helena Bay (*Duncan* 410, in NBG), from which the accompanying new painting by Fay Anderson was produced.

DISTINGUISHING CHARACTERS AND AFFINITIES. *Lachenalia viridiflora* resolved as sister to *L. reflexa* in a phylogenetic analysis (Duncan *et al.*, 2005a) and the distinguishing characteristics are discussed under that species. The two species are geographically isolated, *L. reflexa* having a wider distribution on seasonally wet, flat open ground from Darling to the Cape Peninsula and inland to Franschhoek.

DISTRIBUTION, HABITAT AND CONSERVATION STATUS. *Lachenalia viridiflora* is extremely localised to the north and east of Vredenburg on the Cape west coast (Map 3). It is extinct at its type locality at Witklip Farm south of Vredenburg and the largest populations occur in the vicinity of St Helena Bay at Steenberg's Cove and on privately owned land. These populations are threatened by housing developments; two smaller populations occur to the south-east of St Helena Bay on private land, including Elandskloof Farm (*Duckitt s.n.*, in NBG). The species is confined to Saldanha Granite Strandveld vegetation (Mucina & Rutherford, 2006), occurring in seasonally inundated rock pans and crevices of granite boulders, usually on fairly steep north- and west-facing hillside slopes, or on exposed, more or less flat rock sheets. The bulbs grow in shallow acid humus, exposed or in partial shade, in association with other west coast endemics including *Empodium veratrifolium* and *Pauridia longituba* (Hypoxidaceae), *Pelargonium fulgidum* (Geraniaceae) and various succulent groundcovers. It has a conservation status of Critically Endangered (Victor & Duncan, 2009m).

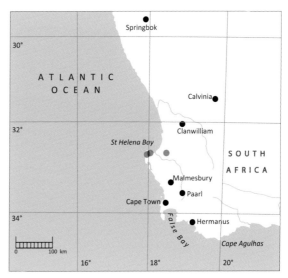

Map 3. Known distribution of *Lachenalia viridiflora*.

NOTES. Leaf upper surface maculation is a polymorphic condition in *L. viridiflora* as expression varies within populations (Duncan, 2005). Most specimens in the wild have plain green leaves and scapes, but some are densely spotted above, the two forms growing side by side. The species is pollinated by Southern Double-collared Sunbirds (*Cinnyris chalybeus*) (Figure 75). The short, rather weak scapes are unable to support the weight of the sunbirds, which probe the flowers from surrounding vegetation or rocks or directly from the ground. In Dirk and Esna Ehlers's private hillside reserve at St Helena Bay, nocturnal Cape porcupines (*Hystrix africaeaustralis* Peters) dig up and feed on the bulbs in summer; bulbs are also scratched out and consumed by helmeted guineafowl (*Numida meleagris* L.). At the same locality, the developing inflorescences and open flowers of *L. viridiflora* are eaten by steenbuck (*Raphicerus campestris* Thunberg).

3. LACHENALIA ORCHIOIDES

Lachenalia orchioides (L.) Aiton, *Hortus Kewensis* 1: 460 (1789); *Hyacinthus orchioides* L., *Species Plantarum* 1: 318 (1753). *Muscari orchioides* (L.) Mill., *The Gardener's Dictionary* 8 (1768).
TYPE: South Africa, Cape, collector and precise locality unknown, figure in Breyne, *Prodromi Fasciculi Rariorum Plantarum*: 24, t. 11, fig. 2 (1739) (lectotype, designated by Stearn (1990)).
(See under subspecies for further synonymy.)

ETYMOLOGY. *orchioides*: orchid-like, with reference to its sweet scent.
DESCRIPTION. *Geophyte*, 55–350 mm high. *Bulb* globose, 10–20 mm in diameter, offset-forming; tunic multilayered, outer layers light to dark brown, spongy; inner cataphyll translucent white, adhering to leaf bases, apex obtuse. *Leaves* (1–)2, lanceolate, 75–120 × 15–35 mm, spreading to suberect,

light to dark green, upper surface plain or heavily marked with darker green, brown or brownish-maroon spots, lower surface plain, suffused with brownish-maroon, or brown-spotted, leathery, flat to slightly canaliculate; leaf margins coriaceous, flat to weakly undulate; leaf bases clasping, 20–40 mm long, lower portion white, upper portion green or suffused with brownish-maroon; primary seedling leaf flat, prostrate. *Inflorescence* spicate, few- to many-flowered, dense, sterile apex short to long; scape erect to suberect, 35–150 mm long, light green, plain or lightly to heavily marked with darker green or brownish-maroon speckles or blotches; lower bracts ovate, becoming lanceolate above, 2–9 × 1–4 mm, white to light green. *Perianth* zygomorphic, tubular, suberect, heavily sweet-scented, maturing to dull reddish-brown; tube cylindrical, 6–8 mm long; cream or greenish-blue to light greenish-yellow with minute blue speckles; outer tepals oblong, 5–15 × 4–5 mm, cream, light blue, greenish-blue, turquoise, violet, purplish-blue or light to dark greenish-yellow, apical gibbosities large, dark green, brown or blue; inner tepals oblong-obovate, 6–22 × 5–7 mm, apices flared, slightly to moderately recurved, translucent white, cream, greenish-yellow, blue or violet, median keels green, dark yellow, blue or bluish-purple, upper two tepals overlapping, lower tepal slightly to distinctly longer, canaliculate. *Stamens* included, rarely shortly exserted, more or less straight; filaments 8–17 mm long, white. *Ovary* ellipsoid, 3–5 × 2–3 mm, light green; style included to shortly exserted, often becoming well exserted as ovary matures, more or less straight, 5–17 mm long, white. *Capsule* ellipsoid, 7–12 × 3–5 mm. *Seed* globose, 0.9–1.1 × 0.8–1.1 mm, glossy, black; strophiole inflated, 0.7–0.8 mm long, smooth. *Chromosome number* variable, see under subspecies below. Plate 2, Figures 82–87.

FLOWERING PERIOD. Late June to late October, with a peak in mid-September.

HISTORY. The plants now known as *L. orchioides* subsp. *orchioides* and *L. orchioides* subsp. *glaucina* were illustrated in watercolour in about 1700 in Jakob and Johann Philipp Breyne's Florilegium *The Flora Capensis of Jakob and Johann Philipp Breyne* (Figure 2), which was published for the first time close to three centuries later by The Brenthurst Press in Johannesburg (Gunn & du Plessis, 1978). Copies of the same two figures and another of *L. contaminata* were used on an engraved plate in a later work of Jakob and Johann Philipp Breyne (1739), their *Prodromi Fasciculi Rariorum Plantarum* (Figure 3), in which figure 2 was described by the phrase name *Hyacinthus orchioides, Africanus, maior, bifolius maculatus; flore sulphureo, obsoleto, majore*. It was cited by Linnaeus as the iconotype of *Hyacinthus orchioides* L. (now *Lachenalia orchioides* subsp. *orchioides*) in his description of the first species included in the genus *Lachenalia* when he introduced his binomial system of classification in his *Species Plantarum* (Linnaeus, 1753). The species was transferred to *Muscari* by Philip Miller (1768), and in 1784, Thunberg described *Phormium orchioides*, citing *Hyacinthus orchioides* L. from the second edition of Linnaeus's *Species Plantarum* (1762), but the five 'varieties' he described under it were all different *Lachenalia* species. The plant was finally transferred to *Lachenalia* in the first volume of *Hortus Kewensis* (Aiton, 1789). The figure in Breyne (1739) was formally lectotypified by William Stearn relatively recently in a paper on the Linnean species of *Hyacinthus* in *Annales Musei Goulandris* (Stearn, 1990). Jakob Breyne (1689) recorded the plant in cultivation in the *Horto Dominae de Flines* in Holland, so it was probably grown in that country before this date. Philip Miller was growing it at the Chelsea Physic Garden in England by 1752 and it was introduced to Kew from seeds sent from the Cape by Francis Masson in 1774 (Gunn & du Plessis, 1978).

The first colour illustration published of the plant now known as *L. orchioides* subsp. *glaucina* appeared in volume two of N. J. Jacquin's *Icones Plantarum Rariorum* in 1793 or 1794, and his description of it followed several years later in volume five of *Collectanea* (Jacquin, 1797a). Originally described by the phrase name *Hyacinthus orchioides, Africanus maior bifolius, flore coeruleo maiore* accompanying

Figure 82 (left). *Lachenalia orchioides* subsp. *orchioides* on a granite outcrop at Brackenfell. Image: Graham Duncan.

Figure 83 (right). *Lachenalia orchioides* subsp. *orchioides* on clay flats, Tokai Forest. Image: Graham Duncan.

figure one in the Breyne's *Prodromi* of 1700, details of its original collector, habitat location and by whom it came to be exported to Europe have been lost; but like *L. orchioides* subsp. *orchioides*, subsp. *glaucina* is known to have been in cultivation in the *Horto Dominae de Flines* in the late 1680s (Breyne, 1689). Two colour forms of the plant were illustrated in *Edwards's Botanical Register* as *L. pallida* Lindl. and *L. pallida* Lindl. var. *coerulescens* Lindl. (Lindley 1830, 1837). Nonetheless, Jacquin's *L. glaucina* was upheld by Baker (1871, 1897a) until the new combination *L. orchioides* (L.) Ait. var. *glaucina* (Jacq.) W. F. Barker was made in the *South African Journal of Botany* (Barker, 1989).

Lachenalia glaucina Jacq. var. *parviflora* W. F. Barker was described in the *Journal of South African Botany* (Barker, 1949). The type material, collected by the Englishman Anthony Wolley-Dod (1861–1948) near Mostert's Ravine on the lower north-east slopes of Devil's Peak in September 1895, is housed in the Bolus Herbarium at the University of Cape Town (holotype) and at the Kew Herbarium (isotype). When Barker (1989) re-assessed *L. glaucina* Jacq. as a variety of *L. orchioides* (L.) Ait., no mention was made of the status of *L. glaucina* var. *parviflora* and the taxon is here formally combined under *L. orchioides*.

DISTINGUISHING CHARACTERS AND AFFINITIES. The three *L. orchioides* taxa always occur and breed true in allopatric populations. Thus, the genes that are specifically associated with each taxon have evolved independently and a step towards speciation is indicated as gene fixation has taken place; subspecific status is therefore attributed to these taxa.

3. LACHENALIA ORCHIOIDES

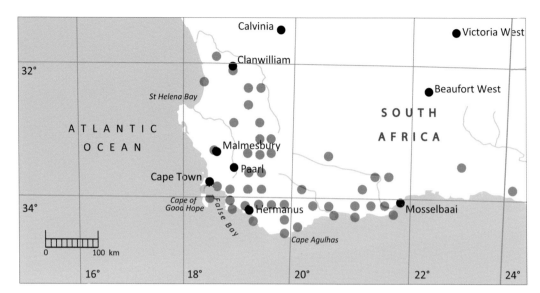

Map 4. Known distribution of *Lachenalia orchioides*: ● = subsp. *orchioides*; ● = subsp. *glaucina*; ● = subsp. *parviflora*.

Lachenalia orchioides resolved adjacent to *L. algoensis* in a phylogenetic analysis of morphological data (Duncan *et al.*, 2005a). Creamy-yellow forms of *L. orchioides* subsp. *orchioides* are frequently confused with *L. algoensis* but the latter differs in its longer perianth tube (9–12 mm), in its longer outer and inner tepals (15–16 mm and 24–30 mm, respectively), and in having the perianth distinctly inflated at the base and becoming narrower towards the mouth, with strongly recurved inner tepal apices and the lower inner tepal distinctly shorter than the upper inner tepals.

DISTRIBUTION AND HABITAT. The species extends from the Nardousberg Mountains near Klawer in the Western Cape to the foothills of the Kouga Mountains north of Haarlem in the Eastern Cape, and inland to Ceres, Montagu and Vanwyksdorp in the Western Cape interior (Map 4). It is usually associated with acid, stony clay substrates in renosterveld but also occurs in moist to arid fynbos in sandy, stony soils, and occasionally on limestone. (See under subspecies for further details.)

NOTES. The heavily sweet-scented flowers are pollinated mainly by honey bees (*Apis mellifera*), and the solitary bee *Anthophora diversipes* has been recorded visiting subsp. *orchioides*. Flowering is profuse in the next spring season immediately after summer bush fires and after clearing of thick undergrowth, but flowering is not dependent on these events. The scapes remain attached to the bulbs following capsule dehiscence and local seed dispersal is effected by the shaking action of wind.

Key to the subspecies

1a Outer tepals 5–6 mm long, violet, Woodstock to Devil's Peak . . . .**c. subsp. parviflora**

1b Outer tepals 8 mm long or more, cream, greenish-blue, greenish-yellow, turquoise, light blue to purplish-blue. 2

2a Inner tepals cream to greenish-yellow, 10–22 mm long, Nardousberg to Kouga Mountains . . . . **a. subsp. orchioides**

2b Inner tepals translucent white to light blue or purplish-blue, 13–15 mm long, Kirstenbosch to Wynberg . . . . **b. subsp. glaucina**

a. subsp. **orchioides**

SYNONYMY: *Orchiastrum aitonii* Lem., *L'Illustration Horticole* 2: 100 (1855). Type: South Africa, Cape, collector and precise locality unknown, figure in *Curtis's Botanical Magazine* 21: t. 854 (1805) (lectotype, designated here).

Orchiastrum virenti-flavum Lem., *L'Illustration Horticole* 2: 100 (1855). Type: South Africa, Cape, collector and precise locality unknown, figure in *Curtis's Botanical Magazine* 31: t. 1269 (1810) (lectotype, designated here).

DESCRIPTION. *Plant* 55–350 mm high. *Bulb* 15–20 mm in diameter *Leaves* 75–120 × 15–35 mm. *Scape* 70–150 mm long; bracts 2–9 × 1–4 mm. *Perianth* suberect; tube 6–8 mm long, cream, greenish-blue or light greenish-yellow with minute blue speckles; outer tepals 8–15 × 4–5 mm, cream, greenish-blue, turquoise or light to dark greenish-yellow; apical gibbosities large, dark green or brown; median keels blue or dark yellow; inner tepals 10–22 × 5–7 mm, cream or greenish-yellow; median keels green. *Filaments* 8–17 mm long. *Ovary* 4–5 × 2–3 mm; style 10–17 mm long. *Capsule* 9–12 × 3–5 mm. *Seed* 0.9–1.1 × 0.9–1.1 mm. *Chromosome number* 2n = 14 (Spies *et al.*, 2008); 2n = 16 (de Wet, 1957); 2n = 16, 17 (Moffett, 1936); 2n = 28 (Hamatani *et al.*, 2007); 2n = 28, 29 (Johnson & Brandham, 1997). Figures 82–85.

FLOWERING PERIOD. Late June to late October, with a peak in mid-September.

Figure 84 (left). *Lachenalia orchioides* subsp. *orchioides* in acid sand near Caledon. Image: Cameron McMaster.

Figure 85 (right). Cultivated specimen of *Lachenalia orchioides* subsp. *orchioides* from Rondebosch. Image: Graham Duncan.

DISTRIBUTION, HABITAT AND CONSERVATION STATUS. The subsp. *orchioides* has the widest distribution of the three taxa, extending from the lower southern slopes of the Nardousberg Mountains near Klawer in the north-western Western Cape to the foothills of the Kouga Mountains north of Haarlem in the south-western Eastern Cape, and inland to Ceres, Montagu and Vanwyksdorp in the Western Cape interior (Map 4). It occurs in 15 vegetation types, Cederberg- and Olifants- Sandstone Fynbos, Kouga Grassy Sandstone Fynbos, Swartland-, Breede-, Ceres-, Montagu-, Western Rûens-, Central Rûens- and Mossel Bay Shale Renosterveld, Boland- and Swartland Granite Renosterveld, Peninsula Granite Fynbos, Elgin Shale Fynbos and Western Gwarrieveld (Mucina & Rutherford, 2006). Its habitat varies from rocky hill and mountain slopes in stony clay (frequently on granite outcrops such as at Perdekop, Bracken Nature Reserve on the Cape Flats) to stony, sandy flats, and it is found in full sun or light to deep shade. It is often encountered in the shade of exotic *Pinus pinea* trees in the southern Cape Peninsula. There, light cream forms predominate in acid, stony soil at Tokai Forest amongst occasional yellow forms, and on the north side of Ou Kaapse Weg. At Caledon, subsp. *orchioides* sometimes grows in association with *L. lutea*; in other renosterveld habitats, it is often seen with *Pterygodium catholicum* (Orchidaceae).

The Nardousberg form (*Duncan 207*, in NBG) is one of the earliest to flower (late June to early July) and has somewhat narrower perianths with turquoise outer tepals and a solitary, broadly lanceolate leaf that is heavily marked with brown spots on the upper surface and with bright magenta bands on the clasping leaf base. This form also occurs on the Pakhuis Pass north-east of Clanwilliam. A similarly early-flowering, dwarf form with two narrowly lanceolate, leathery leaves occurs at Montagu in the south-western part of the Western Cape (*Harrower 3862*, in NBG). The subsp. *orchioides* is recorded from Riviersonderend in the southern Cape, but a collection of *L. mutabilis* (*Zeyher 4289* (K)) from 'mountain ridges by the lower part of Zonder Einde River' (Riviersonderend) was mistakenly cited under *L. orchioides* by Baker in both his monographs (Baker, 1871, 1897a). The taxon is not threatened.

b. subsp. **glaucina** (Jacq.) G. D. Duncan, stat. nov. *Lachenalia glaucina* Jacq., *Collectanea* 5: 59–60 (1797). *Orchiastrum glaucinum* (Jacq.) Lem., *L'Illustration Horticole* 2: 100 (1855). *Lachenalia orchioides* (L.) Aiton var. *glaucina* (Jacq.) W. F. Barker, *South African Journal of Botany* 55 (6): 640–642 (1989).
TYPE: South Africa, Cape, collector and precise locality unknown, figure in N. J. Jacquin, *Icones Plantarum Rariorum* 2(15): t. 391 (1793 or 1794) (neotype, designated here).
SYNONYMY: *Lachenalia pallida* Lindl., *Edwards's Botanical Register* 16: t. 1350 (1830), illegitimate homonym. Type: South Africa, Cape, collector and precise locality unknown, figure in *Edwards's Botanical Register* 16: t. 1350 (1830) (lectotype, designated here).
Lachenalia pallida Lindl. var. *coerulescens* Lindl., *Edwards's Botanical Register* 23: t. 1945 (1837). Type: South Africa, Cape, collector and precise locality unknown, figure in *Edwards's Botanical Register* 23: t. 1945 (1837) (lectotype, designated here).
Lachenalia lilacina Baker, *The Gardeners' Chronicle* 21 (series 2): 668 (1884). Type: South Africa, Cape, collector and precise locality unknown, *ex hort T.S. Ware s.n.*, Tottenham (K!, holo.).

ETYMOLOGY. *glaucina*: blue-grey flowers.
DESCRIPTION. *Plant* 150–400 mm high. *Bulb* 10–20 mm in diameter. *Leaves* 85–150 × 15–30 mm. *Scape* 100–150 mm long; rachis light green at base, mottled with brownish-purple speckles, sterile apex bright blue; bracts 2–8 × 1–4 mm. *Perianth* suberect; tube 6 mm long, turquoise; outer tepals 8–9 × 4–5 mm, light blue to purplish-blue, apical gibbosities dark blue to purplish-blue; inner tepals 13–15 × 4–5 mm, translucent white to light blue or purplish-blue. *Filaments* 10–11 mm long. *Ovary* 4–5 × 3–4

Plate 2. Watercolour painting of *Lachenalia orchioides* subsp. *glaucina* from the Kirstenbosch Estate, Newlands (*Duncan* 131, in NBG), courtesy of Elbe Joubert. Artist: Elbe Joubert.

mm; style 11–12 mm long. *Capsule* 9–12 × 4–5 mm. *Seed* 0.9–1.0 × 0.8–0.9 mm. *Chromosome number* 2n = 28 (Johnson & Brandham, 1997; Hamatani *et al.*, 2007; Spies *et al.*, 2008). Plate 2, Figure 86.

FLOWERING PERIOD. September to October, with a peak in early October.

DISTRIBUTION, HABITAT AND CONSERVATION STATUS. The subsp. *glaucina* has a highly restricted range on the eastern slopes of Table Mountain, comprising three small, isolated populations on the Kirstenbosch Estate, the adjacent Cecilia Forest, and on Zonnestraal Farm on nearby Wynberg Hill (Map 4). It grows in Peninsula Granite Fynbos vegetation (Mucina & Rutherford, 2006), in colonies on lower mountain and hill slopes in stony loam within the semi-shade of surrounding bushy cover or in the deep shade of exotic *Pinus pinea* trees. The plants occur within close proximity of subsp. *orchioides* and subsp. *parviflora* on the Cape Peninsula, subsp. *orchioides* occurring in the southern suburbs of Tokai and Newlands and within the grounds of the University of Cape Town, and subsp. *parviflora* occurring slightly further north towards the city centre. The subsp. *glaucina* is threatened by potential road works and excessive shade caused by *Pinus* plantations, and a conservation status of Endangered is recommended.

Figure 86 (left). *Lachenalia orchioides* subsp. *glaucina* in habitat, Cecilia Forest, southern Cape Peninsula. Image: Graham Duncan.

Figure 87 (right). Cultivated specimen of *Lachenalia orchioides* subsp. *parviflora* from Woodstock, northern Cape Peninsula. Image: Graham Duncan.

c. subsp. parviflora (W. F. Barker) G. D. Duncan, stat. et comb. nov. *Lachenalia glaucina* Jacq. var. *parviflora* W. F. Barker, *Journal of South African Botany* 15 (2): 39 (1949).
TYPE: South Africa, Western Cape, fire guard near Mostert's Ravine, *A.H. Wolley Dod* 380 (BOL!, holo.; K!, iso.).

ETYMOLOGY. *parviflora*: small flowers.
DESCRIPTION. *Plant* 130–200 mm high. *Bulb* 10–15 mm in diameter. *Leaves* 60–100 × 10–15 mm, upper and lower surfaces plain or with brownish-maroon spots. *Scape* 80–100 mm long. *Perianth* spreading to weakly suberect; tube 6 mm long, blue with darker blue speckles; outer tepals 5–6 × 4–5 mm, violet, apical gibbosities large, deep bluish-purple; keels deep blue; inner tepals obovate, 6–7 × 5–6 mm, violet, median keels deep bluish-purple, apices recurved. *Filaments* white, 8–9 mm long. *Ovary* 3 × 2 mm; style 5–6 mm long. *Capsule* 7–8 × 4–5 mm. *Seed* 0.9 × 0.8 mm. *Chromosome number* unknown. Figure 87.
FLOWERING PERIOD. August to October, with a peak in mid-September.
DISTRIBUTION, HABITAT AND CONSERVATION STATUS. The subsp. *parviflora* is restricted to the lower north- and north-east facing slopes of Table Mountain, extending from Woodstock to Devil's Peak (Map 4). It is usually a dwarf plant encountered in full sun, but taller specimens are recorded growing in the shade of pine trees on flatter terrain in the nearby suburb of Observatory, and might still survive in the grounds of the South African Astronomical Observatory. The plants occur in Cape Winelands Shale Fynbos vegetation (Mucina & Rutherford, 2006), flowering in profusion following summer bush fires and growing in large colonies in acidic clay. The subspecies is threatened by housing development and potential road works, and a conservation status of Vulnerable is recommended.

4. LACHENALIA TRICHOPHYLLA

Lachenalia trichophylla Baker, *Journal of Botany (London)* 12: 368 (1874).
TYPE: South Africa, Western Cape, Clanwilliam, *P. A. Mader* in *Herb. MacOwan* 2167 (K!, holo.; SAM!, iso.).
SYNONOMY: *Lachenalia massonii* Baker, *Journal of Botany (London)* 24: 336 (1886). Type: South Africa, Western Cape, Meerhof's Kasteel, figure by Francis Masson (BM!, holo.).

ETYMOLOGY. *trichophylla*: leaf covered with trichomes.
DESCRIPTION. *Geophyte*, 100–200 mm high. *Bulb* subglobose, 15–20 mm in diameter, solitary; tunic multilayered, outer layers dark brown, spongy; inner cataphyll translucent white, adhering to leaf base, apex obtuse. *Leaf* solitary, ovate, 25–60 × 25–50 mm, prostrate, upper surface light to dark green or glaucous, maturing to dull purplish-magenta, trichomes stellate, 0.3–15.0 mm long, dense; lower surface dark green flushed with light to dark maroon; margins entire, trichomes stellate, short to 7 mm long; leaf base clasping, 5–15 mm long, subterranean portion white, aerial portion deep purplish-magenta; primary seedling leaf flat, prostrate. *Inflorescence* spicate or racemose, flowers spirally arranged or in 3-flowered whorls, flaccid to rigid, sterile apex long; scape erect to suberect, 40–120 mm long, rigid, wiry, light green with minute purplish-magenta speckles or light to dark brownish-magenta; rachis green to brownish-magenta, apex light blue to mauve; pedicels absent or suberect, 1–6 mm long, white to magenta; bracts ovate, 4–5 × 1–3 mm, brownish-magenta to white with purplish-magenta speckles, margins translucent white. *Perianth* zygomorphic, tubular, suberect, maturing to dull brownish-pink; tube cylindrical, 6–10 mm long, light creamy-yellow

THE GENUS LACHENALIA
4. LACHENALIA TRICHOPHYLLA

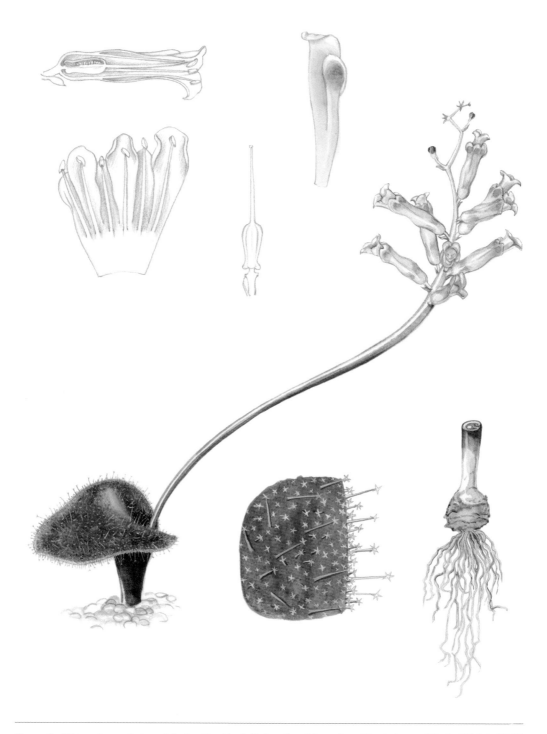

Plate 3. Watercolour painting of *Lachenalia trichophylla* 'northern' form from Kamieskroon (*Hardy 2522*, in PRE) courtesy of The Editor, *The Flowering Plants of Africa*, vol. 46, t. 1808 (1981). Artist: Rhona Collett.

to pinkish-magenta; outer tepals oblong, 6–7 × 3–4 mm, light creamy-yellow, apical gibbosities dark green to brown; inner tepals oblong-obovate, 9–12 × 4–6 mm, creamy-yellow, median keels light green, apices straight to slightly recurved, upper tepals overlapping, lower tepal canaliculate, distinctly longer. *Stamens* included, straight; filaments white, 10–13 mm long. *Ovary* ellipsoid, 5–7 × 2–3 mm, light green; style included, straight, white, 9–12 mm long, included, finally exserted as ovary matures. *Capsule* ellipsoid, 8–13 × 3–5 mm. *Seed* globose, 0.5 × 0.5 mm, matte, black, secondary sculpturing reticulate; strophiole rudimentary, 0.1 mm long, ridged. *Chromosome number* 2n = 14 (Ornduff & Watters, 1978; Johnson & Brandham, 1997). Plate 3, Figures 50, 88–90.

FLOWERING PERIOD. July to September, with a peak in mid-August.

HISTORY. *L. trichophylla* was described by Baker (1874b) in the *Journal of Botany* (London), but the type locality was mistakenly cited as *Caput Bonae Spei in ditione Somerset East, MacOwan* 2197!. The correct locality is Clanwilliam, where the type was collected by P. A. Mader for the MacOwan Herbarium (number 2167). Specimens of this collection have a solitary ovate leaf that is covered with short stellate trichomes on the upper surface and margins, and a spike of tubular, suberect flowers arranged in 3-flowered whorls. Also in the *Journal of Botany* (London), Baker described *L. massonii* from a drawing by Francis Masson that is housed in the British Museum (Natural History), a bulb of which had been collected at Meerhof's Kasteel near Nuwerus on the Knersvlakte in August 1793 (Baker,

Figure 88 (left). *Lachenalia trichophylla* 'northern' form with pedicellate, spirally arranged perianths near Loeriesfontein. Image: Graham Duncan.

Figure 89 (right). *Lachenalia trichophylla* 'southern' form with sessile, whorled perianths near Clanwilliam. Image: Graham Duncan/SANBI.

1886); it had a raceme of shorter tubular flowers and long stellate trichomes on the upper leaf surface. A comparison of herbarium collections of *L. trichophylla* and *L. massonii* resulted in the latter species being placed in synonymy under *L. trichophylla* (Barker, 1980a), mainly on account of intermediate perianth and trichome lengths that exist between the two taxa. The accompanying painting by Rhona Collett illustrates a cultivated specimen from near Kamieskroon in central Namaqualand with pedicellate perianths and both long and short stellate trichomes.

DISTINGUISHING CHARACTERS AND AFFINITIES. In a phylogenetic analysis (Duncan *et al.*, 2005a), *L. trichophylla* resolved adjacent to *L. polypodantha*, with which it shares stellate trichomes on the upper leaf surface and margins, and a small globose seed with reticulate secondary sculpturing and a rudimentary strophiole. *L. polypodantha* differs from *L. trichophylla* in being a dwarf species with widely campanulate, spreading flowers that are borne on long perpendicular pedicels, and in having well-exserted, declinate stamens and a lanceolate, spreading leaf.

DISTRIBUTION, HABITAT AND CONSERVATION STATUS. *L. trichophylla* extends from Garies in southern Namaqualand, south to Hermon in the south-western Cape (Map 5). It occurs in nine vegetation types including Namaqualand Klipkoppe Shrubland, Hantam- and Tanqua Karoo, Bokkeveld, Cederberg- and Olifants Sandstone Fynbos, Agter-Sederberg Shrubland, Atlantis Sand Fynbos and Breede Shale Renosterveld (Mucina & Rutherford, 2006).

Populations can be categorised into 'southern', 'northern' and 'intermediate' forms. Populations of 'southern' forms extend from Hermon north to the Gifberg just south of Vanrhynsdorp and inland to the Tanqua Karoo, usually occurring in fynbos on flats or mountain plateaux in deep sand or on north- or south-east-facing rocky sandstone mountain slopes. The populations found on sandstone slopes have spicate inflorescences with the flowers arranged in distinct verticils or in groups of three, the upper leaf surfaces and margins have very short stellate trichomes that are clearly visible only under magnification.

Populations of 'northern' forms (previously *L. massonii*) occur from Vanrhynsdorp north to Garies, and are usually found on rocky south- and south-west-facing stony shale slopes and granite ridges. The populations found on granite ridges have racemose inflorescences with the flowers spirally arranged, the upper leaf surfaces and margins with both long and short stellate trichomes. Intermediate forms are known from Gifberg, where a population with spicate inflorescences and medium-length trichomes occurs, and also near Tulbagh and at the northern end of the Pakhuis Pass, where populations have inflorescences with three-flowered whorls, very short pedicels and short stellate trichomes. At the northern end of the Pakhuis Pass, an intermediate form that has shortly pedicellate flowers and intermediate-length trichomes occurs. Near Loeriesfontein, both spicate and racemose inflorescences occur within the same population, both forms having long, stellate hairs. The species

Figure 90. Cultivated specimens of a 'southern' form of *Lachenalia trichophylla* from Clanwilliam with very short stellate trichomes. Image: Graham Duncan.

occurs as solitary individuals or in small groups, exposed or within the protection of low scrubby growth. This demography is consistent with wind dispersal of the infructescence. The species is not threatened.
NOTES. Pollination agents are not recorded for *L. trichophylla* in the wild, but Southern Double-collared Sunbirds (*Cinnyris chalybeus*) visit the flowers of potted specimens when displayed in the bulb house within the Conservatory at Kirstenbosch. The flowers of certain forms are partially self-fertile, regularly setting seed under enclosed conditions in the Kirstenbosch Bulb Nursery. The base of the scape detaches shortly after capsule dehiscence and the seeds are dispersed over a wide area as the infructescence is blown about.

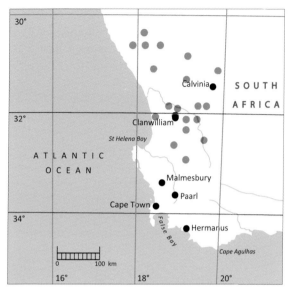

Map 5. Known distribution of *Lachenalia trichophylla*.

5. LACHENALIA MARGINATA

Lachenalia marginata W. F. Barker, *Journal of South African Botany* 45 (2): 204–207 (1979).
TYPE: South Africa, Northern Cape, 4 miles [6.4 km] west of Nieuwoudtville, *W. F. Barker* 6463 (NBG!, holo.).

ETYMOLOGY. *marginata*: thickened leaf margins.
DESCRIPTION. *Geophyte*, 110–300 mm high. *Bulb* subglobose, 10–25 mm in diameter, offset-forming; tunic multilayered, outer layers dark brown, spongy; inner cataphyll translucent white, adhering to leaf bases, apex obtuse. *Leaf* 1(–2), ovate or lanceolate, 45–200 × 18–40 mm, erect to suberect or spreading, dark green to glaucous, upper surface with irregularly scattered, darker green blotches, lamina flat to deeply canaliculate or conduplicate; leaf margin thickened and coriaceous, entire, undulate or crisped, maroon; leaf base clasping, 30–90 mm long, light green with magenta bands and blotches; primary seedling leaf flat, prostrate. *Inflorescence* spicate, few- to many-flowered, dense, sterile apex short to long; scape erect to suberect, 55–100 mm long, light green with dull magenta blotches, upper portion straight to distinctly swollen; rachis light green with dull maroon blotches; lower bracts ovate, becoming lanceolate above, 3–10 × 3–5 mm, light greenish-white, plain or with a brownish-magenta median stripe. *Perianth* zygomorphic, tubular, strongly sweet-scented, ageing to dull red; tube cup-shaped or cylindrical, 3 or 6 mm long, light greenish-yellow with minute light blue speckles; outer tepals oblong, 6–10 × 5–6 mm, apical gibbosities large, dark brown; inner tepals obovate to narrowly obovate, 7–15 × 4–7 mm, median zone green, apical zone brown, upper two tepals overlapping, lower tepal narrower, canaliculate, slightly to distinctly longer, apex straight or recurved. *Stamens* included, straight; filaments white, 6–15 mm long. *Ovary* ellipsoid, 3–6 × 2–4 mm, bright green; style included, 7–15 mm long, white, finally exserted as ovary matures. *Capsule* ellipsoid, 6–8 × 4–5 mm. *Seed* globose, 0.9–1.0 × 0.8–0.9 mm, glossy, black; strophiole inflated, 0.6–0.7 mm long, smooth. *Chromosome number* see under subspecies below. Figures 91, 92.

FLOWERING PERIOD. June to September.

HISTORY. The first record of *L. marginata* is a specimen of unknown provenance that flowered in June 1934 at the Bolus Herbarium and was painted by W. F. Barker. The plant was next collected in July 1940 by Bina Martin in the Matzikamma Mountains south-east of Vanrhynsdorp, and the species has since been recorded mainly around Vanrhynsdorp, to the north-east at Nieuwoudtville, and to the south on the Pakhuis Pass. The type material was collected by W. F. Barker a short distance west of Nieuwoudtville in August 1950, but the species was only formally published 29 years later in the *Journal of South African Botany* (Barker, 1979a). On 12th August 1896, Rudolf Schlechter collected a *Lachenalia* on the farm Zeekoe Vlei west of Clanwilliam. He appended the manuscript name *L. neglecta* Schltr. to this material (*Schlechter* 8490) and distributed it to numerous herbaria in South Africa and abroad. The plant was next collected by W. F. Barker near Clanwilliam in August 1966, but was not recorded again until July 1993 when I came across it between Citrusdal and Clanwilliam, and it was subsequently described as *L. marginata* subsp. *neglecta* in *Bothalia* (Duncan, 1996).

DISTINGUISHING CHARACTERS AND AFFINITIES. *L. marginata* resolved adjacent to *L. marlothii* in a phylogenetic study (Duncan *et al.*, 2005a). *L. marlothii* differs in its oblong-campanulate perianth, longer, broader outer tepals (8–9 × 5–6 mm) and longer filaments (8–10 mm long). It also differs in its globose bulb, the leaf lacks the thickened coriaceous margin and it has a much larger, ovoid seed (1.3–1.4 × 1.8 mm) with a much longer, ridged strophiole. The *L. marginata* taxa breed true in allopatric populations, indicating gene fixation and a step towards speciation.

Figure 91 (left). *Lachenalia marginata* subsp. *marginata* in sandy gravel near Vanrhynsdorp. Image: Graham Duncan.
Figure 92 (right). *Lachenalia marginata* subsp. *neglecta* in fynbos near Clanwilliam. Image: Graham Duncan.

5. LACHENALIA MARGINATA

DISTRIBUTION AND HABITAT. *L. marginata* is confined to the flats and lower to medium slopes of the Bokkeveld Plateau around Nieuwoudtville and the Gifberg Massif, the Cederberg Mountains, and low-lying areas around Clanwilliam (Map 6). It occurs in fynbos in acid, yellowish-brown or red sand or sandy gravel, and the plants occur as scattered individuals or in small groups, often adjacent to or within the protection of restio clumps. The two subspecies are geographically isolated, subsp. *marginata* occurring at relatively high altitude and subsp. *neglecta* being restricted to low-lying areas to the west and south-west of Clanwilliam. For further details, see under subspecies.

NOTES. No known pollinators have been recorded for *L. marginata*, but the species' sweet-scented flowers with extended lower inner tepals indicate that its pollinators are likely to include honey bees (*Apis mellifera*). The subsp. *neglecta* usually commences flowering a few weeks earlier than subsp. *marginata*. The leaves are highly susceptible to the fungus *Uromyces lachenaliae*, both in the wild and in cultivation. The scape detaches once capsule dehiscence has occurred and the seeds are dispersed over a wide area as the infructescence is blown about.

Map 6. Known distribution of *Lachenalia marginata*: ● = subsp. *marginata*; ● = subsp. *neglecta*.

Key to the subspecies

1a Perianth tube 6 mm long; filaments 13–15 mm long; leaf ovate, spreading to suberect, lamina more or less flat, Bokkeveld Plateau to Gifberg and Cederberg. . . . . **a.** subsp. **marginata**

1b Perianth tube 3 mm long; filaments 6–7 mm long; leaf lanceolate, erect to suberect, lamina deeply canaliculate to conduplicate, Clanwilliam . . . . **b.** subsp. **neglecta**

a. subsp. marginata

DESCRIPTION. *Plant* 110–300 mm high. *Bulb* 10–20 mm in diameter. *Leaf* ovate, 85–170 × 30–40 mm, spreading to suberect, lamina more or less flat. *Bracts* 3–10 × 2–4 mm. *Perianth tube* cylindrical, 6 mm long; outer tepals 7–10 × 5–6 mm; inner tepals narrowly obovate, 12–15 × 5–7 mm, lower inner tepal apex recurved. *Filaments* 13–15 mm long. *Ovary* 3–6 × 4 mm; style 11–15 mm long. *Capsule* 7 × 5 mm. Chromosome number 2n = 28, 29 (Johnson & Brandham, 1997). Figure 91.

FLOWERING PERIOD. June to September, with a peak from late July to mid-August.

DISTRIBUTION, HABITAT AND CONSERVATION STATUS. The subsp. *marginata* is distributed from Nieuwoudtville on the Bokkeveld Plateau to the Gifberg south of Vanrhynsdorp and the Cederberg Mountains (Map 6). It occurs in Bokkeveld- and Cederberg Sandstone Fynbos vegetation (Mucina & Rutherford, 2006) at relatively high altitude. At a locality near the Gifberg, it grows in association with *L. mutabilis* and *L. unifolia*. The subsp. *marginata* is not threatened.

b. subsp. **neglecta** Schltr. ex G. D. Duncan, *Bothalia* 26 (1): 7–9 (1996).
TYPE: South Africa, Western Cape, 'Die Berg' Farm, south-west of Clanwilliam, *W. F. Barker* 10428 (NBG!, holo.).

ETYMOLOGY. *neglecta*: neglected; reasons for Schlechter's choice of name are unclear, but possibly point to the taxon having been rarely collected.

DESCRIPTION. *Plant* 110–285 mm high. *Bulb* 11–25 mm in diameter. *Leaf* lanceolate, 45–200 × 18–30 mm, erect to suberect, lamina deeply canaliculate to conduplicate. *Bracts* 3–5 × 3–5 mm. *Perianth tube* cup-shaped, 3 mm long; outer tepals 6–8 × 3–4 mm, minutely blue-speckled; inner tepals obovate, 7–11 × 4–5 mm, lower inner tepal apex straight. *Filaments* 6–7 mm long. *Ovary* 4–5 × 2–3 mm; style 7–8 mm long. *Capsule* 6–8 × 4–5 mm. *Chromosome number* 2n = 14 (Spies *et al.*, 2008). Figure 92.

FLOWERING PERIOD. July to August, with a peak from late July to early August.

DISTRIBUTION, HABITAT AND CONSERVATION STATUS. The subsp. *neglecta* is restricted to the Clanwilliam district (Map 6). It grows in Graafwater Sandstone Fynbos vegetation (Mucina & Rutherford, 2006) in deep, light yellowish-brown acidic sand. The plants occur amongst restios and low shrubs, often in association with *L. mutabilis*, a common species in this area that commences flowering a few weeks later. The subsp. *neglecta* is threatened by extensive clearing of virgin land around Clanwilliam for the rooibos tea industry, and a conservation status of Vulnerable is recommended.

6. LACHENALIA ALGOENSIS

Lachenalia algoensis Schönland, *Transactions of the Royal Society of South Africa* 1: 443 (1910).
TYPE: South Africa, Eastern Cape, Port Elizabeth, *I.L. Drège* 64 (GRA!, lectotype, designated here).

ETYMOLOGY. *algoensis*: after Algoa Bay, Port Elizabeth.

DESCRIPTION. *Geophyte*, 70–300 mm high. *Bulb* subglobose, 10–25 mm in diameter, offset-forming; tunic multilayered, outer layers light to dark brown, spongy; inner cataphyll translucent white, loosely surrounding leaf bases, apex obtuse. *Leaves* (1–)2, lanceolate, 140–200 × 2–20 mm, spreading to suberect, canaliculate, upper surface light green, plain or sporadically marked with dark green blotches, depressed longitudinal grooves prominent; leaf bases clasping, 20–35 mm long, subterranean portion white, aerial portion light green, plain or marked with brownish-magenta blotches; primary seedling leaf flat, prostrate. *Inflorescence* racemose, few- to many-flowered, sterile apex short; scape erect to suberect, 50–85 mm long, light green, plain or with sporadic small to large, dull purple or reddish-brown blotches; rachis light green, plain or with minute purple spots; pedicels suberect, 1–2 mm long, light green; lower bracts ovate, becoming lanceolate above, 2–6 × 1–6 mm, green or translucent white. *Perianth* zygomorphic, tubular, suberect, light yellow to greenish-yellow, narrowing markedly towards apices, sweet-scented; tube cylindrical, 9–12 mm long, base inflated, plain or with sporadic dull blue to purple blotches; outer tepals oblong, 15–16 × 5–7 mm, plain or with small to large dark blue to purple blotches; apical gibbosities light to dark green; inner tepals oblong-obovate, 24–30 × 6–7 mm, apices undulate, recurved, lower inner tepal up to 4 mm shorter than upper two tepals, maturing to dull reddish-pink, median keels green. *Stamens* included, straight; filaments white, 16–18 mm long. *Ovary* ellipsoid, 8–10 × 4–5 mm, bright green; style included, straight, 19–21 mm long, white. *Capsule* ellipsoid, 14–20 × 6–9 mm, suberect. *Seed* globose, 0.9–1.1 × 0.9–1.2 mm, glossy, black; strophiole inflated, 1.1–1.2 mm long, smooth. *Chromosome number* 2n = 14 (Crosby, 1986; Hamatani *et al.*, 2007; Spies *et al.*, 2008). Figures 93, 94.

FLOWERING PERIOD. August to September.

HISTORY. *L. algoensis* was collected for the first time in 1903 at Port Elizabeth by I. L. Drège, son of the German apothecary, naturalist and traveller, C. F. Drège. The plant was described in 1910 by another German, Selmar Schönland, who became Director of the Albany Museum at Grahamstown and first Curator of the Selmar Schönland Herbarium at Rhodes University. In his original description, Schönland cited two collections, one from Port Elizabeth (*I. L. Drège* 64 in GRA) collected on 13th August 1903, and another from nearby Redhouse (*F. Paterson* 92 in BOL), collected in 1916, but did not indicate which of these he considered to be the holotype. The Drège collection is designated here as the lectotype as it better represents the species, showing a whole plant in flower, with the bulb.

DISTINGUISHING CHARACTERS AND AFFINITIES. *L. algoensis* is allied to *L. orchioides* (Duncan *et al.*, 2005a) and is frequently confused with light yellow forms of *L. orchioides* subsp. *orchioides*. The latter taxon is readily set apart by its shorter perianth tube (6–8 mm long), shorter outer tepals (8–15 mm long), and much shorter, flared inner tepals (10–22 mm long) with flat margins, with the lower inner tepal equal to, or slightly longer than, the upper tepals. The flowers of *L. algoensis* are borne on very short, suberect pedicels, whereas those of *L. orchioides* are sessile. Both taxa have similar globose, glossy seeds with a smooth, inflated strophiole, but *L. orchioides* has a much shorter strophiole of 0.7–0.8 mm in length.

Figure 93. *Lachenalia algoensis* with heavily marked outer tepals on grassy dunes at St. Francis Bay. Image: Caryl Logie.

6. LACHENALIA ALGOENSIS

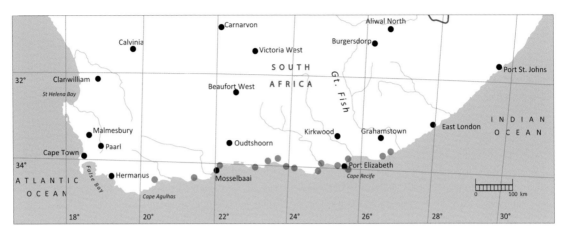

Map 7. Known distribution of *Lachenalia algoensis*.

DISTRIBUTION, HABITAT AND CONSERVATION STATUS. *L. algoensis* is predominantly a coastal species occurring from north-west of Port Beaufort in the southern Cape eastwards and slightly inland to Bathurst in the Eastern Cape (Map 7). The species traverses 10 vegetation types including Knysna Sand Fynbos, Garden Route Shale Fynbos, Groot Brak Dune Strandveld, Blombos Strandveld, Kouga Grassy Sandstone Fynbos, Eastern Coastal Shale Band Vegetation, Algoa Dune Strandveld, Sundays Thicket, Albany Dune Strandveld and Albany Coastal Belt (Mucina & Rutherford, 2006). In coastal areas, it grows in full sun and is most often seen in colonies on sandy flats among thick grass and in low dense groundcover on secondary dunes in openings of dune bush. Inland, it is usually found on dry hillsides and lower mountain slopes in stony ground, in partial shade of surrounding bush. *L. algoensis* has been reduced in number as a result of coastal housing development but it is not currently threatened because it has a relatively wide distribution.

NOTES. The very long perianth tube of *L. algoensis* (9–12 mm) suggests that it is specialised for pollination by sunbirds, but in contrast to most bird-pollinated species, its flowers are sweetly scented. The infructescence remains attached to the bulb for several weeks after capsule dehiscence and the seeds are liberated by the shaking action of wind.

Figure 94. Cultivated specimen of *Lachenalia algoensis* with greenish yellow tepals from St. Francis Bay. Image: Graham Duncan.

7. LACHENALIA VANZYLIAE

Lachenalia vanzyliae (W. F. Barker) G. D. Duncan & T. J. Edwards, stat. nov. *Lachenalia aloides* (L.f.) Engl. var. *vanzyliae* W. F. Barker, *Journal of South African Botany* 50 (4): 543–546 (1984).
TYPE: South Africa, Western Cape, Piketberg Mountain above town, *W. F. Barker* 10349 (NBG!, holo.).

ETYMOLOGY. *vanzyliae*: after Mrs A. van Zyl, a resident of Piketberg.

DESCRIPTION. *Geophyte*, 80–260 mm high. *Bulb* globose, 10–22 mm in diameter, offset-forming; tunic multilayered, outer layers dark brown, spongy; inner cataphyll translucent white, adhering to leaf bases, apex obtuse. *Leaves* (1–) 2, broadly lanceolate, 120–150 × 10–50 mm, spreading, upper surface light green, lightly to densely spotted with purplish-brown, depressed longitudinal grooves distinct, lower surface plain or with small brownish-red spots towards base; leaf margins brownish-red; leaf bases clasping, 20–40 mm long, subterranean, white; primary seedling leaf flat, prostrate. *Inflorescence* racemose, few- to many-flowered, sterile apex short to long; scape erect, 60–90 mm long; light green marked with dull blue or brownish-red blotches; rachis light green in lower half, upper portion turquoise; pedicels suberect, 6–11 mm long, white; lower bracts ovate, becoming lanceolate above, 5–10 × 2–4 mm, white. *Perianth* zygomorphic, tubular, pendulous; tube cylindrical, 6–7 mm long, bright turquoise; outer tepals oblong, 12–14 × 5–6 mm, translucent white, apical gibbosities large, bright green, median keels turquoise, inner tepals oblong-obovate, 17–20 × 7–9 mm, translucent white, median keels bright green, surrounded by broad, bright green zone, apices slightly spreading, white. *Stamens* included to shortly exserted, straight; filaments white, 16–19 mm long. *Ovary* ellipsoid, 4–5 × 3–4 mm, bright green; style included, straight, 17–21 mm long, white, finally protruding beyond perianth as ovary enlarges. *Capsule* ellipsoid, 10–12 × 6–7 mm. *Seed* globose, 1.3–1.4 × 1.2–1.3 mm, glossy, black; strophiole inflated, 1.4–1.5 mm long, smooth. *Chromosome number* 2n = 28 (Crosby, 1986; Hamatani *et al.*, 1998). Plate 4, Figure 95.

FLOWERING PERIOD. September to early November, with a peak from mid-September to mid-October.

HISTORY. Described by W. F. Barker in 1984, this taxon was evidently known to Linnaeus *filius* in the late 18th century, as on page 205 of his *Supplementum Plantarum et Specierum Plantarum* (1781), under the name *Phormium aloides* (later

Figure 95. Cultivated specimen of *Lachenalia vanzyliae* from Piketberg. Image: Graham Duncan.

Plate 4. Watercolour painting of *Lachenalia vanzyliae* from Piketberg [*Thomas s.n.* (697/84), in NBG] courtesy of The Editor, *Flowering Plants of Africa*, vol. 58, t. 2187 (2003). Artist: Fay Anderson.

to become *Lachenalia aloides*), he described the flowers as *Floribus nutantibus cylindrico-infundibuliformis. Variat corollis luteis, corollis croceis, corollis sanguineus apice purpureis, corollis albovirescentibus, et variat proportione inter petala exteriora et interiora. Species difficile determinatur.* A specimen of *L. vanzyliae* collected by Anders Sparrman at an inexact locality at the Cape, probably in 1776, is housed in the *Herbarium Castromii* at the Naturhistoriska Riksmuseet in Stockholm, and a later collection made by Rudolf Schlechter in September 1894 is in the Botanischer Museum der Universitat Zurich. The plant was collected and cultivated by Mrs A. van Zyl, a resident of Piketberg, who presented living plants to Kirstenbosch in 1927. It flowered there in 1929 and was subsequently painted by Barker and given its manuscript name *L. aloides* (L.f.) Engl. var. *vanzyliae* W. F. Barker. Barker later collected the type material on Piketberg Mountain in September 1966. She considered the plant to be sufficiently distinct to be given specific rank, but stated "as *L. aloides* has been found in so many colour varieties and the seed of this plant conforms to the species pattern, it seems best to accommodate it as a variety" (Barker, 1984). Barker's painting was illustrated in monochrome in her 1948 publication, and a watercolour by Fay Anderson was published in *Flowering Plants of Africa* (Duncan & Anderson, 2003).

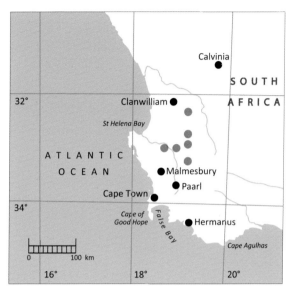

Map 8. Known distribution of *Lachenalia vanzyliae*.

DISTINGUISHING CHARACTERS AND AFFINITIES. *L. vanzyliae* resolved adjacent to *L. thunbergii* in a phylogenetic analysis (Duncan et al., 2005a). The latter differs in its greenish-yellow perianth with shorter outer tepals (10–11 mm long), longer inner tepals (23–24 mm), a bright reddish-orange rachis apex, and a smaller globose seed (1.1–1.2 × 1.2–1.3 mm) with a shorter strophiole (1.1–1.2 mm). The two species are geographically isolated, *L. thunbergii* is restricted to the summit ridge of Riebeek Kasteel Mountain, about 40 km south-west of the nearest *L. vanzyliae* locality in the Twenty Four Rivers mountain range south-east of Porterville. Raising *L. vanzyliae* to specific level is fully justified not only because of its distinct morphological differences but also as it breeds true from seed and rarely forms hybrids in the wild.

Superficially, the flowers of *L. vanzyliae* resemble those of the much earlier-flowering *L. viridiflora*, which differs mainly in its spreading to suberect perianth with much shorter outer (7–9 mm long) and inner tepals (10–13 mm long), with the outer tepals uniformly greenish-turquoise to bright turquoise.

DISTRIBUTION, HABITAT AND CONSERVATION STATUS. *L. vanzyliae* is restricted to the upper reaches of the Piketberg, Porterville, Elandskloof, Cederberg and Twenty Four Rivers mountain ranges of the Western Cape (Map 8). It occurs in four vegetation types, Cederberg-, Piketberg-, Winterhoek- and Hawequas Sandstone Fynbos (Mucina & Rutherford, 2006). The plants are usually encountered on south-facing aspects in light to deep shade, growing in acid humus of sandstone rock crevices or at the base of massive boulders, rarely in full sun. In sunny aspects, the plants are considerably smaller than those in shade, and they grow singly or in small groups. The species is not threatened.

NOTES. The long tubular flowers of *L. vanzyliae* are pollinated by birds, including Orange-breasted Sunbirds (*Anthobaphes violacea*). The scape remains attached well after the capsules have dehisced and the seeds drop to the ground by the shaking action of wind. The only known natural hybrid within the genus is that between *L. vanzyliae* and *L. callista*, recorded from Piketberg Mountain, a specimen of which was collected by M. L. Thomas in 1971 and is housed in NBG.

8. LACHENALIA THUNBERGII

Lachenalia thunbergii G. D. Duncan & T. J. Edwards, sp. nov.
A L. vanzyliae inflorescente breviore, floribus latioribus, tepalis exterioribus brevioribus, viridis citrinus, tepalis interioribus longioribus, viridis citrinus differt.
TYPE: South Africa, Western Cape, 3318 (Cape Town): summit ridge of Riebeek Kasteel Mountain (–BD), *M.C. Botha s.n.* sub. NBG 93.037 (NBG!, holo.).

ETYMOLOGY. *thunbergii*: after C. P. Thunberg (1743–1828), Swedish physician, botanist and naturalist, who made the first collection of plants.

DESCRIPTION. *Geophyte*, 80–120 mm high. *Bulb* globose, 15–18 mm in diameter, offset-forming; tunic multilayered, outer layers light to dark brown, spongy; inner cataphyll translucent white, adhering to leaf bases, apex obtuse. *Leaves* 2, occasionally 1, broadly lanceolate, 70–150 × 15–27 mm, spreading, slightly canaliculate, light green to glaucous, upper surface plain, rarely with dark green or brownish-purple blotches, depressed longitudinal grooves prominent, lower surface plain; leaf margins flat to undulate; leaf bases clasping, 10–20 mm long, subterranean portion white, upper portion light green; primary seedling leaf terete, erect. *Inflorescence* racemose, few- to many-flowered, sterile apex long; scape erect, 60–130 mm long, light green, plain or with light to dark brownish-purple blotches; rachis light green or suffused with brownish-maroon, becoming bright reddish-orange above; lower bracts ovate, green, becoming lanceolate and brownish-orange above, 2–5 × 1–4 mm; pedicels suberect to arcuate, 3–5 mm long, yellowish-green, maturing to brownish-purple in fruit. *Perianth* zygomorphic, tubular, pendulous; tube cylindrical, 6 mm long, bright yellow; outer tepals oblong, 10–11 × 5–6 mm, greenish-

Figure 96. Cultivated specimens of *Lachenalia thunbergii* from Riebeek Kasteel. Image: Graham Duncan/SANBI.

yellow, apical gibbosities large, bright green; inner tepals oblong-obovate, 23–24 × 8–9 mm, greenish-yellow; median keels bright green, apices straight. *Stamens* included, straight; filaments white, 21–24 mm long, pollen yellow at anthesis, maturing to black. *Ovary* ellipsoid, 3–4 × 2–3 mm, light green; style included, straight, 19–20 mm long, white. *Capsule* ellipsoid, 9–12 × 4–5 mm, spreading to suberect. *Seed* globose, 1.1–1.2 × 1.2–1.3 mm, glossy, black; strophiole inflated, 1.1–1.2 mm long, smooth. *Chromosome number* unknown. Figure 96.

FLOWERING PERIOD. August to September.

HISTORY. *L. thunbergii* was collected for the first time by the Swede Carl Peter Thunberg. He found it in early October 1774, at the beginning of his third major journey of botanical exploration into the Cape interior that took him and the Scottish plant collector Francis Masson past Riebeek Kasteel in the south-western Cape. The plants were found at the summit of Riebeek Kasteel Mountain in the remarkable 'chink' formed between massive boulders that provides access from north to south of the mountain. His pressed specimen from this locality (*Thunberg* 8557) is preserved in his herbarium at Uppsala University (Warner & Rourke, 1990). Prof. M. C. Botha visited the site in September 1971 and the specimens he gathered there form the holotype, preserved in the Compton Herbarium at Kirstenbosch. A later visit to the locality by Dr J. P. Rourke in April 1986 resulted in the collection of bulbs for cultivation in the Kirstenbosch Nursery, and the plant was illustrated as an unnamed variety of *L. aloides* in *The Lachenalia Handbook* on plate 18c (Duncan, 1988a).

DISTINGUISHING CHARACTERS AND AFFINITIES. *L. thunbergii* is allied to *L. vanzyliae* (Duncan *et al.*, 2005a), which differs in its turquoise perianth tube, longer, translucent white outer tepals (12–14 mm long) with turquoise median keels, and shorter, translucent white inner tepals (17–20 mm long) with green median keels. It differs further in its shorter filaments (16–19 mm long) and a turquoise upper rachis, and is allopatric, the closest population occurring some 40 km to the north-east in the Twenty Four Rivers mountain range near Porterville. *L. thunbergii* was previously regarded merely as a colour variety of *L. aloides* but is fully deserving of specific rank. In addition to clear morphological differences, it is geographically isolated, breeds true to type from seed and has no known natural hybrids.

DISTRIBUTION, HABITAT AND CONSERVATION STATUS. *L. thunbergii* is restricted to the summit ridge of Riebeek Kasteel, the highest peak of an isolated mountain in the Swartland region of the south-western Cape, above the farm Riebeek River (Map 9). The plants grow in Hawequas Sandstone Fynbos vegetation (Mucina & Rutherford, 2006) and occur as scattered individuals or in small groups in leaf mould in rock cracks, in full sun or within light shade provided by massive sandstone boulders. A conservation status of Rare is recommended.

NOTES. The long tubular flowers with intensely reddish-orange rachis apices are adapted to pollination by sunbirds. The relatively short, weak scapes imply that the sunbirds perch on surrounding vegetation,

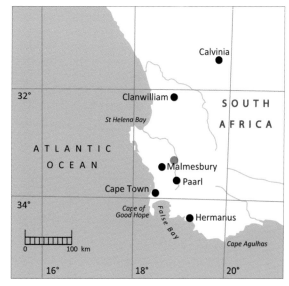

Map 9. Known distribution of *Lachenalia thunbergii*.

rocks or on the ground in order to reach the nectar. The maturing flowers re-orientate to a spreading or suberect position during the fruiting stage. The scapes remain attached following capsule dehiscence and the seeds are dispersed locally.

9. LACHENALIA LUTEOLA

Lachenalia luteola Jacq., *Collectanea* 4: 148–149 (1791). *Lachenalia tricolor* Thunb. var. *luteola* (Jacq.) Baker, in Thiselton-Dyer (ed.), *Flora Capensis* 6: 424 (1897).

TYPE: South Africa, Cape, collector and precise locality unknown, figure in N. J. Jacquin, *Icones Plantarum Rariorum* 2: 12, t. 395 (1792 or 1793) (neotype, designated here).

SYNONYMY: *Lachenalia luteola* Jacq. var. *pallida* Tratt., *Archiv der Gewachskunde* 2: t. 150 (1814). Type: South Africa, Cape, collector and precise locality unknown, figure in *Archiv der Gewachskunde* 2: t. 150 (1814) (lectotype, designated here).

ETYMOLOGY. *luteola*: greenish-yellow, with reference to the outer and inner tepals.

DESCRIPTION. *Geophyte*, 150–340 mm high. *Bulb* globose, 15–20 mm in diameter, offset-forming; tunic multilayered, outer layers dark brown, spongy; inner cataphyll translucent white, adhering to leaf bases, apex obtuse. *Leaves* 2, very rarely solitary, lanceolate, 100–210 × 10–35 mm, spreading, light green, flat to slightly canaliculate, upper surface plain to heavily marked with dark green or brownish-purple blotches, lower surface plain or with sporadic brownish-purple blotches; leaf bases clasping, 20–30 mm long, subterranean portion white, upper portion suffused with dull brownish-magenta on lower surface; primary seedling leaf flat, prostrate. *Inflorescence* racemose, few- to many-flowered, sterile apex short to long; scape erect to suberect, 60–120 mm long, light green or suffused with dull brownish-purple; rachis light green or heavily suffused with brownish-purple in lower half, upper half shading to bright reddish-orange or light yellow; pedicels spreading to suberect, 2–6 mm long, light green; lower bracts ovate, becoming lanceolate above, 1–7 × 1–6 mm, light green to bright red. *Perianth* zygomorphic, tubular, pendulous, becoming spreading to suberect in fruit; tube cylindrical, 5–7 mm long, greenish-yellow, yellowish-green, mustard, bright yellow or reddish-orange; outer tepals oblong, 10–15 × 4–6 mm, greenish-yellow, yellowish-green, mustard, bright yellow or reddish-orange; apical margins plain or magenta, apical gibbosities light green; inner tepals oblong obovate, 17–20 × 5–7 mm, greenish-yellow, yellowish-green, mustard or bright yellow, apices slightly spreading, plain or magenta, median keels light green. *Stamens* included, straight; filaments white, 19–24 mm long. *Ovary* ellipsoid, 3–5 × 3–4 mm, light green; style included, straight, 17–23 mm long, white. *Capsule* ellipsoid, 9–12 × 5–8 mm. *Seed* globose, 1.1–1.2 × 1.1–1.3 mm, glossy, black; strophiole inflated, 1.1–1.2 mm long, smooth. *Chromosome number* $2n = 14$ (Crosby, 1986). Figures 49, 77, 97, 98.

FLOWERING PERIOD. August to November, with a peak from mid-September to mid-October.

HISTORY. *L. luteola* was probably first collected by Thunberg in 1773 at Noordhoek in the southern Cape Peninsula and his specimen (*Thunberg* 8561) is preserved in the Botanical Museum at Uppsala (Thunberg, 1784; Gunn & Codd, 1981). It was described by N. J. Jacquin in 1791 and illustrated on plate 395 of his *Icones Plantarum Rariorum* in 1792 or 1793, although the plant had previously been illustrated on plate 82 of *Curtis's Botanical Magazine*, mistakenly as *L. tricolor* Thunb. (Curtis, 1789). The combination *L. tricolor* var. *luteola* was first made by Ker Gawler (1807a) in *Curtis's Botanical Magazine*, but the plant illustrated on plate 1020 of the latter publication is not the one originally described as *L. luteola* but rather a form of *L. flava*. Baker (1871) placed *L. luteola* in synonymy

Figure 97 (left). Yellowish-green form of *Lachenalia luteola* near Kommetjie. Image: Graham Duncan.

Figure 98 (right). Cultivated specimen of a multicoloured form of *Lachenalia luteola* from the Cape of Good Hope Nature Reserve. Image: Graham Duncan.

under *L. tricolor* Thunb. in his first revision of the genus; in his second and final work (1897a), he afforded it varietal rank as *L. tricolor* Thunb. var. *luteola* (Jacq.) Baker and mistakenly listed Ker Gawler's *L. tricolor* var. *luteola* as synonymous. *L. luteola* was illustrated several times in the early 19th century: an outstanding rendition by Redouté (1809) is included in volume 5 of *Les Liliacées*, and Trattinnick (1814c) described and illustrated a light yellow form as *L. luteola* var. *pallida* in his *Archiv der Gewachskunde*. *L. luteola* was one of the first species used in hybridisation experiments in the United Kingdom in the 1870s and was the female parent of the hybrid *L.* × *nelsonii*.

DISTINGUISHING CHARACTERS AND AFFINITIES. *L. luteola* is very variable with regard to tepal colour. The typical forms flower from August to late September and have greenish-yellow, yellowish-green or mustard outer and inner tepals. Their outer tepals have green apical gibbosities, their inner tepals light green keels and spreading apices, with the rachis apex bright reddish-orange or mustard yellow. Later-flowering forms (late September to early November), such as those occurring in parts of the Cape of Good Hope Nature Reserve, are much more colourful. The outer tepals of these forms are bright reddish-orange at the base, shading to bright yellow above with green apical gibbosities. Their inner tepals are bright yellow or greenish-yellow with broad magenta, slightly spreading apices. Their rachis apex is intensely reddish-orange. It has not been considered feasible to designate taxonomic rank to the various colour forms as, between them, numerous intermediates occur, some with predominantly yellowish-green perianths, the inner tepals plain or with very narrow magenta apices, others with plain mustard yellow perianths; in certain individuals, perianth colour combinations vary considerably within flowers of the same individual.

L. luteola resolved adjacent to *L. thunbergii* in a phylogenetic analysis (Duncan *et al*., 2005a). *L. thunbergii* has greenish-yellow flowers and a bright reddish-orange rachis apex, as do certain forms of *L. luteola*, but it differs in its longer (23–24 mm), broader inner tepals (8–9 mm), and usually smaller stature (85–120 mm high). *L. thunbergii* is confined to relatively high altitude along the summit ridge of Riebeek Kasteel Mountain, 130 km north-east of the closest *L. luteola* populations.

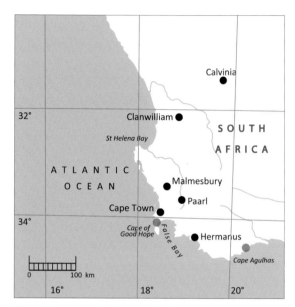

Map 10. Known distribution of *Lachenalia luteola*.

Superficially, the very colourful, late-flowering forms of *L. luteola* from the Cape of Good Hope Nature Reserve resemble *L. quadricolor*, but *L. quadricolor* has much longer inner tepals (27–29 mm, as opposed to 17–20 mm) and its outer tepals have much more prominent green apical gibbosities. Furthermore, *L. quadricolor* is restricted to humus deposits in crevices in granite outcrops of the Cape west coast, and flowers in winter.

DISTRIBUTION, HABITAT AND CONSERVATION STATUS. *L. luteola* is almost restricted to the southern Cape Peninsula, with a single outlying population at Skipskop east of Bredasdorp, a fairly wide disjunction of 235 km to the east (*Admiraal s.n.*, in PRE) (Map 10). The species traverses two vegetation types, Peninsula- and Overberg Sandstone Fynbos vegetation (Mucina & Rutherford, 2006). It occurs at low altitude, singly or in small groups within larger colonies on sharply drained, acid sandstone substrates on flats, rocky outcrops and hill slopes, in full sun or light shade provided by boulders. The species is not threatened.

NOTES. Robust inflorescences are produced in spring immediately following summer or autumn bush fires. The brightly coloured, long tubular flowers are specialised for pollination by sunbirds, the intense reddish-orange rachis apex acting as an additional attraction for the birds. The flowers are pollinated by Orange-breasted Sunbirds (*Anthobaphes violacea*) and Southern Double-collared Sunbirds (*Cinnyris chalybeus*). During fruiting, the orientation of the fertilised flowers changes from pendulous to spreading or suberect. Ripe seeds are dispersed locally from the capsules by the shaking action of wind.

10. LACHENALIA ALOIDES

Lachenalia aloides (L.*f*.) Engl., *Notizblatt des Koniglichen Botanischen Gartens und Museums zu Berlin* 2 (18): 321 (1899); *Phormium aloides* L.*f*., *Supplementum Plantarum et Specierum Plantarum*, ed. 2: 205 (1781); *Lachenalia aloides* (L.*f*.) Pers. (name only), *Synopsis Plantarum* 2: 377 (1805), nom inval.; *Lachenalia aloides* (L.*f*.) Hort. ex Asch. & Graebn., *Synopsis Mitteleuropaischen Flora* 3: 279 (1905).
TYPE: South Africa, Cape, collector and precise locality unknown, figure in N. J. Jacquin, *Icones Plantarum Rariorum* 1 (1): t. 61 (1781) (neotype, designated here).
SYNONYMY: *Lachenalia tricolor* J. Jacq., *Nova Acta Helvetica* 1: 39 (1787). Type: as for *L. aloides* (L.*f*.) Engl.
Lachenalia tricolor Thunb., *Prodromus Plantarum Capensium*: 64 (1794), illegitimate homonym. Type: unknown.

ETYMOLOGY. *aloides*: *Aloe*-like flowers.
DESCRIPTION. *Geophyte*, 90–250 mm high. *Bulb* ovoid, 10–20 mm in diameter, offset- and bulbil-forming, white; tunic multilayered, outer layers dark brown; inner cataphyll translucent green, loosely surrounding leaf bases, apex obtuse. *Leaves* 2, occasionally 1, narrowly lanceolate, 80–200 × 6–40 mm, spreading to suberect, canaliculate, glaucous or bright green, upper surface with faintly depressed longitudinal grooves, plain or heavily marked with green, brownish-magenta or purple blotches, midrib distinct on lower surface; leaf margins slightly coriaceous; leaf bases clasping, 10–60 mm long, white to light green, occasionally speckled with magenta; primary seedling leaf terete, erect. *Inflorescence* racemose, few- to many-flowered, lax, sterile apex short; scape erect, 60–140 mm long, lax to sturdy, light green, plain or heavily marked with large brownish-magenta blotches; rachis light green in lower two thirds, plain or with brownish-magenta blotches, upper third bright red, greenish-yellow or reddish-orange; pedicels 4–5 mm long, spreading or suberect, light green in lower part of inflorescence, reddish-orange or yellowish-orange above; bracts ovate at base of inflorescence, becoming lanceolate above, 2–6 × 1–3 mm, translucent white or light green. *Perianth* zygomorphic, tubular, pendulous; tube cylindrical, bright yellow or deep orange-red, 8–9 mm long; outer tepals oblong, 15–17 × 7–8 mm, bright yellow to greenish-yellow, plain or with bright reddish-orange bases; apical gibbosities prominent, bright green; inner tepals oblong obovate, 26–28 × 8–10 mm, bright yellow to greenish-yellow, apices spreading, broad, deep red or purplish red, margins white. *Stamens* included, straight; filaments yellowish-white, 25–27 mm long, pollen yellow. *Ovary* ellipsoid, 4–5 × 2–3 mm, light green; style straight, yellowish-white, 24–25 mm long, sturdy, thicker than filaments. *Capsule* ellipsoid, 9–14 × 5–9 mm. *Seed* globose, 1.1–1.3 × 1.2–1.4 mm glossy, black; strophiole inflated, 1.1–1.2 mm long, smooth. *Chromosome number* $2n = 14$ (Spies *et al.*, 2008). Plate 5, Figure 99.
FLOWERING PERIOD. Early June to early August, with a peak in mid-July.
HISTORY. Introduced into cultivation at Kew by Francis Masson in 1774 (Aiton,1789), *Phormium aloides* was described by Linnaeus *fil.* in July 1781, in the second edition of his *Supplementum Plantarum et Specierum Plantarum*. There, Linnaeus alluded to several colour 'varieties', all of which are now recognised as species. Two sheets in the Linnean Herbarium could possibly have been used by Linnaeus when he described *P. aloides* (listed in Savage (1945), *Catalogue of the Linnean Herbarium*, as 405-1 and 405-2). The first sheet (405-1), a large plant with two broad leaves and pale flowers, is part of the Banks Herbarium. The second (405-2), a very poor specimen with no leaves, which has dark tips to the inner tepals, has only 'C. B. Sp.' (Cap Bonae Spei) written on the sheet. Although

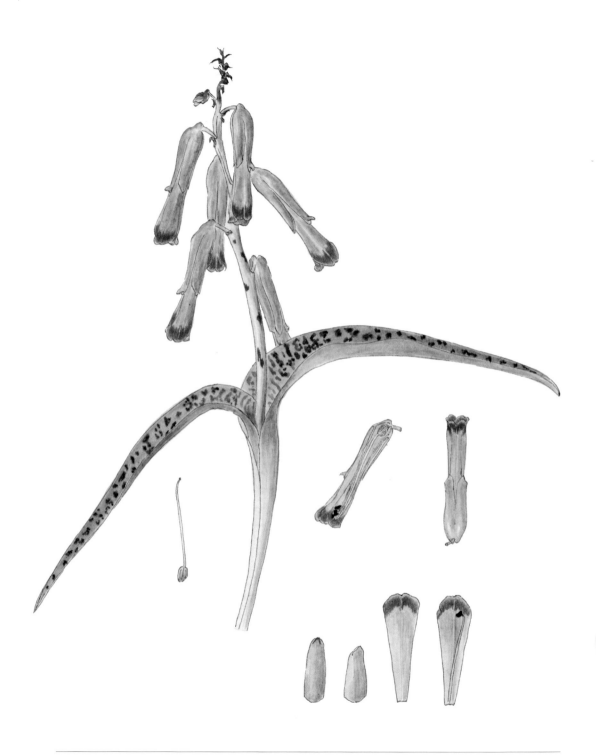

Plate 5. Watercolour painting of *Lachenalia aloides*, courtesy of the Compton Herbarium, Kirstenbosch, South African National Biodiversity Institute. Artist: Winsome Barker.

Linnaeus described the plant in some detail and gave the colour range across several 'varieties', he did not refer to either of the specimens in LINN. At about the same time, J. F. Jacquin described the new genus *Lachenalia*, citing the type species as *L. tricolor*. In 1780, Jacquin sent his manuscript to the editor of *Acta Helvetica*, expecting it to be published that year, but the journal had ceased and his manuscript was only published seven years later in the newly styled *Nova Acta Helvetica* (Jacquin, 1787). In the meantime, J. A. Murray (1784), evidently having seen Jacquin's manuscript and assuming it to have been published, included a short description of *Lachenalia* in volume 14 of *Linnaeus Systema Vegetabilium*, citing Jacquin's *L. tricolor*. In so doing, Murray inadvertently became the first to formally publish the new genus. On page 336 of the same publication, Murray cited *Phormium aloides*, apparently unaware that *L. tricolor* represented the same species. In August 1781, the colour painting of J. F. Jacquin's *L. tricolor*, originally intended to appear in 1780 in volume nine of *Acta Helvetica*, was published on plate 61 of the first volume of his father N. J. Jacquin's *Icones Plantarum Rariorum*; 'Act. Helv. Vol. 9' is printed below the painting.

Phormium aloides L.f. is synonymous with *L. tricolor* J. Jacq. and, in accordance with the *International Code of Botanical Nomenclature*, the earliest available epithet, irrespective of its generic attachment has to be used. Thus Linnaeus's (1781) epithet *aloides* takes precedence. The new combination *Lachenalia aloides* was first made by Persoon (1805), but he merely cited it in synonymy under *L. tricolor*, and thus this combination was not validly published by Persoon. Eventually, H. G. A. Engler (1899) made the combination again, correctly citing the later name *L. tricolor* under *L. aloides*, and the citation for the species became *L. aloides* (L.f.) Engl. (Barker, 1988). A fairly recent watercolour painting of the species by Claire Linder Smith was published in *The Flowering Plants of Africa* (Reid, 1985).

In view of the fact that it is not known which specimens Linnaeus (1781) examined when he described *Phormium aloides*, and the fact that Jacquin's type specimen was lost in a fire between 1939 and 1945, and that Jacquin (1787) described *L. tricolor* as having a red calyx (outer tepals) with greenish tips (gibbosities) and yellow inner tepals with purplish red tips, it has been considered expedient to designate plate 61 of N. J. Jacquin (1781) as the neotype, a view shared by W. F. Barker (Winsome Barker, personal communication).

Thunberg (1784) included three unnamed varieties of *Phormium aloides* L.f. in his *Dissertatio de Novis Generibus Plantarum*. A decade later, in his *Prodromus Plantarum Capensium* (1794), he mistakenly placed one of them (now *L. thunbergii*, Thunberg 8557 in UPS) under *L. pendula* (now *L. bifolia* (Burm. f.) W. F. Barker ex G. D. Duncan). In this later publication,

Figure 99. *Lachenalia aloides* on a granite slope near Atlantis. Image: Graham Duncan.

Thunberg also listed *L. tricolor* under his own name, without any citations, apparently unaware of the publications of Murray (1784) and Jacquin (1787) of the same species under the same name, a mistake perpetuated by several authors including Baker (1871, 1897a). It has not been possible to locate what Thunberg considered to be the type of *L. tricolor* Thunb.. Apart from his specimen collected at Riebeek Kasteel (*L. thunbergii*), there are three other sheets in his herbarium in the Botanical Museum at Uppsala, all of which represent other species: *Thunberg* 8560 (= *L. flava* Andrews, collected on the Paardeberg at Paarl), *Thunberg* 8561 (=*L. luteola* Jacq., collected at Noordhoek in the southern Cape Peninsula) and *Thunberg* 8562 (=*L. quadricolor* Jacq.) from the Swartland.

DISTINGUISHING CHARACTERS AND AFFINITIES. Linnaeus (1781) ended his description of *Phormium aloides* with the words '*Species difficile determinatur*', with reference to the numerous similar-looking 'colour forms' of what he concluded to be a single variable species. N. J. Jacquin and subsequent authors described some of these 'colour forms' as species, two of which (*L. luteola*, *L. quadricolor*) were reduced to synonymy by Baker (1871) and then later reinstated by him (1897a) as varieties of *L. tricolor*. Superficially, the entities previously regarded as varieties of *L. aloides* or *L. tricolor* look similar, but a detailed study of their morphological features, distribution, and breeding behaviour, in which allele fixation results in these entities breeding true (Duncan, 2005; Duncan *et al.*, 2005a), has shown them to constitute eight species: *L. aloides*, *L. quadricolor*, *L. flava*, *L. luteola*, *L. vanzyliae* and three new species, *L. callista*, *L. patentissima* and *L. thunbergii*. Material of these three new species has lain undescribed in herbaria for decades. Apart from morphological differences, all eight species are geographically isolated to a greater or lesser degree, breed true to type from seed, and almost never form hybrids in the wild.

In a phylogenetic analysis of morphological data (Duncan *et al.*, 2005a), *L. aloides* resolved as sister to *L. quadricolor*. The latter differs in its shorter perianth tubes (6–7 mm long), much shorter outer tepals (10–11 mm long) and inner tepals that have deep purplish-magenta apices without white margins. *L. quadricolor* favours slightly different habitat in the crevices of granite rock outcrops and commences flowering several weeks later than *L. aloides*.

DISTRIBUTION, HABITAT AND CONSERVATION STATUS. *L. aloides* is restricted to three locations on granite hillsides; on the Klein Dassenberg Range near Atlantis in the Philadelphia District north of Cape Town, between Wellington and Malmesbury, and near Durbanville to the east of Cape Town (Map 11). It occurs in Swartland Granite Renosterveld vegetation (Mucina & Rutherford, 2006). The plants grow in small clumps within larger colonies on north- and west-facing decomposing granite slopes, in gravelly sand or acid humus, in full sun or light shade, and in association with low succulent scrub and spring-flowering annuals. *L. aloides* is threatened by habitat degradation caused by over-grazing and trampling, and a conservation status of Endangered is recommended.

NOTES. The flowers of *L. aloides* are adapted to bird pollination and are visited by

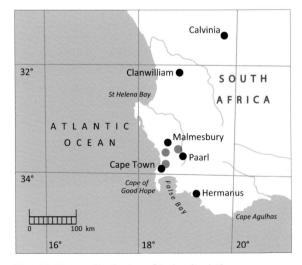

Map 11. Known distribution of *Lachenalia aloides*.

Southern Double-collared Sunbirds (*Cinnyris chalybeus*). These birds gain access to the flowers mainly from the ground or surrounding rocks, or by clinging to the scapes of robust specimens. During the fruiting stage, the ripening capsules re-orientate themselves to a spreading or suberect position and the seeds are dispersed locally by the shaking action of wind. The only member of the genus to have been investigated pharmacologically is *L. aloides* (Watt & Breyer-Brandwijk, 1962), although it is not known which particular 'variety' was used; the bulb gave negative results when tested for the digitalis type of cardiac action, and an aqueous extract of the bulb gave negative antibiotic tests.

11. LACHENALIA QUADRICOLOR

Lachenalia quadricolor Jacq., *Collectanea* 5: 62 (1797). *Lachenalia tricolor* Thunb. var. *quadricolor* (Jacq.) Baker, in Thiselton-Dyer (ed.), *Flora Capensis* 6: 424 (1897). *Lachenalia aloides* (L.f.) Engl. var. *quadricolor* (Jacq.) Engl., *Notizblatt des Koniglichen Botanischen Gartens und Museums zu Berlin* 2 (18): 321 (1899).
TYPE: South Africa, Cape, collector and precise locality unknown, figure in N. J. Jacquin, *Icones Plantarum Rariorum* 2: 16, t. 396 (1795) (neotype, designated here).

ETYMOLOGY. *quadricolor*: having four different colours, with reference to the perianth.
DESCRIPTION. *Geophyte*, 90–200 mm high. *Bulb* ovoid, 15–20 mm in diameter, offset- and bulbil-forming; tunic multilayered, outer tunics dark brown; inner cataphyll translucent green, loosely surrounding leaf bases, apex obtuse. *Leaves* 2, narrowly to broadly lanceolate, 80–200 × 6–40 mm, spreading to suberect, canaliculate, upper surface light green to glaucous, plain to heavily marked with dark green, brownish-magenta or purple blotches, longitudinal grooves faintly depressed; lower surface

Figure 100. *Lachenalia quadricolor* on a granite slope near Brackenfell. Image: Graham Duncan.

plain or flushed with dull maroon at base, midrib distinct; margins slightly coriaceous; bases clasping, 30–60 mm long, subterranean portion white, upper portion light green, plain or with minute magenta speckles; primary seedling leaf terete, erect. *Inflorescence* racemose, few- to many-flowered, sterile apex short to long; scape erect, 50–120 mm long, flaccid to sturdy, light green, plain or heavily marked with large brownish-magenta blotches; rachis light green in lower two thirds, plain to brownish-magenta blotched, upper third bright red to pinkish-red or reddish-orange; pedicels spreading to suberect, 5–7 mm long, light green in lower part of inflorescence, reddish-orange above; lower bracts ovate, becoming lanceolate above, 2–7 × 1–3 mm, translucent white or light green. *Perianth* zygomorphic, tubular, pendulous; tube cylindrical, 6–7 mm long, red to reddish-orange; outer tepals oblong, 10–11 × 7–8 mm, reddish-orange, shading to yellow above, apical gibbosities green; inner tepals oblong obovate, 27–29 × 9–10 mm, greenish-yellow, apices slightly flared, deep purplish-magenta. *Stamens* included, straight; filaments cream to light yellow, 24–26 mm long. *Ovary* ellipsoid, 3–4 × 2–3 mm, light green; style included, straight, 25–27 mm long, sturdy, cream to light yellow. *Capsule* ellipsoid, 10–14 × 5–9 mm. *Seed* globose, 1.1–1.2 × 1.3–1.4 mm, glossy, black; strophiole inflated, 1.1–1.2 mm long, smooth. *Chromosome number* 2n = 14 (Crosby, 1986; Hamatani *et al.*, 1998; Johnson & Brandham, 1997); 2n = 28 (Hancke & Liebenberg, 1990). Plate 6, Figures 40, 100.

FLOWERING PERIOD. Late June to mid-August, with a peak in July.

HISTORY. *L. quadricolor* was illustrated by N. J. Jacquin (1795b) and described by him two years later in *Collectanea* (Jacquin, 1797a). It had, however, been earlier collected by Thunberg in the Swartland and was included as a variety (β) of *Phormium aloides* in his *Nova Genera Plantarum* (Thunberg, 1784). Baker (1871) reduced *L. quadricolor* to synonymy under *L. tricolor* Thunb., but in his second monograph (1897a), he recognised it as *L. tricolor* var. *quadricolor*. When the new combination *L. aloides* (L.f.) Engl. was made by Engler (1899), the taxon became *L. aloides* (L.f.) Engl. var. *quadricolor* (Jacq.) Engl. The taxon superficially resembles *L. aloides* but is restored here to species level as it differs morphologically from the latter and occurs in isolated, morphologically homogeneous populations that breed true to type. It was introduced into horticulture at Kew in 1774 by Francis Masson (Aiton, 1811) and has been in continuous cultivation for well over two centuries. In the late 19th century, *L. quadricolor* was used, in combination with *L. reflexa*, to create the hybrid *L.* × *comesii* Sprenger. It was also used in the breeding programme started at Roodeplaat in South Africa in 1965, and its influence is clearly seen in the popular cultivar 'Namaqua'.

DISTINGUISHING CHARACTERS AND AFFINITIES. Duncan *et al.* (2005a) placed *L. quadricolor* as sister to *L. aloides*. The species share pendulous flowers with bright yellow to greenish-yellow outer and inner tepals, the outer tepals with bright green apical gibbosities. *L. aloides* differs in its longer perianth tube (8–9 mm long), much longer outer tepals (15–17 mm long) and in its inner tepals that have broad red or purplish red apices with white margins. *L. quadricolor* occurs close to *L. aloides* in parts of its range but the two species always occur in discrete populations, the latter confined to three locations, one near Atlantis north of Cape Town, another between Wellington and Malmesbury, the third near Durbanville to the east of Cape Town. *L. aloides* differs in its habitat preference of west-facing, decomposing granite slopes, and the species commences flowering several weeks earlier than *L. quadricolor*.

DISTRIBUTION, HABITAT AND CONSERVATION STATUS. *L. quadricolor* occurs from Langebaan to Darling on the Cape west coast, on the northern slopes of Lion's Head in the northern Cape Peninsula and at Somerset West, Stellenbosch, Brackenfell and Kuilsrivier on the Cape Flats (Map 12). It grows in Saldanha Granite Strandveld, Peninsula Shale Renosterveld and Swartland Granite Renosterveld vegetation (Mucina & Rutherford, 2006). Populations are always associated with granite outcrops,

Plate 6. Watercolour painting of *Lachenalia quadricolor*, courtesy of the Compton Herbarium, Kirstenbosch, South African National Biodiversity Institute. Artist: Winsome Barker.

the bulbs growing in colonies wedged between rock crevices or in depressions of rock sheets, in gravelly sand or acid, humus-rich matter. *L. quadricolor* usually favours north-, north-east and north-west slopes. At Langebaan, it grows in association with the endemic west coast geophytes *Empodium veratrifolium* (Hypoxidaceae), *Gladiolus priorii* and *Watsonia hysterantha* (Iridaceae). A form with narrow, deeply caniculate leaves is rare on Lion's Head in the northern Cape Peninsula. The species is threatened by coastal housing development and a conservation status of Vulnerable is recommended.

NOTES. Southern Double-collared Sunbirds (*Cinnyris chalybeus*) are the primary pollinators of this species; the birds cling to the scapes or, if a delicate specimen is not strong enough to support their weight, access the flowers from the ground, surrounding rocks or vegetation. The flowers are also visited by honey bees (*Apis mellifera*). Developing capsules re-orientate to a spreading or suberect position and the scapes remain attached to the bulbs following capsule dehiscence. The relatively large seeds drop to the ground due to the shaking action of wind, landing close to the mother plant. The species has become naturalised in Western Australia.

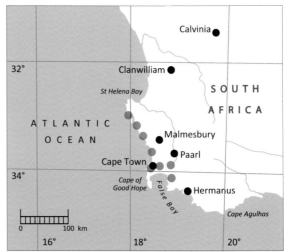

Map 12. Known distribution of *Lachenalia quadricolor*.

12. LACHENALIA FLAVA

Lachenalia flava Andrews, *The Botanist's Repository* 7: t. 456 (1807).
TYPE: South Africa, Cape, collector and precise locality unknown, figure in Andrews *The Botanist's Repository* 7: t. 456 (1807) (lectotype, designated here).
SYNONYMY: *Lachenalia tricolor* Thunb. var. *luteola* (Jacq.) Ker Gawl., *Curtis's Botanical Magazine* 26: t. 1020 (1807). Type: South Africa, Cape, collector and precise locality unknown, figure in *Curtis's Botanical Magazine* 26 t. 1020 (1807) (lectotype, designated here).
Lachenalia quadricolor Jacq. var. *lutea* Sims, *Curtis's Botanical Magazine* 61: t. 1704 (1815). Type: South Africa, Cape, collector and precise locality unknown, figure in *Curtis's Botanical Magazine* t. 1704 (1815) (lectotype, designated here).
Lachenalia macrophylla Lem., *L'Illustration Horticole* 2: 99–100 (1855). Type: South Africa, Cape, collector and precise locality unknown, figure in *Curtis's Botanical Magazine* t. 1704 (1815) (lectotype, designated here).
Lachenalia aurea Lindl., *The Gardener's Chronicle* (first series): 404 (1856). *Lachenalia tricolor* Thunb. var. *aurea* (Lindl.) Hook.f., pro parte, *Curtis's Botanical Magazine* 118: t. 5992 (1872). *Lachenalia tricolor* Thunb. var. *aurea* (Lindl.) Baker, in Thiselton-Dyer, W. T. (ed.), *Flora Capensis* 6: 424–425 (1897). *Lachenalia aloides* (L.f.) Engl. var. *aurea* (Lindl.) Engl., *Notizblatt des Koniglichen Botanischen Gartens und Museums zu Berlin* 2(18): 321 (1899). Type: South Africa, Cape, collector and precise locality unknown, figure in *The Gardener's Chronicle* (first series): 404 (1856) (lectotype, designated here).

Plate 7. Watercolour painting of *Lachenalia flava* from Bainskloof, courtesy of the Compton Herbarium, Kirstenbosch, South African National Biodiversity Institute. Artist: Winsome Barker.

12. LACHENALIA FLAVA

ETYMOLOGY. *flava*: yellow flowers.

DESCRIPTION. *Geophyte*, 190–250 mm high. *Bulb* globose, 15–20 mm in diameter, offset-forming; tunic multilayered, outer layers dark brown, spongy; inner cataphyll translucent white, adhering to leaf bases, apex obtuse. *Leaves* 2, lanceolate, 85–120 × 12–30 mm, spreading to suberect, light green, slightly canaliculate, upper surface plain or light marked with dark green or light to dark brownish-maroon blotches; leaf margins flat to slightly undulate; leaf bases clasping, 10–40 mm long; primary seedling leaf flat, prostrate. *Inflorescence* racemose, many-flowered, sterile apex long; scape erect, 110–170 mm long, dull purple to maroon, or light green with large purple blotches, covered with a delicate waxy bloom; rachis erect, light green, or dull purple to maroon, apex light to bright golden orange, rarely light yellow; pedicels suberect, 5–7 mm long, light golden yellow to greenish-yellow; lower bracts ovate, becoming lanceolate above, 3–7 × 1–4 mm, dark golden yellow, purple or green, apices purple. *Perianth* zygomorphic, tubular, dark golden yellow or orange, rarely light yellow, pendulous; tube cylindrical, 6 mm long; outer tepals oblong, 13–18 × 7–9 mm, apical gibbosities dark golden orange or bright green; inner tepals oblong obovate, 26–28 × 9–12 mm, protruding well beyond outer tepals, apices slightly to widely spreading. *Stamens* included, straight; filaments white to light yellowish-orange, 21–28 mm long. *Ovary* ellipsoid, 4–5 × 3–4 mm, light green; style included, straight, 20–25 mm long, light yellowish-orange. *Capsule* ellipsoid, 9–11 × 4–5 mm. *Seed* globose, 1.1–1.3 × 1.1–1.3 mm, glossy, black; strophiole inflated, 1.1–1.2 mm long, smooth. *Chromosome number* 2n = 14 (Johnson & Brandham, 1997; Hamatani *et al.*, 2004). Plate 7, Figures 44, 101, 102.

FLOWERING PERIOD. July to October, with a peak in early September.

HISTORY. Thunberg found this species for the first time on the Paardeberg at Paarl, probably in early October 1774, and described it as a variety (α) of *Phormium aloides* in his *Nova Genera Plantarum* (Thunberg, 1784); his collection (*Thunberg* 8560) is preserved in the Botanical Museum at Uppsala.

Figure 101. Profuse flowering in *Lachenalia flava* after a fire near Tulbagh. Image: Graham Duncan.

The species was described and illustrated as *L. flava* by H. C. Andrews (1807a) in *The Botanist's Repository*, from a plant cultivated in the nursery of Richard Williams of Turnham Green, London, who noted that he found it 'difficult to flower and slow to increase'. Ker Gawler (1807a) published a plate (t. 1020) of the same species later that year in *Curtis's Botanical Magazine*, erroneously as *L. tricolor* var. *luteola*, a mistake perpetuated by Baker (1897a) who listed it as *L. tricolor* Thunb. var. *luteola* (Jacq.) Baker. The British botanist John Lindley (1799–1865) published *L. aurea*, accompanied by a black-and-white line drawing in *The Gardener's Chronicle*, the specimen having come from an unnamed collector who claimed to have brought it from Natal and grown it to flowering in Chiswick (Lindley, 1856). The line drawing (t. 109) was reproduced again in *The Gardener's Chronicle and Agricultural Gazette*, accompanying an article on the general cultivation of the genus (Anonymous, 1872).

The combination *L. tricolor* var. *aurea* was first made by Hooker (1872) but he cited several other taxa under this name in addition to Lindley's *L. aurea*. Baker (1897a) made the

Figure 102. *Lachenalia flava* in Breede Shale Renosterveld near Tulbagh. Image: Graham Duncan.

combination again, and when Engler (1899) made the new combination *L. aloides*, this entity became *L. aloides* var. *aurea*. However, Andrews's *L. flava* is the same plant as that described by Ker Gawler and Lindley, and being the earliest name for the taxon, Andrews's epithet takes precedence.

L. flava was thought to be restricted to the Bainskloof Mountains west of Worcester, and from early records at Paarl and Wellington. In 2006, Gerard Hansford discovered a further population between Wolseley and Tulbagh, some distance to the north.

DISTINGUISHING CHARACTERS AND AFFINITIES. The species resolved adjacent to *L. luteola* in a phylogenetic analysis (Duncan *et al.*, 2005a). *L. luteola* has similar tubular, pendulous flowers with included, straight stamens, and similar globose seed with inflated strophioles, but differs in its usually plain greenish-yellow to mustard perianth (less frequently bright yellow with magenta apices), narrower outer tepals (4–6 mm wide) and much shorter inner tepals (17–20 mm long). *L. luteola* is confined to the southern Cape Peninsula and a disjunct population east of Bredasdorp.

DISTRIBUTION, HABITAT AND CONSERVATION STATUS. *L. flava* is restricted to the middle and upper slopes of the Bainskloof Mountains west of Worcester, and to a hillside at relatively low altitude (80 m) between Wolseley and Tulbagh in the Tulbagh Valley (Map 13). It is also recorded from Paarl Mountain (*MacOwan* 504 and *MacOwan* 2468, in SAM) where it probably no longer occurs, and from an imprecise locality at Wellington, where a collection was made in 1881 (*J. B. Thompson s.n.*, in PRE). The species is unusual in that it occurs on both sandstone and shale substrates, in Hawequas Sandstone Fynbos and Breede Shale Renosterveld vegetation (Mucina & Rutherford, 2006). On

Bainskloof, the plants grow singly or in small groups on south-facing rocky sandstone slopes, whereas in the Tulbagh Valley, a large, dense population occurs on the north-east side of a low-lying hillside. The latter population is early-flowering, being in full bloom by late July and continuing to flower until mid-August, whereas the Bainskloof plants flower from mid-September to early October. At Tulbagh, flowers of golden yellow predominate, but light yellow forms also occur, and all individuals have unspotted upper leaf surfaces; the plants grow in association with *L. unifolia* and a yellow form of *L. longibracteata*, and alongside numerous spring-flowering geophytes including *Babiana villosa*, *Geissorhiza inflexa* (both Iridaceae) and *Holothrix villosa* (Orchidaceae). In view of the few known populations of *L. flava*, a conservation status of Rare is recommended.

Map 13. Known distribution of *Lachenalia flava*.

NOTES. *L. flava* is geographically isolated from all other taxa previously grouped under *L. aloides*, breeds true to type from seed and has no known hybrids in the wild. It flowers in profusion following bush fires but is not dependent on fire for flowering to occur. The long, tubular, pendulous flowers are adapted to pollination by sunbirds. Malachite sunbirds (*Nectarinia famosa*) were observed pollinating a large stand of the species between Wolseley and Tulbagh in July 2006 (Duncan, Hansford & Cox, pers. obs.) (Figure 76). The sunbirds cling to the scapes while inserting their beaks, and in so doing, lift the flower into a horizontal or suberect position while feeding. The flowers are damaged (splayed open) during the pollination process, the tepals frequently becoming detached and falling to the ground. At the Tulbagh locality, Malachite sunbirds were seen to be most active from 12 noon to 2 pm. Whole flowers are consumed by unknown animals at this locality. The scape remains attached for several weeks following capsule dehiscence and the relatively large seeds are shaken to the ground, close to the mother plant, by wind.

13. LACHENALIA PATENTISSIMA

Lachenalia patentissima G. D. Duncan, sp. nov.
A L. flava folio solitario com marginibus cartilagineus, floribus patentibus suberectisve brevioribus vividus flavis cum pedicellis brevissimis, stylis exsertis differt.
TYPE: South Africa, Western Cape, 3418 (Simonstown): rocky north-facing Table Mountain Sandstone slopes above Smitswinkel Bay, (–CD), *G. D. Duncan* 471 (NBG, holo.; iso.).

ETYMOLOGY. *patentissima*: very widely spreading, descriptive of the orientation of the flowers that diverge from the rachis at almost 90 degrees.

DESCRIPTION. *Dwarf geophyte*, 80–130 mm high. *Bulb* globose, 15–20 mm in diameter, offset-forming; tunic multilayered, outer layers dark brown, spongy; inner cataphyll translucent white, adhering to leaf bases, apex obtuse. *Leaf* solitary, narrowly to broadly lanceolate, 110–150 × 17–25 mm, spreading, upper surface light green, plain or marked with dark green or light brown spots, depressed longitudinal grooves prominent; lower surface suffused with dull magenta in lower part; leaf margins cartilaginous, dull purple; leaf base clasping, 10–20 mm long, subterranean, white; primary seedling leaf flat, prostrate. *Inflorescence* racemose, few- to many-flowered; sterile apex short to long; scape erect, 40–60 mm long, light green, heavily marked with dull purple blotches; rachis light green, plain to lightly or heavily marked with dull purple blotches; pedicels suberect, 2–3 mm long, light green or yellow; lower bracts ovate, becoming lanceolate above, 1–5 × 1–2 mm, white or yellowish-green. *Perianth* zygomorphic, tubular, widely spreading, bright yellow to greenish-yellow, ageing to dull red, slightly sweet-scented; tube cylindrical, 6 mm long, slightly inflated; outer tepals oblong, 7–11 × 4–6 mm, apical gibbosities green; inner tepals oblong-obovate, 14–17 × 6–7 mm, apices slightly spreading; median keels green. *Stamens* included to shortly exserted; filaments straight, 11–15 mm long, white. *Ovary* ellipsoid, 4–5 × 2–3 mm, light green; style included to shortly exserted, becoming well exserted in fruit, straight, 14–20 mm long, white to yellowish-green. *Capsule* ellipsoid, 10–13 × 5–6 mm, suberect. *Seed* globose, 1.1 × 1.1 mm, glossy, black; strophiole inflated, 1.1–1.2 mm long, smooth. *Chromosome number* unknown. Figures 103, 104.

Figure 103 (left). *Lachenalia patentissima* on a sandstone ridge in the Cape of Good Hope Nature Reserve. Image: Graham Duncan.

Figure 104 (right). A cultivated specimen of *Lachenalia patentissima*. Image: Graham Duncan/SANBI.

13. LACHENALIA PATENTISSIMA

FLOWERING PERIOD. September to October, with a peak in late September.

HISTORY. The English soldier and naturalist A. H. Wolley-Dod made the first collection of *L. patentissima* in September 1897 above Smitswinkel Bay in the Cape of Good Hope Nature Reserve. Preserved material of this collection (*Wolley-Dod* 3087) is housed in the Kew Herbarium and at the British Museum (Natural History). The next collection was made by J. B. Gillett in October 1928 above Buffels Bay further south in this Reserve (*Gillett* 730), pressed material of which is housed at the Compton Herbarium at Kirstenbosch, and was tentatively identified as an unnamed variety of *L. aloides* by W. F. Barker. More than 70 years later, in 2002, the plant was found flowering again above Smitswinkel Bay by Kirstenbosch horticulturist Anthony Hitchcock; and the following year, I visited this site with horticulturist Adam Harrower and made the type collection.

DISTINGUISHING CHARACTERS AND AFFINITIES. The species has not been phylogenetically analysed, but appears to be allied to *L. flava*, which has similar globose seeds. Certain forms of *L. flava* have yellow perianths that are similar to those of *L. patentissima*, but *L. flava* differs in its pendulous flowers that are borne on much longer pedicels (5–7 mm long), has much longer outer and inner tepals (13–18 mm; 26–28 mm long, respectively), has a longer, always included style, and has two leaves with non-cartilaginous margins.

DISTRIBUTION, HABITAT AND CONSERVATION STATUS. *L. patentissima* has a highly restricted distribution in the Cape of Good Hope Nature Reserve on the southern Cape Peninsula and is known from just two small populations (Map 14). The plants occur in Peninsula Sandstone Fynbos vegetation (Mucina & Rutherford, 2006), growing as scattered individuals or in small groups in crevices of rocky north- and north-east facing sandstone ledges that are subject to extreme summer heat. The species occurs within several hundred metres of populations of *L. luteola* but retains its identity when grown from seed and does not form hybrid swarms. A conservation status of Rare is recommended.

NOTES. The bright yellow, tubular flowers of *L. patentissima* with their long perianth tubes and copious nectar suggest bird pollination. The short, slender scape is unable to support the weight of a sunbird and suggests that birds must gain access to the flowers by perching on surrounding rocks, nearby vegetation or on the ground, as do birds that visit the flowers of *L. viridiflora*. *L. patentissima* is stimulated to flower profusely in spring following wild fires, but is not dependent on fires for flowering to occur. The tepals age to dull red and the developing capsules become re-orientated to a suberect position. The scape remains attached following capsule dehiscence and the seeds are dispersed locally by the shaking action of wind.

Map 14. Known distribution of *Lachenalia patentissima*.

14. LACHENALIA CALLISTA

Lachenalia callista G. D. Duncan & T. J. Edwards, sp. nov.
A L. aloides tepalis interioribus multo longioribus latioribusque cum apicibus effusis non rubris vel magenteus, filis styloque longioribus, floribus sterilibus apicalis vividis aurantiacis differt.
TYPE: South Africa, Western Cape, 3218 (Clanwilliam): Piketberg, (–DC), *B. E. Martin* 881 (NBG!, holo., iso).

ETYMOLOGY. *callista*: most beautiful, with reference to the flowers.
DESCRIPTION. *Geophyte*, 150–250 mm high. *Bulb* ovoid, 15–25 mm in diameter, offset-forming; tunic multilayered, outer layers dark brown, spongy; inner cataphyll translucent white, longitudinal veins green, loosely surrounding leaf bases, apex obtuse. *Leaves* 2, broadly lanceolate, 90–230 × 9–30 mm, spreading to suberect, bright green, upper surface plain or sparsely to heavily marked with dark green or purplish-brown blotches, depressed longitudinal grooves prominent; leaf margins slightly coriaceous; leaf bases clasping, 10–20 mm long, subterranean portion white, aerial portion light green; primary seedling leaf flat, prostrate. *Inflorescence* racemose, few- to many-flowered, sterile apex long; scape erect, 100–260 mm long, bright green with a delicate powdery bloom, plain or marked with dull purplish-brown blotches; rachis green in lower half, shading to bright orange above; pedicels spreading to suberect, 3–10 mm long, light green in lower half of inflorescence, becoming bright orange in upper half; lower bracts ovate, becoming lanceolate above, 1–5 × 1–4 mm, green. *Perianth* zygomorphic, tubular, pendulous; tube cylindrical, 6–7 mm long, bright yellow; outer tepals oblong, 12–14 × 7–8 mm, bright yellow, uppermost flowers orange-flushed at base, apical gibbosities prominent, bright green; inner tepals oblong-obovate, 32–33 × 11–12 mm, bright yellow, apices tinged with light to bright green, median keels bright green, apices spreading, slightly recurved. *Stamens* included, straight; filaments light yellow, 31–32 mm long. *Ovary* ellipsoid, 5–6 × 3–4 mm, bright green; style included, straight, 29–30 mm long, light yellow, protruding well beyond perianth as it matures. *Capsule* ellipsoid, 10–12 × 4–5 mm. *Seed* globose, 1.1–1.2 × 1.2–1.3 mm, glossy, black; strophiole inflated, 1.1–1.2 mm long, smooth. *Chromosome number* unknown. Plate 8, Figure 105.
FLOWERING PERIOD. August to September, with a peak in late September.
HISTORY. The first collection of *L. callista* was made in September 1949 on Versfeld's Pass above Piketberg by the Kirstenbosch horticulturist and botanical assistant at the Compton Herbarium, Bina Elizabeth Martin. The type material was collected by her at the same location two years later, but the species has been infrequently collected since that time. W. F. Barker had intended to describe it as a variety of *L. aloides*, and the plant was illustrated as '*L. aloides* var.' on plate 18d of *The Lachenalia Handbook* (Duncan, 1988a).
DISTINGUISHING CHARACTERS AND AFFINITIES. The species appears to be related to *L. aloides* (Duncan et al., 2005a), which has similar bright yellow, tubular, pendulous flowers, oblong outer tepals with bright green apical gibbosities and oblong obovate inner tepals with spreading, slightly recurved apices. *L. aloides* differs in its longer outer tepals (15–17 mm long), shorter inner tepals (26–28 mm long) with broad red apices, shorter filaments (25–27 mm long) and shorter style (24–25 mm long), and in its intensely red or yellow rachis apex and glaucous leaves. Superficially, *L. callista* resembles *L. thunbergii*, but the latter differs in its shorter stature (up to 120 mm high) with narrower greenish-yellow flowers and much shorter outer and inner tepals (10–11 mm and 23–24 mm long, respectively). The inner tepal apices of *L. thunbergii* are slightly spreading and not recurved, and it's filaments and style (21–24 mm and 19–20 mm long) are much shorter than those of *L. callista*.

PLATE 8. Watercolour painting of *Lachenalia callista* from Piketberg [*Thomas s.n.* (696/84), in NBG] courtesy of Fay Anderson. Artist: Fay Anderson.

Map 15. Known distribution of *Lachenalia callista*.

Fig. 105. Cultivated specimen of *Lachenalia callista* from Piketberg. Image: Graham Duncan.

DISTRIBUTION, HABITAT AND CONSERVATION STATUS. *L. callista* has a highly restricted distribution and is endemic to the upper reaches of the Piketberg and to an area just north of the nearby farming village of Sauer (Map 15). The plants grow in Piketberg Sandstone Fynbos vegetation (Mucina & Rutherford, 2006), singly or in small groups in humus-rich clefts between sandstone boulders, in full sun or light shade. A conservation status of Rare is recommended.

NOTES. The particularly long tubular flowers of *L. callista* suggest bird pollination. The species grows in close proximity to the turquoise-, green- and white-flowered *L. vanzyliae* but only one natural hybrid has been recorded, in 1971 (*Thomas s.n.*, in NBG). The tepals age to dull red and the developing fruits re-orientate themselves to a spreading or suberect position. The scapes remain attached to the bulbs following capsule dehiscence and the seeds are dispersed locally.

15. LACHENALIA BIFOLIA

Lachenalia bifolia (Burm. *f.*) W. F. Barker ex G. D. Duncan, comb. nov., non *L. bifolia* Ker Gawl., *Curtis's Botanical Magazine* 40: t.1611 (1814). *Aletris bifolia* Burm.*f.*, *Prodromus Florae Capensis*: 10 (1768).
TYPE: South Africa, Cape, collector and precise locality unknown, figure in J. C. Buxbaum, *Plantarum minus cognitarum, Centuria* 3: t. 19 (1729) (neotype, designated here).
SYNONYMY: *Lachenalia bulbifera* (Cirillo) Engl., *Notizblatt des Koniglichen Botanischen Gartens und Museums zu Berlin* 2 (18): 321 (1899). *Phormium bulbiferum* Cirillo, *Plantarum Rariorum Regni Neapolitani* 1:

35, t. 12 (1788). Type: South Africa, Cape, collector and precise locality unknown, figure in D. Cirillo, *Plantarum Rariorum Regni Neapolitani* 1: 35, t. 12 (1788) (lectotype, designated here).

Lachenalia pendula Aiton, *Hortus Kewensis* 1: 461–462 (1789). Type: South Africa, Cape, precise locality unknown, *F. Masson s.n.* (BM!, holo.).

Lachenalia linguiformis Lam., *Encyclopédie Méthodique Botanique* 3 (2): 372–373 (1792). Type: South Africa, Cape, collector and precise locality unknown, figure in J-P. Buc'hoz, *Histoire Universelle Règne Végétal* 2: t. 1 (1773–1778) (lectotype, designated here).

ETYMOLOGY. *bifolia*: having two leaves.

DESCRIPTION. *Geophyte*, 100–300 mm high. *Bulb* subglobose, 10–35 mm in diameter, offset-forming; tunic multilayered, outer layers dark brown, spongy; inner cataphyll translucent white in lower two thirds, tinged with dull magenta in upper third, adhering to leaf bases, apex obtuse. *Leaves* 1–2, lanceolate, 60–200 × 15–90 mm, spreading to suberect, fleshy, slightly to deeply canaliculate, bright to dark green, yellowish-green or glaucous, upper surface with faint depressed longitudinal grooves, plain or heavily marked with few to many dark green, brown or brownish-purple spots or blotches, lower surface plain or tinged with dull maroon; leaf bases clasping, 15–80 mm long, lower portion white, upper portion light green; primary seedling leaf terete, erect. *Inflorescence* racemose, few- to many-flowered, lax to sturdy, sterile apex short; scape erect to suberect, 50–150 mm long, slender to sturdy, light green to greyish-brown, plain or with small to large purplish-brown spots or blotches; rachis uniformly light green, dull red or orange, often light green in lower two thirds, shading to dull red or orange in upper third; pedicels suberect to cernuous, 4–7 mm long, light green, brownish-green, purplish-brown or dull red; bracts ovate, 1–6 × 2–3 mm, translucent white, sometimes tinged with magenta. *Perianth* zygomorphic, tubular, pendulous; tube cylindrical, 6–11 mm long; light to dark orange-red, light to dark red or vermillion; outer tepals oblong, 13–32 × 4–7 mm, light to dark orange-red, plain or mottled with darker orange, light to dark red or vermillion; apical gibbosities light to dark green, greenish-yellow or brown, uppermost outer tepal gibbosity usually conspicuously protruding; inner tepals oblong obovate, 23–37 × 8–11 mm, overlapping, translucent light yellow, orange-red or vermillion, upper margins with broad, dark purple to purplish-magenta zones, apices purplish or green, apical gibbosities prominent, light to bright green, dark magenta or purple, median keels light magenta. *Stamens* included, straight; filaments white, 20–39 mm long; pollen yellow at anthesis, ageing to dull purple. *Ovary* ellipsoid, 4–5 × 2–3 mm, bright green; style included, straight, 25–42 mm long, white, protruding beyond tepals as ovary matures. *Capsule* ellipsoid, 10–20 × 7–10 mm. *Seed* globose, 1.3–1.4 × 1.3–1.4 mm; glossy, black; strophiole inflated, 1.4–1.6 mm long, smooth; raphe translucent, white, joined to strophiole. *Chromosome number* 2n = 14, 56 (Crosby, 1986); 2n = 28, 42 (Moffett, 1936); 2n = 42 (Hamatani *et al.*, 1998; Spies *et al.*, 2008); 2n = 42, 56 (Johnson & Brandham, 1997); 2n = 28, 42, 49, 56 (Kleynhans & Spies, 1999). Figures 21, 26, 27, 106–109.

FLOWERING PERIOD. April to September, with a peak from May to early August.

HISTORY. The species was first described by the phrase name *Orchis hyacinthoides, monophyllos flore coccineo* by the German J. C. Buxbaum (1729) in his *Plantarum minus cognitarum*, and this description was accompanied by a monochrome figure of a stylised, but unmistakeable, flowering plant with a single leaf. Following the binomial system of classification (Linnaeus, 1753), the taxon was later published as *Aletris bifolia* by N. L. Burmann (1768) in his *Prodromus Florae Capensis*. Several years passed before it was again illustrated in monochrome, as *A. bifolia*, in *Histoire Universelle du Règne Vegetal*, by the French physician and naturalist Pierre-Joseph Buc'hoz (Buc'hoz, 1773–1778). It was

Figure 106 (left). Early-flowering form of *Lachenalia bifolia* from Gansbaai. Image: Graham Duncan.
Figure 107 (right). Short-flowered form of *Lachenalia bifolia* from Cape Agulhas. Image: Graham Duncan.

subsequently described in great detail and illustrated in monochrome under the genus *Phormium* as *P. bulbiferum* by the Italian physician and naturalist Domenico Cirillo (1739–1799) in his *Plantarum Rariorum Regni Neapolitani* (Cirillo, 1788). Introduced into cultivation at Kew by Francis Masson in 1774 (Aiton, 1811), it was described under *Lachenalia* first in 1789 and then again in 1792. William Aiton (1789) published it as *L. pendula* in the first volume of *Hortus Kewensis*, apparently unaware that Cirillo had published his own name for the same species the year before. Lamarck (1792) published it as *L. linguiformis*, citing *Aletris bifolia* Burm.f. under it and, mistakenly, using the phrase name *Orchis hyacinthoides, foliis caule & floribus maculates*, which is the species now known as *L. punctata*. More than a century passed before Engler (1899) placed *L. pendula* in synonymy under *L. bulbifera* in the German periodical *Notizblatt des Koniglichen Botanischen Gartens und Museums zu Berlin*. Yet for more than half of the 20th century, the plant continued to be known as *L. pendula* until Ingram (1966) drew attention to the German publication *Synopsis der Mitteleuropäischen Flora* (Ascherson & Graebner, 1905) in which it was listed as *L. bulbifera* (Cirillo) Hort. ex Ascherson & Graebner. Following Ingram's publication, the latter authors were regarded as the legitimate authors of the species until, finally, W. F. Barker drew attention to Engler's 1899 publication and authorship of the species, changing the name once again, to *L. bulbifera* (Cirillo) Engl. In accordance with the *International Code of Botanical Nomenclature*, however, the earliest valid specific epithet available for the plant is N. L. Burmann's (1768) *bifolia*, which W. F. Barker had intended to publish.

15. LACHENALIA BIFOLIA

DISTINGUISHING CHARACTERS AND AFFINITIES. One of the most robust members of the genus, *L. bifolia* is highly variable in inflorescence length, tepal colour and length, capsule size and flowering time, and it forms a polyploid complex with a basic chromosome number of $x = 7$ (see 'Karyology', p. 86). Chromosomal divergence has taken place, but the morphological expression of these changes is still in progress and it is not yet possible to assign taxonomic rank to morphological variants. Collectively, the species has the longest flowering period of all of the species of *Lachenalia*, extending from late April in a short-flowered, orange form from Stilbaai, to early September in a long-flowered, pinkish-red form from Bredasdorp, both in the southern Cape; the longest-flowered forms occur on the Cape west coast and Cape Flats.

L. bifolia is sister to *L. punctata* (Duncan *et al.*, 2005a), with which it shares pendulous, tubular to cernuous flowers with long perianth tubes, oblong outer tepals, oblong obovate inner tepals and globose seeds with inflated strophioles. *L. punctata* is, however, usually a smaller plant with pink, or pink-mottled perianths borne on shorter pedicels (1–4 mm long). With respect to perianth morphology, the lower inner tepal of *L. punctata* is distinctly shorter than its upper inner tepals. It also differs in its globose bulb, in having partially hysteranthous, usually glaucous leaves, a declinate infructescence, shorter strophioles (0.8 mm long) and prostrate, flat primary seedling leaves.

Figure 108 (left). *Lachenalia bifolia* in deep sand at Atlantis. Image: Graham Duncan.
Figure 109 (right). Cultivated specimens of *Lachenalia bifolia* from Koeberg. Image: Graham Duncan.

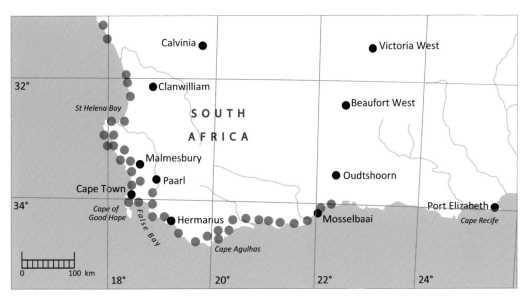

Map 16. Known distribution of *Lachenalia bifolia*.

DISTRIBUTION, HABITAT AND CONSERVATION STATUS. *L. bifolia* has the second-longest coastal distribution of all *Lachenalia* species after *L. punctata*, occurring from Brand-se-Baai west of Bitterfontein in the Northern Cape to George in the southern Cape (Map 16). It is associated primarily with deep sandy, acid or alkaline soils on flats and primary or secondary dunes. It is often encountered just above the high water mark. Populations occur along the margins of coastal dune forest, less frequently in humus in crevices and shallow pans of granite outcrops. *L. bifolia* spans 11 coastal vegetation types: Namaqualand Coastal Duneveld, Namaqualand and Saldanha Flats Strandveld, Saldanha Granite Strandveld, Langebaan-, Cape Flats-, Overberg- and Groot Brak Dune Strandveld, Cape Flats Sand Fynbos, Cape Seashore Vegetation and Blombos Strandveld (Mucina & Rutherford, 2006). It grows in small to large colonies, exposed, in light shade of granite boulders or within the protection of surrounding low shrubby growth. It often occurs in close proximity to populations of *L. punctata*, which flower earlier in the season. The flowering periods of populations in the southern and eastern (early-flowering) parts of the species' range differ markedly from those of populations in the western (mid-season to late-flowering) parts. Numbers of *L. bifolia* plants have been greatly reduced by coastal housing developments, but the species is not considered threatened because of its wide distribution.

NOTES. The species is adapted to pollination by sunbirds (in particular Southern Double-collared Sunbirds (*Cinnyris chalybeus*) and has scapes that are sufficiently sturdy to support their weight. Despite the absence of temperate sunbirds, *L. bifolia* has become naturalised in parts of western and southern Australia, where it is considered to be a weed. Unlike its close relative *L. punctata*, whose scapes change orientation and bend downwards during the fruiting stage, *L. bifolia* has scapes that remain erect to suberect and its seeds are released locally from the ripe capsules by the shaking action of wind. Considerable variation in tepal length and colour (from light orange to deep red) occurs within certain populations of *L. bifolia*, such as those near Atlantis on the Cape west coast. Because *L. bifolia* is the only species that includes red-flowered forms, it has been used extensively in breeding programmes to develop new pot plant cultivars in South Africa.

16. LACHENALIA PUNCTATA

Lachenalia punctata Jacq., *Collectanea* 2: 323 (1788). *Lachenalia rubida* Jacq. var. *punctata* (Jacq.) Baker, in Thiselton-Dyer, W. T. (ed.), *Flora Capensis* 6: 424 (1897).
TYPE: South Africa, Cape, collector and precise locality unknown, figure in N. J. Jacquin, *Icones Plantarum Rariorum* 2: 6, t. 397 (1790) (neotype, designated here).
SYNONYMY: *Aletris linguaeformis* Burm.f., *Prodromus Florae Capensis*: 10 (1768). Type: South Africa, Cape, collector and precise locality unknown, figure in J. C. Buxbaum, *Plantarum minus cognitarum*, *Centuria* 3: t. 20 (1729) (lectotype, designated here).
Lachenalia rubida Jacq., *Collectanea* 5: 60–61 (1797). Type: South Africa, Cape, collector and precise locality unknown, figure in N. J. Jacquin, *Icones Plantarum Rariorum* 2: 15, t. 398 (1793 or 1794), (neotype, designated here).
Lachenalia tigrina Jacq., *Collectanea* 5: 67–68 (1797). *Lachenalia rubida* Jacq. var. *tigrina* (Jacq.) Baker, in Thiselton-Dyer, W. T. (ed.), *Flora Capensis* 6: 424 (1897). Type: South Africa, Cape, collector and precise locality unknown, figure in N. J. Jacquin, *Icones Plantarum Rariorum* 2: 16, t. 399 (1795) (neotype, designated here).

ETYMOLOGY. *punctata*: spotted, with reference to the pinkish-mauve spots on the outer tepals.
DESCRIPTION. *Geophyte*, 60–250 mm high. *Bulb* globose, 10–20 mm in diameter, offset-forming, often deep-seated; tunic multilayered, outer layers light to dark brown, spongy; inner cataphyll translucent white, loosely surrounding leaf bases, apex obtuse. *Leaves* (1)–2, lanceolate, 30–140 × 12–30 mm, partially hysteranthous, erect to suberect or spreading, upper surface glaucous or light to dark green, plain to heavily marked with small to large deep purple or green spots or blotches; lower surface plain or flushed with dull purple, plain or with small to large dull purple or green spots or blotches; leaf bases clasping, 30–70 mm long, lower portion white with light magenta spots, upper portion yellowish-green, plain or purple-spotted; primary seedling leaf flat, prostrate. *Inflorescence* racemose, few- to many-flowered, sterile apex short; scape erect to suberect during flowering stage, usually bending into a horizontal position during fruiting stage, 30–100 mm long, slender to sturdy, light- to yellowish-green with dull purple blotches, or maroonish- to brownish-pink with darker markings; rachis colour as for scape; pedicels suberect, 1–4 mm long, white, light green, pink or brownish-pink; bracts ovate, 1–3 × 3–4 mm, white, apices recurved. *Perianth* zygomorphic, tubular, slightly to distinctly curved, pendulous to cernuous; tube cylindrical, 6–9 mm long, light to bright pink or mottled with pinkish-mauve, rarely light yellow; outer tepals oblong, 17–27 × 4–6 mm, light to bright pink, deep pink to maroonish-pink, or light yellow heavily spotted with pinkish-mauve, apical gibbosities light green, yellowish-pink or brown, upper tepal slightly longer than lower tepals; inner tepals oblong-obovate, 23–33 × 6–9 mm, apices plain or blotched light yellow, margins plain or dark purple, upper two tepals overlapping, lower portion translucent white, upper third heavily marked with light to dark pink, rarely yellow, median keels deep pink, rarely yellow. *Stamens* included to shortly exserted, straight; filaments white, 22–31 mm long. *Ovary* ellipsoid, 3–5 × 2–3 mm, light green or yellow; style shortly- to well exserted, straight, 25–31 mm long, white, apex shading to purple below stigma. *Capsule* ellipsoid, 13–15 × 6–7 mm, infructescence resting near ground level. *Seed* globose, 1.3–1.4 × 1.2–1.4 mm, glossy, black; raphe inflated, translucent white; strophiole inflated, 0.8 mm long, smooth. *Chromosome number* $2n = 14$ (Moffett, 1936; Hamatani *et al.*, 1998); $2n = 14, 28$ (Crosby, 1986). Plate 9, Figures 35, 110–112.
FLOWERING PERIOD. Late March to July, with a peak in May.

THE GENUS LACHENALIA | **161**
16. LACHENALIA PUNCTATA

Plate 9. Watercolour painting of *Lachenalia punctata*, courtesy of the Compton Herbarium, Kirstenbosch, South African National Biodiversity Institute. Artist: Winsome Barker.

16. LACHENALIA PUNCTATA

HISTORY. The earliest published description and illustration appeared in 1729 in volume three of J. C. Buxbaum's *Plantarum minus cognitarum Centuria*, under the phrase name *Orchis hyacinthoides, foliis, caule & floribus maculatis*, illustrated on plate 20 by a stylised fleshy plant with two emerging leaves, with the flowers finely spotted all over. N. L. Burmann (1768) named the plant *Aletris linguaeformis* in *Prodromus Florae Capensis*, citing Buxbaum's plate. Although the earliest applicable name for this species is Burmann's *A. linguaeformis*, in accordance with the *International Code of Botanical Nomenclature*, it cannot be transferred to *Lachenalia* because of Lamarck's *L. linguiformis* Lam. (=*L. bifolia* (Burm.f.) W. F. Barker ex G. D. Duncan), as the two specific epithets are regarded as orthographic variants.

Three *Lachenalia* taxa (*L. punctata*, *L. rubida* and *L. tigrina*) described and illustrated by N. J. Jacquin in the late 18th century are synonymous with Burmann's *Aletris linguaeformis*. *L. punctata* was described in 1788 and illustrated in 1790, *L. rubida* was described in 1797 and illustrated in 1793 or 1794, and *L. tigrina* was described in 1797 and illustrated in 1795. The next earliest available epithet for the species is therefore Jacquin's *L. punctata*. In Baker's first monograph of the genus (1871), *L. punctata* and *L. tigrina* were relegated to varietal status under *L. rubida* as *L. rubida* var. *punctata* and *L. rubida* var. *tigrina*, respectively, a position he maintained in his second and final monograph (Baker, 1897a). The species was introduced into cultivation at Kew by Mr Richard Williams of Turnham Green, London, in about 1803 (Aiton, 1811).

DISTINGUISHING CHARACTERS AND AFFINITIES. *L. punctata* is sister to *L. bifolia* (Duncan et al., 2005a) and the two species are frequently confused as they share similar tubular, pendulous to cernuous flowers with long perianth tubes, oblong outer tepals, oblong obovate inner tepals and globose seeds with inflated strophioles. *L. bifolia* differs in being more robust with straight perianths in which the outer tepals are almost as long as the inner tepals or protrude slightly beyond them,

Figure 110. *Lachenalia punctata* on coastal sand dunes, Saldanha Bay. Image: Graham Duncan.

Figure 111 (left). *Lachenalia punctata* amongst sandstone rocks near Kommetjie. Image: Graham Duncan.
Figure 112 (right). *Lachenalia punctata* with visiting ants, on sandy flats near Gansbaai. Image: Graham Duncan.

and tepal colour varies from light to dark orange-red, red or vermillion, the inner tepals with purplish-magenta and green apices. *L. bifolia* also differs in its intensely orange or red rachis apex, its subglobose bulb and in its often broadly lanceolate leaves, which are usually well developed at flowering. Most forms of *L. bifolia* flower later than *L. punctata* and its seed has a much more prominent strophiole of 1.4–1.6 mm in length.

DISTRIBUTION, HABITAT AND CONSERVATION STATUS. *L. punctata* is a predominantly coastal species with an extensive range from Rooivlei south of Kleinsee in western Namaqualand to Herold's Bay south of George in the southern Western Cape, extending inland in this province to Vanrhynsdorp, Clanwilliam, the Botterkloof Pass between Clanwilliam and Calvinia, and Piketberg (Map 17). Its habitat varies in the Succulent Karoo and Fynbos Biomes; it is mostly encountered in very sandy soils, on flats and primary and secondary dunes, less commonly on hill and mountain slopes. It occurs in 16 vegetation types: Namaqualand Coastal Duneveld, Namaqualand Klipkoppe Shrubland, Namaqualand Strandveld, Van Rhynsdorp Gannabosveld, Bokkeveld-, Piketberg- and Peninsula Sandstone Fynbos, Citrusdal Vygieveld, Lamberts Bay Strandveld, Saldanha Flats Strandveld, Langebaan Dune Strandveld, Cape Flats Sand Fynbos, Cape Flats Dune Strandveld, Overberg Dune Strandveld, Blombos Strandveld and Groot Brak Dune Strandveld (Mucina & Rutherford, 2006).

Along the Namaqualand coastal plain and inland to Vredendal and Vanrhynsdorp, *L. punctata* occurs in deep red sand amongst low succulent vegetation, and at the base of the Matsikammaberg, it grows in brownish-yellow sand within the protection of restio clumps. On the Cape west coast at Saldanha, it occurs in strongly alkaline, deep white sand amongst strandveld vegetation, and at nearby Cape Columbine, in coarse alkaline sand between cracks of granite boulders, within reach of the sea spray. In the southern Cape Peninsula, *L. punctata* grows in fynbos on acid sandy flats and hill slopes, as well as in acid humus in the cracks of sandstone boulders. At Baardscheerdersbos in the southern Cape, it frequents sandy places on limestone hills, and at Herold's Bay near George, it grows in sandy depressions around coastal forest patches. The plants occur as solitary individuals or in small groups within larger colonies. The bulbs are usually deep-seated, sometimes buried to 170 mm. Like *L. bifolia*, with which it sometimes grows, the numbers of *L. punctata* have been greatly reduced as a result of coastal housing development, but it is not regarded as threatened because of its wide distribution.

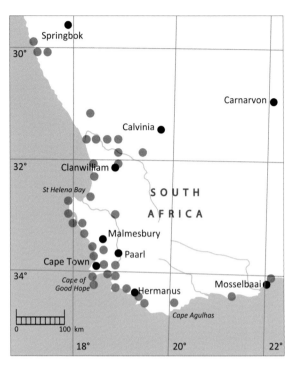

Map 17. Known distribution of *Lachenalia punctata*.

NOTES. The long tubular flowers of *L. punctata* with their long perianth tubes are specialised for pollination by sunbirds, and Southern Double-collared Sunbirds (*Cinnyris chalybeus*) often visit. The slender scapes of this species are not strong enough to support the weight of a sunbird (unlike those of *L. bifolia*), and so the birds must perch on the ground or on surrounding vegetation in order to reach the nectar. The infructescence is unusual in that the weight of the unripe fruits results in the scape bending over and resting on the ground. The scape desiccates and detaches from the bulb before the seeds are fully ripe, thereafter the seeds are shed directly onto the ground and probably dispersed by ants.

The first species of *Lachenalia* to be phytochemically characterised was a collection of *L. punctata* from Bloubergstrand north of Cape Town (*Duncan* 442, in NBG), from which a novel 3-benzyl-chromone was isolated. Although homoisoflavonoids with a 3, 9-double bond are relatively common in the Hyacinthoideae *s.l.*, this was the first report of the isolation of a homoisoflavonoid with a 2, 3-double bond from the Hyacinthaceae *s.l.* (Langlois *et al.*, 2005). A parallel phytochemical investigation of bulbs of the sister taxon *L. bifolia* did not reveal any homoisoflavanones (T. Pohl, 2001, pers. comm.), indicating that homoisoflavonoids with a 2, 3-double bond are not definitive markers for *Lachenalia*.

17. LACHENALIA UNIFOLIA

Lachenalia unifolia Jacq., *Plantarum rariorum horti caesarei Schönbrunnensis* 1: 43, t. 83 (1797). *Scillopsis unifolia* (Jacq.) Lem., *L'Illustration Horticole* 3: 34 (1856).
TYPE: South Africa, Cape, collector and precise locality unknown, figure in *Plantarum rariorum horti caesarei Schönbrunnensis* 1: 43, t. 83 (1797) (lectotype, designated here).

ETYMOLOGY. *unifolia*: single-leafed.
DESCRIPTION. *Geophyte*, 80–350 mm high. *Bulb* subglobose, 12–20 mm in diameter, offset-forming; tunic multilayered, outer layers dark brown, spongy; inner cataphyll translucent white, adhering to leaf base, apex obtuse. *Leaf* solitary, linear, 80–320 × 5–12 mm, suberect, light to dark green, canaliculate, sometimes conduplicate, upper surface plain, lower surface banded with darker green in lower half; leaf base clasping, 55–70 mm long, white, upper part banded with brownish-maroon, subterranean part banded with magenta bands; primary seedling leaf terete, erect. *Inflorescence* racemose, few- to many-flowered, sterile apex short; scape erect to suberect, 100–300 mm long, light green, plain or lightly to heavily mottled with light to dark brownish-maroon; rachis light green, lower half with dull brown or purple mottling, shading to light blue, pink or white above; pedicels

Figure 113 (left). *Lachenalia unifolia* on a south-facing hill slope near Clanwilliam. Image: Graham Duncan.
Figure 114 (right). Cultivated specimens of *Lachenalia unifolia* from Gouda. Image: Graham Duncan.

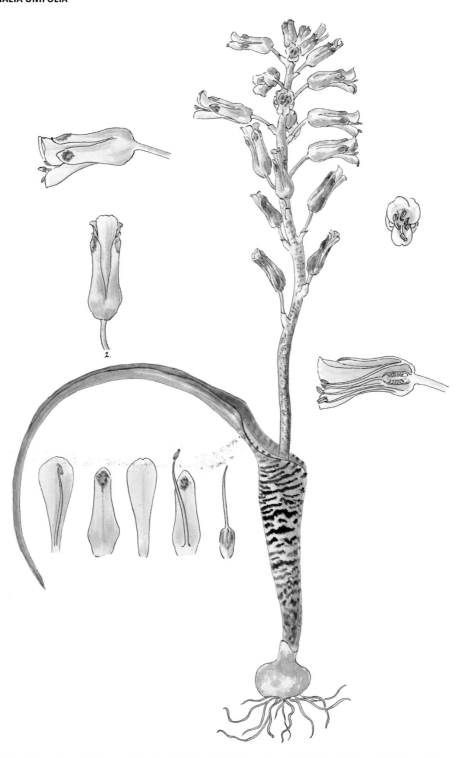

Plate 10. Watercolour painting of *Lachenalia unifolia*, courtesy of the Compton Herbarium, Kirstenbosch, South African National Biodiversity Institute. Artist: Winsome Barker.

suberect at anthesis, 4–15 mm long, increasing in length towards inflorescence apex, white or light green, plain or mottled with brownish-maroon, usually becoming erect in fruit; lower bracts ovate, becoming lanceolate above, 1–3 × 1–3 mm, translucent white or dark brown with white margins. *Perianth* zygomorphic, tubular, spreading to cernuous; tube cylindrical, 6 mm long, white or light to deep blue; outer tepals oblong, 8–9 × 4–5 mm, lower half light to dark blue or pink, upper half white, plain to heavily marked with purplish blotches, apical gibbosity small to large, light to dark green or brown; inner tepals oblong-obovate, 10–16 × 5–6 mm, light blue in lower third, shading to white or yellowish-green in upper two thirds, apices slightly recurved, keels light blue in lower half, shading to light blue or green above, upper two tepals overlapping, lower inner tepal canaliculate, sometimes 1–2 mm longer. *Stamens* included, rarely shortly exserted, straight; filaments white, 11–15 mm long. *Ovary* ellipsoid, 5–6 × 2–4 mm, light green; style included, 9–12 mm long, straight, white. *Capsule* ellipsoid, 9–11 × 4–6 mm, spreading to suberect. *Seed* globose, 0.9–1.0 × 0.9–1.0 mm, glossy, black; strophiole

Figure 115. Cultivated specimens of *Lachenalia unifolia* from Potsdam. Image: Graham Duncan.

rudimentary, 0.1 mm long, ridged. *Chromosome number* 2n = 21, 22, 24, 26 (de Wet, 1957); 2n = 22 (Moffett, 1936; Crosby, 1986; Van Rooyen *et al.*, 2002); 2n = 24 (Hamatani *et al.*, 2004); 2n = 22, 44 (Johnson & Brandham, 1997). Plate 10, Figures 113–115.

FLOWERING PERIOD. August to October, with a peak from mid- to late September.

HISTORY. *L. unifolia* was introduced into cultivation at Kew by Francis Masson in 1795 (Aiton, 1811) and the species was described and illustrated two years later by N. J. Jacquin in the first volume of *Plantarum rariorum horti Schönbrunnensis Descriptiones et Icones* (Jacquin, 1797b). The holotype was lost in a fire in Vienna between 1939 and 1945, and so it has been necessary to designate the colour plate in Jacquin (1797b) as the lectotype. Lemaire (1856) transferred the species to *Scillopsis* as *S. unifolia* (Jacq.) Lem. in *L'Illustration Horticole*, but it was returned to its original name by Baker (1871, 1897a). Baker (1878) described *L. wrightii* in the *Journal of Botany* but later re-assessed it as *L. unifolia* var. *wrightii* in his second monograph (Baker, 1897a). He also described *L. schlechteri* in *Bulletin L'Herbier Boissier* (Baker, 1904), later re-assessed as *L. unifolia* var. *schlechteri* by W. F. Barker (1989) in the *South African Journal of Botany*. Both var. *wrightii* and var. *schlechteri* are here returned to species level as they form distinct evolutionary units and breed true from seed.

DISTINGUISHING CHARACTERS AND AFFINITIES. *L. unifolia* is a variable but distinctive species. The length of the lower inner tepals is variable within some populations: in coastal sandy areas, for example at Mamre, the inner tepals are usually equal, with occasional specimens having slightly longer lower inner tepals; in inland areas in stony clay at Clanwilliam and Gouda, the lower inner tepal is

17. LACHENALIA UNIFOLIA

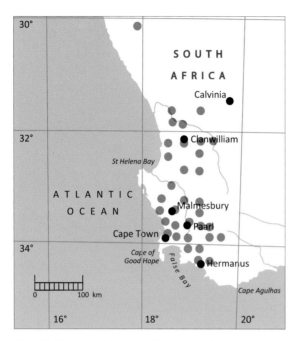

Map 18. Known distribution of *Lachenalia unifolia*.

usually longer than the upper inner tepals, with some individuals having inner tepals that are all the same length. Duncan *et al.* (2005a) considered *L. schlechteri* as sister species of *L. unifolia*, as the two species share solitary linear, canaliculate, banded leaves, usually blue tubular flowers, inner tepals that protrude well beyond the outer tepals, included, straight stamens and globose seeds with rudimentary strophioles. *L. schlechteri* differs in its shorter pedicels (1–2 mm, as opposed to 4–15 mm), suberect perianth (as opposed to a spreading or cernuous perianth), longer perianth tube (9 mm, as opposed to 6 mm), and shorter filaments (9–10 mm, as opposed to 11–15 mm).

DISTRIBUTION, HABITAT AND CONSERVATION STATUS. *L. unifolia* is one of the most common species of *Lachenalia* and is widely distributed within the winter rainfall zone of the Northern and Western Cape, occurring from Kamieskroon in southern Namaqualand to Stanford on the south coast (Map 18). Its habitat is very variable and it traverses 11 vegetation types including Kamiesberg Mountains Shrubland, Bokkeveld-, Graafwater- and Olifants Sandstone Fynbos, Agter-Sederberge Shrubland, Hopefield-, Atlantis- and Cape Flats Sand Fynbos, Swartland Granite- and Breede Shale Renosterveld, and Overberg Dune Strandveld (Mucina & Rutherford, 2006). In coastal areas, *L. unifolia* is usually encountered on flats in deep white sand; whereas inland, it favours lower hill slopes and flats of heavy, stony clay. The plants occur singly, in small groups or dense colonies. At a locality south of Tulbagh, a light blue form grows in large numbers in stony clay, in association with spring-flowering geophytes including *Babiana villosa*, *Geissorhiza inflexa*, *Ixia vinacea* and *Sparaxis grandiflora* (Iridaceae); in the southern Tanqua Karoo, a form that occurs in deep sand on the west coast and in heavy clay south of Tulbagh, grows in association with *L. comptonii* on sandy clay flats. *L. unifolia* is not threatened.

NOTES. The species is pollinated by honey bees (*Apis mellifera*). During the fruiting stage, the pedicels and ripening capsules usually become re-oriented to an erect position. Following capsule dehiscence, the scapes remain attached for several weeks and the seeds are dispersed locally.

18. LACHENALIA SCHLECHTERI

Lachenalia schlechteri Baker, *Bulletin de L'Herbier Boissier* (series 2) 4: 999 (1904). *Lachenalia unifolia* Jacq. var. *schlechteri* (Baker) W. F. Barker, *South African Journal of Botany* 55 (6): 643–644 (1989).
TYPE: South Africa, Western Cape, Piquetberg Road (= Gouda), *F. R. R. Schlechter* 4855 (Z!, holo.).

ETYMOLOGY. *schlechteri*: after Rudolf Schlechter (1872–1925), German botanist and traveller.
DESCRIPTION. *Geophyte*, 120–430 mm high. *Bulb* subglobose, 15–20 mm in diameter, offset-forming; tunic multilayered, outer layers dark brown, spongy; inner cataphyll translucent white, adhering to leaf base, apex obtuse. *Leaf* solitary, linear, 165–310 × 10–12 mm, suberect, light to dark green,

canaliculate, sometimes conduplicate, upper surface plain, lower surface banded with darker green in lower half; leaf base clasping, 55–70 mm long, white, upper part banded with brownish-maroon, subterranean part banded with magenta; primary seedling leaf terete, erect. *Inflorescence* racemose, few-flowered, sterile apex up to 10 mm long; scape erect to suberect, 100–300 mm long, light green, plain or lightly to heavily mottled with light to dark brownish-maroon; rachis light green, lower half with dull brown or purple mottling, shading to light blue or white above; pedicels suberect, 1–2 mm long, white; lower bracts ovate, becoming lanceolate above, 1–3 × 1–3 mm, translucent white. *Perianth* zygomorphic, tubular, suberect; tube cylindrical, 9 mm long, white with light to deep blue blotches; outer tepals oblong, 9–10 × 4–5 mm, white with light to deep blue blotches; apical gibbosities light to dark green or brown; inner tepals oblong obovate, 10–13 × 5–6 mm, light blue below, shading to white or yellowish-green above, apices slightly recurved; keels light blue below, shading to yellowish-green above; upper tepals overlapping, lower inner tepal canaliculate, 1–3 mm longer. *Stamens* included, straight; filaments white, 9–10 mm long. *Ovary* ellipsoid, 6 × 2 mm, light green; style included, 11–12 mm long, straight, white. *Capsule* ellipsoid, 6–11 × 4–6 mm, suberect. *Seed* globose, 0.9 × 0.9 mm, glossy, black; strophiole rudimentary, 0.1 mm long, ridged. *Chromosome number* 2n = 22 (Johnson & Brandham, 1997). Figure 116.

FLOWERING PERIOD. August to September, with a peak in mid-August.

HISTORY. *L. schlechteri* was described from two specimens collected by Rudolf Schlechter at Gouda in the south-western Cape, and the holotype is housed in the Botanical Museum at the University of Zurich (Baker, 1904). More than 80 years later, the taxon was re-assessed as a variety of *L. unifolia*

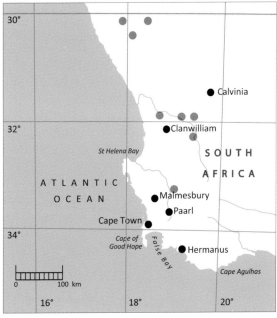

Map 19. Known distribution of *Lachenalia schlechteri*.

Figure 116. Cultivated specimen of *Lachenalia schlechteri* from Kamieskroon. Image: Graham Duncan/SANBI.

(Barker, 1989). The taxon is morpohologically distinct, forms a distinct evolutionary unit, and is returned to species status here.

DISTINGUISHING CHARACTERS AND AFFINITIES. The species is sister to *L. unifolia* (Duncan *et al.*, 2005a), and the relationship is discussed under that species.

DISTRIBUTION, HABITAT AND CONSERVATION STATUS. *L. schlechteri* has a disjunct distribution, occurring on the Kamiesberg in southern Namaqualand, south, east and south-east of Klawer in the Olifants River Valley, and at Gouda in the south-western Cape (Map 19). Its habitat is variable and it is encountered in a variety of vegetation types including Namaqualand Granite Renosterveld, Klawer Sandy- and Agter-Sederberg Shrubland, and Tanqua Karoo and Breede Shale Renosterveld (Mucina & Rutherford, 2006). The plants occur as solitary individuals or in small groups in stony clay or deep sandy soils. The taxon is not threatened.

NOTES. The flowers are pollinated by honey bees (*Apis mellifera*). The scapes remain attached for several weeks following capsule dehiscence and the seeds are locally dispersed by wind.

19. LACHENALIA BRUYNSII

Lachenalia bruynsii G. D. Duncan, sp. nov.

A L. hirta floribus tublatibus longioribus, tepalis exterioribus oblongis, tepalis interioribus oblongis-obovatus cum tepalo interiore infero canaliculato et multo longiore quam tepalis interioribus aliis, filis multo longioribus, capsula ellipsoidea, semini parviore differt.

TYPE: South Africa, Northern Cape, 3119 (Calvinia): 11 km from Moedverloor turnoff on R364 from Clanwilliam to Calvinia (–CC), *G. D. Duncan* 557 (NBG!, holo.).

ETYMOLOGY. *bruynsii*: after Dr P. V. Bruyns (1957–), who made the first documented collection of plants.

DESCRIPTION. *Geophyte*, 100–180 mm high. *Bulb* globose, 10–15 mm in diameter, solitary, light yellow; tunic multilayered, outer layers dark brown, spongy; inner cataphyll translucent white, loosely surrounding leaf base, apex obtuse. *Leaf* solitary, lanceolate, 45–80 × 10–12 mm, widening abruptly at base, spreading, deeply canaliculate, upper surface dark green, glabrous, lower surface dark green in upper half with green transverse bands, lower half white with dark magenta bands, covered with short simple trichomes, midrib distinct; leaf base loosely surrounding base of scape, 20–40 mm long, white, heavily banded with light to dark magenta; leaf margins heavily undulate and crisped for entire length, covered with long green or dark magenta simple trichomes up to 4 mm long; primary seedling leaf terete, suberect. *Inflorescence* racemose, few- to many-flowered, sterile apex long; scape suberect, 40–100 mm long, fairly rigid, slender, light green heavily mottled with maroon, rarely with dark magenta bands at base; lower bracts ovate, becoming lanceolate above, 2–3 × 1–3 mm, white, minutely spotted with dark magenta; rachis light green below, heavily mottled with dark magenta, white to light blue above; pedicels suberect, 3–5 mm long, white, maturing to dull pinkish-red. *Perianth* zygomorphic, tubular, cernuous, ageing to dull pinkish-red; tube cylindrical, 6 mm long, light blue; outer tepals oblong, 10–12 × 4–5 mm, light blue at base, shading to light pink above, apices white, apical gibbosities dark brown; inner tepals oblong obovate, 13–15 × 5–6 mm, light yellowish-white, median keels green, apical marking brown, upper two tepals overlapping, lower tepal 2–3 mm longer, canaliculate. *Stamens* included, straight; filaments white, 13–15 mm long. *Ovary* ellipsoid, 5–6 × 2–3 mm, light green; style included, straight, 10–11 mm long, white, protruding beyond perianth as ovary matures. *Capsule* ellipsoid, 9–10 × 4–5 mm. *Seed* globose, 1.1–1.2 × 0.9–1.0 mm, glossy, black; strophiole rudimentary, 0.1 mm long, ridged. *Chromosome number* unknown. Figure 117.

FLOWERING PERIOD. August to September, with a peak in mid-August.

HISTORY. *L. bruynsii* was illustrated for the first time by Barbara Jeppe, inadvertently as *L. hirta*, on plate 16d of *Spring and Winter Flowering Bulbs of the Cape*, from material of unknown provenance that had been cultivated at Rust-en-Vrede Nursery in Cape Town (Jeppe & Duncan, 1989). Plants were collected by Dr Peter Bruyns north-east of Wupperthal on 17th August 2002, and his pressed specimens (*Bruyns* 9190) were deposited at NBG. In 2004, Gordon Summerfield found the species at a locality some distance north of the Pakhuis Pass, and I visited the site with Summerfield in August 2007 to make the type collection.

DISTINGUISHING CHARACTERS AND AFFINITIES. *L. bruynsii* was not included in the generic phylogeny of Duncan *et al*. (2005a) but appears to be closely related to *L. hirta*, which has a similar leaf and globose bulb with light yellow tissue, but which differs in its much shorter, shallowly cup-shaped perianth tubes (1 mm long), its oblong- campanulate flowers with much shorter, subequal outer and inner tepals (6–8 mm long), and its obovoid capsules and slightly smaller, matt black seeds (0.9 × 0.8–0.9 mm).

DISTRIBUTION, HABITAT AND CONSERVATION STATUS. The species is only recorded from two collections, one from north of the Pakhuis Pass in the south-western Northern Cape and the other from north-east of Wupperthal in the western part of the Western Cape (Map 20). The plants grow as scattered, solitary individuals or in small groups of up to three plants. They occur in Bokkeveld Sandstone Fynbos vegetation (Mucina & Rutherford, 2006) and are confined to flat or slightly sloping sandy plateaux with a slight north-westerly aspect, in reddish-brown sandy substrates among sandstone rocks, exposed or within the protection of low bushes. A conservation status of Rare is recommended.

NOTES. Pollinators for the species are as yet unknown. The inflorescences and leaves are heavily grazed, probably by Cape hares (*Lepus capensis*).

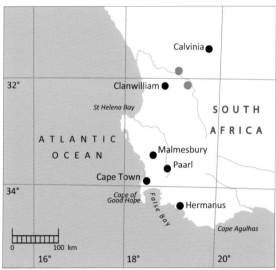

Map 20. Known distribution of *Lachenalia bruynsii*.

Figure 117. *Lachenalia bruynsii* amongst sandstone rocks north-east of Clanwilliam. Image: Graham Duncan.

20. LACHENALIA SARGEANTII

Lachenalia sargeantii W. F. Barker, *Journal of South African Botany* 44 (4): 412–415 (1978).
TYPE: South Africa, Western Cape, Bredasdorp Mountains, in white sand among Table Mountain Sandstone boulders, at upper altitude, *W. F. Barker* 10802 (NBG!, holo., iso.; PRE!, iso.).

ETYMOLOGY. *sargeantii*: after Percival (Percy) A. Sargeant (–1988), mountaineer and expert photographer of Cape flora.

DESCRIPTION. *Geophyte*, 120–300 mm high. *Bulb* globose, 15–20 mm in diameter, offset- and bulbil-forming, neck distinct, 5–8 mm long; tunic multilayered, outer layers cartilaginous, dark brown; inner cataphyll translucent white, adhering to leaf bases, apex obtuse. *Leaves* 2, linear, 200–350 × 4–18 mm, spreading to suberect or arcuate, yellowish-green to olive green, upper surface with depressed longitudinal veins, deeply canaliculate to conduplicate; leaf bases clasping, 40–60 mm long, subterranean, white; primary seedling leaf terete, erect. *Inflorescence* an ordinary or subcorymbose raceme, many-flowered, sterile apex short; scape erect to suberect, 120–280 mm long, rigid, maroonish-purple to bright magenta covered with a powdery bloom; rachis maroonish-purple to bright magenta; pedicels suberect to spreading in bud, becoming declinate at anthesis, 4–15 mm long, bright pinkish-magenta, bending upwards in fruit, increasing in length towards inflorescence apex; lower bracts ovate, becoming lanceolate above, 3–5 × 1–3 mm, translucent white, or white tinged with magenta. *Perianth* zygomorphic, ivory, pendulous; tube cup-shaped, 2–3 mm long, uppermost part flushed with bright magenta; outer tepals oblong, 10–15 × 4–6 mm, apical gibbosities bright green or dark brown; inner tepals oblong obovate, 13–18 × 4–5 mm, apical markings bright green or dark brown. *Stamens* included, straight; filaments white, 11–16 mm long. *Ovary* obovoid, 4–6 × 4–5 mm, bright green; style included, straight, 11–15 mm long, protruding beyond perianth as ovary enlarges. *Capsule* obovoid, 5–7 × 4–6 mm, suberect. *Seed* globose, 1.2–1.3 × 1.3 mm, secondary sculpturing colliculate, matte, black; strophiole rudimentary, 0.2–0.3 mm long, ridged. *Chromosome number* unknown. Plate 11, Figures 24, 29, 51, 118, 119.

FLOWERING PERIOD. October to November, with a peak from late October to early November.

HISTORY. *L. sargeantii* was discovered in full flower in November 1970 by Robert Scott, a visitor from New Zealand who was exploring a mountain at Bredasdorp after an intense fire had swept through the area the previous summer. He photographed the plant in habitat and gave an inflorescence to Percy Sargeant, an ardent naturalist and photographer. Sargeant brought the specimen to the attention of W. F. Barker, who identified it as an undescribed species. The habitat was visited soon afterwards by Barker and Percy, with Robert as their guide, and the type collection was made. A few flowering individuals were found in a subsequent visit to the site in 1971, but none were seen in 1972. *L. sargeantii* was described six years later in the *Journal of South African Botany* (Barker, 1978). For a period of 33 years after the last sighting in 1971, the plants were not recollected. Then, in October 2004, a new colony was discovered in full flower near Napier, north-west of Bredasdorp, by naturalists Cameron and Rhoda McMaster (Duncan *et al.*, 2005c). A painting by Elbe Joubert accompanying an account of the species was published in *Curtis's Botanical Magazine* the following year (Duncan & Edwards, 2005). As a result of more fires in the mountains near Napier in February 2006, two new localities with healthy populations were recorded by the McMasters and a local group of conservationists (Duncan, 2008a).

DISTINGUISHING CHARACTERS AND AFFINITIES. Duncan *et al.* (2005a) placed *L. sargeantii* as sister to *L. montana*. These two species share soboliferous, globose bulbs with linear, deeply canaliculate leaves.

THE GENUS LACHENALIA
20. LACHENALIA SARGEANTII

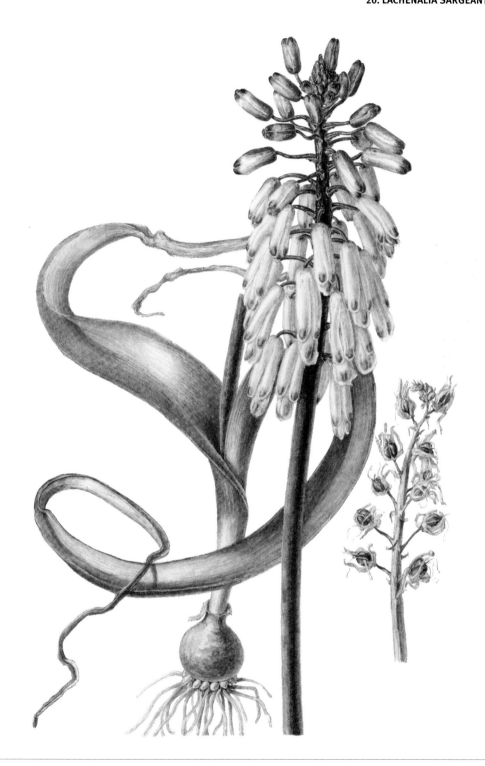

Plate 11. Watercolour painting of *Lachenalia sargeantii* from Napier (*Duncan* 500, in NBG) courtesy of The Editor, *Curtis's Botanical Magazine* vol. 22, t. 534 (2005). Artist: Elbe Joubert.

Figure 118 (left). Form of *Lachenalia sargeantii* with an ordinary raceme after a fire near Napier. Image: Graham Duncan.
Figure 119 (right). Form of *Lachenalia sargeantii* with a subcorymbose raceme from Bredasdorp. Image: Percy Sargeant.

They have rigid scapes with ordinary or subcorymbose racemes and their flowers are usually borne on long, arcuate pedicels. Both produce obovoid capsules containing globose seeds with colliculate secondary sculpturing. *L. montana* differs from *L. sargeantii* mainly in its oblong-campanulate flowers with much shorter outer and inner tepals (6–7 × 4 mm, 8 × 4 mm, respectively), shorter, well-exserted stamens (9–10 mm long) and slightly shorter, rudimentary strophiole (0.1 mm long).

DISTRIBUTION, HABITAT AND CONSERVATION STATUS. *L. sargeantii* is confined to mountainsides in the south-western part of the southern Cape, extending from just north of Napier to Bredasdorp (Map 21). The plants occur in small groups or as scattered individuals in larger colonies on rocky north-facing slopes and summit ridges. Plants occur in Overberg Sandstone Fynbos vegetation (Mucina & Rutherford, 2006) and usually grow wedged between sandstone boulders or among stones in open aspects. They are subject to harsh north-westerly winds in winter and to baking heat during the summer. Notable companion plants that survive fynbos fires by resprouting from subterranean stems include *Protea cynaroides* and *Mimetes cucullatus* (Proteaceae) and several members of the Restionaceae family. Although *L. sargeantii* is only known from a few sites, it occurs in fairly large numbers, and the relatively high altitude at which it grows places most of the plants out of reach of immediate danger. The species is, however, potentially threatened by alien plants, and quarrying for gravel has taken place at one of its lower localities (Duncan *et al.*, 2005). Its conservation status is Vulnerable (Raimondo & Duncan, 2009f).

NOTES. The perianth bases contain nectar and *L. sargeantii* has a rigid scape, long tubular, pendulous, unscented flowers and bright magenta rachis apex and pedicels, all of which imply pollination by sunbirds. Nevertheless, the species has rather short, cup-shaped perianth tubes (2–3 mm long), whereas all other members of the genus that are known or supposed to be sunbird-pollinated have long perianth tubes (6–25 mm long). The insects recorded as visitors to the flowers are the common orange and brown Painted Lady butterfly (*Vanessa cardui*) (Duncan, 2007) and an unidentified species of day-flying moth, both seen in early November 2006 at Fairfield Farm near Napier. The scapes remain attached to the bulbs following capsule dehiscence and the seeds are dispersed locally. The capsules are heavily parasitised by caterpillars in the wild, producing little seed, and this could have led to strong selection pressure for mast-flowering (Sork, 1993).

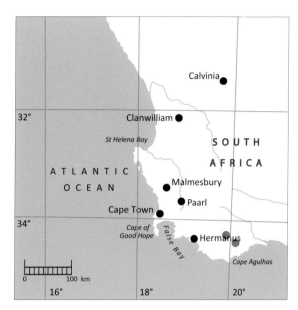

Map 21. Known distribution of *Lachenalia sargeantii*.

L. sargeantii is a true pyrophyte, only flowering and producing leaves in the wild in response to fires during the previous summer or autumn. In exceptional circumstances, a few inflorescences might appear in the second year following a fire. The bulbs of *L. sargeantii* and *L. montana* are deep-seated. In the wild, mature bulbs of *L. sargeantii* remain completely dormant in the period between fires, even though the autumn and winter rains may be sufficient to support leaf growth. Immature bulbs occasionally produce depauperate leaves, as observed at the type locality (Duncan, 2007). A follow-up visit made a year after a number of mature specimens had flowered near Napier in 2005, revealed that all of the plants that had flowered, and all others throughout the population, were completely dormant (McMaster & McMaster, 2006).

21. LACHENALIA VENTRICOSA

Lachenalia ventricosa Schltr. ex W. F. Barker, *Journal of South African Botany* 45 (2): 216–219 (1979).
TYPE: South Africa, Western Cape, Nardousberg Mountain Pass, *T. M. Salter* 3615 (BOL!, holo.; K!, iso.).

ETYMOLOGY. *ventricosa*: descriptive of the inflated scape.
DESCRIPTION. *Geophyte*, 200–400 mm high. *Bulb* subglobose, 15–20 mm in diameter, solitary; tunic multilayered, outer layers light brown, spongy; inner cataphyll translucent white, loosely surrounding leaf base, apex obtuse. *Leaf* solitary, rarely 2, lanceolate, 100–260 × 20–35 mm, suberect to spreading, glaucous, strongly canaliculate; leaf margins undulate; leaf base clasping, 20–60 mm long, subterranean portion white, aerial portion yellowish-green or tinged with light maroon; primary seedling leaf terete, suberect. *Inflorescence* spicate, many-flowered, dense, sterile apex long; scape erect to suberect, 30–160 mm long, yellowish-green, lower portion tinged with light maroon, upper portion strongly

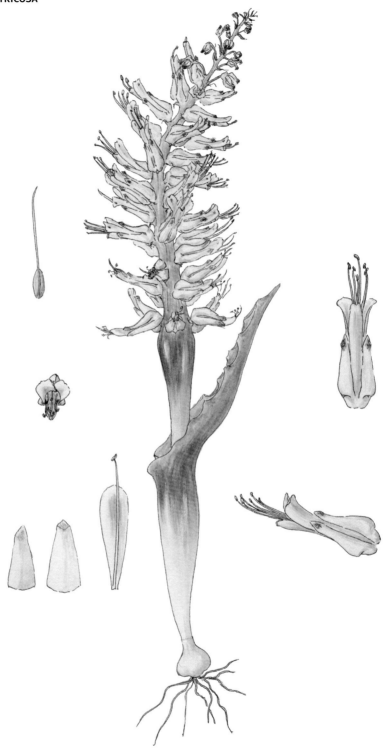

Plate 12. Watercolour painting of *Lachenalia ventricosa* from the Nardousberg (*Salter 3615*, in NBG) courtesy of the Compton Herbarium, Kirstenbosch, South African National Biodiversity Institute. Artist: Winsome Barker.

inflated; rachis yellowish-green, shading to white or light blue above; lower bracts ovate, becoming lanceolate above, 1–4 × 2–3 mm, translucent white. *Perianth* zygomorphic, tubular, spreading to slightly cernuous; tube cup-shaped, 2.0–2.5 mm long, light blue; outer tepals oblong, 10–11 × 5–6 mm, light greenish-yellow, apical gibbosities dark brown; inner tepals oblong-obovate, 12–16 × 5–6 mm, light greenish-yellow, apical marking dark brown. *Stamens* well exserted, more or less straight; filaments white, 15–23 mm long. *Ovary* ellipsoid, 4–5 × 3 mm, light green; style well exserted, straight, 15–17 mm long, white. *Capsule* ellipsoid, 7–8 × 4–5 mm. *Seed* globose, 0.9–1.0 × 0.8–0.9 mm, glossy, black; strophiole inflated, 0.7–0.8 mm long, smooth. *Chromosome number* 2n = 14 (Spies *et al.*, 2008). Plate 12, Figure 120.

FLOWERING PERIOD. July to September, with a peak from mid-August to early September.

HISTORY. Rudolf Schlechter made the first collections of *L. ventricosa* at Lange Kloof north-west of Clanwilliam on 3rd September 1896, and at Brandewynrivier between the Pakhuis Pass and the Doornrivier on 7th September 1897. He distributed material to numerous local and European herbaria under his manuscript name *L. ventricosa* Schltr., but the species remained undescribed for more than 80 years until finally published in the *Journal of South African Botany*, illustrated in monochrome with a painting by W. F. Barker of one of the type specimens (Barker, 1979a). The type collection was made by Hermione Nortier on the Nardousberg Plateau on 6th September 1933 and her specimens were brought to W. F. Barker by T. M. Salter for identification. This collection (*Salter* 3615) is preserved at the Bolus Herbarium in the University of Cape Town, with isotype material at Kew. Annotations to herbarium material show that Barker had initially intended to name the species for Hermione Nortier,

Map 22. Known distribution of *Lachenalia ventricosa*.

Figure 120. Cultivated specimen of *Lachenalia ventricosa* from the Nardousberg. Image: Graham Duncan.

probably before becoming aware of Schlechter's manuscript name.

DISTINGUISHING CHARACTERS AND AFFINITIES. Phylogenetic analysis (Duncan *et al.*, 2005a) suggests that *L. ventricosa* is sister to *L. mutabilis*, sharing subglobose bulbs, lanceolate, canaliculate leaves, and seeds with inflated strophioles. *L. mutabilis* differs from *L. ventricosa* in its urceolate perianths with much shorter, ovate outer tepals (5–8 mm), shorter, obovate inner tepals (7–9 mm), declinate, usually included and much shorter stamens (6–9 mm long), shorter styles (5–7 mm long), and much larger, globose seed (1.2–1.4 × 1.3–1.4 mm).

DISTRIBUTION, HABITAT AND CONSERVATION STATUS. *L. ventricosa* occurs in high altitude parts of the Nardousberg and Cederberg, extending from just south of Klawer south-east to the Biedouw Valley, with a disjunct, relatively low altitude population in the south-eastern part of the Swartruggens Range bordering the Tanqua Karoo Basin (Map 22). The plants occur in Cederberg Sandstone Fynbos and Swartruggens Quartzite Karoo vegetation (Mucina & Rutherford, 2006). They grow singly or in colonies in deep white sand on exposed mountain plateaux, or on south-facing slopes in heavier, stony soil, amongst low bushy growth. In the Swartruggens Range, *L. ventricosa* grows in red quartzitic sand, in association with *L. multifolia* and *L. violacea*. Robust specimens are encountered in years of high rainfall. The species is threatened by agricultural expansion for the rooibos tea industry on the Nardousberg, and a conservation status of Vulnerable is recommended.

NOTES. Pollinators are unknown. Once the capsules have dehisced, the scape detaches at its base and the infructesence is blown away, dispersing the seeds in the process. In seasons of excellent rainfall, the capsules produce so much seed that the bulb is rapidly depleted, often withering completely or failing to flower in subsequent years until sufficient reserves have been accumulated.

22. LACHENALIA BUCHUBERGENSIS

Lachenalia buchubergensis Dinter, *Feddes Repertorium* 30: 84 (1932).
TYPE: Namibia, Buchuberge, *M. K. Dinter* 6558 (B!, holo.; iso.).

ETYMOLOGY. *buchubergensis*: after the Buchuberg Mountains in south-western Namibia.

DESCRIPTION. *Dwarf geophyte*, 70–210 mm high. *Bulb* ovoid, 10–22 mm in diameter, solitary; tunic multilayered, outer layers light brown, spongy; inner cataphyll translucent white to green, adhering to leaf base, apex obtuse. *Leaf* 1, lanceolate to broadly lanceolate, 30–100 × 7–20 mm, suberect to spreading, yellowish-green to glaucous, deeply canaliculate, upper surface plain with distinct depressed longitudinal grooves, lower surface scattered with small to large light green blotches in upper two thirds, shading to brownish-magenta in lower third, midrib distinct; margins undulate or entire; leaf base clasping, 30–80 mm long, subterranean portion white, upper portion light green with distinct magenta to brownish-magenta bands; primary seedling leaf terete, erect. *Inflorescence* racemose, few-flowered, sterile apex short; scape suberect, 20–70 mm long, slender, lax, light green heavily mottled with brownish-magenta; rachis light green, minutely speckled with greyish-magenta; pedicels suberect, 0.5 mm long, light green; lower bracts ovate, becoming lanceolate above, 1.5–3.0 × 1.0–1.5 mm, white, uppermost bracts basally spurred, 0.4–0.6 mm long, white to green. *Perianth* zygomorphic, tubular, spreading to suberect; tube cup-shaped, 2–3 mm long, light blue; outer tepals oblong, 8–9 × 4–5 mm, light greenish-grey, mottled with light blue, apical gibbosities purplish-magenta; inner tepals oblong-obovate, 8–10 × 3–4 mm, light greenish-grey, apices deep magenta, slightly recurved. *Stamens* included to shortly exserted, straight; filaments white, 9 mm long. *Ovary* ellipsoid, 3–4 × 3 mm, bright green; style shortly to well exserted, straight, 10–12 mm long, white. *Capsule* obcordate, 10 × 12 mm, broadly winged. *Seed* oblong, 3.0–3.5 × 1.0–1.1 mm, matte, black, primary sculpturing rugose; strophiole rudimentary, 0.3 mm long, ridged. *Chromosome number* unknown. Figure 121.

FLOWERING PERIOD. May to July, with a peak in late May.

HISTORY. The German botanist and explorer M. K. Dinter (1868–1945) discovered *L. buchubergensis* in July 1929 on the Buchuberg Mountains in south-western Namibia. The holotype (*Dinter* 6558) is preserved at the Botanical Museum at Berlin-Dahlem (B) and the species was described in *Feddes Repertorium* (Dinter, 1932). *L. buchubergensis* was one of five lachenalias included in the treatment of the genus for the *Prodromus einer flora von Südwestafrika* (Sölch & Roessler, 1970), and it was illustrated for the first time in *The Lachenalia Handbook* (Duncan, 1988a).

DISTINGUISHING CHARACTERS AND AFFINITIES. *L. buchubergensis* resolved as sister to *L. nordenstamii* in a cladistic analysis of morphological data (Duncan *et al.*, 2005a). *L. nordenstamii*, a dwarf species restricted to the northern Richtersveld and south-western Namibia, shares a short, few-flowered raceme, a solitary lanceolate leaf, a relatively large, ovoid bulb and large, broadly winged capsules that contain large, oblong seeds with rudimentary, ridged strophioles. It differs from *L.buchubergensis* mainly in its widely campanulate, brownish, cernuous flowers with narrowly spreading, well exserted, relatively thick stamens and a much narrower lanceolate leaf with a prominent midrib.

DISTRIBUTION, HABITAT AND CONSERVATION STATUS. *L. buchubergensis* is restricted to the south-western corner of Namibia and the north-western corner of the Richtersveld. Its distribution extends from a locality between Chamnaib and Bogenfels in the Namib Desert to Skilpad north-east of Alexander Bay in south-western Namibia, and north-east to Numees (Map 23). On the summit of the Buchuberg, *L. buchubergensis* occurs within view of the sea and benefits from drenching sea mists.

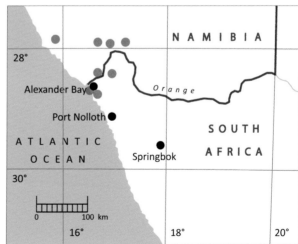

Map 23. Known distribution of *Lachenalia buchubergensis*.

Figure 121. *Lachenalia buchubergensis* on a dolomite slope in the northern Richtersveld. Image: Pieter van Wyk.

In the Richtersveld, this species occurs as widely scattered individuals or in small groups of two to three individuals, in Western Gariep Hills Desert vegetation (Mucina & Rutherford, 2006), usually in crevices between boulders on lower south-facing rocky granite slopes in light shade, or on quartz gravel flats in full sun. Its conservation status is Rare (Victor & Duncan, 2009b). In south-western Namibia it occurs in Desert and Succulent Steppe Vegetation on gentle south-west slopes in sandy washes, in red sand mixed with calcrete, and on quartz gravel plains. Its conservation status there is Rare (Loots, 2005).

NOTES. In common with several other *Lachenalia* species from arid habitats, *L. buchubergensis* responds rapidly to autumn rains, flowering early in the season and entering dormancy in early spring before temperatures rise excessively. Pollinator agents are as yet unknown for the species. Seed dispersal is unusual in that the extraordinarily large, obcordate ripe capsules detach and are blown away.

23. LACHENALIA ISOPETALA

Lachenalia isopetala Jacq., *Collectanea* 5: 68–69 (1797). *Scillopsis isopetala* (Jacq.) Lem., *L'Illustration Horticole* 3: 35 (1856).
TYPE: South Africa, Cape, collector and precise locality unknown, figure in N. J. Jacquin, *Icones Plantarum Rariorum* 2: 15, t. 401 (1793 or 1794) (neotype, designated here).

ETYMOLOGY. *isopetala*: tepals of equal length.

DESCRIPTION. *Geophyte*, 100–300 mm high. *Bulb* globose, 20–30 mm in diameter, solitary; tunic multilayered, outer layers cartilaginous, thick, light to dark brown; neck distinct, 20–80 mm long, tunics thick, apices fasciculate, often protruding above ground; inner cataphyll translucent white, adhering to leaf bases, apex obtuse. *Leaves* 2, lanceolate, 120–200 × 15–25 mm, spreading to arcuate, intensely glaucous, deeply canaliculate, upper surface plain, lower surface with numerous protruding, longitudinal veins; leaf margins undulate and crisped, cartilaginous; leaf bases 25–40 mm long, lower portion white, shading to light green above; primary seedling leaf terete, erect. *Inflorescence* racemose, many-flowered, dense; sterile apex short; scape erect to suberect, 20–50 mm long, glaucous to light brown; rachis light to dark maroonish-brown to greenish-maroon; pedicels suberect, 2–3 mm long, olive green; bracts ovate, 2–6 × 3–4 mm, papery, bright white. *Perianth* zygomorphic, tubular, spreading to suberect; tube shallowly cup-shaped, 1 mm long, light yellowish-maroon; outer tepals

Figure 122. Deep maroon form of *Lachenalia isopetala* on shale flats near Middelpos. Image: Graham Duncan.

THE GENUS LACHENALIA
23. LACHENALIA ISOPETALA

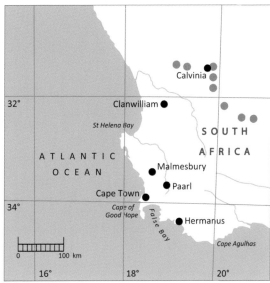

Map 24. Known distribution of *Lachenalia isopetala*.

Figure 123. Cultivated yellowish brown-flowered specimen of *Lachenalia isopetala* from Calvinia. Image: Graham Duncan.

oblong, 12–13 × 4–5 mm, yellowish-maroon to deep brownish-maroon, margins creamish-yellow, broad, apical gibbosities dark maroon to brownish-maroon; inner tepals linear-oblong, 12–13 × 4–5 mm, creamish-yellow to light greenish-yellow, apical marking large, dark maroon to brownish-maroon. *Stamens* included to shortly exserted, straight; filaments light greenish-white, 9–10 mm long. *Ovary* ellipsoid, 5 × 2 mm, light green; style included to shortly exserted, straight, 9–10 mm long, light greenish-white. *Capsule* ellipsoid, 10–11 × 6–7 mm. *Seed* globose, 1.8–2.0 × 1.8–1.9 mm, glossy, black; strophiole rudimentary, 0.1 mm long, ridged. *Chromosome number* 2n = 30 (Johnson & Brandham, 1997); 2n = 40 (Spies *et al.*, 2008). Figures 46, 74, 122, 123.

FLOWERING PERIOD. September to November, with a peak in early October.

HISTORY. *L. isopetala* was first illustrated by N. J. Jacquin in volume two of his *Icones Plantarum Rariorum* in 1793 or 1794 (year uncertain), and a formal description followed in *Collectanea* in 1797. The holotype was lost in a fire in Vienna between 1939–1945. The colour plate is designated here as the neotype as the plate was not referred to in the original protologue. Lemaire (1856) transferred the species to *Scillopsis* as *S. isopetala* in volume three of *L'Illustration Horticole*, but it was returned to *Lachenalia* by Baker (1871). A collection made in Little Brak River in the southern Cape in October 1814 (*Burchell* 6188, in K) was mistakenly cited by Baker (1897a) under *L. isopetala*; it is in fact *L. rosea*.

DISTINGUISHING CHARACTERS AND AFFINITIES. One of the most unusual species, *L. isopetala* stands out amongst the members of section *Lachenalia* as the only species with yellowish- to deep brownish-maroon tepals. It is also unique within the genus in having prominent papery, bright white bracts. Phylogenetic analysis (Duncan *et al.*, 2005) suggests that *L. isopetala* is allied to *L. duncanii*, the two species sharing leathery, canaliculate leaves with undulate and crisped margins and solitary globose bulbs with hard outer tunics. *L. duncanii* differs from *L. isopetala* in having oblong-campanulate, cream flowers with well-exserted, declinate stamens that are light magenta above, sclerotic leaf margins and a much smaller, ovoid seed (1.3–1.4 × 1.1–1.2 mm).

DISTRIBUTION, HABITAT AND CONSERVATION STATUS. *L. isopetala* occurs in the semi-arid south-western Northern Cape, extending from Nieuwoudtville on the Bokkeveld Plateau east to Calvinia and Middelpos in the western Great Karoo and to just east of Sutherland on the Roggeveld Plateau (Map 24). The plants occur in Roggeveld- and Hantam Karoo, and in Nieuwoudtville Shale Renosterveld vegetation (Mucina & Rutherford, 2006). They grow singly or in small groups within larger colonies in red stony clay, on arid flats among karroid scrub, often between small dolerite stones, in full sun. At Sutherland, *L. isopetala* occurs in association with *L. canaliculata* and is subject to snowfalls and sub-zero temperatures for several days at a time in winter. The species is not threatened.

NOTES. The flowers of *L. isopetala* are pollinated by the solitary bee *Amegilla nivea* (Figure 74). Its leaves are proteranthous, having often partially or fully withered by the time the inflorescences emerge in late spring. The scapes remain attached for several weeks after capsule dehiscence and the relatively large seeds drop to the ground due to the shaking action of wind.

24. LACHENALIA SCHELPEI

Lachenalia schelpei W. F. Barker, *Journal of South African Botany* 50 (4): 535–538 (1984).
TYPE: South Africa, Northern Cape, top of Hantamsberg, E. A. *Schelpe s.n.* sub. NBG 127340 (NBG!, holo.; BOL!, iso.).

ETYMOLOGY. *schelpei*: after Prof. E. A. Schelpe (1924–1985), botanist and author.
DESCRIPTION. *Geophyte*, 140–225 mm high. *Bulb* ovoid, 15–25 mm in diameter, offset- and bulbil-forming; tunic multilayered, outer layers spongy, dark brown; neck apex fasciculate; inner cataphyll translucent white, adhering to leaf bases, apex obtuse. *Leaves* 2, broadly lanceolate, 50–180 × 20–40 mm, spreading to suberect, light green, upper and lower surfaces with large, irregularly scattered, bright green to dull maroonish-green blotches, upper surface with distinct depressed longitudinal grooves; leaf bases loosely surrounding base of scape, 10–20 mm long, light green with prominent maroon and magenta bands; primary seedling leaf flat, prostrate. *Inflorescence* racemose, many-flowered, dense, sterile apex short to long; scape suberect, 70–100 mm long, light green, heavily marked with dark maroon blotches; rachis light green, heavily marked with light to dark maroon blotches, apex shading to light electric blue; pedicels suberect, 1.5 mm long, greenish-white; lower bracts ovate, becoming narrowly lanceolate above, 2–5 × 0.5–3.0 mm, translucent white. *Perianth* zygomorphic, tubular, slightly cernuous; tube shallowly cup-shaped, 1 mm long, light bluish-white; outer tepals oblong, 5–7 × 4 mm, light blue at base, shading to greenish-white above, apical gibbosities greenish-brown; inner tepals oblong-obovate, 7–8 × 4–5 mm, overlapping to form a narrow mouth, translucent white, apical marking green, median keels light green. *Stamens* shortly exserted, straight, filaments white, 9–10 mm long. *Ovary* ellipsoid, 4 × 2 mm, yellowish-green; style included to shortly exserted, straight, 7–8 mm long, white. *Capsule* ellipsoid, 7 × 5 mm. *Seed* ovoid, 1.8–1.9 × 1.3–1.4 mm, matte, black; strophiole ridged, 1.2 mm long. *Chromosome number* unknown. Figure 124.

THE GENUS LACHENALIA
24. LACHENALIA SCHELPEI

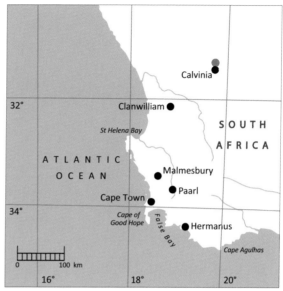

Map 25. Known distribution of *Lachenalia schelpei*.

Figure 124. Cultivated specimen of *Lachenalia schelpei* from Calvinia. Image: Graham Duncan.

FLOWERING PERIOD. June to early August, with a peak in mid-July.

HISTORY. *L. schelpei* is known from very few collections on the Hantamsberg Mountains north of Calvinia. It was collected there for the first time in September 1980 by Prof. E. A. Schelpe, who gathered a mature bulb in fruit and several juvenile bulbs that were subsequently cultivated at Kirstenbosch. The adult bulb flowered in June 1982 and was preserved as the holotype. The species was published in the *Journal of South African Botany* (Barker, 1984) and has since been recorded on only three occasions, twice by Margaret Thomas (*Thomas s.n.*) and once by the author (*Duncan* 401), all in NBG.

DISTINGUISHING CHARACTERS AND AFFINITIES. *L. schelpei* appears to be sister to *L. doleritica* (Duncan et al., 2005a), with which it shares broadly lanceolate leaves and large ovoid seeds with ridged strophioles, but *L. doleritica* differs in its oblong-campanulate flowers borne on longer pedicels (2–5 mm long), longer perianth tubes (2–3 mm long), longer outer and inner tepals (8–9 mm and 9–11 mm long, respectively), declinate stamens and shorter style (6–7 mm long).

DISTRIBUTION, HABITAT AND CONSERVATION STATUS. *L. schelpei* is restricted to the plateau of the Hantamsberg Mountain north of Calvinia in the western Karoo (Map 25). The plants grow in Hantam Plateau Dolerite Renosterveld vegetation (Mucina & Rutherford, 2006) as scattered individuals in seasonally moist, heavy clay in open aspects or within the protection of low bushy scrub. It is potentially threatened by crop cultivation and overgrazing, and its conservation status is Vulnerable (Victor & Duncan, 2009k).

NOTES. *L. schelpei* has no recorded pollinators. The scape of this species remains attached to the bulb for several weeks following capsule dehiscence and the seeds are dispersed locally.

25. LACHENALIA CAPENSIS

Lachenalia capensis W. F. Barker, *Journal of South African Botany* 15: 38–39 (1949).
TYPE: South Africa, Western Cape, Wildschutskraal, *W. F. Barker* 3883 (NBG!, holo.).

ETYMOLOGY. *capensis*: after the Cape Peninsula, where this species is endemic.

DESCRIPTION. *Geophyte*, 120–305 mm high. *Bulb* subglobose, 15–25 mm in diameter, offset-forming; tunic multilayered, outer layers light to dark brown, spongy; inner cataphyll translucent white, loosely surrounding leaf bases, apex obtuse. *Leaves* 1–2, lanceolate, 60–100 × 10–25 mm, light to dark green, upper surface plain or with dark brownish-maroon spots or blotches; leaf margins slightly coriaceous, flat to undulate; leaf bases clasping, 20–40 mm long, lower portion white, upper portion light green to maroon; primary seedling leaf flat, prostrate. *Inflorescence* spicate or racemose, few- to many-flowered, sterile apex short; scape erect, 80–140 mm long, sturdy, light to bright green, plain or with brownish-maroon spots or blotches; rachis light green to glaucous, plain or with small dull brownish-maroon spots, apex light electric blue; pedicels absent or suberect, up to 1 mm long, light green; lower bracts ovate, becoming lanceolate above, 2–5 × 1–3 mm, translucent white. *Perianth* zygomorphic, tubular, heavily sweet-scented, spreading to suberect, cream to white, ageing to dull red; tube cup-shaped, 3 mm long, plain or with light blue tinge; outer tepals oblong, 7–9 × 4–5 mm, apical gibbosities light creamy-yellow, median keels plain to light blue; inner tepals oblong obovate, 10–13 × 3–6 mm, upper inner tepals overlapping, lower inner tepal narrower, up to 2 mm shorter, median keels plain to light blue. *Stamens* included, rarely shortly exserted, straight; filaments cream to white, 10–11

Figure 125 (left). *Lachenalia capensis* with visiting ant on a rocky slope near Simonstown. Image: Graham Duncan.
Figure 126 (right). Cultivated specimens of *Lachenalia capensis* from Ocean View. Image: Graham Duncan.

mm long. *Ovary* ellipsoid, 4 × 3 mm, light to bright green; style included or shortly to well exserted, straight, 10–11 mm long, cream or white. *Capsule* ellipsoid, 8–12 × 3–5 mm. *Seed* globose, 0.9 × 0.8–0.9 mm, glossy, black; strophiole inflated, 0.7–0.8 mm long, smooth. *Chromosome number* 2n = 16 (Hamatani *et al.*, 1998); 2n = 28 (Johnson & Brandham, 1997; Spies *et al.*, 2008). Figures 125, 126.

FLOWERING PERIOD. September to October, with a peak in early October.

HISTORY. The earliest collection of *L. capensis* is that of the Englishman Anthony H. Wolley-Dod, who found it 'on flats near Rondebosch' in August 1895 (*Wolley-Dod* 379, in BOL and NBG) and subsequently in September 1897 on rocky slopes above Smitswinkel Bay in the Cape of Good Hope Nature Reserve (*Wolley-Dod* 3087, in BOL). The type specimens were collected by W. F. Barker in October 1945 at Wildschutskraal in the southern Cape Peninsula and the species was formally published in the *Journal of South African Botany* (Barker, 1949).

DISTINGUISHING CHARACTERS AND AFFINITIES. *L. capensis* is most closely allied with *L. fistulosa* (Duncan, 2005a), an almost exclusively Cape Peninsula species that has similar lanceolate leaves, a heavy sweet scent, straight stamens and globose seeds with inflated strophioles. *L. fistulosa* differs in having urceolate flowers with shorter inner tepals (8–9 mm long) with the apices strongly recurved, the lower inner tepal equal in length to the upper inner tepals, and shorter filaments (5–7 mm long). *L. fistulosa* occurs mainly on south- and north-facing mountain slopes in stony clay soil in renosterveld, and commences flowering a few weeks earlier than *L. capensis*. *L. capensis* resembles cream forms of *L. orchioides* subsp. *orchioides* from the southern Cape Peninsula, but this subspecies differs in having cylindrical perianth tubes 6–8 mm long, much longer tepals, and lower inner tepals that are distinctly longer than the upper inner tepals.

DISTRIBUTION, HABITAT AND CONSERVATION STATUS. Restricted to the Cape Peninsula, *L. capensis* is concentrated in the southern part, occurring from the Cape of Good Hope Nature Reserve to Hout Bay on the western seaboard. It has been known as far east as Steenberg and Rondebosch, but it is certainly extinct at the latter locality (Map 26). The plants occur in Peninsula Granite Fynbos vegetation (Mucina & Rutherford, 2006) on rocky, mainly north- and west-facing slopes and hillsides, growing singly or in small groups in light shade to full sun. On the rocky slopes above Simonstown, it grows in association with *Erica cerinthoides* (Ericaceae) and *Watsonia borbonica* (Iridaceae), and in other parts of the southern Peninsula, it is sometimes encountered within the semi-shade of exotic *Pinus pinea* trees. *L. capensis* is threatened by housing development and encroaching alien vegetation, and has a conservation status of Vulnerable (Helme & Raimondo, 2009b).

NOTES. Flowering in *L. capensis* is profuse during the first spring season following bush fires but the species is not dependent on fire for flowering to occur. The heavily sweet-scented blooms are pollinated by honey bees (*Apis mellifera*). The scapes remain attached to the bulbs for several weeks following capsule dehiscence and the seeds are locally dispersed.

Map 26. Known distribution of *Lachenalia capensis*.

26. LACHENALIA BARKERIANA

Lachenalia barkeriana U. Müll.-Doblies, B. Nord. & D. Müll.-Doblies, *South African Journal of Botany* 53 (6): 486–487 (1987).

TYPE: South Africa, Western Cape, east of Gifkop, between Loeriesfontein and Kliprand, *U. & D. Müll.-Doblies* 80077d (PRE!, holo.; B, BOL!, BTU, G, GRA!, K!, M, NBG!, PRE!, S, STE, Z, iso.).

ETYMOLOGY. *barkeriana*: after Miss W. F. Barker (1907–1994), botanist, *Lachenalia* expert and first Curator of the Compton Herbarium at Kirstenbosch.

DESCRIPTION. *Dwarf geophyte*, 10–20 mm high. *Bulb* globose, 8–15 mm in diameter, solitary; tunic multilayered, outer layers dark brown, spongy; inner cataphyll loosely surrounding leaf bases, subterranean portion translucent white, aerial portion greenish-brown, apex obtuse. *Leaves* 3–12, linear, 35–90 × 1–4 mm, spreading, rosulate, canaliculate, bright green, upper surface plain, lower surface tinged with dull magenta in lower third; primary seedling leaf flat, prostrate. *Inflorescence* a subcapitate raceme, produced at ground level; scape clavate, erect, 5–10 mm long, subterranean during flowering, light green to white, elongating considerably in fruit; rachis light green to white, 5–7 mm long; pedicels erect, 0.5–1.5 mm long, white; lower bracts ovate, becoming lanceolate above, 2–4 × 1–2 mm, translucent white. *Perianth* zygomorphic, tubular, erect, heavily yeast-scented by day and by night; tube shallowly cup-shaped, 0.5–1.0 mm long, greenish-white; outer tepals oblong, 5–6 × 2.0 mm, greenish-white, apical gibbosities and median keels green; inner tepals linear-oblong, 6–8 × 1.0 mm, greenish-white, median keels green. *Stamens* well-exserted, straight, erect; filaments white, 11–14 mm long, inflated above. *Ovary* ellipsoid, 3 × 2 mm, bright green; style well exserted, straight, 7–10 mm long, white. *Capsule* ellipsoid, 7–8 × 6–7 mm. *Seed* globose, 1.1–1.2 × 1.2–1.3 mm, glossy, black; strophiole inflated, 0.6–0.7 mm long, smooth. *Chromosome number* 2n = 14 (Nordenstam, 1982); 2n = 14, 16 (Müller-Doblies *et al.*, 1987). Figure 127.

Figure 127. *Lachenalia barkeriana* in red sand near Bitterfontein. Image: Graham Duncan.

FLOWERING PERIOD. April to June, with a peak in May.

HISTORY. *L. barkeriana* was collected for the first time in 1974 by the Swedish botanists Prof. B. Nordenstam and Dr J. Lundgren on the Knersvlakte north of Bitterfontein. The plant was at first regarded as a narrow-leafed form of *L. pusilla*, but a detailed study of living specimens in the field and in cultivation, and of herbarium material, revealed it to be a distinct but closely related new species.

DISTINGUISHING CHARACTERS AND AFFINITIES. *L. barkeriana* is easily recognised by a rosette of up to 12 linear, canaliculate, spreading leaves and erect filaments that are distinctly swollen in the upper half. The species is sister to *L. pusilla* (Duncan *et al.*, 2005a), which has a similar subcapitate, geoflorous inflorescence of tubular, strongly yeast-scented flowers with well exserted, erect stamens and several to numerous leaves that are produced in a rosette. *L. pusilla* differs from *L. barkeriana* in having lanceolate, prostrate leaves, a terete scape, longer inner tepals (10–13 mm long) and filaments that are not swollen in the upper half.

DISTRIBUTION, HABITAT AND CONSERVATION STATUS. This species occurs mainly between Vanrhynsdorp, Kliprand and Loeriesfontein in the north-western Western Cape and in south-western Namaqualand in the Northern Cape (Map 27). It is found in Namaqualand Strandveld, Namaqualand Klipkoppe Shrubland and Hantam Karoo vegetation (Mucina & Rutherford, 2006), in deep red sand on flats and in shallow pockets of west-facing, gently sloping granite outcrops and in calcrete pans. The plants occur singly or in small groups as part of larger populations. Disjunct populations from Kleinsee at the De Beers Namaqualand Diamond Mine on the south-west coast of Namaqualand (*Duncan 443*, in NBG) and Koingnaas have broader, somewhat fleshier leaves and sometimes occur in association with *L. valeriae*. *L. barkeriana* has a conservation status of Rare (Victor & Duncan, 2009a).

NOTES. *L. barkeriana* has an elongating infructescence, like that of *L. pusilla*. Just prior to capsule dehiscence, the subterranean scape elongates considerably, causing the capsules to be pushed upwards and then to fall to the ground a short distance away from the parent plant, where some of the seeds are dispersed. The scape detaches from the bulb and is carried away by wind, dispersing the remaining seeds (Duncan, 2005). According to Müller-Doblies *et al.* (1987), the prominent strophioles of the seeds act as elaiosomes, facilitating myrmecochory (dispersal by ants). Like those of *L. pusilla*, the flowers of *L. barkeriana* have a very unusual, heavy spicy scent similar to that emitted by *Massonia depressa*, which is rodent-pollinated (Johnson *et al.*, 2001). Preliminary chemical analysis of the scent suggests, however, that both *L. barkeriana* and *L. pusilla* are bee-pollinated (Johnson, pers. comm.). The flowers of *L. barkeriana* are self-fertile and produce copious seeds, regardless of whether they have been visited by animals (Duncan, 2005).

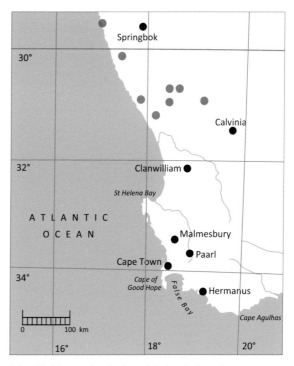

Map 27. Known distribution of *Lachenalia barkeriana*.

27. LACHENALIA PUSILLA

Lachenalia pusilla Jacq., *Collectanea* 5: 71 (1797). *Polyxena pusilla* (Jacq.) Schltr., *Notizblatt des Koniglichen Botanischen Gartens und Museums zu Berlin* 9: 150 (1924).

TYPE: South Africa, Cape, collector and precise locality unknown, figure in N. J. Jacquin, *Icones Plantarum Rariorum* 2: 16, t. 385 (1795) (neotype, designated here).

SYNONYMY: *Massonia undulata* Thunb., *Nova genera plantarum* 2: 41 (1782), non *Lachenalia undulata* Masson ex Baker, *Journal of Botany* (London) 24: 336 (1886). *Brachyscypha undulata* (Thunb.) Baker, *Journal of the Linnean Society* (Botany) 11: 394 (1871). Type: South Africa, Cape, precise locality unknown, *C. P. Thunberg* 7993 (UPS!, holo.).

Lachenalia petiolata Baker, *Bulletin de L'Herbier Boissier* (Ser. 2) 1: 856 (1901). Type: South Africa, Western Cape, Hopefield, *F. E. Bachmann* 1791 (Z!, holo.).

ETYMOLOGY. *pusilla*: descriptive of the very small stature of the plant.

DESCRIPTION. *Dwarf geophyte*, 10–20 mm high. *Bulb* globose, 7–15 mm in diameter, solitary; tunic multilayered, outer layers dark brown; inner cataphyll translucent white, adhering to leaf bases, apex acute. *Leaves* 3–8(–12), narrowly to broadly lanceolate, 20–55 × 5–20 mm, prostrate, rosulate, light to dark green or glaucous, upper surface with small to large, light to dark brown blotches, lower surface plain or centrally tinged with dull purple; leaf margins thickened, dull maroon or yellow; leaf bases clasping, 15–25 mm long, subterranean; primary seedling leaf flat, prostrate. *Inflorescence* a subcapitate raceme, few- to many-flowered, produced at ground level, sterile apex short; scape erect, 10–25 mm long, subterranean, white, terete, lengthening in fruit; rachis white to green; pedicels erect to suberect, 1–3 mm long, white; lower bracts ovate, becoming lanceolate above, 1–4 × 1 mm, translucent white. *Perianth* zygomorphic, tubular, erect, white to cream, heavily yeast-scented; tube shallowly cup-shaped, 1–2 mm long; outer tepals oblong, 5–7 × 0.8–1.0 mm, plain or suffused with dull green or red; inner tepals narrowly oblong obovate, 10–13 × 1–2 mm, plain or suffused with dull green or red. *Stamens* well exserted, straight; filaments white, 11–17 mm long. *Ovary* ellipsoid, 2 × 1 mm, light yellowish-green; style well exserted, white, straight, erect, 8–17 mm long. *Capsule* ellipsoid, 6–7 × 4–6 mm. *Seed* globose, 1.1–1.2 × 1.3 mm, glossy, black; strophiole inflated, 0.6–0.7 mm long, smooth. *Chromosome number* 2n = 14 (Crosby, 1986; Johnson & Brandham, 1997; Hamatani *et al.*, 1998, 2007); 2n = 16 (Nordenstam, 1982). Plate 13, Figures 128, 129.

FLOWERING PERIOD. April to June, with a peak in May.

HISTORY. *L. pusilla* has a complicated history because of its problematic generic placement. It has been shifted between *Massonia*, *Polyxena* and the monotypic *Brachyscypha*, and was first described by C. P. Thunberg (1782) as *Massonia undulata* Thunb.. Baker (1871) placed it in *Brachyscypha* as *B. undulata* (Thunb.) Baker but it was transferred to *Lachenalia* by Bentham and Hooker (1883) under section *Brachyscypha*, and subsequently to subgenus *Brachyscypha* of *Lachenalia* by Baker in his final revision of the genus (Baker, 1897a). Schlechter (1924) mistakenly placed the species in *Polyxena* Kunth on account of its biseriate stamens, unaware that these are also present in *Lachenalia*, and it was again placed in *Brachyscypha* by Hutchinson (1934). Müller-Doblies *et al.* (1987) confirmed Baker's placement of it in subgenus *Brachyscypha*, and at the same time described *L. barkeriana*. The type specimen of *L. pusilla* was destroyed in Vienna during the war years 1939–1945, along with a number of other *Lachenalia* types used in N. J. Jacquin's magnificent work, the *Icones Plantarum Rariorum* (1781–1793). This necessitated the designation of the colour plates as the types of the species. Müller-Doblies *et al.* (1987) mistakenly designated Jacquin's plate 385 as the 'iconotype' of

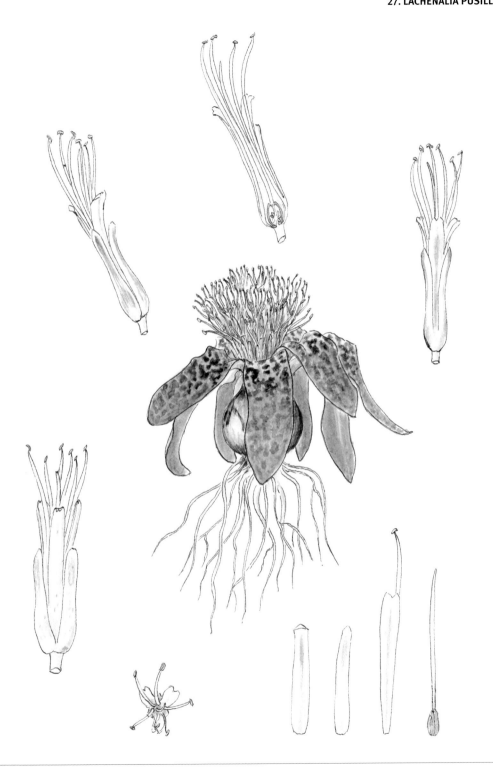

PLATE 13. Watercolour painting of *Lachenalia pusilla* from Citrusdal, courtesy of the Compton Herbarium, Kirstenbosch, South African National Biodiversity Institute. Artist: Winsome Barker.

27. LACHENALIA PUSILLA

Figure 128 (left). Broad-leaved form of *Lachenalia pusilla* on a shale slope near Napier. Image: Graham Duncan.
Figure 129 (right). Narrow-leaved form of *Lachenalia pusilla* from the Cederberg. Image: Graham Duncan/SANBI.

the species, but as no mention of the plate was made in the protologue (Jacquin, 1797a), it is here designated as the neotype. Although Thunberg's (1782) specific epithet *undulata* is the earliest for this species, it had to give way to Jacquin's later epithet *pusilla* because another taxon, *L. undulata* Masson ex Baker, had subsequently been published under that name (Baker, 1886). A collection of *L. pusilla* from Aurora on the Cape west coast (*Duncan* 210, in NBG and K) is the closest match to Jacquin's plate 385, which appeared in volume 2 of his *Icones Plantarum Rariorum* (Jacquin, 1795a).

DISTINGUISHING CHARACTERS AND AFFINITIES. *L. pusilla* is sister to *L. barkeriana* (Duncan *et al.*, 2005a), sharing a subcapitate raceme that is produced at ground level. The latter differs mainly in its linear, deeply channelled, unspotted, spreading to suberect leaves, its shorter, relatively thick pedicels that are as broad as they are long, its clavate scape, and its filaments, which are swollen in the upper half. An unusual feature of the leaves of *L. pusilla* and *L. barkeriana* is the gamophylly (fusion) of the thickened bases of the previous growth cycle within the bulb (Müller-Doblies *et al.*, 1987).

DISTRIBUTION, HABITAT AND CONSERVATION STATUS. *L. pusilla* has a fairly wide distribution in the winter rainfall zone extending from the Bokkeveld Plateau at Nieuwoudtville, south to the Cederberg and Piketberg, to Velddrif on the Cape west coast, inland to the Worcester district, south-east to Bredasdorp and east to the Duiwenhoks River Gorge east of Swellendam in the southern Cape (Map 28). It occurs in ten vegetation types including Nieuwoudtville-Roggeveld Dolerite Renosterveld, Bokkeveld-, Cederberg- and Piketberg Sandstone Fynbos, Leipoldtville- and Hopefield Sand Fynbos, Lamberts Bay Strandveld, Breede Alluvium Fynbos and Central- and Eastern Rûens Shale Renosterveld. It is most frequently encountered in the western and southern parts of the Western

Cape, growing on low-lying sandy and gravelly flats or on rocky east- or south-facing hillsides, ridges, mountain passes and plateaus, amongst boulders and in rock crevices. The leaves of *L. pusilla* show geographic variation, ranging from broadly lanceolate, such as those on plants from Heerenlogement south-west of Vanrhynsdorp, to narrowly lanceolate, such as those on plants near Algeria in the Cederberg. Plants occur singly or in small groups as part of larger colonies, in full sun. The species is not threatened.

NOTES. The infructescence of *L. pusilla* is elongated, like those of several other dwarf lachenalias, including *L. barkeriana*. Before capsule dehiscence, the abbreviated, subterranean scapes of these species elongate before falling into a horizontal position. This results in the base of the scape breaking off, allowing the whole infructescence to be carried away by wind, thereby dispersing seeds over a wide area (Duncan, 2005). Vogel (1954) pointed to the specialised floral ecology of *L. pusilla*, which exhibits the mellitophilous brush mechanism (also seen in *L. barkeriana* and *L. pygmaea*) in which the exserted, erect staminal filaments act as the main visual attractant of the inflorescence.

The flowers of both *L. pusilla* and *L. barkeriana* are very short-lived (up to six days) and emit an unusual, spicy scent by day and by night that is similar to, but stronger than, that emitted by the rodent-pollinated *Massonia depressa* (Johnson *et al.*, 2001). It is possible that *L. pusilla* and *L. barkeriana* might also be pollinated by rodents, although preliminary chemical analysis of the scents of these two species points to bee pollination (Johnson, pers. comm.). The flowers of *L. pusilla* and *L. barkeriana* are self-fertile, producing large numbers of seed regardless of whether they have been visited by animals (Duncan, 2005). Müller-Doblies *et al.* (1987) reported that the prominent strophioles of both of these species act as elaiosomes, promoting dispersal by ants.

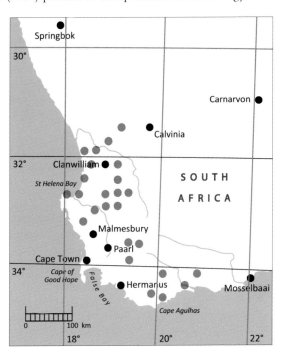

Map 28. Known distribution of *Lachenalia pusilla*.

1.2. Section OBLONGAE

Oblongae G. D. Duncan, sect. nov., *a ceteris sectionibus perianthio oblongo-campanulato, tubo cupulato raro rotato, tepalis interioribus expansis 15–20°, filamentis inclusis usque ad bene exsertis differt.* Type: *Lachenalia mediana* Jacq.

Perianth oblong-campanulate; tube cup- or rarely disc-shaped, up to 3 mm long; outer tepals ovate, straight to weakly convex; inner tepals obovate, radiating 15–20° from the longitudinal axis; filaments straight, recurved or declinate, included to well exserted.

58 species (59 taxa).

28. LACHENALIA MEDIANA

Lachenalia mediana Jacq., *Collectanea* 3: 242–243 (1790). *Scillopsis mediana* (Jacq.) Lem., *L'Illustration Horticole* 3: 35 (1856).
TYPE: South Africa, Cape, collector and precise locality unknown, figure in N. J. Jacquin, *Icones Plantarum Rariorum* 2: 6, t. 392 (1790) (neotype, designated here).

ETYMOLOGY. *mediana*: intermediate, with reference to the intermediate resemblance of the flowers to those of *L. pallida* and *L. orchioides*.

DESCRIPTION. *Geophyte*, 120–400 mm high. *Bulb* globose, 10–15 mm in diameter, offset-forming; tunic multilayered, outer layers light to dark brown, spongy; inner cataphyll translucent white, adhering to leaf bases, apex obtuse. *Leaves* 1–2, lanceolate, 90–300 × 10–25 mm, suberect to spreading, slightly to deeply conduplicate, bright green, upper surface with depressed longitudinal grooves; leaf margins flat to undulate and crisped; leaf base clasping, 20–100 mm long, lower portion white, upper portion light green, plain, minutely maroon speckled, heavily flushed with dark maroon or with transverse magenta bands; primary seedling leaf flat, prostrate. *Inflorescence* racemose, few- to many-flowered, fairly dense, sterile apex short; scape erect, 75–120 mm long, slender to sturdy, light green, plain or spotted to blotched with dark maroon or brown; rachis light bluish-green in lower half, shading to light blue above; pedicels suberect, becoming erect in fruit, 3–10 mm long, white to greenish-white; lower bracts ovate, becoming lanceolate above, 2–6 × 1–4 mm, white to whitish-green. *Perianth* zygomorphic, oblong-campanulate, spreading to slightly suberect or cernuous; tube cup-shaped, 3 mm long, light blue or white; outer tepals ovate, 6–8 × 4–5 mm, grey, light blue or rarely light pink, apices flat to slightly recurved, apical gibbosities light to dark green, greenish-blue or brown, median keels dull blue; inner tepals obovate, 8–10 × 3–6 mm, translucent white with dark blue median keels, apices flat or recurved, lower inner tepal equal to or slightly longer (1 mm) than upper inner tepals. *Stamens* included to shortly exserted, declinate; filaments white, 9–11 mm long. *Ovary* ellipsoid, 3–4 × 2–3 mm, light green; style included to shortly exserted, declinate, 7–9 mm long, white. *Capsule* ellipsoid, 7–9 × 5–7 mm. *Seed* globose, 1.2–1.3 × 1.3 mm, glossy, black; strophiole inflated, 1.1–1.2 mm long, smooth. *Chromosome number*: see under subspecies below. Figures 130–131.

FLOWERING PERIOD. August to November, with a peak from mid-August to early September.

HISTORY. N. J. Jacquin (1790) described *L. mediana* in the third volume of *Collectanea* and, in the same year, published a life-sized colour plate in the second volume of his lavish work, the *Icones Plantarum Rariorum*. Following the loss of Jacquin's holotype sheet in a fire between 1939–1945, the plate is designated here as the neotype, as it did not form part of the protologue when the species was described. Lemaire (1856) transferred the species to *Scillopsis* as *S. mediana* (Jacq.) Lem. in volume three of his *L'Illustration Horticole*, but it was returned to *Lachenalia* by Baker in his first revision of the genus (1871) and upheld in his second and final work (1897a).

The plant now known as *L. mediana* subsp. *rogersii* was originally described as *L. unifolia* var. *rogersii* in *Flora Capensis* (Baker, 1897a), from material collected at an unrecorded locality by the Rev. W. M. Rogers and from another collection made at Tulbagh by the German physician and botanist C. W. L. Pappe. W. F. Barker (1987) considered this taxon more appropriately placed under *L. mediana*, mainly on the grounds of seed morphology, in particular the inflated strophiole that is characteristic of the species, and made the new combination *L. mediana* Jacq. var. *rogersii* (Baker) W. F. Barker.

DISTINGUISHING CHARACTERS AND AFFINITIES. Duncan *et al.* (2005a) suggested that this species is closely allied with *L. thomasiae*, with which it shares oblong-campanulate flowers with declinate, white stamens, lanceolate, canaliculate leaves and a similar globose seed with an inflated strophiole. *L. thomasiae* differs

Figure 130 (left). *Lachenalia mediana* subsp. *mediana* in shade at Observatory, northern Cape Peninsula. Image: Graham Duncan.

Figure 131 (right). *Lachenalia mediana* subsp. *rogersii* in wet clay near Tulbagh. Image: Graham Duncan.

from *L. mediana* in its bright white, strongly cernuous perianth with well-exserted stamens, its pedicels that do not re-orientate to erect in fruit, its shorter strophiole (0.8 mm long) and its subglobose bulbs.

The *L. mediana* taxa (var. *mediana* and var. *rogersii*) are here elevated to subspecific rank as they differ morphologically and breed true in allopatric populations, indicating gene fixation and a step towards speciation. The subsp. *rogersii* differs from subsp. *mediana* in having flat inner tepal apices with the lower inner tepal equal to the upper inner tepals, included filaments, longer styles (8–9 mm long), undulate and crisped leaf margins and banded leaf bases. It has a geographically disjunct distribution, from Hermon to Porterville, and a preference for seasonally wet clay sites.

DISTRIBUTION AND HABITAT. *L. mediana* is endemic to the south-western and southern part of the Western Cape, from Darling on the west coast, south-east to Caledon and inland to Porterville (Map 29). The plants grow in small to large colonies in seasonally moist to wet clay soils. For further details, see under subspecies below.

NOTES. Johnson & Brandham (1997) obtained a chromosome count of $2n = 14$ for a specimen identified as *L. mediana* var. *mediana*, but this may have been the result of a misidentification as all other counts done for the species thus far have given $2n = 26$ (Crosby, 1986; Spies *et al.*, 2008). *L. mediana* is pollinated by honey bees (*Apis mellifera*). During the fruiting stage the pedicels and developing fruits become re-orientated from a suberect to an erect position. The scapes remain attached for several weeks following capsule dehiscence and the seeds are dispersed locally.

28. LACHENALIA MEDIANA

Key to the subspecies

1a Leaf margins flat, leaf bases plain, marked with minute maroon speckles or flushed with dark maroon; perianth spreading to slightly suberect, outer tepals grey, 7–8 mm long, inner tepal apices recurved, lower inner tepal 1 mm longer than upper inner tepals; filaments usually shortly exserted; style 7 mm long; Darling to Caledon . . . . **a.** subsp. **mediana**
1b Leaf margins undulate and crisped, leaf bases banded with magenta; perianth cernuous, outer tepals light blue or light pink, 6–7 mm long, inner tepal apices flat, lower inner tepal equal to upper inner tepals; filaments always included; style 8–9 mm long; Hermon to Porterville . . . . **b.** subsp. **rogersii**

a. subsp. mediana

DESCRIPTION. *Plant* 120–400 mm high. *Bulb* 10–15 mm in diameter. *Leaf* 90–300 × 10–25 mm; leaf bases 40–100 mm long, with minute maroon speckles or heavily flushed with dark maroon. *Scape* 60–150 mm long; rachis light bluish-green in lower half, shading to light blue in upper half; pedicels 3–5 mm long, bracts 3–5 × 1–4 mm. *Perianth* spreading to slightly suberect, outer tepals 7–8 × 4–5 mm, grey with darker blue median keels, apical gibbosities greenish-blue; inner tepals 8–10 × 5–6 mm, apices recurved, lower inner tepal 1 mm longer than upper inner tepals. *Stamens* included to shortly exserted; filaments 9–10 mm long. *Ovary* 3–4 × 2 mm; style 7 mm long. *Capsule* 8–9 × 6–7 mm. *Seed* 1.2–1.3 × 1.3 mm; strophiole 1.1–1.2 mm long. *Chromosome number* 2n = 14 (Johnson & Brandham, 1997); 2n = 26 (Crosby, 1986). Figure 130.

FLOWERING PERIOD. August to October, with a peak in mid-August.

DISTRIBUTION, HABITAT AND CONSERVATION STATUS. Subsp. *mediana* occurs in the south-western and southern Western Cape extending from Darling on the Cape west coast, south-east to Caledon (Map 29). The plants occur in four vegetation types: Swartland Shale Renosterveld, Swartland Granite Renosterveld, Peninsula Shale Renosterveld and Western Rûens Shale Renosterveld (Mucina & Rutherford, 2006), growing in colonies in heavy clay on north-, south-, and west-facing hill slopes and in open aspects. The taxon is threatened by the expansion of vineyards and wheat farming, housing development and encroaching alien plants. A population at the South African Astronomical Observatory in the Cape Town suburb of Observatory still survives in semi-shaded conditions amongst alien grasses underneath alien *Pinus pinea* trees. The subsp. *mediana* has a conservation status of Vulnerable (Raimondo & Helme, 2009a).

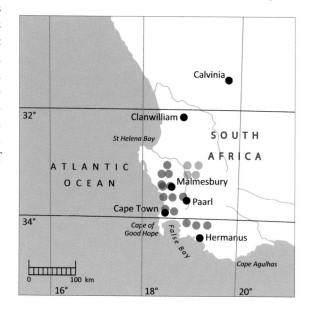

Map 29. Known distribution of *Lachenalia mediana*:
● = subsp. *mediana*; ● = subsp. *rogersii*.

b. subsp. rogersii (Baker) G. D. Duncan, stat. nov. *Lachenalia unifolia* Jacq. var. *rogersii* Baker, in Thiselton-Dyer, W. T. (ed.), *Flora Capensis* 6: 431 (1897); *Lachenalia mediana* Jacq. var. *rogersii* (Baker) W. F. Barker, *South African Journal of Botany* 55(6): 642–643 (1989).

SYNTYPES: South Africa, Cape, precise locality unknown, *W. M. Rogers s.n.* (K!); Tulbagh, *C. W. L. Pappe s.n.* (K!, lecto., designated by Barker (1989); sub SAM!, 23131, isolecto.).

ETYMOLOGY. *rogersii*: after Rev. W. M. Rogers (1835–1920), English plant collector, mainly of the southern Cape.

DESCRIPTION. *Plant* 150–330 mm high. *Bulb* 12–15 mm in diameter. *Leaf* 100–140 × 15–25 mm, lower surface with sporadic purplish-maroon blotches; leaf margins undulate and crisped; leaf bases 20–40 mm long, heavily marked with transverse magenta bands. *Scape* 75–112 mm long, lightly to heavily spotted with dark maroon; pedicels 2–10 mm long; bracts 2–6 × 1–3 mm. *Perianth* slightly cernuous in flower, becoming erect in fruit, outer tepals 6–7 × 4–5 mm, light blue or light pink, apices flat; apical gibbosities light to dark green; inner tepals 9–10 × 3–4 mm, median keels light blue near apex. *Stamens* included; filaments 9–11 mm long. *Ovary* 3–4 × 2–3 mm; style 8–9 mm long. *Capsule* 7–9 × 5–6 mm. *Seed* 1.2 × 1.3 mm; strophiole 1.1 mm long. *Chromosome number* $2n = 26$ (Spies *et al.*, 2008). Figure 131.

FLOWERING PERIOD. August to September, with a peak in early September.

DISTRIBUTION, HABITAT AND CONSERVATION STATUS. The subsp. *rogersii* is restricted to fragmented locations in the Hermon, Saron, Tulbagh and Porterville districts of the south-western Cape (Map 29). It occurs in Breede- and Swartland Shale Renosterveld vegetation (Mucina & Rutherford, 2006), on flats and lower slopes in seasonally wet, heavy clay, often along swamp and stream margins. At Saron north of Tulbagh, it grows close to a large population of *L. bachmannii*, and at a seasonally inundated site south of Tulbagh, it grows in association with *Ornithogalum thyrsoides* (Asparagaceae). Subspecies *rogersii* plant numbers have been severely affected by wheat cultivation, overgrazing and invasive alien plants, and the taxon has a conservation status of Endangered (Raimondo & Helme, 2009b).

29. LACHENALIA PALLIDA

Lachenalia pallida Aiton, *Hortus Kewensis* 1: 460 (1789). *Orchiastrum pallidum* (Aiton) Lem., *L'Illustration Horticole* 2: 100 (1855).

TYPE: South Africa, Cape, collector and precise locality unknown, cultivated specimen (BM!, holo.).

SYNONYMY: *Lachenalia pustulata* Jacq., *Collectanea* 3: 244–245 (1790). *Scillopsis pustulata* (Jacq.) Lem., *L'Illustration Horticole* 3: 34 (1856). Type: South Africa, Cape, collector and precise locality unknown, figure in N. J. Jacquin, *Icones Plantarum Rariorum* 2: 6, t. 386 (1790) (neotype, designated here).

Lachenalia fragrans Jacq., *Plantarum rariorum horti caesarei Schönbrunnensis* 1: 43, t. 82 (1797), non *Lachenalia fragrans* Andrews, *The Botanist's Repository* 5: t. 302 (1803), non *Lachenalia fragrans* Lodd., *Botanical Cabinet*: t. 1140 (1826). *Scillopsis fragrans* (Jacq.) Lem., *L'Illustration Horticole* 3: 34 (1856). *Lachenalia versicolor* Baker var. *fragrans* (Jacq.) Baker, *Journal of the Linnean Society (Botany)* 11: 409 (1871), nom. inval. *Lachenalia unicolor* Jacq. var. *fragrans* (Jacq.) Baker, in Thiselton-Dyer, W. T. (ed.), *Flora Capensis* 6: 435 (1897). Type: South Africa, Cape, collector and precise locality unknown, figure in N. J. Jacquin, *Plantarum rariorum horti caesarei Schönbrunnensis* 1: 43, t. 82 (1797) (lectotype, designated here).

Lachenalia unicolor Jacq., *Collectanea* 5: 61–62 (1797). *Lachenalia versicolor* Baker var. *unicolor* (Jacq.) Baker, *Journal of the Linnean Society (Botany)* 11: 409 (1871), nom. inval. *Scillopsis unicolor* (Jacq.) Lem., *L'Illustration Horticole* 3: 34 (1856). Type: South Africa, Cape, collector and locality unknown, figure in N. J. Jacquin, *Icones Plantarum Rariorum* 2: 16, t. 389 (1795) (neotype, designated here).

Lachenalia purpurea Jacq., *Collectanea* 5: 65–66 (1797). *Scillopsis purpurea* (Jacq.) Lem., *L'Illustration Horticole* 3: 35 (1856); *Lachenalia versicolor* Baker var. *purpurea* (Jacq.) Baker, *Journal of the Linnean Society (Botany)* 11: 409 (1871), nom. inval. *Lachenalia unicolor* Jacq. var. *purpurea* (Jacq.) Baker, in Thiselton-Dyer, W. T. (ed.), *Flora Capensis* 6: 435 (1897). Type: South Africa, Cape, figure in N. J. Jacquin, *Icones Plantarum Rariorum* 2: 12, t. 393 (1792 or 1793) (neotype, designated here).

Lachenalia lucida Ker Gawl., *Curtis's Botanical Magazine* 33: t. 1372 (1811). *Scillopsis lucida* (Ker Gawl.) Lem., *L'Illustration Horticole* 3: 34 (1856). Type: South Africa, Cape, collector and precise locality unknown, figure in *Curtis's Botanical Magazine* 33: t. 1372 (1811) (lectotype, designated here).

Lachenalia racemosa Ker Gawl., *Curtis's Botanical Magazine* 37: t. 1517 (1813). *Scillopsis racemosa* (Ker Gawl.) Lem., *L'Illustration Horticole* 3: 34 (1856). Type: South Africa, Cape, collector and locality unknown, figure in *Curtis's Botanical Magazine* t. 1517 (1813) (lectotype, designated here).

Lachenalia pustulata Jacq. var. *densiflora* Tratt., *Archiv der Gewachskunde* 1: t. 94 (1813). Type: South Africa, Cape, collector and precise locality unknown, figure in *Archiv der Gewachskunde* 1: t. 94 (1813) (lectotype, designated here).

Lachenalia pustulata F. Dietr., *Nachtrag der Lexicon der Gartnerei und Botanik* 4: 291–292 (1818), illegitimate homonym. Type: as for *L. pustulata* Jacq. var. *densiflora* Tratt.

Lachenalia reclinata F. Dietr., *Nachtrag der Lexicon der Gartnerei und Botanik* 4: 292–293 (1818), nom. superfl. Type: as for *L. pustulata* Jacq.

Lachenalia bicolor Lodd., *Loddiges Botanical Cabinet*: t. 1129 (1826). Type: South Africa, Cape, collector and precise locality unknown, figure in *Loddiges Botanical Cabinet*: t. 1129 (1826) (lectotype, designated here).

Lachenalia pyramidalis Dehnh., *Rivista Napolitana* 1(1): 162 (1839). Type: South Africa, Cape, collector and precise locality unknown, figure in N. J. Jacquin, *Plantarum rariorum horti caesarei Schönbrunnensis* 1, t. 82 (1797) (neotype, designated here).

Lachenalia odoratissima Baker, *The Gardener's Chronicle* 21 (Series 2): 668 (1884). Type: South Africa, Cape, collector and precise locality unknown, cultivated specimen, *ex hort T. S. Ware*, April 1884 (K!, holo.).

Lachenalia gillettii W. F. Barker, *The Flowering Plants of South Africa* 13: t. 506 (1933). Type: South Africa, Western Cape, Modderfontein in Olifants River Valley, *J. B. Gillett* 3650 (BOL!, holo.).

ETYMOLOGY. *pallida*: pale, with reference to the tepals.

DESCRIPTION. *Geophyte*, 150–400 mm high. *Bulb* subglobose, 15–20 mm in diameter, offset-forming, light to deep yellow; tunic multilayered, outer layers spongy, dark brown; inner cataphyll translucent white, adhering to leaf bases, apex acute or obtuse. *Leaves* 1–2, narrowly to broadly lanceolate, 60–270 × 15–40 mm, spreading to suberect, bright green to olive green, sometimes suffused with purplish-brown, upper surface smooth or with small to large dome-shaped pustules, depressed longitudinal grooves distinct; leaf bases clasping, 30–70 mm long, green or suffused with maroonish-magenta; leaf margins flat, rarely slightly undulate, white or maroon; primary seedling leaf flat, prostrate. *Inflorescence* racemose, many-flowered, sterile apex short to long; scape erect to suberect,

40–170 mm long, slender, lax, light to dark green, plain or with light purplish-brown blotches, rachis light green, plain or heavily suffused with brownish-magenta, uppermost portion and sterile apex shading to pastel blue, purple or pink; pedicels suberect at anthesis, 1–10 mm long, white, green, blue, mauve or purple, becoming erect in fruit; lower bracts ovate, becoming lanceolate above, 1–4 × 0.8–3.0 mm, light green to white. *Perianth* zygomorphic, oblong-campanulate, spreading, cernuous or suberect, dark purple, mauve, magenta, blue, light to dark yellow, cream or white, yellow forms ageing to dull red, slightly to strongly spice-scented; tube cup-shaped, 1–3 mm long; outer tepals ovate, 4–7 × 3–5 mm, apical gibbosities and median keels dark brown, green, magenta or purple; inner tepals obovate, 6–10 × 4–6 mm, median keels and apical markings dark brown, green, magenta or purple. *Stamens* included to well exserted, declinate; filaments white or light to dark purple, 6–14 mm long. *Ovary* ellipsoid, 2–4 × 2–3 mm, light to dark green; style included to well exserted, declinate, 7–10 mm long, white or light to dark purple. *Capsule* ellipsoid, 7–12 × 4–6 mm. *Seed* globose, 0.9–1.0 × 0.8–1.0 mm, glossy, black; strophiole rudimentary, 0.1–0.2 mm long, ridged. *Chromosome number* 2n = 16 (Moffett, 1936; Crosby, 1986; Johnson & Brandham, 1997; Hamatani *et al.*, 2004; Spies *et al.*, 2000, 2008). Figures 18, 19, 36, 43, 72, 73, 132–137.

Figure 132 (left). Yellow form of *Lachenalia pallida* on gravelly flats near Malmesbury. Image: Graham Duncan.
Figure 133 (right). Cream form of *Lachenalia pallida* on stony flats near Porterville. Image: Graham Duncan.

29. LACHENALIA PALLIDA

Figure 134 (left). Pink form of *Lachenalia pallida* with *Geissorhiza aspera* on sandy flats near Tulbagh. Image: Graham Duncan.

Figure 135 (right). Early-flowering mauve form of *Lachenalia pallida* on a granite outcrop near Vredenburg. Image: Dennis Tsang.

FLOWERING PERIOD. August to November, with a peak in late August.

HISTORY. Extensive synonymy bears testimony to the confusion that has prevailed for more than two centuries over the application of the concept of *L. pallida* Aiton. The species was described by William Aiton (1789) in the first edition of volume one of *Hortus Kewensis*, having been introduced into cultivation at Kew by George Wench in 1782. The holotype sheet in the British Museum (Natural History) comprises a single inflorescence, of which the collector and provenance is unknown. Following Aiton's publication, various forms of this species (as circumscribed here) were described. N. J. Jacquin published three species in *Collectanea*, *L. pustulata* (1790), *L. unicolor* (1797) and *L. purpurea* (1797), and plates of these in volume 2 of his *Icones Plantarum Rariorum* between 1790 and 1795. He also published *L. fragrans* in volume 1 of his *Plantarum rariorum horti caesarei Schönbrunnensis* (Jacquin, 1797b). These were followed by Ker Gawler's *L. lucida* (1811) and *L. racemosa* (1813), Trattinnick's *L. pustulata* var. *densiflora* (1813), Dietrich's *L. reclinata* (1818), Loddiges's *L. bicolor* (1826a), Dehnhardt's *L. pyramidalis* (1839), Baker's *L. odoratissima* (1884) and Barker's *L. gillettii* (1933c). Dehnhardt (1839) published *L. pyramidalis* without any reference to type material or illustration, but his description is comparable to Jacquin's plate 82 (1797) of his white-flowered, heavily scented *L. fragrans*. Baker (1871) reduced Jacquin's *L. fragrans*, *L. purpurea* and *L. unicolor* to varieties of *L. versicolor* Baker, but the latter name is invalid as he based it on the three species (*L. fragrans*, *L. purpurea* and *L. unicolor*), all published much earlier by Jacquin. Baker reversed the latter placements in his 1897 monograph, recognised Jacquin's *L. unicolor* as a species and reduced the latter author's *L. fragrans* and *L. purpurea* to varieties of *L. unicolor*. In the same work, Baker mistakenly cited a collection from Klipfontein near Carnarvon in the Great Karoo (*Burchell* 1534, in K) under *L. pustulata*, this is correctly *L. karooica*.

Aiton's specific epithet *pallida* perfectly describes the light yellow forms of the species, so it is unfortunate that, being the earliest validly published name, it has to take precedence over later epithets that are far more descriptive of the numerous very attractive forms of this species.

DISTINGUISHING CHARACTERS AND AFFINITIES. *L. pallida* is regarded here as a distinct, but highly polymorphic, species complex represented by numerous large, more or less continuously varying populations. Perianth, leaf, bulb and seed shape and stamen orientation are stable across the species, but perianth size and orientation, tepal and stamen colour, degree of stamen exsertion, leaf orientation and the presence or absence of leaf upper surface pustules is highly variable. By way of example, N. J. Jacquin regarded his *L. pustulata* as distinct from his later *L. unicolor* on account of its shorter, white pedicels, white perianth and shortly exserted stamens with white filaments, as opposed to the uniform mauve colouring of the much longer pedicels, perianth and well exserted filaments of *L. unicolor*. *L. pustulata* can have pustulate or smooth upper surfaces and included or shortly exserted stamens, depending on its provenance. Individuals with both smooth and pustulate upper surfaces, and included and shortly exserted stamens, sometimes occur within the same population. Jacquin described *L. purpurea* on account of its purple inner tepals, and distinguished *L. fragrans* because of its sweet-scented flowers, but these too, merely represent forms of an extremely variable species. Similarly, the relatively recently described *L. gillettii* is merely a small-flowered form with very short pedicels, shortly exserted stamens and usually non-pustulate leaves.

Figure 136. Late-flowering deep mauve form of *Lachenalia pallida* on a granite outcrop near Cape Columbine. Image: Graham Duncan.

Figure 137. Purple form of *Lachenalia pallida* en masse near Vanrhynsdorp. Image: Sandra Muller.

Duncan *et al.* (2005a) independently analysed the three species *L. gillettii*, *L. pustulata* and *L. unicolor*, now considered synonymous with *L. pallida*. These three taxa resolved as part of a clade that included *L. purpureo-caerulea*. *L. purpureo-caerulea* resembles purple forms of *L. pallida* in having a similar raceme of oblong-campanulate flowers with declinate, purple stamens and lanceolate leaves, but differs in having widely flared inner tepals that protrude only slightly beyond the outer tepals, which have recurved apices, and in its proteranthous leaves with obtuse apices and seeds with slightly longer strophioles (0.2–0.3 mm long).

DISTRIBUTION, HABITAT AND CONSERVATION STATUS. *L. pallida* occurs in the north-western, western and south-western parts of the Western Cape, extending from just east of Vanrhynsdorp, southwards along the west and south coasts, to just east of Gansbaai, and inland to Worcester (Map 30). Outer tepal colour usually ranges in shades of deep purple or mauve in the northern and western parts, from Vanrhynsdorp to Moorreesburg (deep purple), St Helena Bay to Langebaan (mauve) and Tulbagh (pinkish-mauve). By contrast, in the south-western parts outer tepal colour varies mainly in shades of yellow, cream or white, from Malmesbury and Hopefield (yellow) to Wellington (white), Worcester (cream) and the Cape Peninsula (cream), and south-east to Gansbaai (yellow). The species occurs on a wide variety of substrates, including deep red sand at Leipoldtville south-east of Lambert's Bay. It grows on sandy flats or limestone outcrops at Cape Columbine and Gansbaai, respectively, in humus-rich matter on granite outcrops at Vredenburg, on shale flats in renosterveld around Piketberg and Eendekuil, and in seasonally inundated clay along marshes at Muldersvlei near Stellenbosch.

The species is encountered in at least 11 vegetation types including Vanrhynsdorp Gannabosveld, Leipoldtville- and Hopefield Sand Fynbos, Citrusdal Vygieveld, Breede-, Peninsula- and Swartland Shale Renosterveld, Swartland Granite Renosterveld, Saldanha Granite Strandveld, Peninsula

Granite Fynbos and Overberg Sandstone Fynbos (Mucina & Rutherford, 2006). It has a long flowering period extending from mid-August to early November, three of the latest-flowering forms are a large, mauve-flowered form that grows between crevices of granite outcrops and in rock pans at Cape Columbine, within reach of the sea spray, another a small-flowered, cream form on north-facing clay slopes at Schaapenberg on Vergelegen Farm near Somerset West, and a larger cream-flowered form at Noordhoek in the southern Cape Peninsula.

The species still survives close to Cape Town on south-facing granite slopes on Blaauwberg Hill just north of the city, as well as on the Peninsula itself at Camps Bay, within the grounds of the South African Astronomical Observatory in the suburb of Observatory, and on granite outcrops at Noordhoek. At its most eastern boundary near Gansbaai, it occurs in semi-shade in openings of coastal forest. *L. pallida* usually grows in very large colonies, sometimes numbering thousands of individuals, such as the light yellow form seen around Hopefield, the mauve forms at Saldanha and the deep purple form near Vanrhynsdorp.

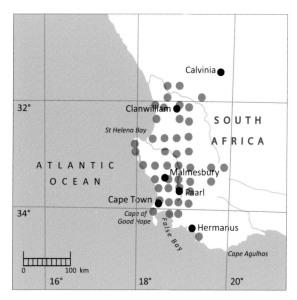

Map 30. Known distribution of *Lachenalia pallida*.

Owing mainly to agricultural extension, *L. pallida* numbers have been greatly reduced in the northern, western and south-western Swartland area of the Western Cape, but the species is still plentiful across its range and by virtue of its wide distribution is not considered threatened.

NOTES. *L. pallida* is pollinated by honey bees (*Apis mellifera*) and the solitary bee *Anthophora diversipes* (Figures 72, 73). It is also visited by the Painted Lady butterfly (*Vanessa cardui*) (Figure 70). The scapes remain attached for several weeks following capsule dehiscence and the seeds are dispersed locally.

30. LACHENALIA PURPUREO-CAERULEA

Lachenalia purpureo-caerulea Jacq. (sphalm. *purpurocaerulea*), *Collectanea* 5: 63–64 (1797). *Scillopsis purpureo-caerulea* (Jacq.) Lem., *L'Illustration Horticole* 3: 35 (1856).
TYPE: South Africa, Cape, collector and precise locality unknown, figure in N. J. Jacquin, *Icones Plantarum Rariorum* 2(14): t. 388 (1793 or 1794) (neotype, designated here).
SYNONYMY: *Lachenalia botryoides* Tratt., *Archiv der Gewachskunde* 2: t. 140 (1814), nom. superfl. Type: South Africa, Cape, collector and precise locality unknown, figure in *Archiv der Gewachskunde* 2, t. 140 (1814) (lectotype, designated here).

ETYMOLOGY. *purpureo-caerulea*: purplish-blue flowers.
DESCRIPTION. *Geophyte*, 100–250 mm high. *Bulb* subglobose, 15–25 mm in diameter, offset-forming, tissue yellow; tunic multilayered, outer layers light brown, spongy; inner cataphyll translucent white, loosely surrounding leaf bases, apex obtuse. *Leaves* 2, lanceolate, 40–100 × 8–20 mm, apices obtuse,

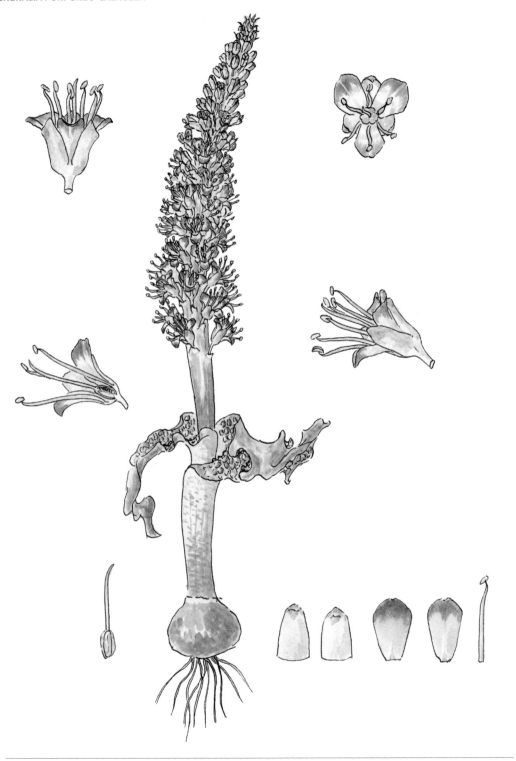

Plate 14. Watercolour painting of *Lachenalia purpureo-caerulea* from Mamre, courtesy of the Compton Herbarium, Kirstenbosch, South African National Biodiversity Institute. Image: Winsome Barker.

spreading to suberect, light to dark green, upper surface densely covered with large flattened pustules, depressed longitudinal grooves prominent; leaf margins maroon, coriaceous; leaf bases clasping, 30–70 mm long, white, subterranean; scape erect to suberect, 50–100 mm long, light green; primary seedling leaf flat, prostrate. *Inflorescence* racemose, many-flowered, dense, sterile apex short; pedicels suberect, 2–4 mm long, mauve; bracts ovate, 1–2 × 1–4 mm, white. *Perianth* zygomorphic, oblong-campanulate, cernuous to suberect, spicy sweet-scented; tube shallowly cup-shaped, 1–2 mm long, white to light blue; outer tepals ovate, 4–5 × 2–3 mm, white to light blue, apices and apical gibbosities deep purple; inner tepals obovate, 5–6 × 3–4 mm, lower half white, upper half purple, apices obtuse, widely flared, median keels deep purple. *Stamens* well exserted, filaments declinate, white or light purple in lower half, shading to purplish-blue above, 6–10 mm long. *Ovary* ellipsoid, 3–4 × 2–3 mm, light green; style well exserted, declinate, 8–10 mm long, white in lower half, deep purple in upper half. *Capsule* ellipsoid, 5–7 × 4–5 mm. *Seed* globose, 0.9–1.0 × 0.8–0.9 mm, glossy, black; strophiole rudimentary, 0.2–0.3 mm long, ridged. *Chromosome number* 2n = 16 (Johnson & Brandham, 1997). Plate 14, Figure 138.

FLOWERING PERIOD. Late September to mid-November, with a peak in late October.

HISTORY. *L. purpureo-caerulea* was illustrated by N. J. Jacquin in the second volume of his *Icones Plantarum Rariorum* in 1793 or 1794 and described in volume 5 of *Collectanea* (Jacquin, 1797a). The colour plate is designated here as the neotype, the holotype material having been destroyed by fire during the period 1939–1945. Richard Williams of Turnham Green, London introduced it into cultivation at Kew in 1798 (Aiton, 1811). Leopold Trattinnick (1814a) described and illustrated *L. botryoides* in volume 2 of his *Archiv der Gewachskunde*, the monochrome plate of which is a replica of

Map 31. Known distribution of *Lachenalia purpureo-caerulea*.

Figure 138. *Lachenalia purpureo-caerulea* in gravelly sand near Darling. Image: Helene Preston.

Andrews's plate 251 of *L. purpureo-caerulea* in volume 4 of *The Botanist's Repository* (Andrews, 1802), but it is synonymous with the latter species and a superfluous name. Lemaire (1856) transferred *L. purpureo-caerulea* to *Scillopsis* as *S. purpureo-caerulea* in volume 3 of his *L'Illustration Horticole*, but it was returned to *Lachenalia* by Baker (1871) who upheld this taxonomy in his final monograph of the genus (Baker, 1897a). An excellent watercolour rendition of the plant by Hilda Mason was published in *Western Cape Sandveld Flowers* (Mason & du Plessis, 1972).

DISTINGUISHING CHARACTERS AND AFFINITIES. Duncan *et al.* (2005) suggested a close alliance between *L. purpureo-caerulea* and *L. pallida*. The latter species resembles *L. purpureo-caerulea* in its similar raceme of oblong-campanulate flowers with declinate stamens and lanceolate leaves, but differs in its narrowly flared inner tepals that protrude well beyond the outer tepals and have non-recurved apices, in its synanthous leaves with acute apices and in its slightly shorter strophiole (0.1–0.2 mm long). Most forms of *L. pallida* flower much earlier than *L. purpureo-caerulea*.

DISTRIBUTION, HABITAT AND CONSERVATION STATUS. *L. purpureo-caerulea* is confined to a small area of the Cape west coast and is currently known from just two severely fragmented locations, one to the west of Darling and another to the north of this town (Map 31). In the 1860s, Prof. Peter Macowan recorded the species at Kalabaskraal, 40 km north of Cape Town (*Macowan* 2604, in K), but it is extinct there. *L. purpureo-caerulea* occurs in Swartland Granite Renosterveld vegetation (Mucina & Rutherford, 2006), on lower hill slopes and seasonally moist or waterlogged flats in deep gravelly sand. At a locality north of Darling, it grows in association with *Ixia curta* (Iridaceae) and the annual *Nemesia strumosa* (Scrophulariaceae) (Helene Preston, pers. comm.). The species' habitat has been severely compromised by wheat cultivation and is also threatened by invasive annual grasses and overgrazing. *L. purpureo-caerulea* has a conservation status of Critically Endangered (Raimondo & Duncan, 2009e).

NOTES. The spicy sweet-scented flowers of *L. purpureo-caerulea* are pollinated by honey bees (*Apis mellifera*) and solitary anthophorid bees (*Anthophora diversipes*). The foliage is heavily grazed, probably by Cape hares (*Lepus capensis*). In the wild, plant growth is erratic, with a percentage of individuals remaining dormant during the growing season, even in years of plentiful rainfall, yet this phenomenon does not occur in cultivation. The scapes remain attached to the bulbs for several weeks following capsule dehiscence and the seeds are locally dispersed.

31. LACHENALIA NARDOUSBERGENSIS

Lachenalia nardousbergensis G. D. Duncan, *Bothalia* 36 (2): 152–155 (2006).

TYPE: South Africa, Western Cape, road to Nardousberge Plateau south-east of Klawer, in deep red sand, *W. F. Barker* 3630 (NBG!, holo.).

ETYMOLOGY. *nardousbergensis*: after the Nardousberg Mountains south-east of Klawer in the Western Cape, where the type collection was made.

DESCRIPTION. *Geophyte*, 150–310 mm high. *Bulb* subglobose, 15–20 mm diameter, dark yellow, solitary, occasionally offset-forming; tunic multilayered, outer layers dark brown, spongy; inner cataphyll translucent white, adhering to leaf bases, apex obtuse. *Leaves* 2, broadly lanceolate, 100–180 x 25–55 mm, prostrate, upper surface olive-green, covered with large purplish flattened pustules, depressed longitudinal grooves distinct; leaf margins coriaceous, white; leaf bases clasping, 10–20 mm long, subterranean, white; primary seedling leaf flat, prostrate. *Inflorescence* racemose, many-flowered, sterile apex short; scape erect, 80–120 mm long, sturdy, light green, heavily marked with small-to-large, light to dark brownish-purple blotches, upper portion often strongly inflated; rachis light green, lower portion often inflated, heavily marked with light to dark brownish-purple blotches; pedicels suberect, 2–6 mm long, brownish-green to brownish-magenta, length increasing towards inflorescence apex; lower bracts ovate, becoming lanceolate above, 1–5 × 3–5 mm, white. *Perianth* zygomorphic, oblong-campanulate, spreading to cernuous; tube cup-shaped, 1–2 mm long, white, or dull blue at base shading to white above; outer tepals ovate, 7–8 × 4–5 mm, light magenta, apical gibbosities and median keels dark- or brownish-magenta or green; inner tepals obovate, 8–9 × 4–5 mm, slightly spreading, light magenta, median keels prominent, dark magenta. *Stamens* well exserted, declinate; filaments white in lower half, shading to deep magenta above, 14–16 mm long. *Ovary* ellipsoid, 3 × 2 mm, light green; style well exserted, declinate, 13–14 mm long, lower half white, shading to light to deep magenta above. *Capsule* ellipsoid, 6–8 × 5–7 mm. *Seed* globose, 1.2 × 1.3 mm, glossy, black; strophiole 0.6–0.7 mm long, smooth. *Chromosome number* unknown. Figure 139.

FLOWERING PERIOD. Late August to early October with a peak in mid-September.

HISTORY. The earliest known collection of *L. nardousbergensis* was made by W. F. Barker on 22nd September 1945 near the Nardousberg Plateau south-east of Klawer (*Barker 3630*, in NBG). This collection serves as the holotype and the species was described more than 60 years later in *Bothalia* (Duncan & Edwards, 2006). No further collections appear to have been made until September 1968, when W. Chater collected it in the same area (*Chater s.n.*, in NBG). It has since been recorded further north at Nieuwoudtville and at several localities further south, in the northern Cederberg and at the northern end of the Piketberg.

Map 32. Known distribution of *Lachenalia nardousbergensis*.

Figure 139. *Lachenalia nardousbergensis* in deep sand, northern Pakhuis Pass. Image: Graham Duncan.

DISTINGUISHING CHARACTERS AND AFFINITIES. Phylogenetic analysis based on morphological data (Duncan *et al.*, 2005a) suggests a close relationship between *L. nardousbergensis* and *L. purpureo-caerulea*. They share oblong- campanulate flowers with well exserted, declinate stamens, lanceolate leaves with heavily pustulate upper surfaces, globose seeds and yellow bulb tissue. *L. purpureo-caerulea* differs in its non-inflated scape, white and purplish-blue, shorter tepals (outer tepals 4–5 mm long; inner tepals flared and 5–6 mm long), much shorter filaments and style (6–10 mm and 8–10 mm, respectively), and smaller seed (0.9–1.0 × 0.8–0.9 mm) with a much shorter strophiole (0.2–0.3 mm). The two species are geographically widely separated: *L. purpureo-caerulea* is confined to sandy or gravelly flats in the Darling District of the south-western Cape, and usually flowers later in the season than *L. nardousbergensis*, from late September to mid-November.

DISTRIBUTION, HABITAT AND CONSERVATION STATUS. *L. nardousbergensis* has a limited distribution in fynbos of the north-western parts of the Northern and Western Cape, extending from the Bokkeveld Plateau at Nieuwoudtville to the Nardousberg Plateau, south-east to the Pakhuis Pass and Middelburg Plateau of the Cederberg. There is an outlying population at the northern end of the Piketberg (Map 32). The plants traverse three vegetation types, Bokkeveld-, Cederberg- and Piketberg Sandstone Fynbos (Mucina & Rutherford, 2006), and grow in areas of fairly level, high-lying ground, in deep red or yellowish-brown sand. They grow as scattered individuals in exposed positions or in small colonies amongst low succulent undergrowth or between restio clumps. The species is not threatened.

NOTES. The flowers are pollinated by honey bees (*Apis mellifera*). The scapes detach from the bulbs shortly after capsule dehiscence and the ripe infructescence is blown away, dispersing the seeds.

32. LACHENALIA THOMASIAE

Lachenalia thomasiae W. F. Barker ex G. D. Duncan, *The Flowering Plants of Africa* 52 (2): t. 2061 (1993).

TYPE: South Africa, Western Cape, rocky ridge near Clanwilliam above Olifants River in dry, thorny succulent scrub, *M. L. Thomas* 271 (NBG!, holo.); *G. D. Duncan* 350 (NBG!, para.).

ETYMOLOGY. *thomasiae*: after Margaret Lilian Thomas (1917–2006), cultivator and collector of South African bulbous plants.

DESCRIPTION. *Geophyte*, 120–450 mm high. *Bulb* subglobose, 10–20 mm in diameter, offset-forming; tunic multilayered, outer layers light to dark brown, spongy; inner cataphyll translucent white, longitudinal veins green, loosely surrounding leaf bases, apex obtuse. *Leaves* 2, lanceolate, 80–300 × 10–40 mm, spreading to suberect, slightly to strongly canaliculate, light to deep green or yellowish-green, upper surface with depressed longitudinal grooves, lower leaf surface tinged with dark maroon in lower portion; margins slightly coriaceous, green or maroon; leaf bases clasping, 30–60 mm long, lower portion white, upper portion deep maroon; primary seedling leaf flat, prostrate. *Inflorescence* racemose, many-flowered, sterile apex long; scape erect to suberect, 50–150 mm long, maroonish-brown, rigid; rachis green to greenish-brown; lower bracts ovate, becoming lanceolate above, 3–5 × 1–3 mm, maroon, light green or brown; pedicels suberect to spreading, 5–15 mm long, lengthening and becoming spreading towards inflorescence apex, green or white. *Perianth* zygomorphic, oblong-campanulate, white, strongly cernuous; tube cup-shaped, 2–3 mm long; outer tepals ovate, 6–7 × 4–5 mm, apical gibbosities bright green, brown or pinkish-maroon; inner tepals obovate, 7–8 × 4–5 mm, apices slightly recurved, median keels bright green, brown or pinkish-maroon. *Stamens* well exserted, declinate; filaments white, 8–12 mm long. *Ovary* ellipsoid,

3–4 × 2–3 mm, bright green; style well exserted, declinate, 7–10 mm long, white, protruding beyond stamens as ovary matures. *Capsule* ellipsoid, 8–10 × 6–7 mm. *Seed* globose, 1.1–1.4 × 1.1–1.4 mm, glossy, black; strophiole inflated, 0.8 mm long, smooth. *Chromosome number* 2n = 14 (Duncan, 1993; Spies *et al.*, 2008). Figures 140, 141.

FLOWERING PERIOD. September to November, with a peak from early to mid-October.

HISTORY. *L. thomasiae* was discovered by Margaret Thomas on 6th October 1986 near Algeria Forest Station south of Clanwilliam. She brought flowering material to W. F. Barker who identified it as undescribed and appended the manuscript name *L. thomasiae* to pressed material in the Compton Herbarium at Kirstenbosch. In October 1991, I visited the type locality with Margaret Thomas to collect additional specimens for cultivation and for illustration by Marieta Visagie, whose colour plate was published in *The Flowering Plants of Africa* (Duncan, 1993).

DISTINGUISHING CHARACTERS AND AFFINITIES. In a phylogenetic analysis of morphological data (Duncan *et al.*, 2005a), *L. thomasiae* resolved adjacent to *L. mediana*, with which it shares oblong-campanulate perianths with declinate, white stamens, lanceolate, canaliculate leaves and a similar globose seed with an inflated strophiole. *L. mediana* differs from *L. thomasiae* in its spreading to slightly suberect perianth with included to shortly exserted stamens, in its pedicels that re-orientate to an erect position in fruit, and in its longer strophiole (1.1–1.2 mm) and globose bulbs.

Figure 140 (left). Cultivated specimen of *Lachenalia thomasiae* from south of Clanwilliam. Image: Graham Duncan.

Figure 141 (right). *Lachenalia thomasiae* in light shade between sandstone boulders south of Clanwilliam. Image: Graham Duncan.

32. LACHENALIA THOMASIAE

DISTRIBUTION, HABITAT AND CONSERVATION STATUS. *L. thomasiae* is confined to a small area of the Olifants River Valley south of Clanwilliam (Map 33). The plants occur in Cederberg Sandstone Fynbos vegetation (Mucina & Rutherford, 2006) amongst thorny, succulent scrub on rocky sandstone ridges with a westerly and south-westerly aspect, and between boulders on hill summits, in full sun or light shade. They grow in a mixture of leaf mould and sand, and occur singly, or in small groups as part of larger colonies. Specimens in full sun have suberect, yellowish-green foliage with maroon margins, whereas those growing in the shade of surrounding shrubs are more robust with broader, flaccid, dark green leaves and longer inflorescences bearing many more flowers. At the type locality, the species grows in association with *L. mutabilis*, *Lapeirousia fabricii* and *Micranthus alopecuroides* (Iridaceae) (Duncan, 1993, 1994). *L. thomasiae* is threatened by invasive alien grasses and agricultural activity; it has a conservation status of Vulnerable (Victor & Raimondo, 2009).

NOTES. *L. thomasiae* is pollinated by honey bees (*Apis mellifera*). Cultivated specimens are at least partially self-fertile, readily forming seeds under isolated, insect-free conditions. The scapes remain attached to the bulbs for several weeks following capsule dehiscence and the seeds drop to the ground due to the shaking action of wind.

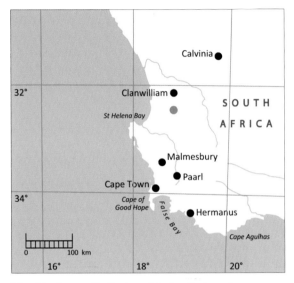

Map 33. Known distribution of *Lachenalia thomasiae*.

33. LACHENALIA MARGARETIAE

Lachenalia margaretiae W. F. Barker (sphalm. *margaretae*), *Journal of South African Botany* 45 (2): 202–204 (1979).
TYPE: South Africa, Western Cape, Pakhuis Pass, *W. F. Barker* 10320 (NBG!, holo.).

ETYMOLOGY. *margaretiae*: after Margaret Lilian Thomas (1917–2006), cultivator and collector of South African bulbous plants.

DESCRIPTION. *Dwarf geophyte*, 35–100 mm high. *Bulb* globose, 10–15 mm in diameter, offset-forming; tunic multilayered, outer layers light to dark brown, spongy; inner cataphyll translucent white, loosely surrounding leaf bases, apex acute. *Leaves* 1(–2), lanceolate, 75–190 × 15–30 mm, spreading, olive green to light green, upper surface with distinct depressed longitudinal grooves, plain or rarely with dark brown spots, lower surface tinged with light to dark maroon; leaf margins slightly coriaceous, light maroon; leaf bases clasping, 20–30 mm long, lower portion white, upper portion light maroon; primary seedling leaf flat, prostrate. *Inflorescence* racemose, few- to many-flowered, fairly dense, sterile apex short; scape erect, 20–70 mm long, light to dark maroon or brownish-maroon, sturdy in full sun, lax in shade; rachis light to dark maroon or brownish-maroon, becoming paler towards apex; pedicels suberect, 3–4 mm long, white; lower bracts ovate, upper bracts lanceolate, 2–3 × 1–4 mm, translucent white, bases dark maroon. *Perianth* zygomorphic, oblong-campanulate, spreading, white; tube cup-shaped, 2 mm long; outer tepals ovate, 4–5 × 3 mm, apical gibbosities large, dark brown to brownish-green; inner tepals obovate, 5–6 × 4 mm, apical marking dark brown. *Stamens* shortly exserted, straight; filaments white, 5–7 mm long. *Ovary* obovoid, 2.5 × 2.0 mm, yellowish-green; style included, 3–4 mm long, white. *Capsule* obovoid, 5–6 × 5 mm. *Seed* globose, 1.3–1.4 × 1.3 mm, glossy, black; strophiole inflated, 0.6 mm long, smooth. *Chromosome number* 2n = 14 (Spies *et al.*, 2008). Figure 142.

FLOWERING PERIOD. October to December, with a peak in early November.

HISTORY. *L. margaretiae* is a very seldom collected species discovered by Margaret Thomas in October 1965 in flower on the Pakhuis Pass east of Clanwilliam. A second collection of leafing specimens was made at the type locality by W. F. Barker in September 1970. The plant was next collected by Fred Paterson in the early 1980s, close to the type locality, following which material became available for cultivation in the nursery at Kirstenbosch. No further records appear to have been made in more than three ensuing decades until November 2006, when Nick Helme found *L. margaretiae* flowering at a new locality north of the Pakhuis Pass summit.

DISTINGUISHING CHARACTERS AND AFFINITIES. *L. margaretiae* resolved adjacent to *L. youngii* in a phylogenetic analysis of morphology (Duncan *et al.*, 2005a). *L. margaretiae* and *L. youngii* share oblong-campanulate perianths with shortly exserted, straight white stamens and a globose seed with an inflated strophiole, but *L. youngii* differs from *L. margaretiae* in its taller stature, its pink, cernuous perianth with longer outer and inner tepals (6–7 mm and 7–8 mm long, respectively), its longer filaments (8–9 mm long), its much longer style (7–8 mm long), and its linear, deeply canaliculate leaves and smaller globose seed (0.9–1.0 × 0.9 mm). The two species are widely separated geographically: *L. youngii* occurs in the southern and south-eastern Western Cape, and in the western Eastern Cape.

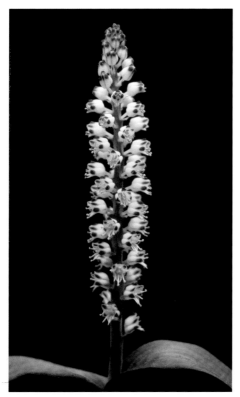

Map 34. Known distribution of *Lachenalia margaretiae*.

Figure 142. Cultivated specimen of *Lachenalia margaretiae* from the Cederberg. Image: Graham Duncan.

Superficially, *L. margaretiae* resembles *L. thomasiae* from rocky sandstone ridges above the Olifants River near Clanwilliam, which has similar white, oblong-campanulate flowers, the outer tepals with dark brown or brownish-green gibbosities and similar lanceolate leaves, but *L. thomasiae* differs from *L. margaretiae* in being much taller with well-exserted, declinate stamens, a subglobose bulb and seed that has a longer strophiole (0.8 mm long).

DISTRIBUTION, HABITAT AND CONSERVATION STATUS. *L. margaretiae* has an extremely limited known distribution in the Cederberg Mountains (Map 34). Three collections are known from the Pakhuis Pass east of Clanwilliam and another a short distance to the north of the Pakhuis Pass summit. The plants grow in Cederberg Sandstone Fynbos vegetation (Mucina & Rutherford, 2006), occurring in light to deep shade in shallow, acid sandy humus on the upper slopes, on ledges and in crevices of the south side of massive sandstone boulders. Although not under direct threat, *L. margaretiae* is estimated to comprise fewer than 1,000 mature plants, although it is likely that it occurs in other parts of the vast Cederberg Range, much of it poorly explored. It has a conservation status of Vulnerable (Helme *et al.*, 2009).

NOTES. Pollination vectors for the species have yet to be recorded, but the oblong-campanulate flower shape suggests pollination by honey bees (*Apis mellifera*). In cultivation, however, the flowers are strongly self-fertile, producing plentiful seeds under enclosed, insect-free conditions, the seedlings always flowering true to type (Duncan, 2008b). The ripe infructescences detach from the bulbs shortly after capsule dehiscence and are blown away, dispersing the seeds.

34. LACHENALIA MATHEWSII

Lachenalia mathewsii W. F. Barker, *The Flowering Plants of South Africa* 11 t. 422 (1931).
TYPE: South Africa, Western Cape, Vredenburg, *J. W. Mathews s.n.* sub. NBG 1646/23 (BOL!, holo.; K!, iso.).

ETYMOLOGY. *mathewsii*: after Joseph William Mathews (1871–1949), first Curator of Kirstenbosch National Botanical Garden, who discovered this species and made the first collection of plants.

DESCRIPTION. *Geophyte*, 100–165 mm high. *Bulb* globose, 12–18 mm in diameter, offset-forming; tunic multilayered, outer layers dark brown, spongy; inner cataphyll translucent white, adhering to leaf bases, apex obtuse. *Leaves* 2, linear, apex attenuate, 20–210 × 5–15 mm, suberect to falcate, glaucous, canaliculate; leaf bases clasping, 30–40 mm long, upper portion glaucous, lower portion white; primary seedling leaf terete, erect. *Inflorescence* racemose, few- to many-flowered, sterile apex short; scape erect to suberect, 80–150 × 3–5 mm, light to dark green with brown speckles, rachis light green with brown speckles; pedicels suberect, 3–4 mm long, light green; lower bracts ovate, becoming lanceolate above, 1–5 × 2–5 mm. *Perianth* zygomorphic, bright yellow, ageing to dull red, spreading, slightly spicy-scented; tube cup-shaped, 2–3 mm long; outer tepals ovate, 7–8 × 5–6 mm, apical gibbosities bright green; inner tepals obovate, 9–10 × 5–6 mm, margins and recurved apices creamy yellow, apical marking bright green, median keels deep yellow. *Stamens* shortly exserted, declinate; filaments white, 8–9 mm. *Ovary* ellipsoid, 3–4 × 2–3 mm, yellowish-green to bright green; style shortly exserted, declinate, 8–9 mm long, white. *Capsule* ellipsoid, 8–9 × 4–5 mm. *Seed* ovoid, 0.9–1.0 × 0.9–1.0 mm, glossy, black; strophiole rudimentary, 0.2 mm long, ridged. *Chromosome number* 2n = 14 (Johnson & Brandham, 1997; Hamatani *et al.*, 1998; Spies *et al.*, 2002, 2008). Plate 15, Figure 143.

FLOWERING PERIOD. September to October, with a peak in late September.

THE GENUS LACHENALIA
34. LACHENALIA MATHEWSII

PLATE 15. Watercolour painting of *Lachenalia mathewsii* from Vredenburg [*Bokelmann & Paine s.n.* (1425/82) in NBG], courtesy of Susan Goldswain. Artist: Ellaphie Ward-Hilhorst.

34. LACHENALIA MATHEWSII

HISTORY. J. W. Mathews discovered *L. mathewsii* close to Vredenburg on the Cape west coast in 1923, and it was described and illustrated in watercolour by W. F. Barker in *The Flowering Plants of South Africa* (Barker, 1931a). A specimen in the Bolus Herbarium at the University of Cape Town (BOL 27878) has a note that the species was exhibited at the Malmesbury Wild Flower Show in October 1926. Mathews made two more collections at Vredenburg in September 1938 and in 1943. It would appear that accurate locality details were not recorded because, despite numerous attempts, *L. mathewsii* was not located again for almost 40 years and was presumed to have become extinct (Duncan, 1986). A collection was finally made in October 1981 between Vredenburg and Velddrift (*L. Hugo* 2956, in PRE) and it was found again in September 1982 by Miss Q. V. Paine and Mrs H. Bokelmann who came across it by chance at another location near Vredenburg, where it grew together with a robust mauve form of *L. pallida*. Material was collected and *L. mathewsii* was painted by Ellaphie Ward-Hilhorst in 1985 from plants grown at Kirstenbosch (Duncan, 1986). It has since been successfully introduced into specialist bulb collections around the world.

DISTINGUISHING CHARACTERS AND AFFINITIES. Duncan *et al.* (2005a) suggest a close alliance with the Namaqualand endemic *L. namaquensis*, which shares linear, canaliculate leaves that taper to a long attenuate apex. *L. namaquensis* differs from *L. mathewsii* in its spike of urceolate, suberect, pinkish- or bluish-magenta flowers with included stamens, and in its seed, which has a longer ridged strophiole (0.5 mm long).

DISTRIBUTION, HABITAT AND CONSERVATION STATUS. *L. mathewsii* is currently known from one large, dense population on private farmland near Vredenburg (Map 35). It grows in Saldanha Granite Strandveld vegetation (Mucina & Rutherford, 2006) in open aspects in seasonally wet, alkaline gravelly clay, unsuitable for cereal crops. Companion species include the succulent groundcover *Drosanthemum*

Figure 143. *Lachenalia mathewsii* in seasonally wet gravelly clay near Vredenburg. Image: Graham Duncan.

hispidum (Mesembryanthemaceae). On the drier margins of the population, a mauve form of *L. pallida* used to occur plentifully, commencing flowering a few weeks earlier than the *L. mathewsii* plants. *L. mathewsii* has been severely affected by overgrazing in winter, which has a detrimental effect on flowering in spring. The plants also have to compete with a number of choking agricultural weeds from surrounding crop fields. Its conservation status is Critically Endangered (Duncan *et al.*, 2009).

NOTES. *L. mathewsii* is pollinated by honey bees (*Apis mellifera*). Following capsule dehiscence, the scapes remain attached to the bulbs for several weeks and the seeds are dispersed locally by the shaking action of wind.

Map 35. Known distribution of *Lachenalia mathewsii*.

35. LACHENALIA VARIEGATA

Lachenalia variegata W. F. Barker, *Journal of South African Botany* 15: 37–38 (1949).
TYPE: South Africa, Western Cape, Paarden Island, *T. M. Salter* 7446 (NBG!, holo.).

ETYMOLOGY. *variegata*: variegated speckles on outer tepals.

DESCRIPTION. *Geophyte* 100–450 mm high. *Bulb* subglobose, 10–20 mm in diameter, offset-forming; tunic multilayered, outer layers dark brown, spongy; inner cataphyll translucent white, tinged with magenta above. *Leaf* solitary, rarely 2, lanceolate, 50–160 × 15–40 mm, suberect to falcate, slightly to strongly canaliculate, coriaceous, glaucous, upper surface plain with depressed longitudinal grooves, lower leaf surface heavily marked with purple, bluish-purple or brownish-purple spots or blotches, leaf margins thickened, undulate and sometimes crisped; leaf base clasping, 20–50 mm long, light greenish-white, heavily marked with purple or purplish-magenta spots or blotches; primary seedling leaf flat, prostrate. *Inflorescence* racemose, many-flowered, fairly dense, sterile apex long; scape erect, 40–100 mm long, sturdy, light green, heavily marked with purple or maroonish-purple spots or blotches; pedicels suberect, 1–2 mm long, white; lower bracts ovate, becoming narrow-lanceolate above, 1–4 × 1–3 mm, light green or white. *Perianth* zygomorphic, oblong-campanulate, spreading to cernuous; tube cup-shaped, 2–3 mm long, light bluish-white; outer tepals ovate, 7–8 × 4–5 mm, light bluish-white, rarely pink, with sporadic light to dark blue, purple or green speckles, apical gibbosities brown or green; inner tepals obovate, 8–10 × 4–5 mm, light yellowish-green, apical zone bright green, margins white, upper two tepals overlapping, lower tepal narrower, canaliculate. *Stamens* declinate, included to shortly exserted; filaments white, 8–9 mm long. *Ovary* ovoid, 3–4 × 3 mm, light to dark green; style included, declinate, 6–7 mm long, white, protruding beyond perianth as ovary enlarges. *Capsule* ovoid, 7–8 × 5 mm. *Seed* globose, 1.2–1.3 × 1.3 mm, glossy, black; strophiole inflated, 1.1 mm long, smooth. *Chromosome number* 2n = 12 (Hamatani *et al.*, 2004); 2n = 14 (Spies *et al.*, 2008). Figures 144, 145.

FLOWERING PERIOD. July to October, with a peak in late September.

HISTORY. The first record of *L. variegata* was that of the Englishman Anthony H. Wolley-Dod, who collected it in September 1897 on sandy flats beyond Paulsberg within the Cape of Good Hope Nature Reserve in the southern Cape Peninsula (*Wolley-Dod* 3191, in BOL). The type collection was made by T. M. Salter at Paarden Island on the northern Cape Peninsula in October 1945, an area now entirely given over to industrial sites, and the species was described in the *Journal of South African Botany* (Barker, 1949). It has frequently been collected along the Cape west coast and on the Cape Flats, but has only recently been recorded on the south coast.

DISTINGUISHING CHARACTERS AND AFFINITIES. Phylogenetic data suggest that *L. variegata* is sister to *L. cernua* (Duncan et al., 2005a). The two species share a many-flowered raceme of spreading to cernuous flowers with included or shortly exserted stamens, usually solitary lanceolate leaves, and globose seeds with inflated strophioles. *L. cernua* differs from *L. variegata* in having urceolate flowers with plain, creamy-white outer tepals, recurved inner tepal apices and more or less straight stamens, and its leaves lack the thickened margin. Both species are represented on the Cape Peninsula, and *L. cernua* also occurs inland at Goudini and Wolseley.

DISTRIBUTION AND HABITAT AND CONSERVATION STATUS. *L. variegata* occurs from Leipoldtville west of Clanwilliam to the Cape Peninsula, and east to Gansbaai (Map 36). Its habitat is variable and it is usually encountered on flats in deep calcareous or acid sand, less frequently on stony lower hill slopes. In the latter habitat, the plants are often more slender and stunted with paler outer tepals and shortly exserted stamens (Barker, 1950). The species occurs in populations of scattered individuals in eight vegetation types: Saldanha Granite- and Cape Flats Dune Strandveld, Peninsula Granite Fynbos,

Figure 144. *Lachenalia variegata* on fynbos flats at Grootbos Private Nature Reserve, Gansbaai. Image: Heiner Lutzeyer.

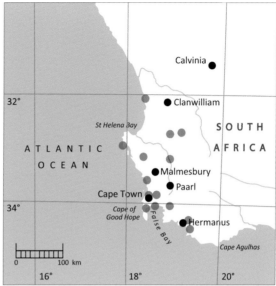

Map 36. Known distribution of *Lachenalia variegata*.

Figure 145. Cultivated specimen of *Lachenalia variegata* from Darling. Image: Graham Duncan.

Leipoldtville- and Cape Flats Sand Fynbos, Graafwater-, Peninsula- and Overberg Sandstone Fynbos (Mucina & Rutherford, 2006). *L. variegata* is still fairly common across its range and is not threatened.
NOTES. At Grootbos Private Nature Reserve near Gansbaai, *L. variegata* flowers prolifically following summer bush fires of the previous season, but the species is not dependent on fire for flowering to occur. It is pollinated by honey bees (*Apis mellifera*). The scape detaches from the bulb shortly after capsule dehiscence and the infructescence is blown about, dispersing the seeds.

36. LACHENALIA MARLOTHII

Lachenalia marlothii W. F. Barker ex G. D. Duncan, *Bothalia* 27 (1): 12–14 (1997).
TYPE: South Africa, Northern Cape, between Vlakkraal and Kalkgat Suid, south of Calvinia, *W. F. Barker* 9330 (NBG!, holo.).

ETYMOLOGY. *marlothii*: after Rudolf Marloth (1855–1931), German chemist and botanist.
DESCRIPTION. *Geophyte*, 90–160 mm high. *Bulb* globose, 15–25 mm in diameter, solitary; tunic multilayered, outer layers dark brown, spongy; inner cataphyll translucent white, loosely surrounding leaf bases, apex obtuse. *Leaf* solitary, broadly lanceolate, 30–60 × 10–25 mm, spreading to suberect, distinctly leathery, dark green, upper surface plain, lower surface with dark purplish-magenta and green transverse bands in lower half; margins slightly undulate or sometimes crisped; leaf base tightly clasping, 30–70 mm long, white, heavily banded with bright magenta or purple; primary seedling leaf flat, prostrate. *Inflorescence* spicate, occasionally racemose, few- to many-flowered, fairly dense, sterile apex short; scape erect to suberect, 40–100 mm long, light green, heavily marked with brownish-

purple blotches; pedicels absent or suberect, 1–2 mm long, white; lower bracts ovate, becoming lanceolate above, 1–2 × 1–2 mm, white. *Flowers* zygomorphic, oblong-campanulate, spreading to suberect, strongly sweet-scented; tube cup-shaped, 2–3 mm long, dull blue; outer tepals ovate, 8–9 × 5–6 mm, dull white or light blue with dark blue or green speckles, apices slightly recurved, apical gibbosities greenish-purple to purplish-brown; inner tepals obovate, 8–11 × 3–6 mm, translucent white or light greenish-yellow, sometimes with brownish margins, median keels purplish-green, apices slightly recurved, upper two tepals overlapping, lower tepal slightly longer, narrower. *Stamens* included to shortly exserted, filaments white, more or less straight, 8–10 mm long. *Ovary* ellipsoid, 4–5 × 3–4 mm, light green; style included, more or less straight, 6–7 mm long, white. *Capsule* ellipsoid, 5–6 × 4–5 mm. *Seed* ovoid, 1.3–1.4 × 1.8 mm, glossy, black; strophiole 1.1 mm long, ridged. *Chromosome number* 2n = 14 (Spies *et al.*, 2008). Figures 146, 147.

FLOWERING PERIOD. July to September, with a peak in late August.

HISTORY. Plants of *L. marlothii* were collected for the first time by Rudolf Marloth in October 1920 at Waterkloof Farm in the Sutherland Roggeveld. W. F. Barker made the next collection at Brandkop north of Nieuwoudtville in August 1950, and made a second collection south of Calvinia in July 1961. She appended the manuscript name *L. marlothii* W. F. Barker to the latter collection, but the species remained undescribed until after her death (Duncan, 1998a). The species has since been found near Springbok in central Namaqualand and at several localities in the northern Richtersveld.

Figure 146 (left). Cultivated specimen of spicate form of *L. marlothii* from Calvinia. Image: Graham Duncan.

Figure 147 (right). Cultivated specimen of racemose form of *L. marlothii* from Steinkopf. Image: Graham Duncan.

DISTINGUISHING CHARACTERS AND AFFINITIES. *L. marlothii* resolved adjacent to *L. marginata* subsp. *neglecta* in a phylogenetic analysis of morphological data (Duncan *et al.*, 2005a). These two taxa share solitary lanceolate leaves that are heavily marked with green and magenta transverse bands on the lower surface and clasping base, undulate and crisped leaf margins, and perianths with greenish-yellow inner tepals. *L. marginata* subsp. *neglecta* differs from *L. marlothii* in its subglobose bulb, narrower, deeply canaliculate, glaucous leaf with a prominent coriaceous margin, shorter, narrower outer tepals (6–8 × 3–4 mm), shorter filaments (6–7 mm), and much smaller, globose seeds (0.9–1.0 × 0.8–0.9 mm) with much shorter, inflated strophioles (0.6–0.7 mm long).

DISTRIBUTION, HABITAT AND CONSERVATION STATUS. *L. marlothii* is currently known from two disjunct centres in the Northern Cape. Its southern boundary is the Roggeveld Plateau near Sutherland, and to the north-west it occurs in the Calvinia district and on the Bokkeveld Plateau. In central and northern

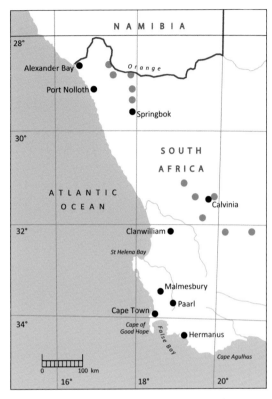

Map 37. Known distribution of *Lachenalia marlothii*.

Namaqualand, it is found near Springbok and Steinkopf, and in the northern Richtersveld, it grows near Eksteenfontein, on the Skimmelberg and just south of the Orange River (Map 37). The species traverses five vegetation types: Central Richtersveld Mountain Shrubland, Namaqualand Klipkoppe Shrubland, Namaqualand Blomveld, Hantam- and Roggeveld Karoo (Mucina & Rutherford, 2006). The plants are encountered as scattered individuals or in small groups, and are usually associated with rocky flats and south-facing hill slopes with clay soil. The species is not threatened.

NOTES. The heavily sweet-scented flowers are pollinated by honey bees (*Apis mellifera*). The base of the scape detaches from the bulb soon after capsule dehiscence and the infructescence is blown away, scattering seed some distance from the parent plant.

37. LACHENALIA LONGIBRACTEATA

Lachenalia longibracteata E. Phillips, *The Flowering Plants of South Africa* 11: t. 405 (1931).
TYPE: South Africa, Western Cape, Lamberts Bay, *J. J. van Nouhuys s.n.* sub. PRE8189 (PRE!, holo.).

ETYMOLOGY. *longibracteata*: descriptive of the long, narrowly lanceolate bracts subtending the upper flowers.

DESCRIPTION. *Geophyte*, 70–380 mm high. *Bulb* globose, 15–20 mm in diameter, offset-forming; tunic multilayered, outer layers dark brown, spongy; inner cataphyll translucent white, longitudinal veins prominent, adhering to leaf bases, apex obtuse. *Leaves* (1–)2, lanceolate, 40–145 × 10–45 mm, suberect to spreading, leathery, light to dark green, upper surface plain or sporadically to heavily marked with light to dark green, brown or purplish-brown spots, rarely with small dome-shaped pustules, with depressed longitudinal grooves; leaf bases 15–40 mm long, white below, shading to light green above; primary seedling leaf flat, prostrate. *Inflorescence* spicate, rarely racemose, few- to many-flowered, dense, sterile apex short to long; scape erect to suberect, 60–150 mm long, slender to sturdy, light green, plain or with scattered light brownish-blue blotches; rachis light green throughout, sometimes shading to light purplish-brown in upper half, plain or with light brownish-blue blotches; pedicels absent or up to 1 mm long; lower bracts ovate, becoming narrowly lanceolate above, 4–16 × 1–4 mm, light green at base, translucent white above. *Perianth* zygomorphic, oblong-campanulate, spreading to suberect, ageing to dull red, sweet-scented; tube cup-shaped, 3 mm long, light blue or yellow, plain or with occasional darker blue speckles; outer tepals ovate, 6–7 × 4–5 mm, translucent light blue, greenish-white or light yellow, plain or with light blue speckles and median keels, apical gibbosities dark brown, brownish-green or bright green; inner tepals obovate, 10–12 × 3–5 mm, translucent white to light greenish-yellow, apices rounded, lower inner tepal slightly shorter; upper inner tepals broader and wider, keels light to bright green. *Stamens* included to shortly exserted, more or less straight; filaments white, 9–11 mm long. *Ovary* ellipsoid, light to dark green, 3–4 × 2–3 mm; style included, straight to declinate, 9–10 mm long, white, protruding well beyond perianth as ovary enlarges. *Capsule* ellipsoid, 8–10 × 4–5 mm. *Seed* globose, 0.9–1.0 × 0.8–0.9 mm, glossy, bla ck; strophiole inflated, 0.8 mm long, smooth. *Chromosome number* 2n = 14 (Crosby, 1986; Hamatani *et al.*, 2007). Figures 148, 149.

Figure 148 (left). Cultivated blue form of *Lachenalia longibracteata* from Vredenburg. Image: Graham Duncan.
Figure 149 (right). Greenish yellow form of *Lachenalia longibracteata* in deep sand near Melkbosstrand. Image: Adam Harrower.

FLOWERING PERIOD. July to September, with a peak from late July to early August.

HISTORY. *L. longibracteata* was described by the Cape Town botanist Dr Edwin Percy Phillips (1884–1967) in volume 11 of *The Flowering Plants of South Africa* (Phillips, 1931). The holotype material, collected at Lamberts Bay in 1926 (*van Nouhuys s.n.*, in PRE) represents the more commonly encountered coastal blue form with very long upper bracts, which is found mainly on granite outcrops of the west coast.

DISTINGUISHING CHARACTERS AND AFFINITIES. *L. longibracteata* is sister to *L. lutea* (Duncan *et al.*, 2005a), sharing spikes of oblong- campanulate flowers with included to shortly exserted, straight stamens, and globose, glossy seeds with inflated strophioles. *L. lutea* differs in its bright yellow, strongly suberect flowers with shorter inner tepals (8–9 mm) with subacute apices and the lower inner tepal longer and slightly recurved, and in having much shorter bracts (2–6 mm long), shorter stamens (5–8 mm long), a shorter style (5–6 mm long) and a shorter inflated strophiole (0.5–0.6 mm long).

The prominent, narrowly lanceolate white bracts (4–16 mm long) of *L. longibracteata* are longest in the upper part of the inflorescence and most conspicuous in the blue forms. The lower inner tepal is usually slightly shorter than the upper inner tepals. *L. longibracteata* is sometimes confused with *L. orchioides* subsp. *orchioides*, but the latter is immediately set apart by its tubular flowers with longer, cylindrical perianth tubes (6–8 mm long) and by its lower inner tepal that is slightly longer than the upper inner tepals and distinctly caniculate. The absence of very long, narrowly lanceolate bracts in the upper part of the inflorescence is also a diagnostic trait for *L. orchioides* subsp. *orchioides*.

DISTRIBUTION, HABITAT AND CONSERVATION STATUS. *L. longibracteata* is confined to the western and south-western parts of the Western Cape, extending from Lambert's Bay south along the Cape west coast to Melkbosstrand and inland to Citrusdal, Piketberg, Tulbagh and Kalbaskraal (Map 38). Its habitat is variable and it occurs mainly in renosterveld, less frequently in coastal fynbos. Blue forms are usually encountered in gravelly clay soils on flats and on the east-facing slopes of granite outcrops, such as at Darling and Vredenburg. Light yellow forms usually occur in stony clay soil (at Tulbagh) and yellowish-green forms in deep, yellow or white sand (at Melkbosstrand and Philadelphia). The plants grow singly, in small groups, or in large colonies. They occur in a variety of vegetation types including Leipoldtville- and Atlantis Sand Fynbos, Piketberg- and Cederberg Sandstone Fynbos, Saldanha Granite Strandveld, Swartland Granite-, Swartland Shale- and Breede Shale Renosterveld. At a locality south of Tulbagh, a yellow form occurs in large numbers in association with *L. flava* and *L. unifolia*, the three species flowering simultaneously. *L. longibracteata* numbers have been much reduced along the Cape west coast owing to agricultural extension and housing development, and the species has a conservation status of Declining (Helme & Raimondo, 2009d).

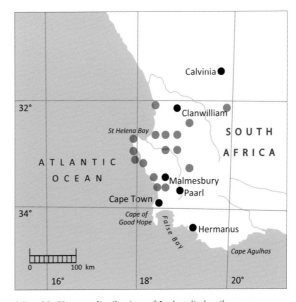

Map 38. Known distribution of *Lachenalia longibracteata*.

NOTES. At a locality near Vredenburg on the Cape west coast, the inflorescences of blue forms of *L. longibracteata* are heavily grazed by steenbuck (*Raphicerus campestris*) and, during the winter months, the bulbs are often unearthed and eaten by Cape porcupines (*Hystrix africaeaustralis*). The longer-flowered, light yellow forms that occur in sandy soils flower profusely following bush fires in the previous summer. The scapes remain attached to the bulbs for several weeks following capsule dehiscence and the seeds are dispersed locally.

38. LACHENALIA LUTEA

Lachenalia lutea G. D. Duncan, *Bothalia* 36 (2): 147–150 (2006).
TYPE: South Africa, Western Cape, Vergelegen Farm, Somerset West, on hillside near reservoir, in heavy clay soil, *W. F. Barker* 9088 (NBG!, holo.).

ETYMOLOGY. *lutea*: yellow flowers.
DESCRIPTION. *Geophyte*, 160–240 mm high. *Bulb* globose, 15–20 mm in diameter, offset-forming; tunic multilayered, outer layers dark brown, spongy; inner cataphyll translucent white, adhering to leaf bases, apex obtuse. *Leaves* 2, occasionally solitary, lanceolate, 90–140 × 12–30 mm, spreading to suberect, flat to weakly canaliculate, bright green to greenish-magenta, rarely glaucous, leathery, upper surface plain or marked with dark green spots, lower surface plain or marked with brownish-purple spots; leaf margins thickened, flat, rarely undulate or crisped; leaf bases clasping, 10–30 mm long, greenish-white in lower half, shading to deep magenta above, plain or occasionally barred or spotted with brownish-purple in upper part; primary seedling leaf flat, prostrate. *Inflorescence* spicate, few- to many-flowered, dense, sterile apex long; scape erect, 70–150 mm long, light green to brownish-green, plain or marked with large, irregularly scattered, brownish-purple blotches; rachis light green, shading to greenish-yellow in upper third; lower bracts ovate, becoming lanceolate above, 2–6 × 1–5 mm, translucent white. *Perianth* zygomorphic, oblong-campanulate, light to bright greenish-yellow, suberect, strongly sweet-scented; tube cup-shaped, 3 mm long; outer tepals ovate, 6–7 × 4 mm, apical gibbosities light to bright green; inner tepals obovate, 8–9 × 4–5 mm, apices subacute, upper two tepals overlapping, lower tepal slightly recurved, longer, median keels bright green. *Stamens* included, rarely shortly exserted; filaments more or less straight, 5–6 mm long, rarely up to 8 mm long, white; pollen ageing to black. *Ovary* ellipsoid, 3.0 × 2.5 mm, light yellowish-green; style included, 5–6 mm long, straight. *Capsule* ellipsoid, 8–9 × 4–5 mm. *Seed* globose, 0.9 × 0.8 mm, glossy, black; strophiole inflated, 0.5–0.6 mm long, smooth. *Chromosome number* 2n = 16 (de Wet, 1957, specimen misidentified as *L. orchioides*). Figure 150.
FLOWERING PERIOD. July to October, with a peak in late August.
HISTORY. The earliest known record of *L. lutea* is that of C. F. Ecklon and C. L. Zeyher who gathered material at Caledon, probably in the early 1830s (*Ecklon & Zeyher* 36, in SAM). The next collection was that of J. F. Solly who collected plants almost a century later, in August 1915, at the foot of Sir Lowry's Pass east of Cape Town. The type collection was made by W. F. Barker at Vergelegen Farm near Somerset West in October 1959. The species has been widely collected in the south-western part of the Western Cape and was overlooked as a distinct taxon until its recent publication in *Bothalia* (Duncan & Edwards, 2006).
DISTINGUISHING CHARACTERS AND AFFINITIES. *L. lutea* is sister to *L. longibracteata* (Duncan et al., 2005a) and distinguishing characters are discussed under that species. *L. lutea* closely resembles *L. arbuthnotiae*,

which has similar oblong-campanulate, greenish-yellow, strongly sweet-scented flowers, but the flowers of *L. arbuthnotiae* differ in being longer and slightly curved, with the inner tepals only slightly longer than the outer ones, and scarcely spreading, and with much longer sexual organs (filaments 8–9 mm long, style 9 mm long).

DISTRIBUTION, HABITAT AND CONSERVATION STATUS. The species is distributed between Strand and Bot River, and from Tulbagh to Villiersdorp, with outliers near Piketberg and Citrusdal (Map 39). It usually occurs in seasonally moist, stony, heavy clay, less frequently in sandy soil. It is encountered in a variety of habitats including low-lying flats and hills, within shale bands of higher mountain slopes or at the edges of quartz patches overlaying shale. It is present in six vegetation types including Cederberg- and Hawequas Sandstone Fynbos, Swartland-, Breede- and Western Rûens Shale Renosterveld, and Elgin Shale Fynbos (Mucina & Rutherford, 2006). The plants grow as scattered individuals or in small to large groups, and the species is common across most of its range. At Caledon, *L. lutea* grows in association with *L. orchioides* subsp. *orchioides*, which flowers later in the season. *L. lutea* is not threatened.

NOTES. The plants flower profusely following bush fires but are not dependent on fire for flowering to occur. The strongly sweet-scented flowers are pollinated by honey bees (*Apis mellifera*). The scapes remain attached for several weeks following capsule dehiscence and the seeds are dispersed locally.

Map 39. Known distribution of *Lachenalia lutea*.

Figure 150. *Lachenalia lutea* on a moist sandstone slope after a fire near Elgin. Image: Graham Duncan.

39. LACHENALIA ARBUTHNOTIAE

Lachenalia arbuthnotiae W. F. Barker, *Journal of South African Botany* 50 (4): 541–543 (1984).
TYPE: South Africa, Western Cape, Isoetes Vlei, Cape Flats, Wetton, in seasonally inundated, deep sandy soil, *M. L. Thomas s.n.* sub. NBG 93529 (NBG!, holo.).

ETYMOLOGY. *arbuthnotiae*: after Miss Isobel A. Arbuthnot (1870–1963), botanical assistant at the Bolus Herbarium until 1939 and at the Compton Herbarium until 1945.

DESCRIPTION. *Geophyte*, 200–400 mm high. *Bulb* subglobose, 15–20 mm in diameter, offset-forming; tunic multilayered, outer layers dark brown, spongy; inner cataphyll translucent white, loosely surrounding leaf bases, apex obtuse. *Leaves* usually 2, occasionally solitary, lanceolate, 70–120 × 10–20 mm, coriaceous, spreading to suberect, upper surface light to dark green or maroon, plain or lightly to heavily blotched with dark purple; margins coriaceous, flat or undulate; leaf bases clasping, 30–50 mm long, lower portion white, shading to green or light magenta above, marked with deep magenta to purplish blotches above; primary seedling leaf flat, prostrate. *Inflorescence* spicate, many-flowered, dense, sterile apex short; scape erect to suberect, 90–140 mm long, light green, plain or heavily marked with small to large dark green or purple blotches; rachis colouring as for scape; lower bracts ovate, becoming lanceolate above, 1–8 × 1–3 mm, white. *Perianth* zygomorphic, narrowly oblong-campanulate, light to bright yellow tinged with green, slightly curved, spreading to slightly suberect, heavily sweet-scented, ageing to dull red; tube cup-shaped, 3 mm long; outer tepals ovate, 5–8 × 4–6 mm apical gibbosities light green; inner tepals obovate, 7–10 × 5–6 mm, slightly spreading, margins light yellow, apices obtuse, undulate and slightly recurved. *Stamens* included to shortly exserted; filaments more or less straight, white, 8–9 mm long. *Ovary* ellipsoid, 3–4 × 2–3 mm, light green; style included, straight, white, 9 mm long, protruding beyond perianth as ovary enlarges. *Capsule* ellipsoid, 9–10 × 5–6 mm. *Seed* globose, 0.9 × 0.8–0.9 mm, glossy, black; strophiole inflated, 0.7–0.8 mm long, smooth. *Chromosome number* 2n = 14 (Crosby, 1986; Johnson & Brandham, 1997; Hamatani *et al.*, 1998; Spies *et al.*, 2008). Plate 16, Figure 151.

FLOWERING PERIOD. Late August to mid-October, with a peak from mid- to late September.

HISTORY. *L. arbuthnotiae* was collected for the first time in August 1931 by the Irishwoman Isobel Arbuthnot, along Blaauwberg Road just north of Cape Town. In September of the same year, she collected the species again, at Plumstead Flats in the southern Cape Peninsula, and in September 1938, she made another collection, this time at Faure on the Cape Flats. A specimen from the Faure collection was painted by W. F. Barker, and this painting was later published in monochrome when the species was described (Barker, 1984). Most records are of plants from Isoetes Vlei in the Edith Stephens Wetland Park in Wetton east of Wynberg in the southern Cape Peninsula; indeed, this was the location for the holotype material, gathered by Margaret Thomas in September 1966 (*Thomas s.n.* sub. NBG 93529). Since its publication, *L. arbuthnotiae* has been very infrequently recorded. In 1986, a collection from the Edith Stephens Wetland Park (*Duncan* 94, in PRE) was painted by Ellaphie Ward-Hilhorst, and this painting appeared in *The Flowering Plants of Africa* (Duncan, 1988b).

DISTINGUISHING CHARACTERS AND AFFINITIES. Phylogenetic evidence (Duncan *et al.*, 2005a) places *L. arbuthnotiae* adjacent to *L. longibracteata*. The latter species has a similar spike of oblong-campanulate, spreading to suberect flowers with more or less straight, included to shortly exserted stamens, but differs from *L. arbuthnotiae* in its prominent long, linear-acuminate bracts in the upper part of the inflorescence, in its non-curved flowers with translucent blue, greenish-white or light yellow outer tepals, and in having inner tepals with flat, non-recurved margins, with the lower inner tepal slightly shorter than the upper inner tepals.

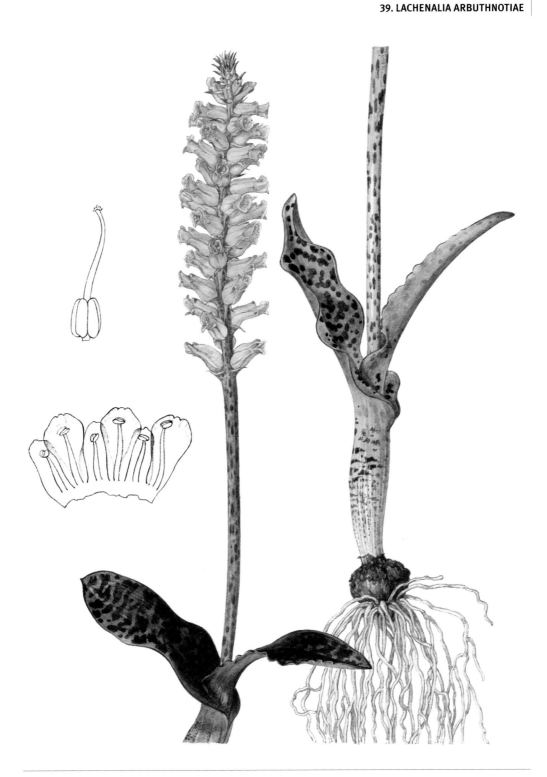

Plate 16. Watercolour painting of *Lachenalia arbuthnotiae* from the Edith Stephens Wetland Park (*Duncan* 94, in PRE) courtesy of The Editor, *Flowering Plants of Africa*, vol. 50, t. 1961 (1988). Artist: Ellaphie Ward-Hilhorst.

39. LACHENALIA ARBUTHNOTIAE

Map 40. Known distribution of *Lachenalia arbuthnotiae*.

Figure 151. *Lachenalia arbuthnotiae* in waterlogged sand in the Edith Stephens Wetland Park. Image: Graham Duncan.

Superficially, *L. arbuthnotiae* resembles *L. lutea*, which has a similar spike of narrowly oblong-campanulate, heavily sweet-scented, yellow flowers, but the latter differs in its non-curved, shorter perianths, with the inner tepal apices acute and flat, and in its flower orientation, which is usually strongly suberect. By comparison with *L. arbuthnotiae*, *L. lutea* has much shorter filaments (5–6 mm) and style (5–6 mm), and its seed has a slightly shorter strophiole (0.5–0.6 mm long).

DISTRIBUTION, HABITAT AND CONSERVATION STATUS. *L. arbuthnotiae* is confined to a small area of the south-western Cape, occurring in low-lying parts of the Cape Flats extending from Faure near Somerset West to the Edith Stephens Wetland Park in Wetton (Map 40). The plants occur in colonies in Cape Flats Sand Fynbos (Mucina & Rutherford, 2006), in seasonally waterlogged depressions and at the edges of seasonal marshes, in full sun. A large colony might still occur within the grounds of Cape Town International Airport. A smaller colony is protected within the Edith Stephens Wetland Park, where it grows in grass thickets in deep sand and in association with *Moraea tripetala*, *Sparaxis bulbifera* (both Iridaceae), *Pelargonium capitatum* and *P. triste* (both Geraniaceae). Most populations of *L. arbuthnotiae* have either been exterminated or are under severe threat because of housing and industrial development, encroaching stands of alien *Acacia* species and eutrophication. Its conservation status is Critically Endangered (Raimondo *et al.*, 2009).

NOTES. The developing inflorescences and scapes of *L. arbuthnotiae* are heavily grazed in the Edith Stephens Wetland Park, probably by vlei rats (*Otomys irroratus*). The heavily sweet-scented flowers are pollinated by honey-bees (*Apis mellifera*). The scapes remain attached to the bulbs for several weeks following capsule dehiscence and the seeds are dispersed locally by the shaking action of wind.

40. LACHENALIA MINIMA

Lachenalia minima W. F. Barker, *South African Journal of Botany* 53 (2): 166–167 (1987).
TYPE: South Africa, Western Cape, 12 km north-east of Bitterfontein, *G. D. Duncan* 172 (NBG! holo., missing).

ETYMOLOGY. *minima*: small, with reference to the plant size.

DESCRIPTION. *Dwarf geophyte*, 20–100 mm high. *Bulb* globose, 10–20 mm in diameter, offset-forming; tunic multilayered, outer layers dark brown, cartilaginous; neck distinct, 5–10 mm long, apex fasciculate; inner cataphyll translucent white, subterranean, loosely surrounding leaf bases, apex acute. *Leaves* 2, lanceolate, 25–90 × 10–25 mm, spreading, dark green, upper surface with depressed longitudinal grooves, smooth to densely covered with small dome-shaped pustules; leaf bases clasping, 20–40 mm long, white; primary seedling leaf terete, erect. *Inflorescence* spicate, few- to many-flowered, dense, sterile apex short; scape green, slender, suberect to erect, 20–60 mm long; rachis light green; lower bracts ovate, becoming lanceolate above, 0.5–1.0 × 1–7 mm, translucent white to light green. *Perianth* zygomorphic, oblong-campanulate, suberect, light greenish-yellow, ageing to dull red, sweet-scented; tube cup-shaped, 3 mm long; outer tepals ovate, 7–9 × 3–4 mm, apical gibbosities green, apices recurved; inner tepals obovate, 9–11 × 3–4 mm, median keels green, apices slightly recurved. *Stamens* included, straight; filaments light yellow, 6–7 mm long. *Ovary* ellipsoid, 2–3 × 2–4 mm, bright green; style shortly exserted, straight, light yellow, 10–11 mm long. *Capsule* ellipsoid, 7–8 × 4–5 mm. *Seed* globose, 0.9 × 1.0 mm, glossy, black; strophiole 0.6–0.7 mm long, ridged. *Chromosome number* 2n = 18 (Spies *et al.*, 2008). Figure 152.

Map 41. Known distribution of *Lachenalia minima*.

Figure 152. *Lachenalia minima* in heavy clay near Bitterfontein. Image: Graham Duncan/SANBI.

FLOWERING PERIOD. June.

HISTORY. Paymaster-Captain T. M. Salter was the first to collect this interesting dwarf species. He found a single flowering plant south of Bitterfontein on the Knersvlakte on 16th June 1931, and in July 1935, he collected the species again (this time in seed) in the same area. In August 1974, Drs B. Nordenstam and J. Lundgren found specimens in seed at a locality north-west of Bitterfontein. In order for W. F. Barker to confirm its identity as a new species and formally describe the plant, I made an expedition to the Bitterfontein area in June 1985, and was rewarded by finding good flowering material south and north-west of this town. *L. minima* was subsequently described in the *South African Journal of Botany* (Barker, 1987). The holotype material (*Duncan* 172) has been missing for many years, but cultivated pressed specimens from this collection are preserved in NBG as an interim measure.

DISTINGUISHING CHARACTERS AND AFFINITIES. Phylogenetic analysis of morphological traits placed the species close to *L. undulata* (Duncan *et al.*, 2005a), which shares a spike of oblong-campanulate flowers, included, straight stamens, a deep-seated bulb covered with hard outer tunics, and an ovoid seed. *L. undulata* differs in its light green and white, usually spreading flowers with flat tepal apices, longer filaments (8–10 mm), a shorter style (7–9 mm), much larger, broadly lanceolate leaves (60–200 × 20–60 mm) and a matte black seed.

DISTRIBUTION, HABITAT AND CONSERVATION STATUS. *L. minima* is confined to the Knersvlakte in the north-western part of the Western Cape, occurring near Bitterfontein and Vanrhynsdorp (Map 41). The plants grow in Namaqualand Klipkoppe Shrubland and Vanrhynsdorp Gannabosveld vegetation (Mucina & Rutherford, 2006), in small colonies on seasonally moist flats in heavy clay that bakes rock hard in summer. At a locality north-west of Bitterfontein, *L. minima* grows in association with *L. inconspicua* and *L. splendida*, which flower in August. *L. minima* is threatened by overgrazing and potentially by road widening, and has a conservation status of Vulnerable (Victor & Duncan, 2009g).

NOTES. The flowers are pollinated by honey bees (*Apis mellifera*). The scapes detach from the bulbs shortly after capsule dehiscence and the infructescence is blown away, during which time the seeds are scattered.

41. LACHENALIA UNDULATA

Lachenalia undulata Masson ex Baker, *Journal of Botany* 24: 336 (1886).
TYPE: South Africa, Cape, precise locality unknown, drawing by Francis Masson in British Museum (Natural History), *circa* 1793 (BM!, holo.).

ETYMOLOGY. *undulata*: wavy leaf margins.

DESCRIPTION. *Geophyte*, 50–300 mm high. *Bulb* globose, 15–25 mm in diameter, solitary; tunic multilayered, outer layers dark brown, cartilaginous; neck distinct, 10–20 mm long, apex fasciculate; inner cataphyll translucent white to light green, longitudinal veins dark green, loosely surrounding leaf bases, apex obtuse. *Leaves* 2, occasionally solitary, broadly lanceolate, 50–200 × 20–60 mm, suberect to spreading, light to dark green or intensely glaucous, upper surface with depressed longitudinal grooves, plain or with dark brownish-purple spots; leaf margins slightly to distinctly thickened, light reddish-brown to green, entire or slightly to heavily undulate and crisped; leaf bases clasping, 15–25 mm long, white, usually subterranean; primary seedling leaf terete, erect. *Inflorescence* spicate, sparse to dense, few- to many-flowered, sterile apex short; scape erect to suberect, sturdy, light yellowish-green to glaucous, 20–150 mm long; rachis light yellowish-green to glaucous; lower bracts ovate at base of inflorescence, becoming lanceolate above, 3–7 × 1–4 mm, white to light green. *Perianth* zygomorphic, narrowly oblong-campanulate, spreading to slightly suberect; tube cup-shaped, 1–2 mm long, dull white tinged with light blue; outer tepals ovate, 6–7 × 4–5 mm, light greenish-white to yellowish-green, apical gibbosities dark brown to greenish-brown; inner tepals obovate, 8–10 × 4–5 mm, translucent white, caniculate, median keels light to dark brown. *Stamens* included to just emerging, more or less straight; filaments white, 8–10 mm long. *Ovary* ellipsoid, 4–5 mm long, light to bright green; style included, 7–9 mm long, white, protruding well beyond stamens as ovary enlarges. *Capsule* ellipsoid, 5–7 × 4 mm. *Seed* ovoid, 1.1–1.2 × 1.0–1.1 mm; matte, black, strophiole 0.5–0.6 mm long, ridged. *Chromosome number* $2n = 20$ (Johnson & Brandham, 1997). Figures 153, 154.

FLOWERING PERIOD. May to June, with a peak in early June.

HISTORY. *L. undulata* was described by J. G. Baker in volume 24 of the *Journal of Botany*, based on a drawing by Francis Masson that was completed in about 1793 and is currently housed in the British Museum (Natural History). Precise locality details that describe the origin of Masson's specimen are unknown, but it could have come from the vicinity of Nuwerus as the drawing closely matches a specimen collected near Nuwerus by W. F. Barker in June 1963 (*Barker* 9937, in NBG) (Barker, 1966). Before seeing a photograph of Masson's type drawing, Barker mistakenly published one of her own paintings of what she considered to be *L. undulata* in *The Flowering Plants of South Africa* (Barker, 1931c). This painting actually depicted a different species with greenish-yellow tepals that have purplish-magenta tips and often undulate leaf margins. The species in Barker's painting, later described as *L. framesii*, had been collected by P. Ross-Frames north of Vanrhynsdorp and this collection is housed in the Bolus Herbarium at the University of Cape Town (Barker, 1966).

Figure 153 (left). Glaucous-leafed form of *Lachenalia undulata* near Bitterfontein. Image: Graham Duncan.
Figure 154 (right). Robust green-leafed form of *Lachenalia undulata* south of Vanrhynsdorp. Image: Graham Duncan.

41. LACHENALIA UNDULATA

DISTINGUISHING CHARACTERS AND AFFINITIES. Analysis of morphological data (Duncan et al., 2005a) suggests that *L. undulata* is sister to *L. duncanii*. The latter differs in its racemose inflorescence, and cream outer and inner tepals with green apical gibbosities and median keels, and in its well exserted, declinate, bicoloured filaments that are light magenta in the upper half, cream below. The leaf margins of *L. duncanii* are sclerotic, and the species has an ovoid seed (1.3–1.4 × 1.1–1.2 mm) that is larger than that of *L. undulata* and has a slightly longer strophiole (0.6–0.7 mm long).

Perianth shape and colour is consistent across the many forms of *L. undulata*, but this species shows great variation in leaf size and colour and in the degree of undulation of leaf margins. The leaf upper surfaces of *L. undulata* have prominent depressed grooves and are plain or lightly to heavily marked with brownish-purple spots. Leaf colour varies from light to dark green, or the leaves can be intensely glaucous. Glaucous-leaved forms usually have heavily undulate and crisped, often dull maroon margins, whereas the green leaves vary from flat to moderately undulate and have green margins. Leaf size decreases and leaves become increasingly glaucous, with increased upper surface spotting, as habitat becomes drier (from Nuwerus northwards). Southern populations tend to have much larger leaves (especially in the Vanrhynsdorp District) that are various shades of green, usually without any upper surface spotting.

DISTRIBUTION, HABITAT AND CONSERVATION STATUS. *L. undulata* is widely distributed from Khubus in the northern Richtersveld to Clanwilliam in the Olifants River Valley, with a disjunct record at Aggeneys in north-western Bushmanland (Map 42). The plants are usually encountered as scattered, solitary individuals on flats and lower hill and mountain slopes in stony, heavy red clay.

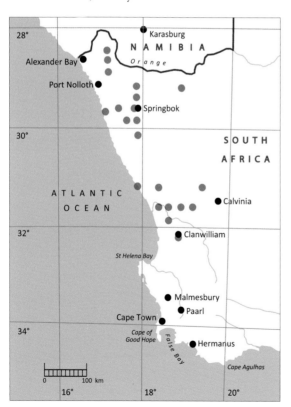

This species occurs in ten vegetation types: Central Richtersveld Mountain Shrubland, Namaqualand Strandveld, Namaqualand Blomveld, Namaqualand Klipkoppe- and Namaqualand Shale Shrubland, Riethuis-Wallekraal Quartz Vygieveld, Bushmanland Sandy Grassland, Klawer Sandy Shrubland, Nieuwoudtville Shale Renosterveld and Citrusdal Vygieveld (Mucina & Rutherford, 2006). Near Kleinsee on the west coast of Namaqualand, a form of *L. undulata* with intensely glaucous leaves and heavily undulate, crisped margins grows between low succulents on quartzite flats in association with *L. framesii*, which flowers two months later. *L. undulata* is fairly common throughout its range, especially around Vanrhynsdorp, and is not threatened.

NOTES. Pollinators are as yet unknown for *L. undulata* but its oblong-campanulate perianth suggests that these are likely to include honey bees (*Apis mellifera*). The bulbs of *L. undulata* enter dormancy in early spring before those of most other members of the genus. At this time, the scapes detach from the bulbs and the ripe infructescences are blown away, scattering seed over relatively large distances.

Map 42. Known distribution of *Lachenalia undulata*.

42. LACHENALIA DUNCANII

Lachenalia duncanii W. F. Barker, *South African Journal of Botany* 55 (6): 630–631 (1989).
TYPE: South Africa, Western Cape, near Kliprand on Pofadder Road to Loeriesfonein, *G. D. Duncan* 189 (NBG!, holo., iso.).

ETYMOLOGY. *duncanii*: after Graham Duncan (1959–).

DESCRIPTION. *Geophyte*, 120–180 mm high. *Bulb* ovoid, 20–30 mm in diameter, solitary; tunic multilayered, outer layers dark brown, cartilaginous, thick, neck distinct, 10–20 mm long, apex fasciculate; inner cataphyll translucent green, adhering to leaf bases, apex obtuse. *Leaves* 2, lanceolate, 85–110 × 15–25 mm, spreading to suberect, slightly to deeply canaliculate, coriaceous, upper surface glaucous to light green, plain or marked with irregularly scattered large brownish-purple blotches, depressed longitudinal grooves prominent, lower surface glaucous to light green; margins sclerotic, undulate and often heavily crisped, light maroon to brownish-white; leaf bases clasping, 10–20 mm long, lower half white, upper half light green to glaucous above; primary seedling leaf terete, erect. *Inflorescence* racemose, many-flowered, dense, sterile apex short; scape erect to suberect, 30–70 mm long, light green to brownish-purple, straight or swollen above; rachis light green to brownish-purple; pedicels suberect, 2–4 mm long, light green to light brownish-magenta; lower bracts ovate, becoming narrowly lanceolate above, 1–3 × 1–2 mm, light green. *Perianth* zygomorphic, oblong-campanulate, spreading to slightly cernuous, cream, ageing to light brownish-pink; tube cup-shaped, 2 mm long;

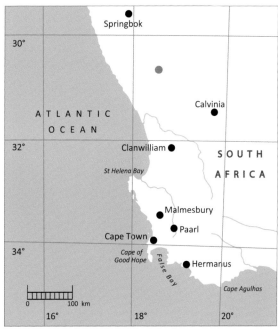

Map 43. Known distribution of *Lachenalia duncanii*.

Figure 155. *Lachenalia duncanii* near Kliprand.
Image: Graham Duncan/SANBI.

THE GENUS LACHENALIA
42. LACHENALIA DUNCANII

PLATE 17. Watercolour painting of *Lachenalia duncanii* from near Kliprand (*Duncan 189*, in NBG) courtesy of The Editor, *Flowering Plants of Africa*, vol. 56, t. 2143 (1999). Artist: Claire Linder Smith.

outer tepals narrowly ovate, 7–8 × 3–4 mm, apices slightly recurved, apical gibbosities and median keels green; inner tepals obovate, 9–10 × 2–3 mm, apices slightly recurved, median keels light green, shading to darker green at apex. *Stamens* well exserted, declinate; filaments cream in lower half, light magenta above, 11–12 mm long. *Ovary* obovoid, 4 × 3 mm, light green; style well exserted, declinate, 11–12 mm long, cream in lower half, light magenta above. *Capsule* obovoid, 5–6 × 3–4 mm. *Seed* ovoid, 1.3–1.4 × 1.1–1.2 mm, glossy, black; strophiole 0.6–0.7 mm long, ridged. *Chromosome number* $2n = 18$ (Spies *et al.*, 2008). Plate 17, Figure 155.

FLOWERING PERIOD. August to September, with a peak in late August.

HISTORY. I first collected this very distinctive species, which was in flower near Kliprand in western Bushmanland, on 10th September 1985. It was described from this material by W. F. Barker in the *South African Journal of Botany* (Barker, 1989), and is only known from the type locality. A watercolour painting by Claire Linder Smith was published in *Flowering Plants of Africa* (Duncan & Linder Smith, 1999a).

DISTINGUISHING CHARACTERS AND AFFINITIES. *L. duncanii* is recognised by the undulate and crisped, sclerotic margins of its leaves, which are only present in one other *Lachenalia* species, *L. nervosa*. The green apical gibbosities and median keels stand out against the cream background of the outer and inner tepals and the declinate, well-exserted filaments are prominent in their light magenta upper parts. The solitary bulb is distinctive in having numerous cartilaginous outer tunics that are produced into a thick, fasciculate neck. *L. duncanii* is sister to *L. undulata* (Duncan *et al.*, 2005a) and the relationship is discussed under that species.

DISTRIBUTION, HABITAT AND CONSERVATION STATUS. The species is currently known only from near Kliprand in south-western Bushmanland, in the extreme north-western part of the Western Cape (Map 43). The plants occur in Namaqualand Klipkoppe Shrubland vegetation (Mucina & Rutherford, 2006). They grow as exposed, scattered individuals in stony, deep red sand amongst low succulent scrub, in an area no larger than 10 square metres. *L. duncanii* may well occur elsewhere in this poorly botanised, arid region, but in view of the only known population's position near a road, it has a conservation status of Vulnerable (Victor & Duncan, 2009d).

NOTES. Although *L. duncanii* occurs in a transitional rainfall zone, receiving both winter and summer rain, it follows a winter growth cycle and undergoes a pronounced dormant period from mid-September to mid-April. The bulbs are adapted to remain dormant for one or more years during periods of insufficient rain. The oblong-campanulate flowers are pollinated by honey bees (*Apis mellifera*). The scape remains attached to the bulb for several weeks following capsule dehiscence and the seeds are dispersed locally by the shaking action of wind.

43. LACHENALIA SUMMERFIELDII

Lachenalia summerfieldii G. D. Duncan, sp. nov.
A L. neilii foliis vividus viridibus, floribus effusis cum staminibus declinatis, tepalis brevioribus, semini longiore angustiore cum strophiolo longiore differt.

TYPE: South Africa, Northern Cape, 3120 (Williston), 23.3 km north-west of Middelpos, on Leliekrantz Road (–CC), *G. D. Duncan* 562 (NBG!, holo., iso).

ETYMOLOGY. *summerfieldii*: after Gordon Summerfield (1941–), collector and expert cultivator of South African bulbous plants, who discovered this species and made the first collection of plants.

DESCRIPTION. *Geophyte*, 90–180 mm high. *Bulb* globose, 15–17 mm in diameter, solitary; tunic multilayered, outer layers dark brown, cartilaginous, neck apices fasciculate; inner cataphyll translucent white to light brown, longitudinal veins magenta, adhering to leaf bases, apex obtuse. *Leaves* 2, lanceolate, 100–120 × 12–20 mm, spreading to suberect, bright green, upper surface plain, with prominent depressed longitudinal grooves, lower surface heavily marked with brownish-magenta blotches; leaf bases clasping, 10–30 mm long, light green with dark magenta blotches or bands; primary seedling leaf terete, erect. *Inflorescence* racemose, many-flowered, sterile apex short; scape erect, 55–65 mm long, light green, heavily marked with dark brownish-purple blotches; pedicels suberect, 3–9 mm long; light green to white; lower bracts ovate, becoming lanceolate above, 1–3 × 1–4 mm, white to light green with dull maroon speckles. *Perianth* zygomorphic, oblong-campanulate, spreading to slightly suberect; tube cup-shaped, 3 mm long, white speckled with light blue; outer tepals ovate, 5–6 × 2–4 mm, light blue at base, shading to white or greenish-white above, apical gibbosities dark brownish-maroon or green; inner tepals obovate, 7–8 × 5–6 mm, translucent white, median keels dark brownish-maroon or dull yellow. *Stamens* shortly exserted, declinate; filaments white, 8–9 mm long. *Ovary* obovoid, 3–4 × 2–3 mm, light green; style shortly exserted, weakly declinate, 6–8 mm long, white, protruding well beyond perianth as ovary matures. *Capsule* ellipsoid, 6–7 × 5 mm. *Seed* ovoid, 1.7–2.0 × 1.2–1.3 mm, matte, black; strophiole 0.7–0.8 mm long, ridged. *Chromosome number* unknown. Figure 156.

FLOWERING PERIOD. September to October, with a peak in late September and early October.

HISTORY. *L. summerfieldii* was discovered by Gordon Summerfield just north of Middelpos in the spring of 2005. Suspecting it to be an undescribed species, he cultivated a few plants and the following

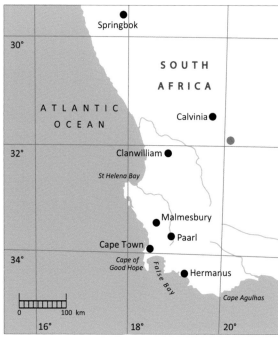

Map 44. Known distribution of *Lachenalia summerfieldii*.

Figure 156. Cultivated specimen of *Lachenalia summerfieldii* from Middelpos. Image: Graham Duncan.

year, I accompanied him to the locality, but almost all the plants had been heavily grazed, and only one individual was seen in fruit. On our return to the locality in September 2007, a few flowering specimens in good condition were collected to serve as the type material, although most of the plants had again been heavily grazed.

DISTINGUISHING CHARACTERS AND AFFINITIES. The species has not yet been included in a phylogenetic analysis. Morphologically, it appears to be allied to *L. neilii*, which has similar oblong-campanulate flowers with shortly exserted stamens, lanceolate, suberect, glaucous leaves and ovoid seeds. *L. neilii* differs in its slightly cernuous flower orientation with longer, recurved, greenish-white outer tepals (7–8 mm) and longer inner tepals (9–10 mm). The filaments of *L. neilii* are recurved and its seed is smaller (1.4 × 1.8 mm) with a shorter strophiole (0.6 mm long).

DISTRIBUTION, HABITAT AND CONSERVATION STATUS. *L. summerfieldii* is currently only known from the Roggeveld Plateau just north of Middelpos (Map 44). The plants occur in Roggeveld Karoo vegetation (Mucina & Rutherford, 2006), growing singly or in small groups, in seasonally moist, red, stony doleritic clay that bakes rock hard in summer. The plants occur on flats close to a large population of *Moraea marlothii* (Iridaceae). The species is threatened by overgrazing and a conservation status of Vulnerable is recommended.

NOTES. Heavy grazing of the leaves and developing inflorescences is probably the work of Cape hares (*Lepus capensis*). The oblong-campanulate flower shape suggests pollination by honey bees (*Apis mellifera*). The scapes remain attached to the bulbs for several weeks following capsule dehiscence and the seeds are dispersed locally.

44. LACHENALIA DOLERITICA

Lachenalia doleritica G. D. Duncan, *Bothalia* 28 (2): 134–135 (1998).
TYPE: South Africa, Northern Cape, Akkerendam Nature Reserve, Calvinia, M.L. Thomas *s.n.* sub. NBG 140212 (NBG!, holo.).

ETYMOLOGY. *doleritica*: occurring in doleritic clay.
DESCRIPTION. *Geophyte*, 120–240 mm high. *Bulb* ovoid, 15–50 mm in diameter, solitary; tunic multilayered, outer layers dark brown, cartilaginous; inner cataphyll translucent white, adhering to leaf bases, apex obtuse. *Leaves* 2, broadly lanceolate, 70–90 × 20–35 mm, spreading to slightly arcuate, apices sometimes resting at ground level, bright green to yellowish-green, upper surface with prominent depressed longitudinal grooves; leaf margins coriaceous, light to dark maroon; leaf bases clasping, 30–40 mm long, subterranean portion white, upper portion light green; primary seedling leaf terete, erect. *Inflorescence* racemose, many-flowered, sterile apex short; scape erect, 40–130 mm long, light green, sturdy, plain or with light brown speckles or dull purple blotches; rachis light green or heavily tinged with brown, apex brownish-pink to white; pedicels suberect to spreading, 2–5 mm long, increasing in length towards apex; lower bracts ovate, becoming lanceolate above, 3–5 × 2–4 mm, white. *Perianth* oblong-campanulate, spreading to suberect, greenish-yellow, ageing to dull pinkish-brown; tube cup-shaped, 2–3 mm long; outer tepals ovate, 8–9 × 4–5 mm, apical gibbosities bright green, brown or brownish-mauve, median keels green or brown; inner tepals obovate, 9–11 × 4–5 mm, apices slightly recurved, lower inner tepal slightly longer than upper tepals; apical marking and median keels green or dark brown. *Stamens* included to shortly exserted; filaments declinate, 9–11 mm long, white. *Ovary* ellipsoid, 4–5 × 3 mm, yellowish-green; style included, more or less straight, 6–7 mm long, white, protruding beyond perianth as ovary matures.

44. LACHENALIA DOLERITICA

Figure 157. *Lachenalia doleritica* on dolerite flats near Calvinia. Image: Graham Duncan.

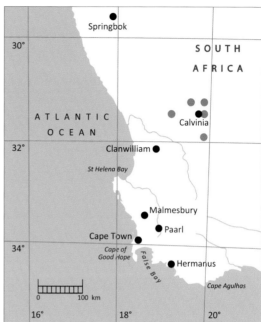

Map 45. Known distribution of *Lachenalia doleritica*.

Capsule ellipsoid, 8–9 × 5–6 mm. *Seed* ovoid, 1.8–1.9 × 1.8 mm, glossy, black; strophiole 1.1–1.2 mm long, ridged. *Chromosome number* 2n = 18 (Spies *et al.*, 2008). Figures 28, 157.

FLOWERING PERIOD. September to October, with a peak in early September.

HISTORY. The first collection of *L. doleritica* was made by the South African pasture ecologist and vegetation specialist J. P. H. Acocks (1911–1979) on 28th September 1953 at the farm Perdefontein in the Calvinia District (*Acocks* 17310, in PRE). It was subsequently collected on numerous occasions in the environs of Calvinia, but the species was only formally described 45 years later, from a collection made by Margaret Thomas in the Akkerendam Nature Reserve at Calvinia, in September 1984 (Duncan, 1998a).

DISTINGUISHING CHARACTERS AND AFFINITIES. Duncan *et al.* (2005a) suggested an alliance between *L. doleritica* and *L. schelpei*; the latter also has two broadly lanceolate leaves and large, ovoid seeds with long, ridged strophioles. *L. schelpei* differs in having tubular, slightly cernuous flowers borne on very short pedicels (1.5 mm long). Its flowers have very short perianth tubes (1 mm long), shorter outer and inner tepals (5–7 mm, 7–8 mm, respectively) and longer styles (7–8 mm).

DISTRIBUTION AND HABITAT AND CONSERVATION STATUS. *L. doleritica* is confined to the Calvinia District, occurring in Hantam Karoo vegetation (Mucina & Rutherford, 2006), singly or in large colonies on flats and between dolerite ridges, in doleritic clay that bakes rock hard over summer (Map 45). The bulbs are adapted to remain dormant for one or more years until favourable conditions return. *L. doleritica* is threatened by road widening and overgrazing, and has a conservation status of Vulnerable (Victor & Duncan, 2009c).

NOTES. *L. doleritica* is pollinated by honey bees (*Apis mellifera*). The scapes remain attached to the bulbs for several weeks following capsule dehiscence and the seeds are dispersed locally by the shaking action of wind.

45. LACHENALIA DASYBOTRYA

Lachenalia dasybotrya Diels, *Botanische Jahrbücher für Systematik Pflanzengeschichte und Pflanzengeographie* 44: 116 (1909).
TYPE: South Africa, Northern Cape, 'Karooflache sudostlich von Calvinia' [Karoo flats south-east of Calvinia], *F. L. E. Diels* 699 (B!, holo.).

ETYMOLOGY. *dasybotrya*: shaggy inflorescences.
DESCRIPTION. *Dwarf geophyte*, 80–130 mm high. *Bulb* globose, 20–30 mm in diameter, solitary, deep-seated; tunic multilayered, outer layers dark brown, cartilaginous, neck apex fasciculate; inner cataphyll translucent white, adhering to leaf bases, apex obtuse; leaf bases clasping, 30–80 mm long, subterranean, white. *Leaves* 2, broadly lanceolate, 50–90 × 25–40 mm, spreading, upper surface light green to glaucous with depressed longitudinal grooves, lower surface light green; leaf bases 10–25 mm long, subterranean, white; leaf margins flat to undulate, cartilaginous; primary seedling leaf terete, erect. *Inflorescence* racemose, many-flowered, dense, sterile apex short; scape erect, 20–60 mm long, light green, plain or with brownish-purple blotches; rachis light green or dull maroon in lower part, shading to mauvish-white above; lower bracts ovate, becoming lanceolate above, 1–3 × 1–3 mm, white; pedicels perpendicular to suberect, 3–6 mm long, white. *Perianth* zygomorphic, oblong-campanulate, white to cream, ageing to dull pink, spreading to suberect; spicy-scented; tube cup-shaped, 3 mm long; outer tepals ovate, 5–7 × 4–5 mm, apical gibbosities dull green or brown,

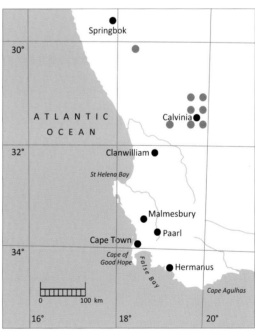

Map 46. Known distribution of *Lachenalia dasybotrya*.

Figure 158. Cultivated specimen of *Lachenalia dasybotrya* from Calvinia. Image: Graham Duncan/SANBI.

median keels absent or pinkish-brown; inner tepals obovate, 8–9 × 4–5 mm, apices slightly recurved, median keels green or brown. *Stamens* included to shortly exserted; filaments declinate, white, 6–10 mm long. *Ovary* ellipsoid, 3–4 × 2–3 mm, light green; style included to shortly exserted, declinate, 7–8 mm long, white. *Capsule* ellipsoid, 6–7 × 4–5 mm. *Seed* ovoid, 1.3–1.4 × 1.4 mm, matte, black; strophiole 0.5 mm long, ridged. *Chromosome number* unknown. Figure 158.

FLOWERING PERIOD. August to October, with a peak in late September.

HISTORY. *L. dasybotrya* was collected for the first time south-east of Calvinia in the western Great Karoo on 15th September 1900 by F. L. E. Diels, German Professor of Botany at the University of Berlin and Director of the Berlin-Dahlem Botanical Garden. The holotype material (*Diels* 699) is housed in the Botanical Museum of Berlin-Dahlem and the species was published nine years after its discovery, in volume 44 of *Engler's Botanische Jahrbücher* (Diels, 1909). *L. dasybotrya* has seldom been collected because of its extended periods of dormancy in arid environments.

DISTINGUISHING CHARACTERS AND AFFINITIES. A relationship exists between *L. dasybotrya* and *L. doleritica* (Duncan et al, 2005a) which share racemose, oblong-campanulate, spreading to suberect flowers with declinate stamens. Both have light green to glaucous, unmarked leaves, and ovoid, matte black seeds. *L. doleritica*, however, is a taller plant (up to 240 mm high), differing in its smaller, greenish-yellow perianths with longer outer and inner tepals (8–9 mm and 9–11 mm long, respectively). In addition, *L. doleritica* produces larger ovoid seeds (1.8–1.9 × 1.8 mm) with bigger strophioles (1.1–1.2 mm).

DISTRIBUTION, HABITAT AND CONSERVATION STATUS. *L. dasybotrya* is centred around Calvinia on the western edge of the Great Karoo, with an outlier south-east of Gamoep in western Bushmanland (Map 46). Its deep-seated bulbs grow in colonies, mainly in shale or stony red doleritic clay on open flats, in Hantam Karoo or Vanrhynsdorp Shale Renosterveld vegetation (Mucina & Rutherford, 2006). In common with numerous other lachenalias from arid habitats, the bulb is protected by cartilaginous outer tunics and is adapted to remain dormant for one or more years during periods of poor rainfall. The species is threatened by overgrazing, and a conservation status of Vulnerable is recommended.

NOTES. The oblong-campanulate flower shape of *L. dasybotrya* is indicative of pollination by honey bees (*Apis mellifera*). The scapes detach from the bulbs shortly after capsule dehiscence and the ripe infructescences are blown away, dispersing the seeds over a wide area.

46. LACHENALIA ALBA

Lachenalia alba W. F. Barker ex G. D. Duncan, *Bothalia* 26 (1): 6–7 (1996).
TYPE: South Africa, Northern Cape, Charlies Hoek, Nieuwoudtville, *W. F. Barker* 10888 (NBG!, holo.).

ETYMOLOGY. *alba*: white flowers.

DESCRIPTION. *Geophyte*, 100–330 mm high. *Bulb* globose, 15–20 mm in diameter, offset-forming; tunic multilayered, outer layers dark brown, spongy; inner cataphyll translucent white, longitudinal veins green, adhering to leaf bases, apex obtuse. *Leaves* 2, lanceolate, 120–230 × 15–25 mm, suberect, canaliculate, glaucous; upper surface with depressed longitudinal grooves, plain, lower surface plain or suffused with dull maroon towards base; margins slightly coriaceous, flat to slightly undulate; leaf bases clasping, 20–80 mm long; primary seedling leaf terete, erect. *Inflorescence* racemose, many-flowered, sterile apex short; scape erect to suberect, 50–150 mm long, light green; rachis light green below, apex white, speckled with dull maroonish-brown; pedicels suberect, 0.5–2.0 mm long, suberect, white; lower bracts ovate, becoming lanceolate above, 2–3 × 2–3 mm, translucent white.

THE GENUS LACHENALIA
46. LACHENALIA ALBA

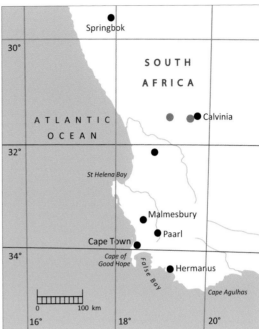

Map 47. Known distribution of *Lachenalia alba*.

Figure 159. *Lachenalia alba* at Nieuwoudtville. Image: Graham Duncan.

Perianth oblong-campanulate, suberect; tube cup-shaped, 2–3 mm long, white, or dull blue at base shading to white above; outer tepals ovate, 7–8 × 3–4 mm, white, apical gibbosities light brownish-green or dull red; inner tepals obovate, 10–11 × 4–6 mm, white, keels dull red, brownish-green or blue, apices undulate, recurved. *Stamens* recurved, shortly exserted; filaments white, 10–13 mm long. *Ovary* ellipsoid, 3–4 × 2–3 mm, light green; style included to shortly exserted, more or less straight, 8–9 mm long, white. *Capsule* ellipsoid, 6–7 × 3–4 mm, light green. *Seed* ovoid, 1.2–1.4 × 1.2–1 4 mm, glossy, black; strophiole 0.5–0.6 mm long, ridged. *Chromosome number* $2n = 18, 20, 40$ (Johnson & Brandham, 1997). Figure 159.

FLOWERING PERIOD. September to October, with a peak in late September.

HISTORY. *L. alba* was collected for the first time by W. F. Barker in September 1930 at Klipkoppies, also known as Charlie's Hoek, just east of Nieuwoudtville. The holotype material (*Barker* 10888, in NBG) was also collected at this location by Barker, in September 1973, and numerous subsequent collections have been made in the vicinity of this town. The manuscript name *L. alba* W. F. Barker was appended to pressed material at NBG, but the species was only formally described more than 60 years after its discovery, in *Bothalia* (Duncan, 1996).

DISTINGUISHING CHARACTERS AND AFFINITIES. *L. alba* and *L. neilii* are sister species (Duncan *et al.*, 2005a), sharing similar racemes of oblong-campanulate flowers with shortly exserted stamens, two glaucous, suberect, unmarked leaves and ovoid seeds with ridged strophioles. *L. neilii* differs in its smaller, more numerous, spreading, greenish-white perianths that have less flared inner tepals and prominent dull blue perianth tubes, larger capsules (8–9 × 6–7 mm) and larger seeds (1.4 × 1.8 mm).

DISTRIBUTION, HABITAT AND CONSERVATION STATUS. *L. alba* is confined to the Bokkeveld Plateau at Nieuwoudtville, with a disjunct record on Merweshoek Farm at the south-western end of the Hantamsberg near Calvinia (Map 47). The plants grow in Nieuwoudtville Shale Renosterveld and Hantam Karoo vegetation (Mucina & Rutherford, 2006), on flats in large colonies amongst low scrub and grasses. The seasonally moist, heavy red doleritic clays of these areas bake rock hard during summer. *L. alba* sometimes occurs in association with *L. suaveolens*. *L. alba* is threatened by agricultural expansion, overgrazing and trampling by livestock, and the bulbs are consumed by Cape porcupines (*Hystrix africaeaustralis*). It has a conservation status of Vulnerable (Duncan & Raimondo, 2009a).

NOTES. The flowers of *L. alba* are pollinated by honey bees (*Apis mellifera*). The scapes remain attached to the bulbs for a number of weeks following capsule dehiscence and the seeds are dispersed locally.

47. LACHENALIA NEILII

Lachenalia neilii W. F. Barker ex G. D. Duncan, *Bothalia* 26 (1): 5–6 (1996).
TYPE: South Africa, Northern Cape, Nieuwoudtville–Calvinia Road (R27), at turnoff to Rondekop, in ditches on both sides of road, *G. D. Duncan* 196 (NBG!, holo.).

ETYMOLOGY. *neilii*: after Neil McGregor (1936–2010), farmer, naturalist and former owner of Glen Lyon Farm, Nieuwoudtville.

DESCRIPTION. *Geophyte*, 120–230 mm high. *Bulb* globose, 12–25 mm in diameter, offset-forming; tunic multilayered, outer layers dark brown, spongy; inner cataphyll translucent white, adhering to leaf bases, apex obtuse. *Leaves* 2, lanceolate, 55–145 × 7–22 mm, suberect, canaliculate, glaucous, upper surface with depressed longitudinal grooves; leaf bases clasping, 40–70 mm long, light yellowish-green, sometimes speckled with maroon; primary seedling leaf terete, erect. *Inflorescence* racemose, many-flowered, sterile apex short; scape erect, 40–110 mm long, sturdy, light green, plain or with maroon speckles; rachis light green; pedicels suberect, 3–5 mm long, white; lower bracts ovate becoming lanceolate above, 2–3 × 1–2 mm, white. *Perianth* oblong-campanulate, ageing to dull red, spreading to slightly cernuous; tube cup-shaped, 2–3 mm long, dull blue; outer tepals ovate, 7–8 × 3–4 mm, greenish-white, bases dull blue; apical gibbosities green or brown; inner tepals obovate, 9–10 × 4–5 mm, translucent white, apices slightly recurved, median keels green. *Stamens* recurved, shortly exserted; filaments white, 9–11 mm long. *Ovary* ellipsoid, 3–4 × 2–3 mm, bright green; style included to shortly exserted, more or less straight, 8–11 mm long, white. *Capsule* ellipsoid, 8–9 × 6–7 mm. *Seed* ovoid, 1.4 × 1.8 mm, matte, black; strophiole 0.6 mm long, ridged. *Chromosome number* 2n = 18 (Spies *et al.*, 2008). Figure 160.

FLOWERING PERIOD. August to October, with a peak in mid-September.

HISTORY. The first collection of *L. neilii* was made by staff of the National Botanical Gardens, Kirstenbosch at Klipkoppies in the Nieuwoudtville area in September 1930. W. F. Barker collected it several times during the 1960s and appended her manuscript name *L. neilii* W. F. Barker to pressed specimens at NBG, but the species was only formally described many years later in *Bothalia* (Duncan, 1996). The type material (*Duncan* 196, in NBG) was collected in a ditch along the Nieuwoudtville–Calvinia Road in September 1985, but that population is no longer extant as a result of road verge clearing and smothering alien grasses.

DISTINGUISHING CHARACTERS AND AFFINITIES. Duncan *et al.* (2005a) consider this species to be sister to *L. alba* and the morphological similarities are discussed under that taxon.

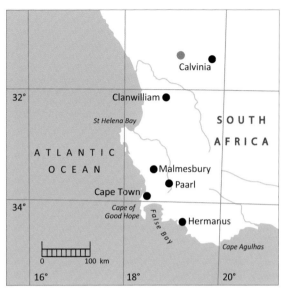

Map 48. Known distribution of *Lachenalia neilii*.

Figure 160. *Lachenalia neilii* in habitat near Nieuwoudtville. Image: Graham Duncan.

DISTRIBUTION, HABITAT AND CONSERVATION STATUS. *L. neilii* has a highly restricted distribution on the Bokkeveld Plateau immediately east of Nieuwoudtville (Map 48). The plants grow in Nieuwoudtville-Roggeveld Dolerite Renosterveld vegetation (Mucina & Rutherford, 2006) and are limited to a few severely fragmented subpopulations in open aspects on flats in seasonally moist, heavy red doleritic clay that bakes rock hard in summer. The species is extinct at its type locality and is in decline due to road widening and overgrazing. It has a conservation status of Endangered (Duncan & Raimondo, 2009c).
NOTES. The flowers are pollinated by honey bees (*Apis mellifera*). The scape remains attached to the bulb for several weeks following capsule dehiscence and the seeds are released locally.

48. LACHENALIA BOWKERI

Lachenalia bowkeri Baker, in Thiselton-Dyer, W. T. (ed.), *Flora Capensis* 6: 427 (1897).
TYPE: South Africa, Eastern Cape, Somerset Division, *J. H. Bowker s.n.* (K!, holo.).
SYNONYMY: *Lachenalia subspicata* Fourc., *Transactions of the Royal Society of South Africa* 21 (1): 79 (1934). Type: South Africa, Eastern Cape, Zuur Anys Flats on road to Kouga, *H. G. Fourcade* 3044 (BOL!, holo.).

ETYMOLOGY. *bowkeri*: after Col. J. H. Bowker (1822–1900), collector of plants and butterflies from the Eastern Cape and KwaZulu-Natal.
DESCRIPTION. *Geophyte*, 50–300 mm high. *Bulb* globose, 10–15 mm in diameter, offset-forming; tunic multilayered, outer layers dark brown, cartilaginous, neck fasciculate; inner cataphyll translucent white,

48. LACHENALIA BOWKERI

Figure 161 (left). Cultivated specimen of *Lachenalia bowkeri* from Port Elizabeth. Image: Graham Duncan.
Figure 162 (right). Cultivated specimen of *Lachenalia bowkeri* from near Oudtshoorn. Image: Graham Duncan.

adhering to leaf bases, apex obtuse. *Leaves* 1(–2), lanceolate, 80–190 × 10–15 mm, canaliculate, light green to glaucous, plain, upper surface with depressed longitudinal grooves; leaf bases clasping, 30–90 mm long, white below, shading to light green above, plain or with minute magenta speckles; primary seedling leaf terete, suberect. *Inflorescence* spicate or racemose, few- to many-flowered, moderately dense, sterile apex short; scape erect, 100–200 mm long, light green in lower half, upper half minutely speckled with dull magenta; rachis light green speckled with dull magenta; pedicels absent or up to 2 mm long, suberect, white; lower bracts ovate, becoming lanceolate above, 1–3 × 2–3 mm, white with light green or magenta speckles. *Perianth* zygomorphic, oblong-campanulate, spreading to cernuous, lightly to moderately sweet-scented, ageing to dull brownish-magenta; tube cup-shaped, 2–3 mm long, light greenish-white; outer tepals narrowly ovate, 5–6 × 4 mm, glaucous or light purple, speckled with dull blue at base, apical gibbosities dark brown to brownish-magenta; inner tepals obovate, 7–9 × 4–5 mm, apices rounded, flat or slightly recurved, translucent white, median keels light to dark green or purple. *Stamens* included or shortly to well exserted; filaments declinate, 7–10 mm long, white; pollen ageing to black. *Ovary* ellipsoid, 3 × 2 mm, light green; style included to well exserted, 8–12 mm long, white. *Capsule* ellipsoid, 5–6 × 4–5 mm. *Seed* globose, 0.9–1.0 × 0.9–1.0 mm, matte, black, secondary sculpturing reticulate; strophiole 0.6–0.7 mm long, ridged. *Chromosome number* $2n = 16$ (Johnson & Brandham, 1997). Figures 25, 161, 162.

FLOWERING PERIOD. July to September, with a peak in late August.

HISTORY. James G. Baker (1897a) described *L. bowkeri* in volume 6 of *Flora Capensis*. The type specimen, a single sheet housed in the Kew Herbarium, was collected by the South African naturalist Col. J. H. Bowker at an inexact locality in the Eastern Cape in the 'Somerset Division', presumed to be in the vicinity of Somerset East. Forms of *L. bowkeri* with sessile flowers are sometimes misidentified as *L. orchioides*, and even Baker (1897a) mistakenly cited a specimen of *L. bowkeri* collected near Grahamstown (*MacOwan* 1337, in GRA) under *L. orchioides*.

L. subspicata was described by H. G. Fourcade (1934) in volume 21 of *Transactions of the Royal Society*, from specimens he collected in September 1925 on the Zuur Anys flats on the road to Kouga in the Eastern Cape, where *L. bowkeri* has been collected several times. Evidently Fourcade did not consider it comparable with *L. bowkeri* as his specimens were shortly pedicellate and Baker's description stated that the species had sessile flowers. Consequently, Fourcade compared his specimens with *L. orchioides*, intimating that *L. orchioides* has broader leaves, more numerous flowers and shorter tepals, and with *L. youngii*, which has longer pedicels. Reid (1993) listed *L. subspicata* under *L. bowkeri* and it was formally placed in synonymy by Dold & Phillipson (1998).

DISTINGUISHING CHARACTERS AND AFFINITIES. *L. bowkeri* is a variable species as regards inflorescence type, flower orientation and degree of stamen exsertion. Most forms are readily identified by their sessile or shortly pedicellate, spreading flowers that have included stamens, whereas others have much longer pedicels (up to 3 mm long) and are cernuous with shortly to well-exserted stamens. Outer tepal colour varies in shades of light green to grey-green or light purple, speckled with light blue at the base, with dark brown or brownish-magenta gibbosities. Inner tepals are translucent white with rounded, slightly recurved apices and light to dark green or purple median keels. The identification of pressed material with faded flowers is problematic as the stamens of forms with exserted filaments tend to recede into the perianth and the normally spreading inner tepal apices become narrower upon pressing.

Duncan *et al*. (2005a) placed the species as sister to *L. dehoopensis*, which has similar oblong-campanulate flowers, and globose seeds with reticulate secondary sculpturing. *L. dehoopensis* differs in its smaller stature and flat, acute inner tepal apices. It has two linear, deeply caniculate leaves that

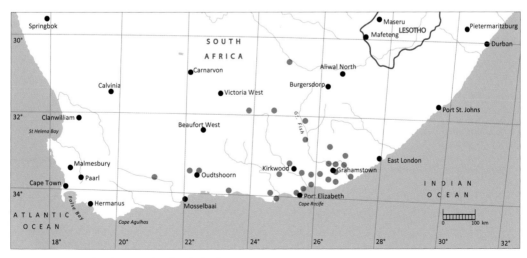

Map 49. Known distribution of *Lachenalia bowkeri*.

are heavily banded with dark green on the lower leaf surfaces and with dark maroon to magenta on the clasping bases. The species is restricted to De Hoop Nature Reserve east of Bredasdorp.

Superficially, *L. bowkeri* resembles *L. karooica*, which has similar oblong-campanulate flowers with light green or glaucous outer tepals. *L. karooica* also occurs in the Great Karoo and Eastern Cape interior but differs from *L. bowkeri* in having smaller flowers with strongly recurved inner tepals and recurved stamens, flowers that are arranged in whorls of 3, and a much larger, ovoid seed (1.3–1.4 × 1.2–1.3 mm) with a longer, ridged strophiole (1.1 mm long).

DISTRIBUTION, HABITAT AND CONSERVATION STATUS. *L. bowkeri* is a near endemic of the Eastern Cape, occurring from Muiskraal in the Little Karoo in the Western Cape, east to the Peddie District, inland to the eastern foothills of the Sneeuberg east of Murraysberg, and to Bell Rock Farm west of Colesberg in the eastern Great Karoo (Map 49). It occurs in 11 vegetation types: Eastern Upper Karoo, Bedford Dry Grassland, Western- and Eastern Little Karoo, Kouga-, Algoa- and Kouga Grassy Sandstone Fynbos, Kowie- and Great Fish Thicket, Suurberg Shale Fynbos and Humansdorp Shale Renosterveld (Mucina & Rutherford, 2006). *L. bowkeri* grows singly or in small groups within larger colonies in a variety of aspects, including exposed flats, shaded to partially shaded south- or south-west-facing mountain and hill slopes, or hot west-facing situations. Its substrates include alluvial sand, loose sandy gravel, stony clay and limestone. The species is associated with karroid scrub, succulent ground cover, dune vegetation and rocky grassland. It is fairly common on the Swartkops River floodplain near Port Elizabeth and most collections have been made around Grahamstown in the Albany District. *L. bowkeri* is not threatened.

NOTES. The foliage of *L. bowkeri* is heavily grazed by herbivores (Dold & Phillipson, 1998). The sweet-scented flowers are pollinated by honey bees (*Apis mellifera*). The scapes remain attached to the bulbs following capsule dehiscence and the seeds are dispersed locally.

49. LACHENALIA DEHOOPENSIS

Lachenalia dehoopensis W. F. Barker, *South African Journal of Botany* 53 (2): 167–168 (1987).
TYPE: South Africa, Western Cape, De Hoop-Potberg Nature reserve, on De Hoop flats, *A. Scott s.n.* sub. NBG 129587 (NBG!, holo.).

ETYMOLOGY. *dehoopensis*: after De Hoop Nature Reserve in the southern Cape.
DESCRIPTION. *Dwarf geophyte*, 80–170 mm high. *Bulb* globose, 10–15 mm in diameter, offset-forming; tunic multilayered, outer layers dark brown, cartilaginous, neck fasciculate; inner cataphyll translucent white, adhering to leaf bases, apex obtuse. *Leaves* 2, linear, 80–130 × 4–7 mm, bifacial, spreading to suberect, canaliculate, upper surface dark green, plain, lower surface dark green in upper half, shading to whitish-green below, with darker green, slightly raised fleshy bands in lower half, midrib distinct; leaf bases clasping, 15–40 mm long, white with dark maroon to magenta bands; primary seedling leaf flat, prostrate. *Inflorescence* racemose, few- to many-flowered, lax, sterile apex short; scape erect, 50–100 mm long, slender, light green, plain or marked with minute greenish-grey, dark maroon or purple blotches; rachis colouring as for scape; pedicels suberect, 2–3 mm long, white; lower bracts ovate, becoming narrow-lanceolate above, 1–2 × 1–3 mm, white to greenish-white. *Perianth* zygomorphic, oblong-campanulate, spreading to cernuous; tube shallow cup-shaped, 1–2 mm long, white; outer tepals ovate, 5–7 × 3–4 mm, bases light blue, becoming light grey above, apical gibbosities green to brownish-purple; inner tepals obovate, 7–10 × 4–6 mm, translucent white, median keels broad, green or brownish-purple, apices acute. *Stamens* included to shortly exserted, declinate; filaments white,

THE GENUS LACHENALIA
49. LACHENALIA DEHOOPENSIS

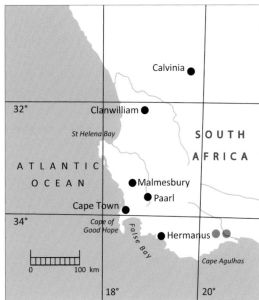

Map 50. Known distribution of *Lachenalia dehoopensis*.

Figure 163. Cultivated specimens of *Lachenalia dehoopensis* from De Hoop Nature Reserve. Image: Graham Duncan.

9–10 mm long. *Ovary* ellipsoid, 2–3 × 2 mm, light green; style included, more or less straight, 8–9 mm long, white, becoming well exserted as ovary matures. *Capsule* ellipsoid, 5–6 × 4–5 mm. *Seed* globose, 0.9–1.0 × 0.8 mm, matte, black, secondary sculpturing reticulate; strophiole 0.6–0.7 mm long, ridged. *Chromosome number* unknown. Figure 163.

FLOWERING PERIOD. August to September, with a peak in early September.

HISTORY. The South African botanist J. P. H. Acocks (1911–1979) was the first to collect *L. dehoopensis* in August 1963, at Kathoek near Bredasdorp in the southern Cape (*Acocks* 22596, in PRE); however, the specimen was initially mis-identified as *L. subspicata* Fourc. Two further collections were made by C. J. Burgers in 1979 and 1980, both in the De Hoop Nature Reserve east of Bredasdorp. In 1984, Ann Scott collected the species on the De Hoop flats. Living specimens of this collection were studied by W. F. Barker and material became available for cultivation at Kirstenbosch. *L. dehoopensis* was described in the *South African Journal of Botany* (Barker, 1987) with the Scott collection as the holotype (*Scott s.n.*, in NBG). Flowering plants have since been observed on several occasions within the De Hoop Nature Reserve.

DISTINGUISHING CHARACTERS AND AFFINITIES. Duncan *et al.* (2005a) placed *L. dehoopensis* as sister to *L. bowkeri*, which has a similar inflorescence of oblong-campanulate flowers and globose seeds with reticulate secondary sculpturing, but which differs in its more robust stature with one or two lanceolate, usually plain leaves. The outer tepals of *L. bowkeri* are glaucous or light purple and its inner tepals are rounded, with slightly recurved apices. *L. bowkeri* is allopatric with a relatively wide distribution that extends from the southern Little Karoo in the west to the southern Eastern Cape, and inland to west of Colesberg in the central Great Karoo.

49. LACHENALIA DEHOOPENSIS

DISTRIBUTION, HABITAT AND CONSERVATION STATUS. *L. dehoopensis* is confined to just a few sites within the De Hoop Nature Reserve (Map 50). It grows in De Hoop Limestone Fynbos vegetation (Mucina & Rutherford, 2006) and is nowhere plentiful, occurring as scattered individuals or in small groups. This species favours open flats in dark loam overlaying limestone deposits, and also occurs on hill slopes in hard, gravelly substrates. The species is potentially threatened by alien invader plants and has a conservation status of Vulnerable (Raimondo & Duncan, 2009a).

NOTES. The species is stimulated to profuse flowering following bush fires but is not dependent on fire for flowering to occur. The oblong-campanulate flowers suggest pollination by honey bees (*Apis mellifera*). The scapes remain attached to the bulbs following capsule dehiscence and the seeds are dispersed locally.

50. LACHENALIA OBSCURA

Lachenalia obscura Schltr. ex G. D. Duncan, *Bothalia* 27 (1): 9–11 (1997).
TYPE: South Africa, Northern Cape, Vogelstruis Vlakte, Calvinia district, *R. H. Compton* 11174 (NBG!, holo.).

ETYMOLOGY. *obscura*: obscure flower colouring.
DESCRIPTION. *Geophyte*, 55–400 mm high. *Bulb* globose, 10–25 mm in diameter, offset-forming; tunic multilayered, outer layers dark brown, cartilaginous, apex fasciculate; inner cataphyll translucent white, loosely surrounding leaf bases, apex obtuse. *Leaves* 1–2, lanceolate, 25–280 × 10–25 mm,

Figure 164 (left). *Lachenalia obscura* on stony clay flats at Calvinia. Image: Graham Duncan.
Figure 165 (right). *Lachenalia obscura* on grassy flats at Nieuwoudtville. Image: Graham Duncan.

suberect, canaliculate, yellowish-green to dark green, upper surface plain, with depressed longitudinal grooves, lower surface usually heavily banded with bright green; leaf bases clasping, 10–20 mm long, heavily banded with dull brownish-purple or magenta; primary seedling leaf terete, erect. *Inflorescence* spicate, erect to suberect, many-flowered, sterile apex short; scape erect to suberect, 25–100 mm long, light to dark green, blotched with light to dark purple; rachis light green, mottled with light bluish-purple; lower bracts ovate, becoming lanceolate above, 2–4 × 1–3 mm, greenish-white. *Perianth* oblong-campanulate, spreading to slightly cernuous, arranged in 3-flowered whorls; tube cup-shaped, 2–3 mm long, dull white with dull blue speckles; outer tepals ovate, 6–9 × 4–5 mm, dull white with dull blue speckles; apical gibbosities greenish-brown; inner tepals obovate, 8–10 × 5–7 mm, translucent white, apices obtuse, apical margins bright to deep magenta, median keels green. *Stamens* included to shortly exserted, more or less straight; filaments white, 8–11 mm long, white. *Ovary* ellipsoid, 3–4 × 2–3 mm, light green; style straight, 8 mm long, protruding beyond perianth as ovary enlarges. *Capsule* ellipsoid, 7–8 × 4–6 mm. *Seed* ovoid, 1.2–1.3 × 1.3–1.4 mm, glossy, black; strophiole 0.6–0.7 mm long, ridged. *Chromosome number* 2n = 36 (Spies *et al.*, 2008). Figures 164, 165.

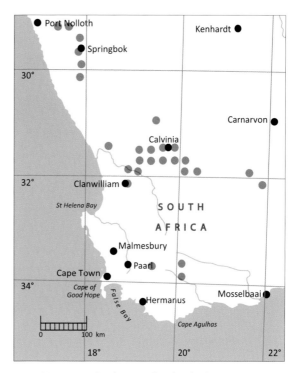

Map 51. Known distribution of *Lachenalia obscura*.

FLOWERING PERIOD. June to October, with a peak in late August.

HISTORY. The German traveller Rudolf Schlechter first collected this species in August 1897, at Papkuilsfontein Farm south-east of Vanrhynsdorp. He distributed pressed material under his manuscript name *L. obscura* Schltr. to nine local and foreign herbaria, where it languished for a century until its formal publication in *Bothalia* (Duncan, 1997). The type material was collected by Prof. R. H. Compton at Vogelstruis Vlakte in July 1941 in the Calvinia district, where the species is common.

DISTINGUISHING CHARACTERS AND AFFINITIES. *L. obscura* is allied to *L. canaliculata* (Duncan *et al.*, 2005a). These species share several characters including spikes of oblong-campanulate flowers that are arranged in three-flowered whorls, perianths with dull blue tubes, and ovoid seeds with ridged strophioles. *L. canaliculata* differs in its narrower perianths, acute inner tepals, the lower tepal being distinctly canaliculate and longer. It also differs in its declinate stamens, intensely glaucous leaves and narrower seed (1.0–1.1 mm wide). Collections from the vicinity of Sutherland (*Hall 3252*, *Hall 3287* in NBG), originally included under *L. obscura* (Duncan, 1997) have been assigned to *L. canaliculata*.

DISTRIBUTION, HABITAT AND CONSERVATION STATUS. *L. obscura* is a variable, arid habitat species with a relatively wide distribution extending from Steinkopf in the southern Richtersveld, south to the Montagu District in the Little Karoo (Map 51). The species is encountered in eight vegetation types: Namaqualand Blomveld, Namaqualand Klipkoppe Shrubland, Hantam- and Roggeveld Karoo,

Nieuwoudtville-Roggeveld Dolerite Renosterveld, Citrusdal Vygieveld, Western Upper Karoo and Montagu Shale Renosterveld (Mucina & Rutherford, 2006). It is common in and around Calvinia and Nieuwoudtville, and throughout its range, it occurs as solitary individuals or in small groups as part of larger populations. *L. obscura* is usually associated with hard, dry stony red clay soils, growing on flats, lower hill slopes and rocky outcrops, amongst karroid vegetation. The species is not threatened.

NOTES. The flowers are pollinated mainly by honey bees (*Apis mellifera*) and by the solitary bee *Anthophora diversipes*. Shortly after capsule dehiscence, the scapes detach and the seeds are dispersed over a wide area as the infructescence is blown about.

51. LACHENALIA CANALICULATA

Lachenalia canaliculata G. D. Duncan, sp. nov.

A L. obscura foliis glaucis, tepalis angustioribus cano-azureis, tepalo interiore infero canaliculato et ultra superioribus tepalis interioribus protruso, staminibusque declinatis differt.

TYPE: South Africa, Northern Cape, 3220 (Sutherland): Voëlfontein Farm, Sutherland (–AD), H. Hall 3252 (NBG!, holo.; PRE!, iso.).

ETYMOLOGY. *canaliculata*: channelled, descriptive of the lower inner tepal.

DESCRIPTION. *Geophyte*, 150–450 mm high. *Bulb* globose, 15–20 mm in diameter, offset-forming; outer tunic multilayered, outer layers dark brown to black, cartilaginous, neck apex fasciculate; inner cataphyll translucent white, adhering to leaf bases, apex obtuse. *Leaves* (1)–2, lanceolate, 100–150 × 13–35 mm, spreading to suberect, dark green to glaucous, upper surface plain, lower surface plain or marked with light magenta spots or banded with bright green; leaf bases clasping, 30–60 mm long, plain or banded with purplish-maroon; primary seedling leaf terete, erect. *Inflorescence* spicate, few- to many-flowered; scape erect, 80–120 mm long, light green with darker green or dull brownish-magenta blotches; rachis light green, plain or with brownish-magenta blotches; lower bracts ovate, becoming lanceolate above, 1–2 × 1–3 mm. *Perianth* zygomorphic, oblong-campanulate, suberect, arranged in 3-flowered whorls, faintly spicy-scented; tube cup-shaped, 3 mm long, cream mottled with dull blue; outer tepals ovate, 5–6 × 4–5 mm, translucent white or cream with dull blue speckles, apical gibbosities dark brown or green; inner tepals obovate, 8–11 × 3–4 mm, translucent white to yellowish-green, fading to light magenta, apices acute, median keels light to bright green, upper two tepals overlapping, lower inner tepal narrower, 2 mm longer, strongly canaliculate. *Stamens* included to shortly exserted, declinate; filaments white, 8–10 mm long. *Ovary* ellipsoid, 3–4 × 2–3 mm, light green; style declinate, 7–8 mm long, white. *Capsule* ellipsoid, 6–8 × 5–6 mm. *Seed* ovoid, 1.1–1.2 × 1.0–1.1 mm, glossy, black; strophiole 0.7–0.8 mm long, ridged. *Chromosome number* unknown. Figures 166–168.

FLOWERING PERIOD. August to October, with a peak in early September.

HISTORY. A collection made by Rudolf Marloth at an unrecorded locality on the Roggeveld near Sutherland in October 1920 (*Marloth* 9716, in NBG) appears to be the first collection of *L. canaliculata*. The type material was collected in October 1968 by Harry Hall, a former Kirstenbosch horticulturist, at Voëlfontein Farm west of Sutherland, the location of several other collections of this species. Specimens from the Sutherland District (*Hall* 3252, *Hall* 3287 in NBG) recognised here as *L. canaliculata* were originally included under *L. obscura* when the latter species was published in *Bothalia* (Duncan, 1997).

THE GENUS LACHENALIA | 247
51. LACHENALIA CANALICULATA

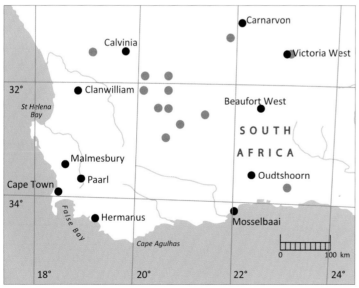

Map 52. Known distribution of *Lachenalia canaliculata*.

Figure 166 (left). *Lachenalia canaliculata* in moist clay near Nieuwoudtville. Image: Graham Duncan.

Figure 167 (centre). *Lachenalia canaliculata* in stony clay near Sutherland. Image: Graham Duncan.

Figure 168 (right). *Lachenalia canaliculata* amongst dolerite stones at Victoria West, central Great Karoo. Image: Graham Duncan.

DISTINGUISHING CHARACTERS AND AFFINITIES. Duncan *et al.* (2005a) suggested an alliance with *L. obscura*, and the relationship is discussed under that species.

DISTRIBUTION, HABITAT AND CONSERVATION STATUS. *L. canaliculata* is distributed from the Bokkeveld Plateau at Nieuwoudtville south-east to the Sutherland district and north-east to near Carnarvon and Victoria West in the central Great Karoo (Map 52). A disjunct record exists north of Uniondale in the extreme east of the Western Cape. The plants occur in five vegetation types: Nieuwoudtville-, Roggeveld- and Uniondale Shale Renosterveld, Roggeveld Karoo and Eastern Upper Karoo (Mucina & Rutherford, 2006). They grow as solitary individuals or in small to large groups. *L. canaliculata* is common in and around Nieuwoudtville, where it flowers in early September, occurring in moist doleritic clay and along the edges of seasonally inundated sandy depressions. At Papkuilsfontein Farm south of Nieuwoudtville, it grows in large numbers, adjacent to large populations of *L. mutabilis* that favour drier, sandier terrain. On the outskirts of Sutherland on the Roggeveld Plateau, *L. canaliculata* occurs on flats and on the lower slopes of ridges in doleritic clay; here, it flowers in early October and grows in association with *L. isopetala*. At Victoria West in the central Great Karoo, *L. canaliculata* occurs in arid conditions amongst dolerite stones. In this eastern part of its range, rainfall is mainly in summer, but the species nevertheless follows a winter growth cycle. *L. canaliculata* is not threatened.

NOTES. The flowers are pollinated by honey bees (*Apis mellifera*). The scape remains attached to the bulb for several weeks following capsule dehiscence and the seeds are dispersed locally. The foliage is heavily grazed by Cape hares (*Lepus capensis*).

52. LACHENALIA INCONSPICUA

Lachenalia inconspicua G. D. Duncan, *Bothalia* 27 (1): 11–12 (1997).
TYPE: South Africa, Northern Cape, 500 m beyond Gamoep, on Springbok road to Gamoep, in deep gravelly red sand, *G. D. Duncan* 259 (NBG!, holo.).

ETYMOLOGY. *inconspicua*: inconspicuous flowers.

DESCRIPTION. *Geophyte*, 120–160 mm high. *Bulb* globose, 15–20 mm in diameter, solitary; tunic multilayered, outer layers thick, dark brown, spongy; inner cataphyll translucent white, adhering to leaf bases, apex obtuse. *Leaf* solitary, rarely 2, lanceolate, 85–150 × 15–20 mm, spreading, deeply canaliculate, coriaceous, upper surface glaucous with distinct depressed, longitudinal grooves, plain or with irregularly scattered purplish-brown spots, lower surface glaucous, plain or with brownish bands; leaf base clasping, 10–15 mm long, upper portion light green to brownish-magenta, lower portion white, plain or with maroon bands; primary seedling leaf terete, erect. *Inflorescence* spicate, erect to suberect, few- to many-flowered, fairly dense, sterile apex short; scape erect, 55–80 mm long, sturdy, light green with large purplish-brown blotches; lower bracts ovate, becoming lanceolate above, 0.5–1.0 × 0.5–1.0 mm. *Perianth* zygomorphic, oblong-campanulate, suberect, usually arranged in distinct three-flowered whorls, ageing to dull brownish-purple; tube cup-shaped, 1–2 mm long, light blue; outer tepals ovate, 7–8 × 4–5 mm, apices slightly recurved, light greenish-blue, base marked with darker blue, apical gibbosities large, dull purplish-brown to brownish-green; inner tepals obovate, 8–11 × 4–5 mm, translucent white, median keels brownish-green, apices slightly recurved, upper two tepals overlapping, lower tepal 2 mm longer, canaliculate. *Stamens* included to shortly exserted, more or less straight; filaments white, 7–9 mm long. *Ovary* ellipsoid, 2–3 × 1–2 mm, light green; style included, straight, 8–9 mm long. *Capsule* ellipsoid, 9–10 × 5–6 mm. *Seed* globose, 1 × 1 mm, matte, black; strophiole ridged, 1.1 mm long. *Chromosome number* 2n = 18 (Spies *et al.*, 2008). Figure 169.

52. LACHENALIA INCONSPICUA

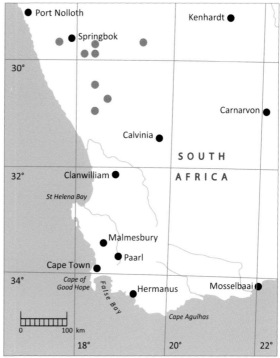

Map 53. Known distribution of *Lachenalia inconspicua*.

Figure 169. *Lachenalia inconspicua* in stony red sand at Gamoep. Image: Graham Duncan.

FLOWERING PERIOD. July to August, with a peak in late July.

HISTORY. The first collection of *L. inconspicua* was made in August 1982 by Fiona Archer, north-east of Gamoep in western Bushmanland. The holotype material (*Duncan 259*, in NBG) was collected in 1986 just beyond Gamoep, on the road to Springbok, and the species was published in *Bothalia* (Duncan, 1997).

DISTINGUISHING CHARACTERS AND AFFINITIES. The phylogeny of Duncan *et al.* (2005a) suggests that *L. inconspicua* is closely related to *L. karooica*. They have similar glaucous leaves and oblong-campanulate perianths, but *L. karooica* differs in its spirally arranged flowers, narrower tepals (2–3 mm wide) (the inner whorl of which are strongly recurved), and in its longer, well exserted, recurved filaments. By comparison with *L. inconspicua*, *L. karooica* has much larger, ovoid seeds (1.3–1.4 × 1.2–1.3 mm) and a much wider distribution, mainly in summer rainfall parts of the Great Karoo.

DISTRIBUTION, HABITAT AND CONSERVATION STATUS. *L. inconspicua* occurs in central, western and southern Bushmanland at Middeldeurvlei, Gamoep and Kliprand Farms, respectively, and on the Kamiesberg east of Leliefontein. An outlying population occurs on the Spektakelberg west of Springbok (Map 53). This species grows in Namaqualand Klipkoppe Shrubland and Bushmanland Arid Grassland (Mucina & Rutherford, 2006) and is locally common. Plants occur as scattered individuals or in groups of two to three, in red or reddish-brown, stony or gravelly sand, on calcrete flats and at the base of hill slopes. *L. inconspicua* is not threatened.

NOTES. *L. inconspicua* is pollinated by honey bees (*Apis mellifera*). The scapes detach from the bulbs shortly after capsule dehiscence in September and the infructescences are tumbled about by wind, dispersing the seeds. The bulbs are adapted to remain dormant during prolonged periods of drought.

53. LACHENALIA KAROOICA

Lachenalia karooica W. F. Barker ex G. D. Duncan, *Bothalia* 26 (1): 1–3 (1996).
TYPE: South Africa, Free State, 0.5 km on road from Fauresmith to Koffiefontein, *P. Chaplin s.n.* (NBG!, holo.).

ETYMOLOGY. *karooica*: after the Great Karoo, where the largest number of collections have been made.
DESCRIPTION. *Geophyte*, 45–220 mm high. *Bulb* ovoid, 10–20 mm in diameter; tunic multilayered, outer layers dark brown, spongy; inner cataphyll translucent white, adhering to leaf bases, apex acute. *Leaves* 1–2, lanceolate, 40–200 × 5–20 mm, suberect to spreading, canaliculate to conduplicate, glaucous, upper surface with depressed longitudinal grooves, upper and lower surfaces with irregularly scattered brown, green or maroon blotches; leaf bases clasping, 5–30 mm long, upper portion glaucous, plain or banded with purplish-maroon, lower portion white, plain or banded with magenta; primary seedling leaf terete, erect. *Inflorescence* spicate or racemose, few- to many-flowered, sterile apex short; scape erect to suberect, 20–80 mm long, slender to sturdy, light green mottled with light maroon to dull purple; rachis colouring as for scape; pedicels absent or up to 2 mm long, suberect, white; lower bracts ovate, becoming lanceolate above, 1–2 × 1–3 mm. *Perianth* zygomorphic, oblong-campanulate, spreading to suberect, ageing to dull purple or red; tube cup-shaped, 1–2 mm long, dull white or blue; outer tepals ovate, 5–7 × 2–3 mm, light grey, apices slightly recurved, apical gibbosities brownish-green, median keels dull blue or green; inner tepals obovate, 7–10 × 2–3 mm, light grey, apices strongly recurved, median keels dull blue or green. *Stamens* well exserted, recurved; filaments 10–11 mm long, white. *Ovary* ellipsoid, 2–3 × 1–2 mm, light green; style well exserted, 10–14 mm long, straight, white, protruding well beyond stamens as ovary matures. *Capsule* ellipsoid, 5–7 × 3–4 mm. *Seed* ovoid, 1.3–1.4 × 1.2–1.3 mm, glossy, black, secondary sculpturing reticulate; strophiole 1.1 mm long, ridged. *Chromosome number* 2n = 16 (Duncan, 1996); 2n = 20 (Johnson & Brandham, 1997). Figure 170.

Figure 170. Cultivated specimen of *Lachenalia karooica* from Fauresmith. Image: Graham Duncan/SANBI.

FLOWERING PERIOD. June to September, with a peak in early August.

HISTORY. The naturalist, traveller and author W. J. Burchell made the first known collection of *L. karooica* at Klipfontein in the Carnarvon district of the Northern Cape in 1811. This specimen, which is preserved in the Kew Herbarium (*Burchell* 1534) was, however, mistakenly cited under *L. pustulata* Jacq. by Baker (1871, 1897a), and more than a century and a half elapsed before the species *L. karooica* was finally published (Duncan, 1996). D. F. Gilfillan collected this species in August 1899 near Middelburg in the Eastern Cape, and in 1925, I. B. Pole Evans found it at Fauresmith in the south-western Free State. *L. karooica* has been collected several times at the Fauresmith location and the collection made there by P. Chaplin, former Curator of the Free State National Botanical Garden, in 1976 is preserved as the holotype (*Chaplin s.n.*, in NBG).

DISTINGUISHING CHARACTERS AND AFFINITIES. *L. karooica* is distinctive in its light grey, spreading to suberect, oblong-campanulate flowers with strongly recurved inner tepals, well-exserted, recurved filaments and relatively large, ovoid seeds. Duncan *et al.* (2005a) suggested an alliance with *L. inconspicua*, and the relationship is discussed under that species.

DISTRIBUTION, HABITAT AND CONSERVATION STATUS. The species is recorded mainly from the Great Karoo, most collections having been made in the central and north-eastern Northern Cape, south-western Free State and north-western Eastern Cape. Currently, the known range of *L. karooica* extends from Carnarvon in the central Great Karoo north-east to Prieska and Barkly West, south-east to Fauresmith in the Free State, south to Colesberg, Hanover, Middelburg and Nieu Bethesda, and to the Mountain Zebra National Park west of Cradock in the Eastern Cape interior (Map 54). A single record from Karoopoort north of Ceres in the Western Cape is a wide disjunction. *L. karooica* occurs in exceptionally arid conditions and is usually encountered on south-facing slopes of dolomitic limestone outcrops. There, it grows in sandy or gravelly soils, in a variety of vegetation types including Kimberley- and Queenstown Thornveld, Lower Gariep Broken Veld, Northern- and

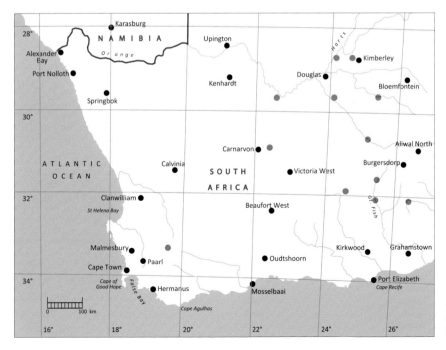

Map 54. Known distribution of *Lachenalia karooica*.

Eastern Upper Karoo, Xhariep Karroid Grassland, Karoo Escarpment Grassland and Tanqua Karoo. At Fauresmith, *L. karooica* grows within the protection and partial shade of *Rhus burchellii* shrubs (Anacardiaceae), in association with the geophytes *Freesia andersoniae* and *Moraea pallida* (Iridaceae). *L. karooica* is probably widespread throughout the Great Karoo; the remote and arid nature of this vast area, coupled with the plant's inconspicuous appearance probably accounts for the relatively small number of scattered specimens recorded.

Although most of the distribution range of *L. karooica* falls within areas of predominantly summer rainfall, it receives sufficient autumn rain. This allows *L. karooica* to follow the typical pattern of winter growth and summer dormancy characteristic of almost all lachenalias. The species is not threatened.

NOTES. *L. karooica* is pollinated by honey bees (*Apis mellifera*). Shortly after capsule dehiscence, the scapes detach from the bulbs and the infructescence is blown away, dispersing the seeds.

54. LACHENALIA CONCORDIANA

Lachenalia concordiana Schltr. ex W. F. Barker, *Journal of South African Botany* 44 (4): 407–409 (1978).
TYPE: South Africa, Northern Cape, 15 miles [24 km] north-east of Springbok, *W. F. Barker* 6762 (NBG!, holo.; iso.).

ETYMOLOGY. *concordiana*: after Concordia, a town in central Namaqualand.
DESCRIPTION. *Geophyte*, 70–250 mm high. *Bulb* globose, 12–20 mm in diameter, solitary; neck distinct, up to 30 mm long; tunic multilayered, outer layers dark brown, hard; inner cataphyll translucent green, loosely surrounding leaf base, apex acute. *Leaf* solitary, lanceolate, 100–120 × 8–10 mm, spreading to suberect, dark green, upper surface plain or with sporadic dark green spots, lower surface heavily marked with dark green blotches, midrib prominent; leaf base tightly clasping, 30–70 mm long, white, heavily banded with dark magenta; primary seedling leaf terete, erect. *Inflorescence* spicate, few-flowered, fairly dense, erect to suberect, sterile apex short; scape erect to suberect, 40–50 mm long, slender, light yellowish-green, heavily suffused with dull brown; rachis light green, heavily suffused with maroonish-brown; lower bracts ovate, becoming lanceolate above, translucent white, 1–3 × 1 mm. *Perianth* zygomorphic, oblong-campanulate, translucent white to cream, spreading to suberect, arranged in 3-flowered whorls; tube cup-shaped, 3 mm long, plain or with small light blue speckles; outer tepals ovate, 6–7 × 3–4 mm, plain or with light blue speckles, apices recurved, apical gibbosities bright green; inner tepals obovate, 8–11 × 3–4 mm, apices recurved, median keels bright green, upper inner tepals overlapping, lower inner tepal longer, caniculate. *Stamens* included, declinate; filaments white, 5–7 mm long; pollen yellow at anthesis, ageing to black. *Ovary* ellipsoid, 3–4 × 2–3 mm, dark green; style included, slightly declinate, 5 mm long, white. *Capsule* ellipsoid, 7–8 × 3–4 mm. *Seed* ovoid, 1.3–1.4 × 0.8–0.9 mm; matte, black, strophiole 0.8 mm long, ridged. *Chromosome number* 2n = 14 (Spies *et al.*, 2008). Figure 171.

FLOWERING PERIOD. July to September, with a peak in late August.
HISTORY. Rudolf Schlechter collected *L. concordiana* in September 1897 near the small town of Concordia in central Namaqualand. He appended his manuscript name, *L. concordiana* Schltr., to the pressed material and distributed it to seven local and foreign herbaria, but the species lay undescribed for more than 80 years until finally published in the *Journal of South African Botany* (Barker, 1978). The holotype collection (*Barker* 6762, in NBG) was made north-east of Springbok in September

Map 55. Known distribution of *Lachenalia concordiana*.

Figure 171. Cultivated specimen of *Lachenalia concordiana* from the Calvinia district. Image: Graham Duncan/SANBI.

1950. Very few subsequent collections have been made, but this species' range is now known to extend to the southern Richtersveld, western Bushmanland and Hantam Karoo.

DISTINGUISHING CHARACTERS AND AFFINITIES. Duncan *et al.* (2005a) regarded this species as sister to *L. verticillata*. Both have flowers arranged in whorls of three, with included, declinate filaments. *L. verticillata* differs in its taller stature, broadly lanceolate, intensely glaucous leaf, urceolate flowers with strongly recurved, purple apices and a smaller ovoid seed (1.1–1.2 × 1.3 mm).

DISTRIBUTION, HABITAT AND CONSERVATION STATUS. The species is confined to the north-western Northern Cape and is recorded from three disjunct centres: Steinkopf, Springbok and Garies in Namaqualand; Aggeneys in western Bushmanland; and the Calvinia district in the western Karoo (Map 55). The plants grow singly in full sun on flats and in dry river beds in red sandy soil. *L. concordiana* has been recorded from four vegetation types: Eenriet Plains Succulent Shrubland, Bushmanland Sandy Grassland, Namaqualand Klipkoppe and Hantam Karoo (Mucina & Rutherford, 2006). The species occurs sporadically and has a conservation status of Rare (Victor, 2009).

NOTES. Pollination vectors for *L. concordiana* are as yet unknown but the oblong-campanulate flower shape indicates these are likely to include honey bees (*Apis mellifera*). The scape detaches from the bulb shortly after capsule dehiscence and the seeds are widely scattered.

55. LACHENALIA MAXIMILIANI

Lachenalia maximiliani Schltr. ex W. F. Barker, *Journal of South African Botany* 45 (2): 209–212 (1979).
TYPE: South Africa, Western Cape, Wupperthal, *W. F. Barker* 8993 (NBG!, holo.).

ETYMOLOGY. *maximiliani*: after Maximilian Schlechter (1874–1960), trader and plant collector of the Port Nolloth district, and brother to Rudolf Schlechter.

DESCRIPTION. *Geophyte*, 100–200 mm high. *Bulb* globose, 10–17 mm in diameter, offset- and bulbil-forming, neck distinct, fasciculate; tunic multilayered, outer layers dark brown, hard; inner cataphyll translucent white, adhering to leaf bases, apex obtuse. *Leaf* solitary, lanceolate, 30–90 × 10–30 mm, erect to suberect, slightly to deeply canaliculate, yellowish-green, upper surface plain or spotted greenish-brown, with depressed longitudinal grooves; lower surface plain, midrib distinct; leaf bases 25–40 mm long, lower portion white, light green above, plain or brownish-maroon tinged; primary seedling leaf terete, erect. *Inflorescence* almost fully racemose (lowermost flowers sessile), erect to suberect, few- to many-flowered, fairly dense, sterile apex short; scape erect, 35–60 mm long, light green; rachis light green in lower third, shading to light whitish-blue in upper two thirds; lower bracts ovate, becoming lanceolate above, 1–3 × 1–2 mm, translucent white; pedicels suberect, up to 1 mm long, white. *Perianth* zygomorphic, oblong-campanulate, suberect; tube cup-shaped, 2 mm long, light blue; outer tepals ovate, 6–7 × 3–4 mm, apices recurved, whitish-grey, apical gibbosities brown or brownish-green to reddish-brown; inner tepals obovate, 7–9 × 4 mm, translucent white, apices recurved, ageing to pinkish-magenta, lower inner tepal narrower, deeply canaliculate, slightly

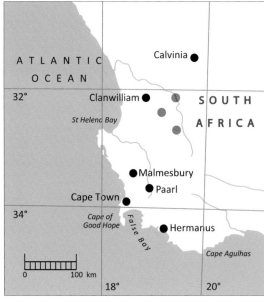

Map 56. Known distribution of *Lachenalia maximiliani*.

Figure 172. Cultivated specimen of *Lachenalia maximiliani* from Wupperthal. Image: Graham Duncan.

longer than upper inner tepals. *Stamens* included, straight; filaments translucent whitish-grey, 5–7 mm long. *Ovary* ellipsoid, 3 × 2 mm, light green; style included, straight, 3–4 mm long, whitish-grey. *Capsule* ellipsoid, 5–6 × 3–4 mm. *Seed* ovoid, 1.2 × 1.3 mm; strophiole 0.6 mm long, ridged. *Chromosome number* 2n = 16 (Spies *et al*., 2009). Figure 172.

FLOWERING PERIOD. July to August, with a peak in late July.

HISTORY. The German botanist and traveller, Rudolf Schlechter (1872–1925) collected *L. maximiliani* for the first time in August 1896 at Koudeberg, at the top of the steep pass that runs down to the missionary town of Wupperthal south-east of Clanwilliam. He distributed material under his manuscript name, *L. maximiliani* Schltr., to numerous herbaria locally and abroad, but the species was only validly published 83 years later in the *Journal of South African Botany* (Barker, 1979a). It was collected again by Wagener in 1943 at Matjiesrivier in the Cederberg and has since been found in several localities in the Biedouw Valley.

DISTINGUISHING CHARACTERS AND AFFINITIES. In a phylogenetic analysis of morphological data (Duncan *et al*., 2005a), *L. maximiliani* resolved adjacent to *L. mathewsii*. These species share oblong-campanulate perianths and ovoid seeds, but *L. mathewsii* differs markedly in its bright yellow, spreading perianth with bright green apical gibbosities, shortly exserted, declinate stamens, two narrowly lanceolate, glaucous leaves that taper to long terete apices, and smaller seeds (0.9–1.0 × 0.9–1.0 mm) with much shorter strophioles (0.2 mm long).

DISTRIBUTION, HABITAT AND CONSERVATION STATUS. *L. maximiliani* is restricted to the area around Wupperthal and northwards to the Biedouw Valley in the Cederberg Mountains (Map 56). It grows in Agter-Sederberg Shrubland vegetation (Mucina & Rutherford, 2006) in large, dense colonies. Substrates are seasonally wet, heavy stony clay or sandy soil on moderate to steep north- and south-facing mountain slopes or along river banks. It has a conservation status of Rare (Raimondo & Duncan, 2009b).

NOTES. The oblong-campanulate flowers are pollinated by honey bees (*Apis mellifera*). The mother bulb produces numerous compressed bulbils at its base, between the bulb scales. The scape remains attached for several weeks after the capsules have dehisced and the seeds are dispersed locally by the shaking action of wind.

56. LACHENALIA PERRYAE

Lachenalia perryae G. D. Duncan, *Bothalia* 26 (1): 3–5 (1996).
TYPE: South Africa, Western Cape, Karoo Desert National Botanical Garden veld reserve, Worcester, *P. L. Perry s.n.*, *sub*. NBG 142275 (NBG!, holo.; PRE!, iso.).

ETYMOLOGY. *perryae*: after Pauline L. Perry (1927–), botanist, formerly of the Karoo Desert National Botanical Garden, Worcester.

DESCRIPTION. *Geophyte*, 100–170 mm high. *Bulb* globose, 12–15 mm in diameter, offset-forming, often deep-seated up to 50 mm; tunic multilayered, outer layers dark brown, spongy; inner cataphyll translucent white, loosely surrounding leaf base, apex optuse. *Leaf* solitary, narrowly lanceolate, 100–150 × 15–22 mm, canaliculate, slightly spreading to suberect, yellowish-green, upper surface plain or with light green blotches, lower surface with darker green transverse bands; leaf base 20–40 mm long, white, plain or with dark magenta transverse bands; primary seedling leaf terete, erect. *Inflorescence* racemose, few- to many-flowered, sterile apex short; scape erect to suberect, 40–50 mm long, light green to brownish-magenta with darker magenta blotches and speckles; rachis light green to brownish-magenta, plain or densely marked with darker brownish-magenta speckles, apex light blue; lower bracts ovate, becoming lanceolate above, 1–3 × 1–3 mm, white; pedicels suberect, 1–3 mm long, white or light to dark brownish-purple, increasing in length towards inflorescence apex. *Perianth* zygomorphic, oblong-campanulate, spreading to slightly cernuous; tube cup-shaped, 1–2 mm long, light bluish-grey or cream; outer tepals ovate, 5–7 × 3–5 mm, light bluish-grey or cream, apical gibbosities dark purplish-brown or green; inner tepals obovate, 8–9 × 4–6 mm, flared, apices rounded, translucent white to light greyish-blue, median keels dark purplish-brown or green, upper tepals overlapping, lower inner tepal narrower, canaliculate. *Stamens* included to shortly exserted, more or less straight; filaments white, 7–10 mm long. *Ovary* obovoid, 2–3 × 1–2 mm, yellowish-green; style included, 5–8 mm long, white. *Capsule* obovoid, 5–7 × 5 mm. *Seed* ovoid, 1.2–1.3 × 1.3–1.4 mm, matte, black, strophiole ridged, 0.5 mm long. *Chromosome number* unknown. Figure 173.

Figure 173. *Lachenalia perryae* on renosterveld flats at Worcester. Image: Graham Duncan.

FLOWERING PERIOD. July to September, with a peak in mid-August.

HISTORY. The first known record of *L. perryae* is that made by W. F. Barker near Robertson in the Little Karoo, in August 1949. The species has been collected on numerous occasions during more than four ensuing decades and was finally published in *Bothalia* (Duncan, 1996). *L. perryae* has been recorded numerous times in the Worcester District and the type collection was made by P. L. Perry in August 1985, within the grounds of the natural area of the Karoo Desert National Botanical Garden at Worcester.

DISTINGUISHING CHARACTERS AND AFFINITIES. *L. perryae* is a variable species and is recognised by a many-flowered raceme of oblong-campanulate, spreading to slightly cernuous flowers in shades of light bluish-grey or rarely cream. The inner tepals have rounded apices, and the straight stamens are usually included, or rarely, shortly exserted. The solitary, narrowly lanceolate, canaliculate leaf is heavily marked with transverse bands on the lower surface and clasping base.

In a phylogenetic analysis of morphological data (Duncan *et al.*, 2005a), *L. perryae* resolved adjacent to *L. latimeriae*. Both have a usually solitary lanceolate leaf, and a raceme of oblong-campanulate flowers with straight filaments and obovoid capsules. *L. latimeriae* differs in its strongly cernuous, light pink to lilac flowers, which are borne on much longer pedicels (4–5 mm long), and in its shorter, globose seeds (0.9–1.0 × 1.3–1.4 mm) with slightly longer strophioles (0.6 mm long).

DISTRIBUTION, HABITAT AND CONSERVATION STATUS. *L. perryae* is native to the south-western and southern Western Cape, extending from the Hex River Mountains north of Worcester, south-east

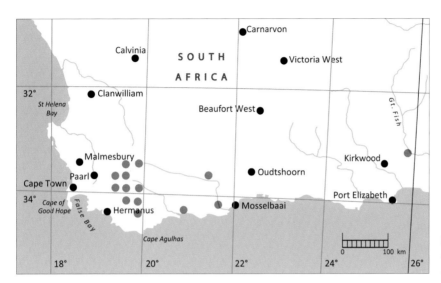

Map 57. Known distribution of *Lachenalia perryae*.

to Port Beaufort, east to Riversdale and inland to Vanwyksdorp in the central Little Karoo, with an isolated record (*Bayliss* 5919 in NBG, PRE) near Kommadagga in the western Eastern Cape (Map 57). The species grows in a number of vegetation types including Northern Inland Shale Band Vegetation, Breede Alluvium Renosterveld, Western- and Eastern Rûens Shale Renosterveld, Breede- and Mossel Bay Shale Renosterveld and Albany Broken Veld (Mucina & Rutherford, 2006). It occurs almost exclusively on shale substrates and the plants occur on moist, stony flats and south-facing mountain slopes in small groups, exposed or in light shade of surrounding bush. At a locality near Worcester, *L. perryae* occurs in association with the geophytes *Massonia depressa* (Asparagaceae) and *Oxalis obtusa* (Oxalidaceae). The species is declining in the vicinity of Worcester as the result of agricultural expansion and housing development, but overall, it is still frequent across its range and not threatened.

NOTES. The flowers of *L. perryae* are pollinated by honey bees (*Apis mellifera*). Following capsule dehiscence, the scapes remain attached for several weeks and the seeds are dispersed locally. In the Worcester district, the leaves and developing inflorescences are heavily grazed by Cape hares (*Lepus capensis*).

57. LACHENALIA LATIMERIAE

Lachenalia latimeriae W. F. Barker (sphalm. *latimerae*), *Journal of South African Botany* 45 (2): 196–199 (1979).

TYPE: South Africa, Eastern Cape, Ferndale Farm, Patensie, *M. Courtenay-Latimer s.n.* sub. NBG 72287 (NBG!, holo.).

ETYMOLOGY. *latimeriae*: after Dr Marjorie Courtenay-Latimer (1907–2004), former Director of the East London Museum, who made the type collection, in recognition of her active role in the conservation of the flora of the southern and Eastern Cape.

DESCRIPTION. *Geophyte*, 90–280 mm high. *Bulb* globose, 10–20 mm in diameter, offset-forming; tunic multilayered, outer layers light brown; inner cataphyll translucent white, adhering to leaf bases, apex obtuse. *Leaves* 1–2, lanceolate, 110–200 × 4–20 mm, suberect to falcate, canaliculate, light green,

258 THE GENUS LACHENALIA
57. LACHENALIA LATIMERIAE

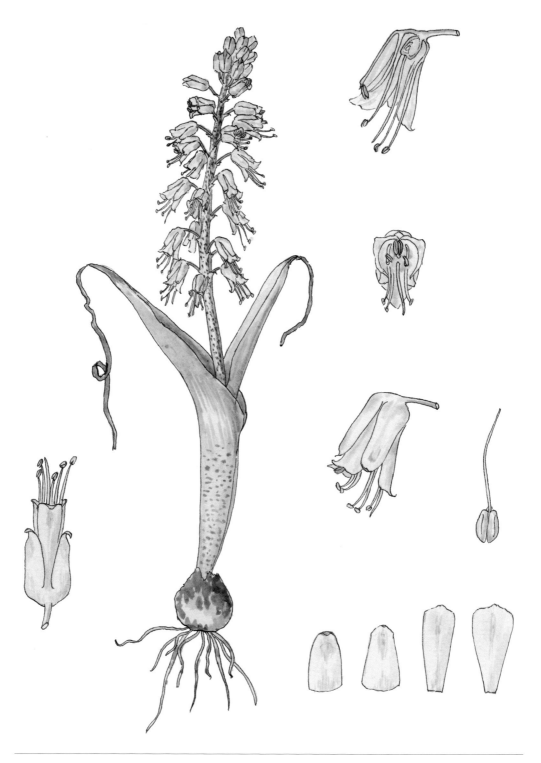

Plate 18. Watercolour painting of *Lachenalia latimeriae* from Cango Caves [*Salter s.n.* (1987/36), in NBG] courtesy of the Compton Herbarium, Kirstenbosch, South African National Biodiversity Institute. Artist: Winsome Barker.

upper surface plain, with depressed longitudinal grooves, lower surface plain, minutely spotted or heavily flushed with brownish-magenta at base; leaf bases clasping, 20–50 mm long, greenish-white, plain or minutely spotted with brownish-magenta; primary seedling leaf terete, erect. *Inflorescence* racemose, erect, lax, few- to many-flowered, sterile apex short; scape slender, 30–60 mm long, light green, plain or with small brownish-maroon blotches or spots; rachis light green in lower half, shading to light brownish-pink and to light electric blue in sterile apex; pedicels spreading to suberect, 4–5 mm long, white to light pink; lower bracts ovate, becoming lanceolate above, 1–3 × 0.5–2.0 mm, light green or white. *Perianth* zygomorphic, oblong-campanulate, light pink to lilac, cernuous; tube cup-shaped, 2 mm long; outer tepals ovate, 5–6 × 3–4 mm, apical gibbosities pinkish- to greenish-brown; inner tepals obovate, 6–7 × 4–5 mm, median keels brownish-pink, apices obtuse. *Stamens* well exserted, more or less straight; filaments white, maturing to lilac, 10–13 mm long. *Ovary* obovoid, 2.0 × 1.5 mm, yellowish-green; style well exserted, 9–10 mm long, white, maturing to lilac. *Capsule* obovoid, 6–7 × 5 mm. *Seed* globose, 0.9–1.0 × 1.3–1.4 mm; strophiole 0.6 mm long, ridged.

Figure 174. Cultivated specimens of *Lachenalia latimeriae* from the Baviaanskloof. Image: Graham Duncan.

Chromosome number 2n = 14 (Spies *et al.*, 2008); 2n = 18 (Hamatani *et al.*, 2007). Plate 18, Figure 174.

FLOWERING PERIOD. July to September, with a peak in early September.

HISTORY. *L. latimeriae* was collected for the first time in August 1936 by Paymaster Captain T. M. Salter near the Cango Caves. The bulbs were cultivated in the nursery at Kirstenbosch, and when they flowered in 1941, a specimen was painted by W. F. Barker and pressed material was preserved in the Compton Herbarium. In 1948, Marjorie Courtenay-Latimer collected this species at Ferndale Farm near Patensie, and the species was described in the *Journal of South African Botany*, with the Courtenay-Latimer specimens serving as the holotype (Barker, 1979). *L. latimeriae* has been infrequently recorded and most collections are from the Baviaanskloof Mountains and the Patensie District in the southern Eastern Cape. In December 1938, Dr Courtenay-Latimer brought to the attention of the world the discovery of the Coelacanth, a fish thought to have been extinct for 65 million years until it was found at the mouth of the Chalumna River in the Eastern Cape. The fish was later named *Latimeria chalumnae* Smith in her honour.

DISTINGUISHING CHARACTERS AND AFFINITIES. Phylogenetic analysis of *Lachenalia* placed *L. latimeriae* adjacent to *L. haarlemensis* (Duncan *et al.*, 2005a). These species share similar many-flowered racemes of oblong-campanulate, cernuous flowers with well exserted, straight stamens and obovoid capsules. The latter differs in its smaller, greenish-grey flowers with shorter pedicels (2–3 mm), shorter outer

57. LACHENALIA LATIMERIAE

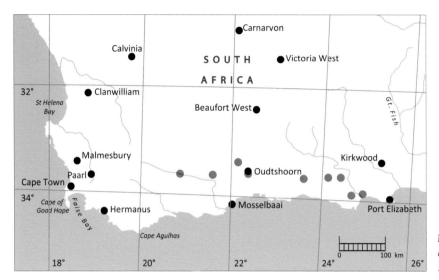

Map 58. Known distribution of *Lachenalia latimeriae*.

tepals (4–5 mm) and filaments that are bright mauve in the upper half. Vegetatively, *L. haarlemensis* differs from *L. latimeriae* in its lower leaf surfaces and clasping bases, which are heavily banded with purplish-maroon and magenta.

Superficially, *L. latimeriae* resembles certain forms of *L. juncifolia* that have similar light pink oblong-campanulate flowers with well exserted, straight stamens, but *L. juncifolia* is immediately set apart by its two linear leaves that are often terete in the uppermost part.

DISTRIBUTION, HABITAT AND CONSERVATION STATUS. *L. latimeriae* is uncommon over its distribution range, occurring from the Touwsberg mountains in the Little Karoo, east to Ferndale Farm south of Patensie (Map 58). It traverses four vegetation types: Western- and Eastern Little Karoo, Kouga Sandstone Fynbos and Gamtoos Thicket (Mucina & Rutherford, 2006). The plants occur on seasonally moist ridges and south-facing hills, and on mountain slopes. Niches include moss between rock cracks in semi shade, in open aspects in loose gravel, and along river banks support scattered individuals or small colonies. The species has been recorded growing in association with the grass *Cynodon dactylon* (Poaceae) and the shrub *Polygala myrtifolia* (Polygalaceae). It is not threatened.

NOTES. Pollinators are as yet unrecorded for *L. latimeriae* but are likely to include honey bees (*Apis mellifera*). The scapes remain attached to the bulbs for several weeks following capsule dehiscence and the seeds are dispersed locally.

58. LACHENALIA HAARLEMENSIS

Lachenalia haarlemensis Fourc., *Transactions of the Royal Society of South Africa* 21 (1): 79 (1934). **TYPE:** South Africa, Western Cape, 12 miles (19 km) from Avontuur, north of Haarlem, *G. H. Fourcade* 4345 (BOL!, holo.).

ETYMOLOGY. *haarlemensis*: after Haarlem, a small town in the southern Western Cape, near the type locality.

DESCRIPTION. *Geophyte*, 120–300 mm high. *Bulb* globose, 100–200 mm in diameter, offset-forming; tunic multilayered, outer layers dark brown, spongy; inner cataphyll translucent white, loosely surrounding leaf bases, apex acute. *Leaf* solitary, lanceolate, 100–150 × 3–22 mm, erect to suberect or falcate, deeply canaliculate, upper surface yellowish-green to glaucous or dark green, plain with depressed longitudinal grooves, lower surface light green, upper half heavily marked with darker green spots, lower half with purplish-maroon to magenta bands; leaf base 20–40 mm long, white to light green, heavily marked with transverse magenta bands; primary seedling leaf terete, erect. *Inflorescence* racemose, many-flowered, moderately dense, sterile apex short; scape erect, 100–180 mm long, slender, light green to glaucous, heavily marked with dull maroon or purple blotches and speckles; rachis light green to glaucous, marked with dull maroon or purple speckles; pedicels suberect, 2–3 mm long, short at base of inflorescence, lengthening above, light green to glaucous; lower bracts ovate, becoming lanceolate above, 1–4 × 1–3 mm, green to white. *Perianth* zygomorphic, oblong-campanulate, light greenish-grey, spreading to slightly cernuous; tube cup-shaped, 1–2 mm long; outer tepals ovate, 4–5 × 3–4 mm, apical gibbosities light brown; inner tepals obovate, 6–7 × 3–4 mm, median keels light blue to mauve, apical zone dark mauve. *Stamens* well exserted, more or less straight; filaments bright mauve in upper portion, white below, 11–12 mm long. *Ovary* obovoid, 2 × 2 mm, light green; style well exserted, straight, apex light to dark mauve, white below, 10–12 mm long. *Capsule* obovoid, 3–4 × 4 mm. *Seed* globose, 1.2–1.3 × 1.3–1.4 mm, glossy, black; strophiole 0.5–0.6 mm long, ridged. *Chromosome number* 2n = 18 (Johnson & Brandham, 1997). Figure 175.

Figure 175. Cultivated specimens of *Lachenalia haarlemensis* from near Haarlem. Image: Graham Duncan.

FLOWERING PERIOD. September to October, with a peak in early October.

HISTORY. The first collection of *L. haarlemensis* was made in October 1930 between the southern Cape towns of Haarlem and Avontuur by the French forest officer and botanical collector, Henri Georges Fourcade (1866–1948). The species was published by him in *Transactions of the Royal Society of South Africa* (Fourcade, 1934). *L. haarlemensis* has since been infrequently collected in southern parts of the Western Cape, and twice near Onder Kouga in the western Eastern Cape. The accompanying photograph is of wild-collected specimens gathered by Pauline Perry between Avontuur and Haarlem in November 1984, cultivated in the Kirstenbosch Nursery.

DISTINGUISHING CHARACTERS AND AFFINITIES. Phylogenetic analysis of morphological data (Duncan *et al.*, 2005a) placed *L. haarlemensis* adjacent to *L. latimeriae*. These species share oblong-campanulate perianths with well-exserted, straight stamens and usually a solitary lanceolate leaf, but *L. latimeriae* differs in its larger, light pink, lilac or cream flowers that are borne on longer pedicels (4–5 mm long), its white filaments, and its spotted lower leaf surfaces and leaf bases.

58. LACHENALIA HAARLEMENSIS

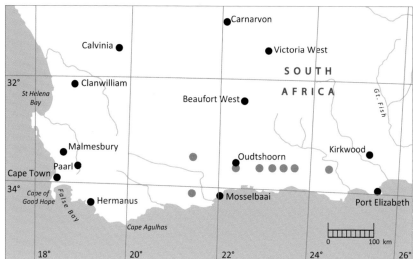

Map 59. Known distribution of *Lachenalia haarlemensis*.

DISTRIBUTION, HABITAT AND CONSERVATION STATUS. *L. haarlemensis* extends from Seweweekspoort near Ladismith in the southern Cape, south to Riversdale and east to Onder Kouga in the western Eastern Cape (Map 59). The plants occur in five vegetation types: Montagu-, Langkloof- and Eastern Rûens Shale Renosterveld, North Kammanassie Sandstone Fynbos and Willowmore Gwarrieveld (Mucina & Rutherford, 2006). They grow as scattered individuals or in small groups in deep loam or stony soils among low scrub, in grassy areas on flats, or on south- or south-east-facing hillsides and lower mountain slopes, in full sun or light shade. *L. haarlemensis* has been recorded growing in association with *Aloe ferox* (Xanthorrhoeceae) in the Riversdale District. The species is threatened by crop cultivation and overgrazing, and has a conservation status of Vulnerable (Vlok & Raimondo, 2009).

NOTES. The oblong-campanulate perianth shape of *L. haarlemensis* suggests that it is pollinated by honey bees (*Apis mellifera*). The scapes remain attached to the bulbs for several weeks following capsule dehiscence and the seeds are dispersed locally.

59. LACHENALIA LACTOSA

Lachenalia lactosa G. D. Duncan, *Bothalia* 28 (2): 135–137 (1998).
TYPE: South Africa, Western Cape, Honingklip Farm near Bot River, *W. F. Barker* 10510 (NBG!, holo.; PRE!, iso.).

ETYMOLOGY. *lactosa*: descriptive of the milky white outer tepals of certain forms of this species.

DESCRIPTION. Geophyte, 100–265 mm high. *Bulb* globose, 8–15 mm in diameter, offset-forming; tunic multilayered, outer layers dark brown, spongy; inner cataphyll translucent white, loosely surrounding leaf bases, apex obtuse. *Leaves* 1–2, lanceolate, 40–200 × 5–18 mm, spreading to suberect, upper surface light green with distinct longitudinal grooves, plain or occasionally marked with purplish-brown spots, lower surface tinged with purplish-maroon; leaf bases clasping, 10–20 mm long, subterranean, white; primary seedling leaf flat, prostrate. *Inflorescence* racemose, few- to many-flowered, fairly dense, sterile apex short; scape erect to suberect, 50–130 mm long, light green, heavily marked with small to large purplish-maroon spots or blotches; rachis erect, light green marked with small dull purplish blotches; pedicels suberect at anthesis, 2–4 mm long, becoming erect in fruiting stage, light green; lower bracts ovate, becoming lanceolate above, 1–3 × 1–3 mm, purplish-maroon to white. *Perianth* zygomorphic, oblong-campanulate, suberect to spreading; tube shallowly cup-shaped, 0.5–1.0 mm long, light blue to white; outer tepals ovate, 3–4 × 5 mm, white or cream, or light blue at base shading to white or cream above; apical gibbosities and keels dark magenta or greenish-brown; inner tepals obovate, 5–7 × 3 mm, translucent white to cream, median keels dark magenta or greenish-brown. *Stamens* shortly exserted, straight; filaments 4–6 mm long, white. *Ovary* ellipsoid, 2–3 × 2 mm, bright green; style more or less straight, included, 4–5 mm long, white. *Capsule* ellipsoid, 5–6 × 4–5 mm, erect. *Seed* globose, 0.9–1.0 × 1.2–1.3 mm, glossy, black; strophiole inflated, 0.5 mm long, smooth. *Chromosome number* 2n = 14 (Spies *et al.*, 2008). Figures 176, 177.

FLOWERING PERIOD. September to October, with a peak in early October.

HISTORY. W. F. Barker first collected material of *L. lactosa* in October 1933 near Kleinmond in the southern Western Cape. Occasional records were made in the ensuing three decades but the species remained poorly known until October 1996, when Louis Mostert found it on land set aside for housing development at Kleinmond, and material was donated to the Kirstenbosch Bulb Collection for cultivation. The plant was subsequently studied in habitat at Kleinmond, and its publication followed in *Bothalia* (Duncan, 1998a).

DISTINGUISHING CHARACTERS AND AFFINITIES. Duncan *et al.* (2005a) consider *L. peersii* to be the closest relative of *L. lactosa*. The two species share straight stamens and globose seeds with inflated

Figure 176 (left). *Lachenalia lactosa* on sandy flats at Kleinmond. Image: Graham Duncan.
Figure 177 (right). Cultivated specimen of *Lachenalia lactosa* from Kleinmond. Image: Graham Duncan.

strophioles, but *L. peersii* differs in its usually taller stature, urceolate, strongly carnation-scented perianths with recurved inner tepals, a plain, uniformly light green to brownish-maroon scape, capsules that retain the same orientation as the flowers, and narrower seeds (0.8–0.9 mm) with longer strophioles (0.6–0.7 mm).

DISTRIBUTION, HABITAT AND CONSERVATION STATUS. *L. lactosa* is confined to a short stretch of coastal fynbos in the southern Western Cape from Houw Hoek Pass eastwards to just beyond Baardscheerdersbos (Map 60). It occurs in Kogelberg- and Overberg Sandstone Fynbos vegetation (Mucina & Rutherford, 2006), on acid sandy flats and lower mountain slopes in full sun, growing singly or in small clumps. The plants are not easily detected in their natural surroundings, not even when in flower, because of their small stature and excellent camouflage against the white sandy soil and surrounding fynbos vegetation (Duncan, 1999b). *L. lactosa* used to be quite common around Kleinmond but is now under severe threat from housing development and has a conservation status of Endangered (Victor *et al.*, 2009b).

NOTES. Pollinators of *L. lactosa* are as yet unknown but are likely to include honey bees (*Apis mellifera*). The species is strongly self-fertile in cultivation, consistently producing copious seed under isolated, insect-free conditions, and the seedlings flower true to type. The pedicels and developing fruits re-orientate to an erect position and the scape remains attached to the bulb following capsule dehiscence, the seeds being dispersed locally.

Map 60. Known distribution of *Lachenalia lactosa*.

60. LACHENALIA ROSEA

Lachenalia rosea Andrews, *The Botanist's Repository* 5: t. 296 (1803). *Scillopsis rosea* (Andrews) Lem., *L'Illustration Horticole* 3: 35 (1856).

TYPE: South Africa, Cape, collector and precise locality unknown, figure in Andrews, *The Botanist's Repository* 5: t. 296 (1803) (lectotype, designated here).

SYNONYMY: *Lachenalia bifolia* Ker Gawl., *Curtis's Botanical Magazine* 40: t.1611 (1814), non *L. bifolia* (Burm. *f.*) W. F. Barker ex G. D. Duncan. *Scillopsis bifolia* (Ker Gawl.) Lem., *L'Illustration Horticole* 3: 35 (1856). Type: South Africa, Cape, collector and precise locality unknown, figure in *Curtis's Botanical Magazine* 40: t. 1611 (1814) (lectotype, designated here).

Lachenalia sanguinolenta Willd. ex Kunth, *Enumeratio Plantarum* 4: 286 (1843). Type: South Africa, Cape, precise locality unknown, *C. L. Willdenow* 6745 (B!, holo.).

Lachenalia unifolia Jacq. var. *pappei* Baker, in Thiselton-Dyer, W. T. (ed.), *Flora Capensis* 6: 431 (1897). Type: South Africa, Western Cape, Cape Flats, *C. W. Pappe s.n.* (K!, holo.).

ETYMOLOGY. *rosea*: rose-pink, referring to the tepals.

Plate 19. Watercolour painting of *Lachenalia rosea* from Cape Agulhas (*Crous* 240, in NBG) courtesy of The Editor, *Flowering Plants of Africa*, vol. 55, t. 2126 (1997). Artist: Fay Anderson.

60. LACHENALIA ROSEA

DESCRIPTION. *Geophyte*, 60–360 mm high. *Bulb* subglobose, 10–25 mm in diameter, offset-forming; tunic multilayered, outer layers dark brown, spongy; inner cataphyll translucent white, adhering to leaf bases, apex obtuse. *Leaves* 1–2, lanceolate, 80–250 × 16–22 mm, suberect, light to dark green, slightly to strongly canaliculate, leathery, upper surface with depressed longitudinal grooves, plain or marked with light to dark brown blotches, lower surface plain or flushed with maroon near base; leaf margins coriaceous, white to greenish-yellow; leaf bases clasping, 10–40 mm long, upper portion green or deep maroonish-magenta, lower portion white; primary seedling leaf flat, prostrate. *Inflorescence* racemose, many-flowered, sterile apex short; scape erect, 55–120 mm long, light green or tinged with dark pinkish-maroon, sturdy; rachis tinged with light to dark pinkish-maroon; pedicels suberect at anthesis, 2–4 mm long, white, green or light blue, becoming erect in fruit; lower bracts ovate, becoming lanceolate above, 2–4 × 1–3 mm, yellowish-green or maroonish-magenta. *Perianth* zygomorphic, oblong-campanulate, suberect to spreading at anthesis; tube cup-shaped, 3 mm long, light blue to bluish-grey; outer tepals ovate, 8–9 × 4–5 mm, light blue at base, shading to rose-pink or bluish-grey above, apical gibbosities dark rose-pink or bluish-grey; inner tepals obovate, 9–10 × 3–4 mm, light rose-pink or bluish-grey, median keels darker pink or bluish-grey, lower inner tepal very slightly longer than upper two tepals. *Stamens* included, rarely slightly exserted; filaments more or less straight, white, 9–10 mm long. *Ovary* obovoid, 3–4 × 2.0–2.5 mm, light green; style included to well exserted, 8–10 mm long, straight, white. *Capsule* obovoid, 8–10 × 5–6 mm, erect. *Seed* globose, 0.9–1.0 × 0.9 mm, glossy, black; strophiole inflated, 0.7–0.8 mm long, smooth. *Chromosome number* 2n = 14 (Johnson & Brandham, 1997; Hamatani *et al.*, 2007; Spies *et al.*, 2008). Plate 19, Figure 178.

Figure 178. *Lachenalia rosea* in fynbos after a fire at Grootbos Private Nature Reserve near Gansbaai. Image: Heiner Lutzeyer.

FLOWERING PERIOD. August to December, with a peak from late September to mid-November.

HISTORY. H. C. Andrews (1803b) described and illustrated *L. rosea* in volume 5 of *The Botanist's Repository*. The plant used for this purpose was collected by James Niven, resident collector of one of the most ardent English gardeners of his time, George Hibbert of Clapham. Niven collected this plant at an unspecified locality during his visit to the Cape between 1798 and 1803. *L. rosea* has three synonyms, the first of which, *L. bifolia* Ker Gawl., was published in 1814 accompanying plate 1611 of *Curtis's Botanical Magazine*. The second is that of C. S. Kunth (1843), who published a Willdenow specimen (*Willdenow* 6745) as *L. sanguinolenta* Willd. ex Kunth in his *Enumeratio Plantarum*. In his second revision of the genus, Baker (1897a) described the new variety *L. unifolia* Jacq. var. *pappei* Baker from material collected by Ludwig Pappe at an unrecorded locality on the Cape Flats. The holotype sheet is housed in the Kew Herbarium and clearly matches *L. rosea*. A watercolour painting by Fay Anderson of an especially attractive form from near Cape Agulhas (*Crous* 240, in NBG), which has spotted upper leaf surfaces, was reproduced in *Flowering Plants of Africa* (Duncan & Anderson, 1997).

DISTINGUISHING CHARACTERS AND AFFINITIES. *L. rosea* is allied to *L. salteri* (Duncan et al., 2005a), which shares oblong-campanulate flowers, the pink forms of which are similar to the usual pink forms of *L. rosea*, more or less straight stamens, lanceolate, leathery leaves with cartilaginous margins and similar globose seeds with inflated strophioles. *L. salteri* differs in its spreading inner tepals with recurved apices, shortly to well-exserted stamens, shorter outer tepals (6–8 mm long) and longer styles (11–13 mm).

DISTRIBUTION, HABITAT AND CONSERVATION STATUS. *L. rosea* is confined to the southern coastal part of the Western Cape, occurring from the Cape of Good Hope Nature Reserve on the southern Cape Peninsula (where it is rare) east to Knysna (Map 61). The plants grow as scattered individuals or in small groups, often along swamp margins in peaty acid sand, less frequently in acid clay-loam on low hills and steep rocky mountain slopes. It is encountered in five vegetation types: Peninsula, Kogelberg-, Overberg- and South Outeniqua Sandstone Fynbos, and Elgin Shale Fynbos (Mucina & Rutherford, 2006). At Grootbos Private Nature Reserve near Gansbaai, *L. rosea* grows within clumps of *Chondropetalum tectorum* (Restionaceae) and in close association with the pyrophyte *L. lutzeyeri*. *L. rosea* is still commonly encountered across its range and is not threatened.

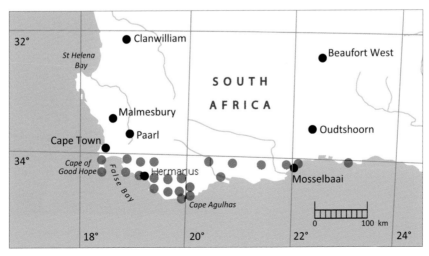

Map 61. Known distribution of *Lachenalia rosea*.

NOTES. *L. rosea* plants flower prolifically following bush fires but are not dependent on fire for flowering to occur. Honey bees (*Apis mellifera*) are the primary pollinators. The scapes remain attached to the bulb for several weeks after the capsules have dehisced and the seeds are dispersed locally.

61. LACHENALIA JUDITHIAE

Lachenalia judithiae G. D. Duncan, sp. nov.
A L. rosea *tepalis interioribus protrusis, tepalo interiore inferiore canaliculato et longiore quam tepalis interioribus aliis, staminibus declinatis, semini longiore et latiore differt.*
TYPE: South Africa, Eastern Cape, 3323 (Willowmore): 1.5 miles [2.4 km] west of Groot Bo-Kouga (–CB), *M.L. Thomas s.n.* sub. NBG 98476 (NBG!, holo.).

ETYMOLOGY. *judithiae*: after Vivienne Judith van Warmelo (1933–), mother of the author.

DESCRIPTION. *Geophyte* 150–300 mm high. *Bulb* subglobose, 15–20 mm in diameter, offset-forming; tunic multilayered, outer layers dark brown, spongy; inner cataphyll translucent white, subterranean, adhering to leaf bases, apex acute. *Leaves* 1–2, narrowly to broadly lanceolate, 200–300 × 30–35 mm, suberect, slightly canaliculate, coriaceous, leathery, light to dark green, upper surface heavily marked with large brownish-purple blotches, rarely plain, lower surface plain; margins entire, distinctly cartilaginous, white to dull red; leaf bases clasping, 15–30 mm long; primary seedling leaf flat, prostrate. *Inflorescence* racemose, many-flowered, sterile apex short; scape erect to suberect, 70–150 mm long, light green, plain or heavily marked with large brownish-purple blotches; rachis dull brownish-purple; pedicels suberect, 2–4 mm long, white, lengthening towards inflorescence apex; lower bracts ovate, becoming lanceolate above, 2–4 × 1–4 mm, light green to white. *Perianth* zygomorphic, oblong-campanulate, spreading to suberect, ageing to dull red or purple, becoming erect in fruit; tube cup-shaped, 3 mm long, metallic blue; outer tepals narrowly ovate, 5–10 × 4–5 mm, metallic blue, apices translucent white, apical gibbosities yellowish-green to greenish-brown, median keels bluish-green; inner tepals obovate, 7–12 × 4–6 mm, translucent white, median keels green, apices slightly to strongly recurved, upper two tepals overlapping, lower tepal canaliculate, equal to or 1–2 mm longer than upper tepals, diverging markedly from upper tepals. *Stamens* included, declinate; filaments white, 5–9 mm long. *Ovary* ellipsoid, 3–4 × 2–3 mm, bright green to

Figure 179. Cultivated specimen of *Lachenalia judithiae* from Greyton. Image: Graham Duncan.

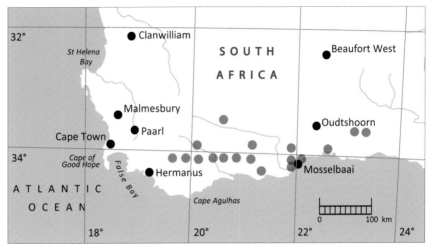

Map 62. Known distribution of *Lachenalia judithiae*.

yellowish-green; style included, declinate, 4–7 mm long, white. *Capsule* ellipsoid, 8–10 × 6–7 mm. *Seed* globose, 1.1–1.3 × 1.2–1.3 mm, glossy, black; strophiole inflated, 0.7–0.8 mm long, smooth. *Chromosome number* unknown. Figure 179.

FLOWERING PERIOD. Early August to late October.

HISTORY. The first collection of *L. judithiae* was made by Prof. R. H. Compton, who found it at Keur Kloof near Montagu in August 1940 (*Compton s.n.*, in NBG). The holotype collection in NBG was made by Margaret Thomas just west of Groot Bo-Kouga in the Kouga Mountains in the extreme south-eastern part of the Western Cape, on 23rd August 1973. Material of this species was previously included under *L. rosea*.

DISTINGUISHING CHARACTERS AND AFFINITIES. *L. judithiae* is identified by a heavily blotched scape, metallic blue or occasionally light pink perianth tubes and outer tepals, the latter with translucent white, slightly recurved apices, and green or brown apical gibbosities. The translucent white inner tepals have bluish-green keels and strongly recurved apices, with the deeply canaliculate lower inner tepal usually being distinctly longer than, and diverging strongly from, the upper inner tepals. Flower size varies considerably, and the lanceolate leaves are heavily marked with brownish-purple blotches on the upper surface and have prominent cartilaginous, hyaline margins.

The species' ally, *L. rosea*, has similar oblong-campanulate flowers with blue perianth tubes and leathery, lanceolate leaves with cartilaginous margins, but differs in its light to bright pink outer and inner tepals, with the inner tepals being all the same length and only slightly longer than the outer ones, and in its declinate stamens and smaller globose seed (0.9–1.0 × 0.9 mm).

L. judithiae is superficially similar to *L. martleyi*, which has similar metallic blue perianth tubes and outer tepals, a heavily blotched scape, leathery leaves and globose seeds with inflated strophioles, but the latter differs in its longer inner tepals (12–13 mm long), with the lower inner tepal equal to the upper inner tepals and all with flat apices, and in its longer stamens (10–12 mm).

DISTRIBUTION, HABITAT AND CONSERVATION STATUS. *L. judithiae* extends from Greyton in the west, inland to Montagu in the Little Karoo, south to Swellendam and along the coast and inland to the Kouga Mountains south of Willowmore in the east (Map 62). It occurs mainly in heavy, stony clay ground in various renosterveld vegetation types, including Montagu-, Eastern Rûens- and Mossel Bay Shale Renosterveld, and in fynbos types, including Garden Route Shale Fynbos, Garden Route

Granite Fynbos and Kouga Sandstone Fynbos, as well as in Canca Limestone Fynbos overlaying sandstone (Mucina & Rutherford, 2006). The plants occur in colonies in open aspects or on south-facing hillsides and lower mountain slopes. Near Greyton, *L. judithiae* occurs in association with *L. lutea*. *L. judithiae* is not threatened.

NOTES. Honey bees (*Apis mellifera*) pollinate this species. After flowering, the pedicels and developing fruits become re-orientated to an erect position. The scapes remain attached to the bulbs for several months following capsule dehiscence and the seeds are dispersed locally.

62. LACHENALIA MARTLEYI

Lachenalia martleyi G. D. Duncan, sp. nov.

A L. mediana, bulbo subgloboso, marginibus foliorum cartilagineis, atroguttatis scapis, tepalis exterioribus anguste ovatis et interioribus oblongis-obovatis, staminibus rectis et stylis brevioribus distinguenda est.

TYPE: South Africa, Western Cape, 3418 (Simonstown): Lourensford Farm, Somerset West (–BB), *W. D. Baxter s.n.* (BOL!, holo.).

ETYMOLOGY. *martleyi*: after Commander J. F. Martley, naval officer and bulb-grower.

DESCRIPTION. *Geophyte*, 300–340 mm high. *Bulb* subglobose, 15–20 mm in diameter, offset-forming; tunic multilayered, outer layers dark brown, spongy; inner cataphyll translucent white, apex obtuse. *Leaves* (1–)2, lanceolate, 150–210 × 10–20 mm, suberect, canaliculate, light green, upper surface heavily marked with purplish-brown blotches, lower surface plain; leaf margins cartilaginous; leaf bases clasping, 60–120 mm long, subterranean portion white, light green above. *Inflorescence* racemose, many-flowered, dense, erect to suberect, sterile apex short; scape suberect, 70–120 mm long, light green, heavily marked with purplish-brown blotches; rachis light green; pedicels suberect, 1–5 mm long, becoming erect in fruit, light green, length increasing towards apex; lower bracts ovate, becoming lanceolate above, 2–5 × 1–4 mm, translucent white. *Perianth* zygomorphic, oblong-campanulate, suberect, becoming erect in fruit, ageing to dull reddish-pink; tube cup-shaped, 3 mm long, metallic blue; outer tepals narrowly ovate, 9–11 × 3–4 mm, metallic blue, apices whitish-yellow, apical gibbosities greenish-yellow; inner tepals oblong-obovate, 12–13 × 4–5 mm, white, apices flat; median keels bluish-green. *Stamens* included, declinate; filaments white, 10–12 mm long. *Ovary* ellipsoid, bright yellowish-green, 5 × 3 mm; style included, declinate, 5 mm long. *Capsule* ellipsoid, 5–6 × 2–3 mm. *Seed* globose, 1.2 × 1.2 mm, glossy, black; strophiole inflated, 1.0–1.1 mm long, smooth. *Chromosome number* unknown. Plate 20.

FLOWERING PERIOD. September to November.

HISTORY. The first collection of *L. martleyi* was made by William D. Baxter (*Baxter s.n.*, in BOL) at Lourensford Farm, Somerset West in October 1929. Six years later, in September 1935, a cultivated specimen of unknown origin flowered in the Stellenbosch garden of a friend of Commander J. F. Martley, who brought the plant to W. F. Barker at the Bolus Herbarium. She painted it in watercolour, compiled a brief description, and preserved the inflorescence, leaves and a floral dissection (specimen in BOL 21947). No recent collections have been made.

DISTINGUISHING CHARACTERS AND AFFINITIES. *L. martleyi* appears to be related to *L. mediana*, which has similar oblong-campanulate, suberect flowers with blue perianth tubes and outer tepals, and globose seeds with inflated strophioles. *L. mediana* differs in its slender, unmarked scapes, shorter, ovate outer tepals (6–8 mm long), shorter, obovate inner tepals (8–10 mm long), longer style (7–9 mm long), non-leathery leaves, unmarked upper leaf surfaces and soft leaf margins.

THE GENUS LACHENALIA | 271
62. LACHENALIA MARTLEYI

Plate 20. Watercolour painting of *Lachenalia martleyi* (*ex hort*, in BOL 21947) courtesy of the Compton Herbarium, Kirstenbosch, South African National Biodiversity Institute. Artist: Winsome Barker.

Superficially, *L. martleyi* resembles *L. judithiae*, which shares metallic blue perianth tubes and outer tepals, included stamens, a heavily blotched scape, similar leathery lanceolate leaves, usually with blotched upper surfaces, and globose seeds with inflated strophioles. *L. judithiae* differs in its shorter inner tepals (7–12 mm long) with strongly recurved apices, the lower inner tepal diverging strongly from and usually protruding well beyond, the upper inner tepals, and in its shorter filaments (5–9 mm long).

DISTRIBUTION, HABITAT AND CONSERVATION STATUS. *L. martleyi* is currently only known from the Somerset West District, occurring in acidic, seasonally moist, loamy-clay soils, in Cape Winelands Shale Fynbos (Map 63). Populations are threatened by agricultural activity, and a conservation status of Vulnerable is recommended.

NOTES. *L. martleyi* forms part of a group of species that includes *L. mediana*, *L. judithiae*, *L. rosea* and *L. salteri* whose pedicels and flowers change orientation from a suberect to an erect position during the fruiting period. The oblong-campanulate flower shape suggests pollination by honey bees (*Apis mellifera*).

Map 63. Known distribution of *Lachenalia martleyi*.

63. LACHENALIA SALTERI

Lachenalia salteri W. F. Barker, *The Flowering Plants of South Africa* 13: t. 505 (1933).
TYPE: South Africa, Western Cape, near Cirkels Vlei, Cape of Good Hope Nature Reserve, *T. M. Salter* 1861 (BOL!, holo.; K!, iso.).

ETYMOLOGY. *salteri*: after Captain Terence Macleane Salter (1883–1969), author and botanical collector.

DESCRIPTION. *Geophyte*, 150–350 mm high. *Bulb* globose, 15–20 mm in diameter, solitary; tunic multilayered, outer layers dark brown, spongy; inner cataphyll translucent white, adhering to leaf bases, apex obtuse. *Leaves* 1–2, lanceolate, 70–170 × 10–25 mm, suberect to spreading, yellowish-green to dark green, slightly to deeply canaliculate, upper surface plain or marked with dark green or purplish-brown blotches, with slightly depressed longitudinal grooves, lower surface tinged with dark maroon in lower portion; leaf margins coriaceous, maroon; leaf bases clasping, 15–55 mm long, white, plain or with sporadic maroon or magenta blotches in lower half, densely blotched in upper half; primary seedling leaf flat, prostrate. *Inflorescence* racemose, 60–120 mm long, many-flowered, moderately dense, erect, sterile apex short; scape erect, 50–150 mm long, dull brownish-green, plain or with small to large purplish- or light brownish-maroon blotches, slender to sturdy; rachis light green in lower half, shading to light blue in upper half, plain or with minute blue speckles; pedicels suberect at anthesis, 2–4 mm long, elongating towards inflorescence apex, white or light blue, becoming erect in fruit; lower bracts ovate, becoming lanceolate above, 3–7 × 1–5 mm white to light green, plain or with purplish-brown markings. *Perianth* zygomorphic, oblong-campanulate,

PLATE 21. Watercolour painting of *Lachenalia salteri* from Elim (*Nicklin* 136, in NBG), courtesy of The Editor, *Curtis's Botanical Magazine* vol. 20, t. 477 (2003). Artist: Fay Anderson.

suberect to slightly spreading in the flowering stage, becoming erect in fruit, spicy sweet-scented; tube cup-shaped, 2–3 mm long, white to light blue; outer tepals ovate, 6–8 × 3–4 mm, white or light to deep maroonish-purple, or light blue shading to whitish-pink above, apical gibbosities dark purple or brownish-pink, apices flat to slightly recurved; inner tepals obovate, 9–11 × 4 mm, white or maroonish-purple to light pink, median keels dark purple or blue, apices slightly to well recurved. *Stamens* shortly to well exserted, straight; filaments white, 10–12 mm long. *Ovary* ellipsoid, 3–5 × 1.5–2.0 mm, light green with minute blue speckles; style well exserted, straight, 11–13 mm long, white. *Capsule* ellipsoid, 7–10 × 3–5 mm. *Seed* globose, 0.9–1.0 × 0.8–0.9 mm, glossy, black; strophiole inflated, 0.6–0.7 mm long, smooth. *Chromosome number* unknown. Plate 21, Figure 180.

FLOWERING PERIOD. September to early January, with a peak from mid-November to mid-December.

HISTORY. The type collection was made by Paymaster-Captain T. M. Salter near Cirkels Vlei in the Cape of Good Hope Nature Reserve in November 1931, and the species was described in *The Flowering Plants of South Africa*, with the description accompanied by a watercolour plate of a maroonish-purple form from this Reserve (Barker, 1933d). *L. salteri* has since been collected rather infrequently, probably because of its very late flowering period and its appearance in large numbers only after wild fires, mainly in the vicinity of Betty's Bay, Kleinmond and Elim.

DISTINGUISHING CHARACTERS AND AFFINITIES. In a phylogenetic analysis of unweighted morphological data (Duncan *et al.*, 2005a), *L. salteri* resolved as sister to *L. rosea*, with which it is often confused. The distinguishing characteristics are discussed under *L. rosea*.

DISTRIBUTION, HABITAT AND CONSERVATION STATUS. *L. salteri* is confined to the south-western and southern Cape, extending from the Cape of Good Hope Nature Reserve and Red Hill in the southern

Map 64. Known distribution of *Lachenalia salteri*.

Figure 180. *Lachenalia salteri* in seasonally wet fynbos at Betty's Bay. Image: Graham Duncan.

Cape Peninsula (where it is rare) east along the coastal belt to Elim (Map 64). The species traverses three vegetation types: Peninsula-, Kogelberg- and Overberg Sandstone Fynbos (Mucina & Rutherford, 2006). Plants occur singly or in small colonies in acid, moist or seasonally wet peaty soil at the edges of swamps and in cleared areas among short grass. In Jill Attwell's wild garden at Betty's Bay, a pink form grows in close association with dense clumps of the amaryllid *Haemanthus canaliculatus*, within metres of the rocky coastline. *L. salteri* is threatened by coastal housing development and has a conservation status of Endangered (Helme & Raimondo, 2009h).

NOTES. *L. salteri* plants are stimulated to profuse flowering from late spring to midsummer following the clearing of thick undergrowth by bush fires in the previous summer and autumn, which provide the plants with adequate sunlight. The species is, however, not dependent on fire for flowering to occur (Duncan, 2003d). The flowers of *L. salteri* emit a pleasant, spicy-sweet scent and are pollinated by honey bees (*Apis mellifera*). The scapes remain attached to the bulbs for several weeks following capsule dehiscence and the seeds are dispersed locally.

64. LACHENALIA LUTZEYERI

Lachenalia lutzeyeri G. D. Duncan, *Bothalia* 37 (1): 31–34 (2007).
TYPE: South Africa, Western Cape, Witkransberg area of Grootbos Private Nature Reserve, northwest of Gansbaai, on south-west-facing mountain slope in Table Mountain Sandstone, *H. Lutzeyer s.n.* (NBG!, holo.!; PRE!, iso.).

ETYMOLOGY. *lutzeyeri*: after Heiner Lutzeyer (1928–2012) of Grootbos Private Nature Reserve near Gansbaai, who discovered this species and made the first collection, in recognition of the contribution he has made to nature conservation in the southern Cape.

DESCRIPTION. *Geophyte*, 200–420 mm high. *Bulb* globose, 20–25 mm in diameter, solitary, deep-seated; roots mainly contractile; tunic multilayered, outer layers dark brown, spongy; inner cataphyll translucent white, adhering to leaf bases, apex acute. *Leaves* 1–2, narrowly lanceolate, 85–290 × 4–13 mm, spreading to suberect, deeply canaliculate to conduplicate, yellowish-green or light to deep maroon, upper surface plain or with large dark green, brown or maroon flattened pustules, lower surface plain or with dark green blotches; leaf margins cartilaginous, flat to slightly undulate; leaf bases clasping, 30–85 mm long, subterranean portion white, aerial portion yellowish-green to deep maroon; primary seedling leaf terete, erect. *Inflorescence* racemose, erect to suberect, many-flowered, moderately dense, sterile apex short to long; scape erect to suberect, 120–200 mm long, rigid, light green to yellowish-green, plain to heavily marked with minute maroon speckles or small to large maroon blotches; rachis light green to dull maroon, plain or with darker maroon spots; pedicels suberect at anthesis, 3–4 mm long, green, erect in fruit; lower bracts ovate, becoming lanceolate above, 1–2 × 0.5–2.0 mm, translucent white. *Perianth* zygomorphic, oblong-campanulate, cernuous, ageing to dull maroon, soapy sweet-scented; tube cup-shaped, 2–3 mm long, light yellowish-cream; outer tepals ovate, 5–6 × 3–4 mm, light yellowish-cream, apical gibbosities olive green to brown; inner tepals obovate, 6–7 × 4–5 mm, translucent white, apical marking dark green to brown, median keels yellowish-green. *Stamens* well exserted, declinate; filaments white, 9–10 mm long. *Ovary* obovoid, 2 × 3 mm, bright green; style well exserted, straight, 9 mm long, white. *Capsule* obovoid, 6–7 × 6–7 mm. *Seed* globose, 1.1–1.2 × 1.2–1.3 mm, glossy, black; strophiole inflated, 0.8–0.9 mm long, smooth. *Chromosome number* unknown. Plate 22, Figures 34, 181.

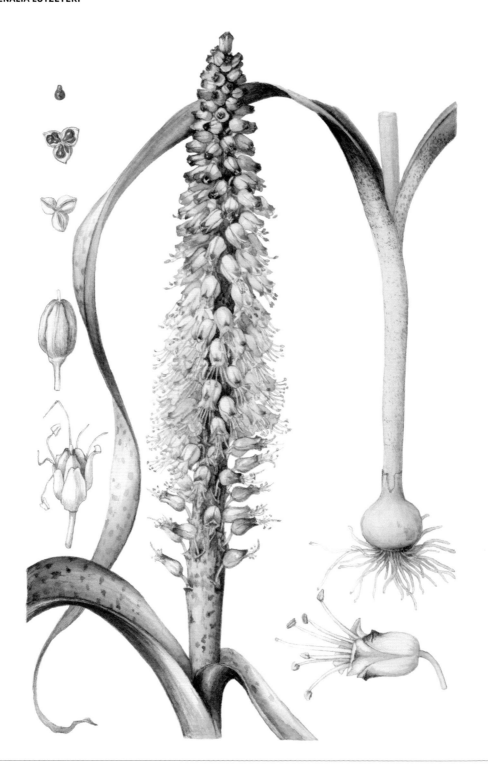

Plate 22. Watercolour painting of *Lachenalia lutzeyeri* from Grootbos Private Nature Reserve near Gansbaai (*Lutzeyer s.n.*, in NBG), courtesy of Vicki Thomas. Artist: Vicki Thomas.

FLOWERING PERIOD. November to December, rarely to late January, with a peak in mid-November.

HISTORY. *L. lutzeyeri* was first collected by Heiner Lutzeyer in late March 2004, following an extensive wild fire that burnt over the Witkransberg area of Grootbos Private Nature Reserve north-west of Gansbaai. Large numbers of plants flowered in November 2004, but during the corresponding period in November and December 2005, the number of flowering plants had decreased drastically to just a few individuals. Possible reasons for the species having remained undetected until recently are that flowering is extremely erratic because of its dependence on summer or early autumn fires, coupled with its very late flowering period and slim chance of being recognised as a distinct taxon because of its superficial similarity to other lachenalias. In February 2006, the *L. lutzeyeri* type population was burnt again as a result of a devastating fire that raged across most of Grootbos Private Nature Reserve, and this resulted in the appearance of large numbers of flowering individuals in November of that year. The subsequent absence of fires has resulted in the plants remaining completely dormant. The species was described in *Bothalia* (Duncan & Edwards, 2007).

DISTINGUISHING CHARACTERS AND AFFINITIES. *L. lutzeyeri* appears to be most closely allied to *L. youngii*, another late-flowering species that occurs much further east, from Mossel Bay in the southern part of the Western Cape to Jeffreys Bay in the Eastern Cape. *L. youngii* has similar oblong-campanulate, cernuous flowers that are produced on suberect white pedicels that re-orientate to an erect position in fruit. It differs mainly in having linear leaves without cartilaginous margins, a lax scape, shortly exserted, straight stamens and light to deep pink perianths. The shallow-seated, offset-forming bulbs of *L. youngii* are not dependent on fire for flowering to occur (Duncan & Edwards, 2007).

Map 65. Known distribution of *Lachenalia lutzeyeri*.

Figure 181. *Lachenalia lutzeyeri* flowering after fire at Grootbos Private Nature Reserve. Image: Graham Duncan.

DISTRIBUTION, HABITAT AND CONSERVATION STATUS. *L. lutzeyeri* is presently only known from Grootbos Private Nature Reserve north-west of Gansbaai in the southern Cape (Map 65). Recorded subpopulations are restricted to Overberg Sandstone Fynbos vegetation (Mucina & Rutherford, 2006). The species is plentiful and grows in full sun in open aspects between Table Mountain Sandstone boulders, on moderate slopes with south-western, southern and northern aspects. It may well be recorded in other parts of the southern Cape in due course. Although fully protected, a conservation status of Rare is recommended because of the species' limited distribution within the Reserve.

NOTES. The flowers of *L. lutzeyeri* are pollinated by honey bees (*Apis mellifera*) and are also visited by the Painted Lady butterfly (*Vanessa cardui*). *L. lutzeyeri* is one of only three members of the genus that are entirely dependent on fire for flowering to occur. The other two pyrophytes are *L. montana* and *L. sargeantii* (Duncan & Edwards, 2007). *L. lutzeyeri* and *L. rosea* grow in close association at Grootbos Private Nature Reserve and flower simultaneously; *L. rosea* has very similar leaves but is not pyrophytic. Other companion species in this Reserve include *Mimetes cucullatus* (Proteaceae) and various genera of Restionaceae.

65. LACHENALIA LEOMONTANA

Lachenalia leomontana W. F. Barker, *South African Journal of Botany* 53 (2): 169–170 (1987).
TYPE: South Africa, Western Cape, kloof at south base of Leeurivierberg, NW of Swellendam, E. E. *Esterhuysen* 33322 (NBG!, holo.; BOL!, K!, iso.).

ETYMOLOGY. *leomontana*: after the Leeurivierberg (Lion River Mountain), the Afrikaans name of the mountain north-west of Swellendam in the southern Cape where the type collection was made.

DESCRIPTION. *Geophyte*, 100–300 mm high. *Bulb* globose, 10–15 mm in diameter, offset- and bulbil-forming; tunic multilayered, outer layers light brown, papery; inner cataphyll translucent white, adhering to leaf bases, apex obtuse. *Leaf* solitary, 70–120 × 8–15 mm, lanceolate, spreading, slightly canaliculate, olive green, upper surface plain to heavily marked with dark purple spots, depressed longitudinal grooves prominent, lower surface tinged with maroon towards base; leaf base 10–25 mm long, clasping, lower half white, shading to maroon above; primary seedling leaf flat, prostrate. *Inflorescence* racemose, erect, few- to many-flowered, lax, sterile apex short; scape erect to suberect, 70–180 mm long, light green, plain to heavily marked with small light purple blotches or spots, slender; rachis light green, plain or with small light purple spots, apex white; pedicels suberect at anthesis, 4–7 mm long, becoming erect in fruit; light green to white; lower bracts ovate, becoming lanceolate above, 1–2 × 1–2 mm, brown at base, shading to white above. *Perianth* zygomorphic, oblong-campanulate, white, cernuous; tube cup-shaped, 2 mm long; outer tepals ovate, 6–7 × 4–5 mm, apices recurved, apical gibbosities light green or yellowish-brown; inner tepals obovate, 7–8 × 5–6 mm, apices recurved, median keels light yellow. *Stamens* included, straight; filaments white, 7–8 mm long. *Ovary* obovoid, 3 × 3 mm, light green; style included, straight, 4 mm long, white. *Capsule* obovoid, 8 × 5 mm. *Seed* globose, 1.3–1.4 × 1.3 mm, glossy, black; strophiole inflated, 0.5–0.6 mm long, smooth. *Chromosome number* 2n = 14 (Spies et al., 2008). Figure 182.

FLOWERING PERIOD. October to November, with a peak in early November.

HISTORY. *L. leomontana* was initially collected by Elsie Esterhuysen north-west of Swellendam in November 1973. The only other collection is that of Prof. M. C. Botha, made at the same locality in 1977. The accompanying photograph is of a plant from the latter collection, in cultivation in the Kirstenbosch Bulb Nursery. The species was described in the *South African Journal of Botany* (Barker,

1987). The holotype sheet (*Esterhuysen* 33322) is housed in the Compton Herbarium at Kirstenbosch (NBG), with isotype material at the Kew and Bolus Herbaria.

DISTINGUISHING CHARACTERS AND AFFINITIES. An analysis of unweighted morphological data suggested a close alliance between *L. leomontana* and *L. youngii* (Duncan *et al.*, 2005a). *L. youngii* has a similar many-flowered raceme of oblong-campanulate, cernuous flowers borne on long suberect pedicels that bend upwards during the fruiting stage, and straight stamens. It differs from *L. leomontana* in its pink tepals with flat inner tepal apices, its shortly exserted stamens, paired, linear, deeply canaliculate leaves and smaller globose seeds (0.9–1.0 × 0.9 mm). It is geographically isolated from *L. leomontana*, having a more easterly distribution from George in the southern Cape to Jeffrey's Bay in the Eastern Cape.

DISTRIBUTION, HABITAT AND CONSERVATION STATUS. *L. leomontana* is confined to the southern base of the Leeurivierberg north-west of Swellendam (Map 66). It grows in colonies in South Langeberg Sandstone Fynbos vegetation (Mucina & Rutherford, 2006), in shallow sandy soil between large sandstone boulders, in semi-shade or full sun. The species is not under threat, but because of its isolation at a single locality, it has a conservation status of Critically Rare (Helme, 2009).

NOTES. The oblong-campanulate flower shape of *L. leomontana* suggests pollination by honey bees (*Apis mellifera*). Although the open flowers have a cernuous orientation, developing capsules become erect and the seeds are dispersed locally from the capsules by the shaking action of wind.

Map 66. Known distribution of *Lachenalia leomontana*.

Figure 182. Cultivated specimen of *Lachenalia leomontana* from the Leeuwrivierberg. Image: Graham Duncan/SANBI.

66. LACHENALIA YOUNGII

Lachenalia youngii Baker, in Thiselton-Dyer, W. T. (ed.), *Flora Capensis* 6: 433 (1897).
TYPE: South Africa, Western Cape, Montagu Pass, *E. W. Young* 5545 (K!, holo., iso.; BOL!, iso.).

ETYMOLOGY. *youngii*: after E. W. Young, who discovered this species and made the first collection of plants.

DESCRIPTION. *Geophyte*, 150–250 mm high. *Bulb* globose, 15–20 mm in diameter, offset-forming; tunic multilayered, outer layers dark brown, spongy; inner cataphyll translucent white, adhering to leaf bases, apex acute. *Leaves* 2, linear, 140–300 × 5–10 mm, suberect to spreading, light to dark green, deeply canaliculate, adaxial surface plain, abaxial surface tinged with brownish-maroon; leaf bases loosely surrounding base of scape, 10–30 mm long, mostly subterranean, lower portion white, upper portion light green; primary seedling leaf terete, erect. *Inflorescence* racemose, many-flowered, erect, sterile apex short; scape erect to suberect, 90–150 mm long, light green, densely mottled or tinged with brownish-pink, lax; rachis densely mottled or tinged with brownish-pink; pedicels suberect at anthesis, 3–5 mm long, white to light brownish-pink, becoming erect in fruit; lower bracts ovate, becoming lanceolate above, 2–5 × 0.5–3.0 mm, white. *Perianth* zygomorphic, oblong-campanulate, light to deep pink, cernuous, slightly spicy-scented; tube cup-shaped, 2 mm long; outer tepals ovate, 6–7 × 3–4 mm, apical gibbosities greenish-brown to brownish-pink, apices slightly recurved; inner tepals obovate, 7–8 × 3–4 mm, median keels dark brownish-pink. *Stamens* shortly exserted, straight; filaments white, 8–9 mm long, white. *Ovary* obovoid, 3 × 3 mm, light yellowish-green; style included, straight, 7–8 mm long, white. *Capsule* obovoid, 5–6 × 5–6 mm. *Seed* globose, 0.9–1.0 × 0.9 mm, glossy, black; strophiole inflated, 0.6 mm long, smooth. *Chromosome number* 2n = 16 (Spies *et al.*, 2008). Figure 183.

Figure 183. Cultivated specimens of *Lachenalia youngii* from Knysna. Image: Graham Duncan.

FLOWERING PERIOD. September to November, with a peak in late October and early November.

HISTORY. *L. youngii* was discovered by E. W. Young in October 1880 on Montagu Pass, and was described in *Flora Capensis* (Baker, 1897a). In addition to Young's holotype collection (*Young* 5545, in K), an isotype sheet at Kew was mistakenly cited under *L. unicolor* by Baker, and a second isotype is housed in the Bolus Herbarium at the University of Cape Town. The plant has been infrequently collected and was illustrated in colour for the first time in *The Lachenalia Handbook*, almost a century after its original publication (Duncan, 1988a).

DISTINGUISHING CHARACTERS AND AFFINITIES. Duncan *et al.* (2005a) placed *L. youngii* adjacent to *L. leomontana* because these two species share similar many-flowered racemes of oblong-campanulate,

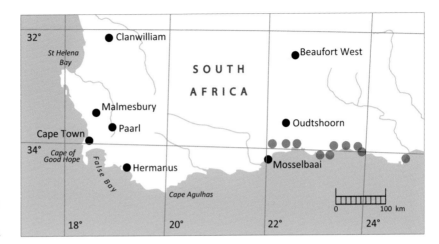

Map 67. Known distribution of *Lachenalia youngii*.

cernuous flowers with straight stamens, which are borne on fairly long, suberect pedicels. *L. leomontana* differs in its pure white perianth with distinctly recurved inner tepal apices, and included stamens. It has a solitary lanceolate leaf that is heavily spotted on the upper surface and globose seeds that are much larger than those of *L. youngii* (1.3–1.4 × 1.3 mm).

DISTRIBUTION, HABITAT AND CONSERVATION STATUS. *L. youngii* extends from west and north of George in the southern Cape to Jeffrey's Bay just east of Humansdorp in the south-western Eastern Cape (Map 67). It occurs in three vegetation types: Tsitsikamma-, South Outeniqua-, and Kouga Grassy Sandstone Fynbos (Mucina & Rutherford, 2006). The plants form colonies in acid, seasonally wet, sandy-loam soils. The habitat terrain varies from flats in fynbos clearings and forest marginal swamps to mountain and hill slopes and sandy river banks. *L. youngii* is threatened by coastal housing development in the southernmost parts of its range and a conservation status of Vulnerable is recommended.

NOTES. *L. youngii* plants flower profusely following bush fires but are not dependent on fire for flowering to occur. Pollinators for this species are not known but are likely to include honey bees (*Apis mellifera*). During the fruiting stage, the pedicels and developing fruits re-orientate to an erect position. The scape remains attached to the bulb for several weeks following capsule dehiscence and the seeds are dispersed locally by wind.

67. LACHENALIA MONTANA

Lachenalia montana Schltr. ex W. F. Barker, *Journal of South African Botany* 44 (4): 415–418 (1978).
TYPE: South Africa, Western Cape, Onrus, Hermanus, *W. F. Barker* 10485 (NBG!, holo.).

ETYMOLOGY. *montana*: from mountainous terrain.
DESCRIPTION. *Geophyte*, 100–330 mm high. *Bulb* globose, 15–20 mm in diameter, offset- and bulbil-forming, deep-seated; tunic multilayered, outer layers dark brown, cartilaginous; inner cataphyll translucent white, adhering to leaf bases, apex obtuse. *Leaves* 2, linear, 150–350 × 5–15 mm, spreading to suberect or arcuate, deeply canaliculate, adaxial and abaxial surfaces bright to dark green, plain; leaf margins light maroon, leaf bases clasping, 20–50 mm long, upper portion heavily tinged with magenta, subterranean portion white; primary seedling leaf terete, erect. *Inflorescence* racemose, many-flowered, dense, erect, sterile apex short; scape erect to suberect, 80–230 mm long,

67. LACHENALIA MONTANA

rigid, bright green tinged with light magenta; rachis light green, tinged with light magenta; pedicels arcuate, becoming suberect in fruit, 4–15 mm long, length increasing from base of inflorescence to apex, cream to light magenta; lower bracts ovate, becoming lanceolate above, 3–5 × 1–3 mm, cream tinged with light magenta. *Perianth* zygomorphic, oblong-campanulate, cream, cernuous to pendulous, spice-scented; tube shallow cup-shaped, 1–2 mm long; outer tepals ovate, 6–7 × 4 mm, apical gibbosities bright green to greenish-brown; inner tepals obovate, 8 × 4 mm, median keels bright green to greenish-brown. *Stamens* shortly exserted, straight; filaments white, 9–10 mm long. *Ovary* obovoid, 3 × 3 mm, bright green, stipitate, stipule 0.5 mm long; style shortly exserted, straight, 6–7 mm long, white. *Capsule* obovoid, 4–6 × 4–5 mm. *Seed* globose, 1.2–1.3 × 1.3 mm, matte, black, secondary sculpturing colliculate; strophiole rudimentary, 0.1 mm long, ridged. *Chromosome number* unknown. Figures 184, 185.

FLOWERING PERIOD. October to December, with a peak in mid-October.

HISTORY. *L. montana* was recorded for the first time in December 1831 by the German botanical collector Carl Zeyher at Houw Hoek in the south-western Cape. His preserved specimens (*Zeyher 4294*) are lodged in the South African Museum collection at Kirstenbosch and at the Kew Herbarium. In November 1896, Rudolf Schlechter collected a large number of specimens at Houw Hoek and distributed them to numerous local and foreign herbaria under his manuscript name *L. montana* Schltr.. A number of other collections were subsequently made by other collectors but pressed material lay unnamed in herbaria for decades before the species was finally described in the *Journal of*

Figure 184 (left). *Lachenalia montana* in fynbos after a fire at Hermanus. Image: Graham Duncan/SANBI.
Figure 185 (right). Detail of *Lachenalia montana* from Hermanus. Image: Graham Duncan/SANBI.

South African Botany; the holotype material was collected by W. F. Barker at Onrus near Hermanus in October 1966 (Barker, 1978).

DISTINGUISHING CHARACTERS AND AFFINITIES. *L. montana* is sister to *L. sargeantii* (Duncan et al., 2005a), with which it shares soboliferous, globose bulbs with linear, deeply caniculate leaves. They have rigid scapes and their flowers are usually borne on long, arcuate pedicels. Both produce obovoid capsules containing globose seeds that have colliculate secondary sculpturing. The flowers of *L. sargeantii* differ from those of *L. montana* in being tubular and pendulous with oblong outer tepals and oblong-obovate inner tepals. Its stamens are longer (11–16 mm) than those of *L. montana* but remain within the perianth.

Map 68. Known distribution of *Lachenalia montana*.

DISTRIBUTION, HABITAT AND CONSERVATION STATUS. *L. montana* is confined to the south-western Cape, extending from Louwshoek Peak east of Franschhoek, south to Houw Hoek Pass and Hangklip, and east along the coast to Hermanus (Map 68). The plants grow in Hawequas-, Kogelberg- and Overberg Sandstone Fynbos vegetation (Mucina & Rutherford, 2006), in large colonies on acid sandstone, usually on south-facing, lower mountain slopes from just above sea level to about 600 m. At Hermanus, a large population grows with *L. peersii*, and the two species flower simultaneously in October and November after fire (Duncan, 1988a). *L. montana* is not threatened.

NOTES. *L. sargeantii*, *L. lutzeyeri* and *L. montana* are the only pyrophytic members of the genus, only flowering in late spring and early summer after bush fires (Duncan & Edwards, 2007). During their first post-fire season, the plants flower in profusion, but flowering is drastically reduced or completely absent in the second year after a fire, following which the plants remain dormant until the next fire. During the fruiting stage, the pedicels re-orientate from spreading to suberect (Duncan et al., 2005). The scapes remain attached for several weeks after capsule dehiscence and the seeds are dispersed locally. The bulbs are deep-seated through the action of contractile roots. Pollinators have not been recorded for *L. montana*, although the oblong-campanulate flower shape and episodic flowering suggests generalist pollination. The flowers are at least partially self-fertile in cultivation and set seed under isolated conditions.

68. LACHENALIA SPLENDIDA

Lachenalia splendida Diels, *Botanische Jahrbücher für Systematik Pflanzengeschichte und Pflanzengeographie* 44: 116 (1909).
TYPE: South Africa, Western Cape, Vanrhynsdorp district, between Windhoek and Troe-Troe Rivers, *F. L. E. Diels* 444 (B!, lectotype, designated here).
SYNONYMY: *L. roodeae* E. Phillips, *The Flowering Plants of South Africa* 3: t. 91 (1923). Type: South Africa, Western Cape, Vanrhynsdorp, *E. Rood s.n.* sub. PRE 1461 (PRE!, holo.; K!, iso.).

68. LACHENALIA SPLENDIDA

ETYMOLOGY. *splendida*: splendid, bright lilac flowers.

DESCRIPTION. *Geophyte*, 60–350 mm high. *Bulb* subglobose, 12–20 mm in diameter, offset-forming, yellow; tunic multilayered, outer layers light to dark brown, hard; inner cataphyll translucent white, loosely surrounding leaf bases, apex obtuse. *Leaves* 2, lanceolate, 60–150 × 7–35 mm, spreading to suberect, bright green, canaliculate, upper surface with depressed longitudinal grooves; leaf bases clasping, 10–20 mm long, aerial portion green or tinged with lilac, subterranean portion white; primary seedling leaf terete, erect. *Inflorescence* spicate, many-flowered, dense, sterile apex short; scape erect to suberect, 40–120 mm long, light green, sturdy, upper portion sometimes slightly inflated, light green; rachis light green to brownish-lilac, apex shading to bright electric blue or lilac; lower bracts ovate, becoming lanceolate above, 2–4 × 2–4 mm, light green or white. *Perianth* zygomorphic, oblong-campanulate, spreading to cernuous; tube cup-shaped, 2–3 mm long, whitish-blue; outer tepals ovate, 6–8 × 3–5 mm, whitish-blue at base, shading to lilac above, apical gibbosities brownish-green to lilac-brown; inner tepals obovate, 9–12 × 5–6 mm, bright lilac, median keels dark brownish- to greenish-lilac. *Stamens* well exserted, declinate; filaments light to dark lilac, 11–13 mm long. *Ovary* ellipsoid, 3 × 2 mm, light green; style well exserted, declinate, 12–13 mm long, light lilac. *Capsule* broadly ellipsoid, 7–11 × 5–8 mm. *Seed* globose, 1.1–1.2 × 1.1–1.3 mm, glossy, black; strophiole rudimentary, 0.3 mm long, ridged. *Chromosome number* 2n = 16 (Crosby, 1986; Johnson & Brandham, 1997; Hamatani *et al.*, 1998). Figures 42, 186, 187.

FLOWERING PERIOD. July to August, with a peak in late August.

HISTORY. Material of *L. splendida* was collected for the first time by the German F. L. E. Diels (1874–1945), Professor of Botany at the University of Berlin and Director of the Berlin-Dahlem Botanic Garden. He found it near Vanrhynsdorp on 7th September 1900, during a short stay at the Cape on his way to Australia. Diels made a second collection two days later (*Diels* 517, in B) at

Figure 186. *Lachenalia splendida* flowering en masse near Vredendal. Image: Graham Duncan.

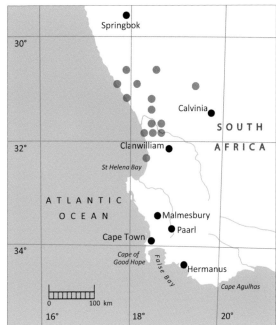

Map 69. Known distribution of *Lachenalia splendida*.

Figure 187. *Lachenalia splendida* on red clay flats near Vredendal. Image: Graham Duncan.

Ebenezer Farm near Vanrhynsdorp and the species was described by him in volume 44 of Engler's *Botanischer Jahrbücher fur Systematik Pflanzengeschichte und Pflanzengeographie* (Diels, 1909). Diels cited both specimens in his publication but did not state which of these he considered the holotype; his collection *Diels* 444 is designated here as the lectotype as it better represents the species, including one complete plant in flower and with its bulb attached. E. P. Phillips (1923) described *L. roodeae* in *The Flowering Plants of South Africa* from material collected at another locality near Vanrhynsdorp, but his species matches the earlier *L. splendida* and was reduced to synonymy (Barker, 1989).

DISTINGUISHING CHARACTERS AND AFFINITIES. Duncan *et al.* (2005a) considered *L. splendida* to be allied to *L. framesii*, which shares a spicate inflorescence with unmarked, bright green leaves and declinate stamens. Leaves of *L. framesii* differ from those of *L. splendida* in often being undulate, and *L. framesii* also differs markedly in having urceolate flowers with distinctive greenish-yellow outer tepals, magenta-tipped, recurved inner tepals, and included stamens.

DISTRIBUTION, HABITAT AND CONSERVATION STATUS. *L. splendida* extends from Garies south to Heerenlogement south-west of Vanrhynsdorp, with disjunct populations near Loeriesfontein in the western Karoo and at Elandsbaai on the Cape west coast (Map 69). It occurs in four vegetation types: Namaqualand Klipkoppe Shrubland, Knersvlakte Shale Vygieveld, Hantam Karoo and Leipoldtville Sand Fynbos (Mucina & Rutherford, 2006). This species is common around Vanrhynsdorp, Vredendal and Bitterfontein, and the plants are gregarious, usually occurring in large colonies numbering hundreds or sometimes thousands of individuals, on open flats, stony hilltops, at the base of rocky outcrops or along river banks, in seasonally moist clay or red sand. *L. splendida* is not threatened.

NOTES. The oblong-campanulate flowers of *L. splendida* are pollinated by honey-bees (*Apis mellifera*). The scapes detach from the bulbs shortly after capsule dehiscence and the infructescence is blown about, dispersing the seeds.

69. LACHENALIA NUTANS

Lachenalia nutans G. D. Duncan, *Bothalia* 28 (2): 131–132 (1998).
TYPE: Namibia, 2.5 km north-east of Schlafkuppe, on sandy gravel flats, *N. J. van Berkel* 563 (NBG!, holo.).

ETYMOLOGY. *nutans*: nodding flowers.
DESCRIPTION. *Geophyte*, 35–110 mm high. *Bulb* globose, 10–20 mm in diameter, solitary; tunic multilayered, outer layers reddish-brown, spongy; inner cataphyll translucent white, adhering to leaf bases, apex shape unknown. *Leaf* solitary, lanceolate, 30–60 × 4–20 mm, suberect, upper surface dark green, plain, with depressed longitudinal grooves, lower surface glaucous, plain or barred with transverse dark green bands below; margins dark maroon; leaf base clasping, 20–55 mm long, white, subterranean. *Inflorescence* racemose, erect, sterile apex short; scape erect, 25–40 mm long, light green, mottled with dull purplish-red, uppermost portion slightly to strongly inflated; rachis marked with dull purplish-red spots, lower half light green, slightly to strongly inflated, upper half white; pedicels erect to suberect at flowering, 2–5 mm long, strongly declinate in fruit; lower bracts ovate, becoming lanceolate above, 1.0–1.5 × 1–2 mm, white. *Perianth* zygomorphic, oblong-campanulate, strongly nodding; tube cup-shaped, 3 mm long, white; outer tepals ovate, 4–6 × 2.5 mm, white, base greenish, apical gibbosities yellowish- to brownish-green; inner tepals obovate, 4–6 × 3–4, white, apical zone light greenish-yellow. *Stamens* declinate, well exserted; filaments white, 6–10 mm long. *Ovary* obovoid, 2–3 mm long, light green; style well exserted, declinate, 7–9 mm long, white. *Capsule* obovoid, 4–6 × 4 mm. *Seed* globose, 1.4 × 1.3 mm, glossy, black; strophiole rudimentary, 0.2 mm long, ridged. *Chromosome number* unknown. Figure 188.
FLOWERING PERIOD. July to August.
HISTORY. In late August 1929, the German botanical collector M. K. Dinter (1868–1945) collected what he thought to be *L. klinghardtiana* at Haalenberg east of Lüderitz in south-western Namibia, and deposited pressed material (*Dinter* 6666) at B, BM, BOL, K, M and PRE. The specimens on the sheet at BOL, and those on one of the two sheets at B, were fruiting, and were not *L. klinghardtiana* but represented the first collection of a new species, *L. nutans*. In their account of *Lachenalia* for the *Prodromus einer flora von Südwestafrika*, Sölch & Roessler (1970) mistakenly listed Dinter's sheet 6666 of fruiting specimens in B (described much later as *L. nutans*) as *L. anguinea*. In July 1986, more than half a century after its discovery, Nicky van Berkel collected flowering material of *L. nutans* further south at a locality north-east of Schlafkuppe. The species was described in *Bothalia* with the van Berkel material serving as the holotype (Duncan, 1998a). The year 2001 was a favourable one for *L. nutans* as three records by different collectors were made, all within the diamond mining area north of the Klinghardt Mountains and south-east of Lüderitz; these specimens are housed in the Windhoek Herbarium, Namibia.
DISTINGUISHING CHARACTERS AND AFFINITIES. Duncan *et al.* (2005a) treated *L. nutans* as sister to *L. xerophila*, the two species sharing oblong-campanulate flowers with well-exserted, declinate stamens and a solitary lanceolate leaf. *L. xerophila* differs in its more numerous, slightly cernuous flowers, with the inner tepals distinctly longer than the outer ones. The outer tepals of *L. xerophila* have large, dark

THE GENUS LACHENALIA
69. LACHENALIA NUTANS

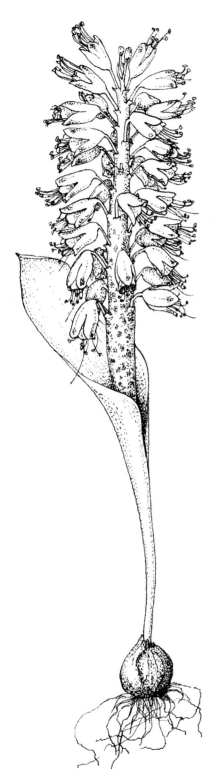

brown apical gibbosities and the inner tepals have broad, dark brown keels. *L. xerophila* also differs in its longer filaments (8–15 mm), a more strongly inflated scape, ellipsoid capsules and an oblong seed.

DISTRIBUTION, HABITAT AND CONSERVATION STATUS. *L. nutans* is confined to south-western Namibia in the area to the east and south-east of Lüderitz, extending from Haalenberg in the north to the Klinghardt Mountains in the south (Map 70). The plants grow in Desert and Succulent Steppe vegetation in colonies on lower hill slopes and red sandy or gravelly quartzite plains in full sun. They typically have deep-seated bulbs occurring up to 100 mm deep. *L. nutans* is not threatened but has a conservation status of Rare (Loots, 2005).

NOTES. Bulbs of *L. nutans* are adapted to remain dormant for many years until rains that are adequate to support vigorous growth have fallen. Pollinating agents are as yet unknown but the oblong-campanulate flower shape of *L. nutans* suggests bee pollination. The capsules become strongly nodding during the late fruiting stage and, as with other *Lachenalia* species that have inflated scapes, the base of the scape probably breaks off shortly after capsule dehiscence, allowing the whole infructescence to be taken away by wind and thus the seeds to be dispersed over a wide area.

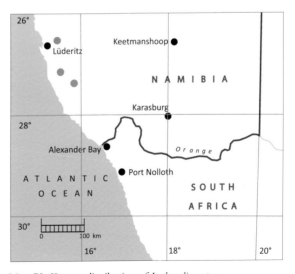

Map 70. Known distribution of *Lachenalia nutans*.

Figure 188. *Lachenalia nutans* from Schlafkuppe, south-western Namibia (*van Berkel* 563, in NBG), life size. Drawn by Vicki Thomas.

70. LACHENALIA KLINGHARDTIANA

Lachenalia klinghardtiana Dinter, *Feddes Repertorium* 16: 341–342 (1920).
TYPE: Namibia, Klinghardt Mountains, *F. Schäfer* 554 (B!, holo.).

ETYMOLOGY. *klinghardtiana*: after the Klinghardt Mountain range in south-western Namibia, where the holotype collection was made.

DESCRIPTION. *Geophyte*, 80–300 mm high. *Bulb* globose, 10–20 mm in diameter, offset-forming; tunic multilayered, outer layers light to dark brown, spongy; inner cataphyll translucent white to light green, loosely surrounding leaf base, apex obtuse. *Leaves* usually 1, rarely 2, broadly lanceolate, 80–150 × 15–50 mm, spreading to suberect or arcuate, glaucous, leathery, flat to deeply canaliculate, upper surface plain or marked with sporadic green or brown spots, lower surface plain or heavily blotched with dark magenta in lower third; leaf margins flat to undulate and crisped, thickened, dull maroon; leaf base clasping, 15–45 mm long, greenish-white at base, shading to glaucous above, plain or with small to large, light to deep magenta blotches; primary seedling leaf terete, erect. *Inflorescence* racemose, few- to many-flowered, fairly dense, sterile apex short; scape erect to suberect, 50–180 mm long, light green to glaucous, plain or marked with maroon spots or large blotches, upper half slightly to conspicuously swollen; rachis light green to glaucous, plain or marked with small to large, light maroon or purple blotches, slightly to conspicuously swollen in lower half; pedicels suberect, 1–7 mm long, white to light green, lengthening towards inflorescence apex; lower bracts ovate,

Figure 189. *Lachenalia klinghardtiana* on a stony hillside near Alexander Bay. Image: Pieter van Wyk.

becoming lanceolate above, 1–4 × 2–5 mm, translucent white. *Perianth* zygomorphic, oblong-campanulate, spreading to cernuous, ageing to dull red; tube shallow cup-shaped, 1 mm long, white, cream, light green or blue; outer tepals ovate, 6–8 × 3–4 mm, translucent white, cream, greenish-yellow to light green, apical gibbosities dark green, brownish-green or purple; inner tepals obovate, 7–9 × 4–5 mm, translucent white, cream or light green, median keels dark green to brown, apices spreading. *Stamens* declinate, well-exserted; filaments white, 10–11 mm. *Ovary* ellipsoid, 2–4 × 2–3 mm, bright green, plain or with purple blotches; style exserted, declinate, 10–12 mm long, white. *Capsule* ellipsoid, 9–10 × 7–8 mm. *Seed* globose, 1.2–1.3 × 1.2–1.3 mm, glossy, black; strophiole 0.5–0.6 mm long, ridged. *Chromosome number* 2n = 14 (Spies *et al.*, 2008). Figures 189, 190.

FLOWERING PERIOD. June to July, with a peak in early July.

HISTORY. The German Dr Fritz Schäfer, who was medical practitioner on the Lüderitz-Keetmanshoop Railway in Namibia, first collected this species on 14th August 1913 in the Klinghardt Mountains. It was described by his German compatriot and botanical explorer M. K. Dinter (1868–1945) in *Feddes Repertorium*, and the holotype sheet is housed in the Botanical Museum of Berlin-Dahlem (Dinter, 1920). *L. klinghardtiana* was next collected by Dinter on 30th August 1929 at Haalenberg, some distance to the north. Subsequently, the species has been infrequently collected, probably because of its desolate habitat and the relatively early appearance of the flowers in mid-winter.

DISTINGUISHING CHARACTERS AND AFFINITIES. Phylogenetic analysis of morphological data (Duncan *et al.*, 2005a) suggests an alliance of *L. klinghardtiana* with *L. dasybotrya*, which shares a many-flowered raceme of white, oblong-campanulate flowers that have declinate stamens. *L. dasybotrya* differs in always having two spreading, flat, light green leaves with subterranean bases and dense inflorescences. Its perianth tubes are longer (3 mm) than those of *L. klinghardtiana* and its stamens are included or shortly exserted.

Map 71. Known distribution of *Lachenalia klinghardtiana*.

Figure 190. *Lachenalia klinghardtiana* on a sandy slope near Alexander Bay. Image: Pieter van Wyk.

DISTRIBUTION, HABITAT AND CONSERVATION STATUS. *L. klinghardtiana* stretches from north and east of Lüderitz in Namibia to the Holgat River in the southern Richtersveld, South Africa (Map 71). It is fairly common east of Alexander Bay in western Richtersveld, growing as scattered individuals within larger colonies. *L. klinghardtiana* is encountered in three vegetation types in South Africa: Richtersveld Coastal Duneveld, Western Gariep Lowland Desert and Western Gariep Hills Desert (Mucina & Rutherford, 2006) and in south-western Namibia it occurs in Desert and Succulent Steppe vegetation (Loots, 2005). It occurs on exposed sandy plains and lower dune slopes in deep red sand, on stony quartz flats, and in crevices of rocky hillsides in heavier soils derived from mica schist, often growing amongst low succulents. In common with several other lachenalias from arid south-western Namibia and the Richtersveld, *L. klinghardtiana* flowers relatively early, from early to mid-winter, and its bulbs enter dormancy early in the spring, well before the onset of intense summer heat. *L. klinghardtiana* is not threatened but in Namibia it has a conservation status of Rare (Loots, 2005).

NOTES. *L. klinghardtiana* is pollinated by honey bees (*Apis mellifera*). Its fruits develop rapidly and its scape detaches from the bulb shortly after capsule dehiscence, after which the infructescence is blown away and the seeds are widely dispersed. The bulbs are adapted to remain dormant during drought cycles.

71. LACHENALIA PHYSOCAULOS

Lachenalia physocaulos W. F. Barker, *Journal of South African Botany* 50 (4): 538–541 (1984).
TYPE: South Africa, Western Cape, Erfdeel Farm between Robertson and McGregor, *G. J. Lewis 5635* (NBG!, holo.).

ETYMOLOGY. *physocaulos*: inflated scapes.

DESCRIPTION. *Geophyte*, 100–300 mm high. *Bulb* globose, 10–30 mm in diameter, solitary; tunic multilayered, outer layers dark brown, spongy; inner cataphyll translucent white, adhering to leaf base, apex acute. *Leaves* 1–2, narrowly lanceolate, 120–180 × 10–20 mm, spreading to suberect, glaucous, upper surface canaliculate, lower surface dark maroon at base; leaf base divergent, 40–60 mm long, subterranean, widening abruptly, white with minute magenta speckles; leaf margins flat to slightly undulate; primary seedling leaf terete, suberect. *Inflorescence* racemose, many-flowered, dense, sterile apex short; scape erect, 60–180 mm long, upper portion strongly inflated, light green, upper two-thirds densely marked with dark maroon spots and blotches, shading to light magenta in lower third; rachis erect, light green, inflated at base, densely marked with dull maroon blotches; pedicels suberect, 1 mm long, white; lower bracts ovate, becoming lanceolate above, 1–2 × 1–4 mm, white. *Perianth* zygomorphic, oblong-campanulate, spreading to slightly cernuous; tube cup-shaped, 2 mm long, light blue; outer tepals ovate, 5–6 × 3–4 mm, light blue at base shading to white, apical gibbosities greenish-brown to brownish-magenta; inner tepals obovate, 6–8 × 4 mm, light pinkish-magenta, margins white, median keels dark magenta. *Stamens* well exserted, declinate; filaments white in lower half, shading to magenta above, 9–13 mm long. *Ovary* ellipsoid, 3–4 × 2 mm, style well exserted, declinate, 9–11 mm, white in lower half, shading to light magenta in upper half, apex white. *Capsule* ellipsoid, 5–7 × 4–5 mm. *Seed* globose, 0.9 × 0.9 mm, glossy, black; strophiole inflated, 0.5 mm long, smooth. *Chromosome number* $2n = 14$ (Spies *et al.*, 2008). Figures 191, 192.

FLOWERING PERIOD. August to September, with a peak in mid-September.

HISTORY. This beautiful and unusual species was collected from two different locations between Robertson and Swellendam in the south-western Cape in September 1962: the first collection by the

THE GENUS LACHENALIA
71. LACHENALIA PHYSOCAULOS

Figure 191 (left). *Lachenalia physocaulos* with divergent spreading leaf (at left) in deep sand near Robertson. Image: Graham Duncan.

Figure 192 (right). Cultivated specimen of *Lachenalia physocaulos* from Robertson. Image: Graham Duncan.

Cape Town botanist Dr G. J. Lewis (*Lewis* 5635, in NBG), the other by Liebenberg (*Liebenberg* 6467, in NBG). It was described more than 20 years later in the *Journal of South African Botany* (Barker, 1984) and has since been occasionally collected only in these two areas.

DISTINGUISHING CHARACTERS AND AFFINITIES. Duncan *et al.* (2005a) considered *L. physocaulos* to be sister to *L. leipoldtii*, the species sharing shortly pedicellate racemes of oblong-campanulate flowers that have well-exserted, declinate stamens and slightly to strongly inflated scapes. *L. leipoldtii* differs in its smaller, light creamy-white or yellowish-white flowers with white filaments. It has light green leathery leaves with purplish-magenta-spotted lower surfaces and distinctive coriaceous leaf margins, and its leaf bases clasp the scape along their entire length. By contrast, the highly divergent leaf bases of *L. physocaulos* are subterranean and emerge at a distance from the scape. The two species occur in close proximity in the Little Karoo north of Montagu, but *L. leipoldtii* has a wider distribution that extends from the Olifants River Valley to Karoopoort, and east to a locality between Touws Rivier and Montagu (Duncan, 1998a).

DISTRIBUTION, HABITAT AND CONSERVATION STATUS. *L. physocaulos* is recorded from Erfdeel Farm between Robertson and McGregor, and from the Bontebok National Park near Swellendam (Map 72). The species grows as solitary individuals within larger colonies on south-facing lower slopes and flats in deep brownish-white sand. It occurs in Robertson Karoo and Swellendam Silcrete Fynbos vegetation (Mucina & Rutherford, 2006), sometimes in transition vegetation between renosterveld and fynbos. The farm populations have been largely extirpated by vineyard expansion and no plants were recorded there in September 2007. *L. physocaulos* has a conservation status of Endangered (Raimondo & Duncan, 2009d).

NOTES. The flowers are pollinated by honey bees (*Apis mellifera*). Shortly after capsule dehiscence, the scape detaches from the bulb and the infructescence is blown about, during which process the seeds are dispersed.

Map 72. Known distribution of *Lachenalia physocaulos*.

72. LACHENALIA LEIPOLDTII

Lachenalia leipoldtii G. D. Duncan, *Bothalia* 28 (2): 137–139 (1998).
TYPE: South Africa, Western Cape, 4.8 km north of Citrusdal, *T. M. Salter* 3608 (BOL!, holo.; BM!, iso.).

ETYMOLOGY. *leipoldtii*: after Dr C. Louis Leipoldt (1880–1947), celebrated South African naturalist, physician and poet, whose interest in lachenalias from the Clanwilliam-Citrusdal area is reflected in his famous Afrikaans poem 'Oktobermaand'.

DESCRIPTION. *Geophyte*, 60–280 mm high. *Bulb* globose, 10–15 mm in diameter, solitary; tunic multilayered, outer layers dark brown, spongy; inner cataphyll translucent white, adhering to leaf bases, apex acute. *Leaves* 1–2, lanceolate, 30–120 × 3–20 mm, spreading to suberect, yellowish- to bright green, flat to canaliculate, leathery, upper and lower surfaces plain or sporadically spotted or blotched with dark green or purplish-magenta; margins cartilaginous, flat to slightly undulate; leaf bases clasping, 10–20 mm long, subterranean portion white, aerial portion yellowish-green with purplish-magenta spots or blotches; primary seedling leaf terete, suberect. *Inflorescence* racemose, many-flowered, erect to suberect, sterile apex short to long; scape erect to suberect, 45–100 mm long, light green with dark magenta blotches, slightly to conspicuously inflated in upper half; pedicels suberect, 1–2 mm long, white to light green; lower bracts ovate, becoming lanceolate above, 2–3 × 1–2 mm, white. *Perianth* zygomorphic, oblong-campanulate, spreading to slightly cernuous; tube cup-shaped, 1–2 mm long, light blue; outer tepals ovate, 4–6 × 3–4 mm, creamy-white, apical gibbosities and median keels greenish- or pinkish-brown; inner tepals obovate, 6–7 × 4–5 mm, translucent creamy-white. *Stamens* well-exserted, declinate; filaments white, 8–10 mm long. *Ovary* ellipsoid, 2.0 × 1.5 mm, light green; style well exserted, declinate, 8–10 mm long, white. *Capsule* ellipsoid, 6–8 × 4–5 mm. *Seed* globose, 0.9–1.0 × 0.9 mm, glossy, black; strophiole inflated, 0.5 mm long, smooth. *Chromosome number* unknown. Figure 193.

THE GENUS LACHENALIA
72. LACHENALIA LEIPOLDTII

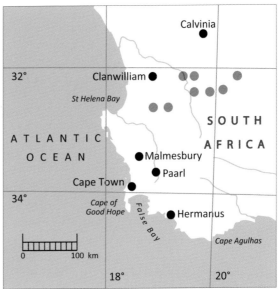

Map 73. Known distribution of *Lachenalia leipoldtii*.

Figure 193. Cultivated specimen of *Lachenalia leipoldtii* from De Doorns. Image: Graham Duncan/SANBI.

FLOWERING PERIOD. August to September with a peak in late August.

HISTORY. The German botanical collector C. L. Zeyher was the first to collect this species, at Brakfontein Farm near Citrusdal in the Olifants River Valley, in 1831 (Gunn & Codd, 1981). The species was described more than 150 years later in *Bothalia* (Duncan, 1998a). Dr C. Louis Leipoldt collected *L. leipoldtii* at Piekenierskloof in 1931 and it has since been recorded from a number of inland localities in the Western Cape. The holotype collection, made by T. M. Salter just north of Citrusdal in September 1933, is housed in the Bolus Herbarium at the University of Cape Town, with an isotype in the British Museum (Natural History).

DISTINGUISHING CHARACTERS AND AFFINITIES. *L. leipoldtii* appears to be the sister species to *L. physocaulos* (Duncan *et al.*, 2005a), and the relationship is discussed under that species.

DISTRIBUTION, HABITAT AND CONSERVATION STATUS. Confined to the Western Cape, *L. leipoldtii* is recorded from the Olifants River Valley to the southern Tanqua Karoo and the western edge of the Little Karoo, extending from just west of Citrusdal to between Touwsrivier and Montagu (Map 73). The plants occur in Cederberg Sandstone Fynbos and Breede Alluvium Fynbos, and Western Little Karoo vegetation types (Mucina & Rutherford, 2006), growing in colonies on sandy flats in full sun and on south-facing inclines. The species is threatened by vineyard expansion and habitat degradation through trampling by livestock near De Doorns in the Hex River Valley. It has a conservation status of Vulnerable (Harrower & Raimondo, 2009).

NOTES. Pollinators have yet to be recorded for *L. leipoldtii* but will probably include honey bees (*Apis mellifera*). The scapes detach from the bulbs shortly after capsule dehiscence and the seeds are dispersed from the infructescences as they are blown away.

73. LACHENALIA XEROPHILA

Lachenalia xerophila Schltr. ex G. D. Duncan, *Bothalia* 27 (1): 14–15 (1997).
TYPE: South Africa, Northern Cape, Kouberg Farm, off R355 from Springbok to Gamoep, western Bushmanland, *M. C. Botha s.n.* sub. NBG 95451 (NBG!, holo.).

ETYMOLOGY. *xerophila*: growing in dry places.
DESCRIPTION. *Geophyte*, 100–370 mm high. *Bulb* globose, 15–25 mm in diameter, solitary, deep-seated (up to 70 mm); tunic multilayered, outer layers dark brown, spongy; inner cataphyll translucent white, subterranean, adhering to leaf base, apex acute. *Leaves* 1–2, lanceolate, 85–220 × 16–45 mm, spreading to suberect or erect, deeply canaliculate, glaucous; margins coriaceous, dull maroon, flat to slightly undulate and crisped; leaf base clasping, 45–70 mm long, shortly aerial, white below shading to glaucous above; primary seedling leaf terete, erect. *Inflorescence* racemose, erect to suberect, dense, sterile apex short; scape erect, 30–150 mm long, light green to glaucous, lower portion plain or flushed with dull purple, upper portion flushed with brownish-maroon, distinctly swollen; rachis light green, 70–200 mm long, flushed with brownish-maroon; pedicels suberect, 1–5 mm long, white, short at base of inflorescence, lengthening towards apex; lower bracts ovate, becoming lanceolate above, 0.5–1 × 1–2 mm, translucent white. *Perianth* zygomorphic, oblong-campanulate, cernuous, faintly spice-scented; tube cup-shaped, 1–2 mm long, white or dull blue; outer tepals ovate, 5–7 × 3–5 mm, white to light mauve, apical gibbosities dark brown or mauve, prominent; inner tepals obovate, 7–9 × 4–6 mm, spreading, translucent white or light mauve, median keels dark brown or mauve, broad. *Stamens* well exserted, declinate; filaments white, 8–15 mm long. *Ovary* ellipsoid, 2–3 × 2 mm, dull green; style well exserted, straight, 7–16 mm long, white. *Capsule* ellipsoid, 8–9 × 5–6 mm. *Seed* oblong, 1.8–1.9 × 0.9–1.0 mm, glossy, black; strophiole rudimentary, 0.3 mm long; raphe inflated. *Chromosome number* unknown. Figure 194.

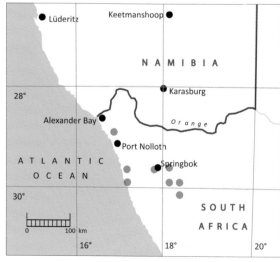

Map 74. Known distribution of *Lachenalia xerophila*.

Figure 194. *Lachenalia xerophila* in alluvial sand near Gamoep. Image: Graham Duncan.

FLOWERING PERIOD. August to September, with a peak in early August.

HISTORY. Rudolf Schlechter (1872–1925) was the first to collect material of *L. xerophila*, which he discovered at Leeuwpoort north of Concordia in central Namaqualand on 21st September 1897. He distributed pressed material to seven local and foreign herbaria under his manuscript name *L. xerophila* Schltr. (Barker, 1983a), but these specimens remained undescribed for just over a century before finally being published in *Bothalia* (Duncan, 1997). The collection that serves as the holotype was collected by Prof. M. C. Botha at Kouberg Farm on the road between Springbok and Gamoep in western Bushmanland in August 1972. The species has since been collected on several occasions in the southern Richtersveld, Namaqualand and western Bushmanland.

DISTINGUISHING CHARACTERS AND AFFINITIES. Phylogenetic analysis of morphological data (Duncan *et al.*, 2005a) suggests that *L. nutans* is sister to *L. xerophila*. The species share oblong-campanulate, white flowers with well-exserted, declinate stamens, a deep-seated, globose bulb and glaucous leaves. *L. nutans* differs in its strongly cernuous flowers with the inner tepals being scarcely longer than the outer ones, pedicels that bend downwards during the fruiting stage, and obovoid capsules.

DISTRIBUTION, HABITAT AND CONSERVATION STATUS. *L. xerophila* is confined to the dry southern Richtersveld, western and central parts of Namaqualand and western Bushmanland, extending from the Holgat River west of Eksteenfontein south-east to Varsputs Farm in western Bushmanland (Map 74). It is frequent in the area around Kleinsee and Gamoep, and in seasons of excellent rainfall it is seen in large populations sometimes numbering hundreds of individuals. The plants favour deep red gravelly sand and are frequent in open aspects on flats and along the edges of watercourses; they are often encountered within the protection of grass tufts or low bushes. *L. xerophila* occurs in five vegetation types: Richtersveld- and Namaqualand Coastal Duneveld, Namaqualand Klipkoppe and Platbakkies Succulent Shrubland, and Bushmanland Arid Grassland vegetation (Mucina & Rutherford, 2006). At Varsputs Farm east of Springbok, *L. xerophila* grows in association with *L. polypodantha* subsp. *polypodantha*, and at Rietfontein Farm near Gamoep, it grows with *L. inconspicua* and *L. polypodantha* subsp. *eburnea*. *L. xerophila* is not threatened.

NOTES. The oblong-campanulate flowers of *L. xerophila* are pollinated by honey bees (*Apis mellifera*). The scapes detach from the bulbs shortly after capsule dehiscence and the ripe infructescence is blown away, scattering the seeds. The leaf fungus *Uromyces lachenaliae* infests the leaves in winter, both in the wild and in cultivation.

74. LACHENALIA VIOLACEA

Lachenalia violacea Jacq., *Collectanea* 4: 147–148 (1791). *Scillopsis violacea* (Jacq.) Lem., *L'Illustration Horticole* 3: 35 (1856).

TYPE: South Africa, Cape, collector and precise locality unknown, figure in N. J. Jacquin, *Icones Plantarum Rariorum* 2: 12, t. 394 (1792 or 1793) (neotype, designated here).

ETYMOLOGY. *violacea*: violet, with reference to the inner tepal apices.

DESCRIPTION. *Geophyte*, 150–400 mm high. *Bulb* ovoid, 15–30 mm in diameter, offset-forming; tunic multilayered, outer layers dark brown, spongy; inner cataphyll translucent white, loosely surrounding leaf bases, apex acute. *Leaves* 1–2, broadly lanceolate, 120–270 × 20–50 mm, spreading to suberect, canaliculate to conduplicate, glaucous to light green, upper surface plain or with sporadic dark green blotches, lower surface plain or sporadically to densely marked with green or maroon spots; leaf bases clasping, 50–70 mm long, light green to glaucous, plain or lightly to heavily marked with

Plate 23. Watercolour painting of *Lachenalia violacea* from north of Clanwilliam (*Duncan* 190, in NBG) courtesy of The Editor, *Curtis's Botanical Magazine* vol. 16, t. 373 (1999). Artist: Claire Linder Smith.

magenta bands; leaf margins moderately to strongly coriaceous, green to brownish-maroon, slightly to strongly undulate and/or crisped; primary seedling leaf terete, erect. *Inflorescence* racemose, many-flowered, dense, sterile apex short; scape erect to suberect, 150–200 mm long, light green, plain or heavily marked with dark purplish-brown blotches, upper portion slightly to strongly inflated; rachis light green, plain or heavily mottled with small to large, light to dark brownish-maroon blotches, lowermost portion inflated; pedicels spreading to suberect, 3–9 mm long, white, light green or dull purplish-brown; lower bracts ovate, becoming lanceolate above, 1–4 × 1–3 mm, white to green. *Perianth* zygomorphic, oblong-campanulate, slightly to strongly cernuous, spicy sweet-scented; tube cup-shaped, 1–3 mm long, turquoise to dull blue; outer tepals ovate, 5–7 × 4–5 mm, light green to bluish-green, apical gibbosities purple, green or brown; inner tepals obovate, 6–7 × 4–5 mm, bluish-green, apices or upper halves dark purple, margins flat to slightly recurved. *Stamens* well exserted; filaments straight, dark purple in upper two thirds, white below, 10–13 mm long. *Ovary* ellipsoid, 3–5 × 3–4 mm, light green; style well exserted, straight, 9–12 mm long, bright purple in upper two thirds, white below. *Capsule* ellipsoid, 5–9 × 4–7 mm. *Seed* globose, glossy, black, 1.2–1.4 × 1.3–1.4 mm; strophiole inflated, 0.7–0.8 mm long, smooth. *Chromosome number* 2n = 14 (Ornduff & Watters, 1978; Johnson & Brandham, 1997; Hamatani *et al.*, 1998). Plate 23, Figures 195–197.

FLOWERING PERIOD. August to October, with a peak in mid-September.

HISTORY. *L. violacea* was introduced into cultivation at Kew in 1795 (Aiton, 1811). It was described by N. J. Jacquin (1791) in volume 4 of *Collectanea* and later illustrated in the second volume of his *Icones Plantarum Rariorum* (1792–1793). During a visit to the herbarium at Kew in 1937, W. F. Barker

Figure 195 (left). *Lachenalia violacea* on a granite outcrop near Garies. Image: Graham Duncan.

Figure 196 (right). *Lachenalia violacea* in deep sand in the Tanqua Karoo. Image: Graham Duncan.

examined the preserved specimen of *L. violacea* that had been sent there from Vienna specially on her account, together with many other Jacquin specimens used to depict the species in the *Icones*. The Vienna specimen proved to be an exact match of the painting of *L. violacea* in the *Icones*, but was tragically lost in a fire in Vienna between 1939–1945 (Duncan, 1999a). The painting was mistakenly designated as the lectotype for the species by Barker (Barker, 1989); it is here designated as a neotype as it was published after the description and was not referred to in the original protologue.

J. G. Baker (1871, 1897a) mistakenly placed *L. bicolor* Lodd. in synonymy under *L. violacea*; the former is correctly a synonym of *L. pallida* Aiton. Lemaire (1856) transferred *L. violacea* to *Scillopsis* as *S. violacea* (Jacq.) Lem. in his *L'Illustration Horticole*, but it was returned to *Lachenalia* by Baker (1897a). Two duplicate collections in K and BOL (*W. Morris* in Herb. Bol. 5804, collected near O'kiep in central Namaqualand, and *Bolus* 6591 from near Modderfontein in the Olifants River Valley), were mistakenly cited by Baker (1897a) under *L. unicolor*, and are correctly *L. violacea*. W. F. Barker (1989) described *L. violacea* var. *glauca* in the *South African Journal of Botany*, distinguishing it mainly by the colour of the tepals that 'all shade to pale magenta in the upper half, giving the inflorescence an overall glaucous appearance'; the taxon is here upgraded to species rank as it is morphologically distinct and because it does not form hybrid swarms at the contact zones with *L. violacea*, and is thus in a different gene pool.

DISTINGUISHING CHARACTERS AND AFFINITIES. *L. violacea* is a distinctive but highly polymorphic species. Duncan *et al.* (2005a) treated *L. glauca* as sister in a phylogenetic analysis of morphology. Like *L. violacea*, *L. glauca* has cernuous, oblong-campanulate perianths with well exserted, straight stamens

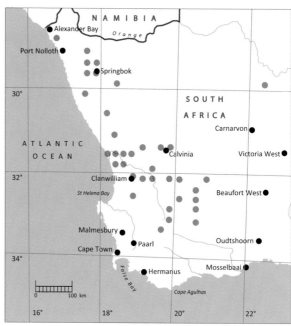

Map 75. Known distribution of *Lachenalia violacea*.

Figure 197. Cultivated specimens of *Lachenalia violacea* from the Nardousberg. Image: Graham Duncan.

and inflated scapes; but *L. glauca* differs from *L. violacea* in its shorter perianth tube (1 mm long), greyish-mauve tepals, white or greyish-mauve filaments, heavy coconut scent, unmarked, usually narrower leaves and smaller globose seeds (0.9–1.0 × 0.9–1.0 mm), which have shorter inflated strophioles (0.5–0.6 mm long).

DISTRIBUTION, HABITAT AND CONSERVATION STATUS. *L. violacea* extends from the Holgat River in the southern Richtersveld to Whitehill Station east of Matjiesfontein in the southern Great Karoo. A collection at Prieska in the northern Great Karoo is a wide disjunction (Map 75). It occurs in 15 vegetation types including Southern Richtersveld Yellow Duneveld, Namaqualand Blomveld, Namaqualand Shale Shrubland, Namaqualand Heuweltjieveld, Namaqualand Klipkoppe Shrubland, Spinescent Grassland, Lower Gariep Broken Veld, Vanrhynsdorp Gannabosveld, Bokkeveld Sandstone Fynbos, Nieuwoudtville-Roggeveld Dolerite Renosterveld, Leipoldtville Sand Fynbos, Agter-Sederberg Shrubland, and Hantam-, Roggeveld- and Tanqua Karoo (Mucina & Rutherford, 2006). Its habitat is very varied and mostly arid, including sandy or stony red clay flats, crevices of granite outcrops or south-facing mountain slopes. The plants grow singly, in small groups or in colonies amongst karroid or succulent scrub, and within restio tufts. *L. violacea* has a long flowering period; an early-flowering form in the Tanqua Karoo National Park flowers in mid-August, another at Bitterfontein flowers in mid-September and a late-flowering form near Clanwilliam extends its flowering into early October. *L. violacea* is sympatric with *L. glauca* to the north and west of Springbok, with *L. aurioliae* in the central Tanqua Karoo, and with *L. martiniae* at Clanwilliam. *L. violacea* is not threatened.

NOTES. *L. violacea* is pollinated by honey bees (*Apis mellifera*). Following capsule dehiscence, the scapes remain attached to the bulbs for several weeks and the seeds are dispersed locally.

75. LACHENALIA GLAUCA

Lachenalia glauca (W. F. Barker) G. D. Duncan, stat. nov. *Lachenalia violacea* Jacq. var. *glauca* W. F. Barker, *South African Journal of Botany* 55 (6): 639–640 (1989).
TYPE: South Africa, Northern Cape, Grootvlei, west of Kamieskroon, *J. P. Rourke* 812 (NBG!, holo.).

ETYMOLOGY. *glauca*: greyish-mauve, with reference to the colour of the outer and inner tepals.
DESCRIPTION. *Geophyte*, 150–350 mm high. *Bulb* ovoid, 15–25 mm in diameter, offset-forming; tunic multilayered, outer layers dark brown, spongy; inner cataphyll translucent white, loosely surrounding leaf bases, apex acute. *Leaves* 1–2, narrowly to broadly lanceolate, 100–250 × 20–50 mm, suberect, glaucous to light green, plain, leaf bases clasping, 50–70 mm long, light green to glaucous, plain or tinged with dull maroon; leaf margins slightly to moderately coriaceous, flat or slightly to moderately undulate; primary seedling leaf terete, erect. *Inflorescence* racemose, many-flowered, dense, sterile apex short; scape erect to suberect, 150–200 mm long, light green, upper portion slightly to moderately inflated; rachis light green, plain, lowermost portion slightly inflated; pedicels spreading to suberect, 4–8 mm long, white to light green; lower bracts ovate, becoming lanceolate above, 1–3 × 1–5 mm. *Perianth* zygomorphic, oblong-campanulate, slightly to strongly cernuous, heavily coconut-scented; tube cup-shaped, 1 mm long, dull blue or light to dark greyish-mauve; outer tepals ovate, 6–8 × 4–5 mm, greyish-mauve, apical gibbosities green or brown; inner tepals obovate, 7–8 × 5–6 mm, light to dark greyish-mauve, apical margins flat to slightly recurved, plain or white. *Stamens* well exserted; filaments straight, white, or light mauve in upper two thirds, white below, 10–11 mm long. *Ovary*

PLATE 24. Watercolour painting of *Lachenalia glauca* from Kamieskroon (*Struben s.n.*, in BOL) courtesy of the Compton Herbarium, Kirstenbosch, South African National Biodiversity Institute. Artist: Winsome Barker.

Figure 198 (left). *Lachenalia glauca* in Namaqualand Blomveld vegetation near Springbok. Image: Graham Duncan.
Figure 199 (right). Cultivated specimens of *Lachenalia glauca* from near Hondeklipbaai. Image: Graham Duncan.

ellipsoid, 2–3 × 2–3 mm, light green; style well exserted, straight, 9–10 mm long, white, or light mauve in upper two thirds, white below. *Capsule* ellipsoid, 5–7 × 4–6 mm. *Seed* globose, glossy, black, 0.9–1.0 × 0.9–1.0 mm; strophiole inflated, 0.5–0.6 mm long, smooth. *Chromosome number* unknown. Plate 24, Figures 198, 199.

FLOWERING PERIOD. September to October, with a peak in late September.

HISTORY. *L. glauca* was first collected in September 1925 by the University of Cape Town botanist Dr M. R. Levyns (1890–1975) at Zeekoe Vlei near Clanwilliam (*Levyns* 1177, in BOL). More than 60 years elapsed before it was finally described as *L. violacea* var. *glauca* (Barker, 1989). The taxon is morphologically distinct from *L. violacea* and is here upgraded to species level.

DISTINGUISHING CHARACTERS AND AFFINITIES. In a phylogenetic analysis of unweighted morphological data (Duncan *et al.*, 2005a), *L. glauca* resolved as sister to *L. violacea*, and the distinguishing characteristics are discussed under that species.

DISTRIBUTION, HABITAT AND CONSERVATION STATUS. *L. glauca* has a limited range from just north of Springbok to Spektakel Mountain west of this town and south-east to Kamieskroon. A collection from Zeekoe Vlei near Clanwilliam is a wide disjunction (Map 76). It occurs singly or in small to large colonies sometimes numbering thousands of individuals, and is common on the Kamiesberg Pass and at Skilpad west of Kamieskroon. The species traverses six vegetation types including Namaqualand Blomveld, Namaqualand Klipkoppe Shrubland, Granite Renosterveld, Kamiesberg Granite

Fynbos, Kamiesberg Mountains Shrubland and Graafwater Sandstone Fynbos (Mucina & Rutherford, 2006). On the Kamiesberg and to the north and west of Springbok, it is sympatric with *L. violacea*, but the two species always occur in discrete populations and no hybrids are known. *L. glauca* grows in mixed populations with *L. namaquensis* on Spektakel Mountain west of Springbok.

Plants grow on flats, granite outcrops or mountainsides in sandy granitic soils or sandy clay. They generally favour substrates that are wetter than those preferred by their close relative *L. violacea* and they commence flowering later than *L. violacea*, from mid-September to early October. *L. glauca* is not threatened.

NOTES. The species is pollinated by honey bees (*Apis mellifera*). In years of excellent rainfall, specimens are robust, producing dense, many-flowered racemes of up to 350 mm high. Following capsule dehiscence, the scapes remain attached for several weeks and the seeds are dispersed locally by the shaking action of wind.

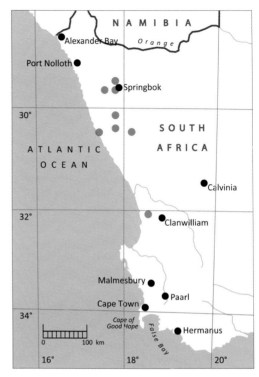

Map 76. Known distribution of *Lachenalia glauca*.

76. LACHENALIA WHITEHILLENSIS

Lachenalia whitehillensis W. F. Barker, *Journal of South African Botany* 49 (4): 432–434 (1983).
TYPE: South Africa, Western Cape, Whitehill, Laingsburg Division, *R. H. Compton* 14845 (NBG!, holo.).

ETYMOLOGY. *whitehillensis*: after Whitehill Railway Station near Matjiesfontein, where the type collection was made.

DESCRIPTION. *Geophyte*, 150–360 mm high. *Bulb* subglobose, 15–25 mm in diameter, solitary; tunic multilayered, outer layers dark brown, spongy; inner cataphyll translucent white, adhering to leaf bases, apex obtuse. *Leaves* 1–2, narrowly lanceolate, 70–140 × 10–20 mm, suberect, canaliculate, conduplicate, upper surface dark green to olive green with depressed longitudinal grooves, lower surface light green, heavily marked with dark green transverse bands; leaf margins flat to undulate and crisped in lower portion; leaf bases clasping, 40–80 mm long, white, aerial portion heavily banded with dark maroon shading to magenta below; primary seedling leaf terete, erect. *Inflorescence* racemose, many-flowered, sterile apex short; scape erect, 100–150 mm long, light green, heavily marked with dark maroon to dull purple blotches; rachis colouring as for scape; pedicels suberect, 4–6 mm long, light green to white, speckled light blue to purple; bracts ovate, 1–4 × 2–4 mm, light greenish-white. *Perianth* zygomorphic, oblong-campanulate, cernuous; tube cup-shaped, 1 mm long, light greenish-blue; outer tepals ovate, 5–7 × 4–5 mm, dull blue to light greenish-blue in lower half, shading to translucent white above, apical gibbosities dull reddish- to greenish-brown; inner tepals obovate, 6–7 × 5–6 mm, light greenish-blue in lower half, shading to translucent white above, median keels light blue, apical marking brownish-purple. *Stamens* well exserted, declinate; filaments white, 10–17 mm long. *Ovary* obovoid, 2–3 × 2 mm, light green; style well exserted, declinate, 8–10 mm long, white. *Capsule* obovoid, 7–8 × 5–6 mm. *Seed* ovoid, matte, black, 1.4 × 1.3–1.4 mm; strophiole 0.5–0.6 mm long, ridged. *Chromosome number* unknown. Figures 200, 201.

FLOWERING PERIOD. September to October, with a peak in late September.

HISTORY. *L. whitehillensis* was described from specimens collected by Prof. R. H. Compton at the original site of the Karoo National Botanical Garden at Whitehill Station east of Matjiesfontein (Barker, 1983b). The earliest known collection of *L. whitehillensis* is that of Dr Rudolf Marloth, who found it on Uitkyk Valley Farm west of Sutherland in October 1920. It has since been found by Gordon Summerfield to the north-west and south of Sutherland.

DISTINGUISHING CHARACTERS AND AFFINITIES. *L. whitehillensis* resolved adjacent to *L. attenuata* in a phylogenetic analysis of morphological data (Duncan *et al.*, 2005a); the species share light blue oblong-campanulate, cernuous flowers, declinate stamens and ovoid, matte black seeds, but *L. attenuata* differs mainly in having a broadly linear leaf, shorter, included to shortly exserted stamens (6–7 mm long) and seeds that have much longer strophioles (1.1 mm long).

DISTRIBUTION, HABITAT AND CONSERVATION STATUS. *L. whitehillensis* has a narrow range extending from Quaggasfontein Farm in the Roggeveld Mountains north-west of Sutherland to Matjiesfontein in the southern Great Karoo (Map 77). It occurs along river beds in deep alluvial reddish-yellow sand, in shallow sand on adjacent rocky outcrops, or sometimes among stones within river beds that

Figure 200 (left). *Lachenalia whitehillensis* in stony alluvial sand on the Komsberg. Image: Graham Duncan.
Figure 201 (right). Cultivated specimen of *Lachenalia whitehillensis* from Matjiesfontein. Image: Graham Duncan.

remain moist for short periods. Plants traverse Roggeveld Karoo and Matjiesfontein Shale Fynbos vegetation (Mucina & Rutherford, 2006). They are solitary or occur in small groups. In seasons of adequate rainfall, they can become robust, reaching up to 360 mm high. In the wild, the leaves are proteranthous, having partially to fully withered by the time the plants flower. North-west of Sutherland and on the Komsberg, *L. whitehillensis* grows in association with *L. canaliculata*; *L. comptonii* and *L. juncifolia* also occur at the latter locality. *L. whitehillensis* is not threatened.

NOTES. The oblong-campanulate flowers of *L. whitehillensis* are pollinated by honey bees (*Apis mellifera*). The scapes detach from the bulbs shortly after capsule dehiscence and the infructescences are blown away, scattering seed in the process. During periods of insufficient winter rainfall, the bulbs remain dormant.

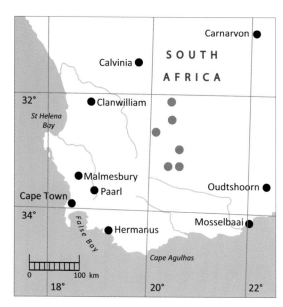

Map 77. Known distribution of *Lachenalia whitehillensis*.

77. LACHENALIA ZEBRINA

Lachenalia zebrina W. F. Barker, *Journal of South African Botany* 49 (4): 424–428 (1983).
TYPE: South Africa, Western Cape, Whitehill, *R. H. Compton* 17392 (NBG!, holo.; BOL!, iso.).
SYNONYMY: *Lachenalia zebrina* W. F. Barker forma *densiflora* W. F. Barker, *Journal of South African Botany* 49 (4): 426 (1983). Type: South Africa, Cape, 23 miles [37 km] N of Downes on Klipwerf Road, *M. L. Thomas s.n.* sub. NBG 105714 (NBG!, holo.).

ETYMOLOGY. *zebrina*: leaf markings reminiscent of a zebra's hind leg.
DESCRIPTION. *Geophyte*, 150–300 mm high. *Bulb* ovoid, 10–30 mm in diameter; tunic multilayered, outer layers dark brown, thick, spongy; inner cataphyll translucent white, adhering to leaf base, apex obtuse. *Leaf* solitary, lanceolate, 60–220 × 20–60 mm, weakly to strongly curved, glaucous, canaliculate, often conduplicate, upper surface plain, lower surface heavily marked with dark green blotches in upper two thirds, shading to maroon blotches in lower third; leaf margins flat to undulate and crisped; leaf base clasping, 40–70 mm long, silvery-white, heavily banded with maroon, shading to magenta; primary seedling leaf terete, erect. *Inflorescence* racemose, many-flowered, dense, lax to sturdy, sterile apex short; scape erect to suberect, 50–120 mm long, sturdy, light green to glaucous, heavily blotched with light to dark maroon; rachis light green to glaucous, heavily marked with dull maroon; pedicels suberect to arcuate, 2–20 mm long, shortest at base of inflorescence, lengthening towards apex, greenish-brown to purplish-grey; lower bracts ovate, becoming narrow-lanceolate above, 1–4 × 0.5–2.0 mm, brown at base, shading to white above. *Perianth* zygomorphic, oblong-campanulate, cernuous, creamy-green to light brown, ageing to dull maroon, foetid-scented; tube a flat disc 4–5 mm in diameter, outer tepals ovate, 4–6 × 4 mm, apical gibbosities greenish-brown, apices translucent white, slightly recurved; inner tepals obovate, 5–6 × 2 mm, median keels brownish-green, apices

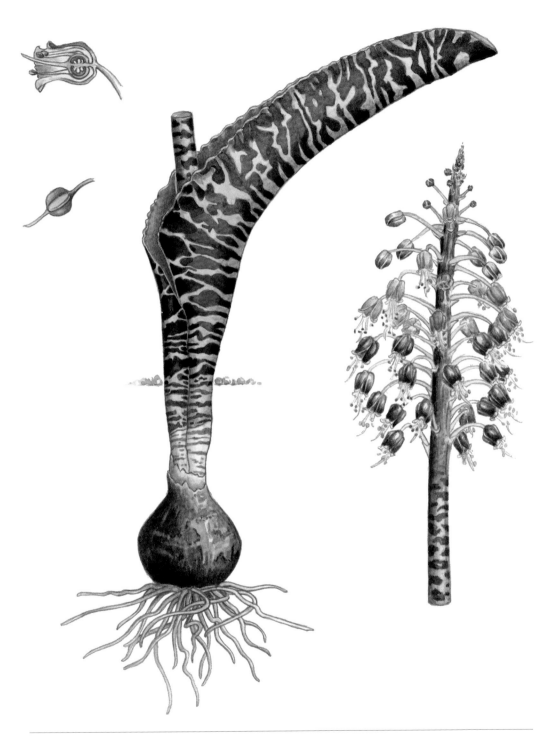

PLATE 25. Watercolour painting of *Lachenalia zebrina* from Calvinia (*Thomas* 272, in NBG), courtesy of The Editor, *Flowering Plants of Africa*, vol. 57, t. 2166 (2003). Artist: Marieta Visagie.

slightly recurved, translucent white. *Stamens* well exserted, straight; filaments white, 8–15 mm long. *Ovary* ellipsoid, 3–4 × 2 mm, light to bright green; style well exserted, straight, 5–8 mm long, white. *Capsule* broadly obcordate, 8–10 × 8–9 mm, broadly winged. *Seed* ovoid, 1.8–1.9 × 1.8–2.0 mm, matte, black; strophiole 0.5–0.6 mm long, ridged, micropyle 0.5 mm long, extruded. *Chromosome number* 2n = 30 (Johnson & Brandham, 1997; Spies *et al.*, 2008). Plate 25, Figures 41, 202, 203.

FLOWERING PERIOD. August to October, with a peak in early September.

HISTORY. A fruiting specimen collected in October 1905 at Grootfontein, one of several farms by this name in the Great Karoo (*Lamb s.n.* sub. BOL10030), appears to be the earliest recorded collection, but insufficient information exists to determine precisely which Grootfontein it was collected at. *L. zebrina* was long considered an inland form of *L. anguinea* (Barker, 1930b). Following an examination of the seed characters and other morphological traits from both coastal and inland populations, Barker described the inland taxon *L. zebrina*, recognising the forms from coastal areas as *L. anguinea*. *L. zebrina* was described in the *Journal of South African Botany*, the description was accompanied by a watercolour painting that was reproduced in monochrome (Barker, 1983b). Another watercolour of the species by Marieta Visagie was subsequently published in *Flowering Plants of Africa* (Duncan & Visagie, 2001).

DISTINGUISHING CHARACTERS AND AFFINITIES. *L. zebrina* is a very variable, but distinctive species. Barker (1983) designated the two extremes of the graduated morphological sequence of the species as forma *zebrina* and forma *densiflora*. She recognised forma *zebrina* for specimens with lax inflorescences

Figure 202 (left). *Lachenalia zebrina* on stony clay flats near Loeriesfontein. Image: Graham Duncan.

Figure 203 (right). Fruiting specimen of *Lachenalia zebrina* on a shale hillside near Middelpos. Image: Graham Duncan.

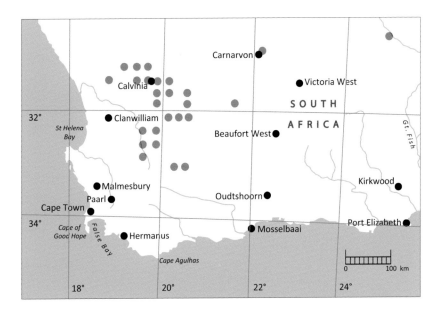

Map 78. Known distribution of *Lachenalia zebrina*.

and long pedicels, and forma *densiflora* for those having more numerous flowers that are borne on shorter pedicels, forming dense, elongated inflorescences. As intermediate forms occur between the two extremes, it is considered expedient to recognise a single, very variable taxon.

Duncan *et al.* (2005a) considered *L. zebrina* to be allied to *L. buchubergensis* and *L. nordenstamii*. These species share solitary lanceolate leaves, which are heavily banded on the lower surface, and broadly winged, obcordate capsules. *L. buchubergensis* differs from *L. zebrina* mainly in its tubular, spreading, light greenish-grey flowers, the inner tepals with deep magenta apices, with included stamens and a much longer, oblong seed (3.0–3.5 × 1.0–1.1 mm), whereas *L. nordenstamii* differs in its few-flowered inflorescence of widely campanulate, deep maroon flowers that have narrowly spreading, stout filaments and that produce oblong seeds (2.5 × 0.8–0.9 mm). *L. buchubergensis* and *L. nordenstamii* also differ from *L. zebrina* in being dwarf species. Both have a much more northerly distribution in the north-western Richtersveld in South Africa and in the south-western corner of Namibia.

DISTRIBUTION, HABITAT AND CONSERVATION STATUS. *L. zebrina* is widely distributed in the Western Karoo extending from Loeriesfontein south to the Tanqua Karoo. It extends further south-east to Grootvlakte Farm in the western Little Karoo and inland to Laingsburg and Fraserburg, with wide disjunctions at Carnarvon in the central Karoo, and Bell Rock Farm west of Colesberg (Map 78). *L. zebrina* occurs in nine vegetation types: Knersvlakte Shale Vygieveld, Roggeveld Shale Renosterveld, Hantam-, Roggeveld-, Tanqua-, Koedoesberge-Moordenaars-, Western Upper-, Eastern Upper- and Northern Upper Karoo (Mucina & Rutherford, 2006). It is common in the Calvinia and Loeriesfontein districts. On flats of the Tanqua Karoo, *L. zebrina* occurs in exceptionally harsh conditions in hard, pebbly clay, and the plants tend to be stunted. It is usually encountered in full sun on stony shale flats or rocky south-facing hill slopes in clay or very hard, sandy soil, exposed or within the protection of karroid scrub. *L. zebrina* is not threatened.

NOTES. The flowers of *L. zebrina* emit an unusual, somewhat foetid scent, but pollinating agents are as yet unknown for this species. The scape remains attached to the bulb following capsule dehiscence and the broadly winged, aerodynamic capsules detach and are blown away, dispersing the seeds.

78. LACHENALIA MARTINIAE

Lachenalia martiniae W. F. Barker (sphalm. *martinae*), *Journal of South African Botany* 45 (2): 207–209 (1979).
TYPE: South Africa, Western Cape, Alpha Farm, Olifants River Valley, B. E. Martin *s.n.* sub. hort. NBG 126/37 (NBG!, holo.).

ETYMOLOGY. *martiniae*: after Miss Bina E. Martin (1900–), a former Kirstenbosch horticulturist.
DESCRIPTION. *Geophyte*, 100–300 mm high. *Bulb* globose, 10–22 mm in diameter, solitary; tunic multilayered, outer layers dark brown to grey-brown, hard; inner cataphyll creamy-white, loosely surrounding leaf bases, apex obtuse. *Leaf* solitary, narrowly to broadly lanceolate, 60–120 × 20–35 mm, spreading to suberect, bright green, slightly canaliculate, upper surface plain, with strongly depressed longitudinal grooves, lower surface heavily banded with green in central part, shading to maroon towards base; leaf margins flat or slightly to heavily undulate and crisped; leaf bases clasping, 30–70 mm long, white, aerial portion heavily banded with maroon shading to magenta in subterranean portion; primary seedling leaf terete, erect. *Inflorescence* racemose, erect to suberect, few- to many-flowered, sterile apex short; scape erect, 65–160 mm long, light green with maroon or dull brownish-blue speckles or blotches, slender; rachis light green in lower two thirds, shading to light greenish-blue in upper third, with maroon or dull brownish-blue speckles; pedicels suberect, 1–2 mm long, white; lower bracts ovate, becoming narrow-lanceolate above with long, filiform apices, 1–5 × 1–4 mm. *Perianth* zygomorphic, narrowly oblong-campanulate, whitish-grey, spreading to slightly cernuous; tube cup-shaped, 1–3 mm long; outer tepals ovate, 7–8 × 4–5 mm, apical gibbosities brownish-green; inner tepals obovate, 10–11 × 4–6 mm, whitish-grey, apical marking dull reddish-brown to brownish-blue. *Stamens* included to shortly exserted, more or less straight; filaments white, 7–10 mm long. *Ovary* ellipsoid, 3–5 × 2–3 mm, bright green; style included, straight, 7–9 mm long, white. *Capsule* ellipsoid, 7–9 × 4–5 mm. *Seed* globose, 1.3–1.4 × 1.3–1.4 mm, glossy, black; strophiole inflated, 0.5–0.8 mm long, smooth. *Chromosome number* $2n = 26$ (Spies *et al.*, 2008). Figure 204.
FLOWERING PERIOD. July to August, with a peak in mid-July.
HISTORY. The first specimens of *L. martiniae* were collected by Miss Bina Martin in August 1937 at Alpha Farm in the Olifants River Valley, just north of Citrusdal (Martin *s.n.* sub. hort. NBG 126/37, in NBG). The species has since been collected sporadically around Clanwilliam and once near Vanrhynsdorp. Bina Martin was a horticulturist at Kirstenbosch from 1934–1938, returning after WW2 in 1945 to work as secretary to the Director until her retirement in 1965.
DISTINGUISHING CHARACTERS AND AFFINITIES. Phylogenetic analysis of morphological data placed *L. martiniae* adjacent to *L. isopetala* (Duncan *et al.*, 2005a). These two species share undulate leaf margins, short pedicels and straight, included to shortly exserted stamens. *L. isopetala* differs in its papery bracts and tubular perianth; its outer tepals are oblong and its inner tepals are linear-oblong. The seeds of *L. isopetala* are considerably larger than those of *L. martiniae* (1.8–2.0 × 1.8–1.9 mm) and have rudimentary strophioles (0.1 mm long).
DISTRIBUTION, HABITAT AND CONSERVATION STATUS. *L. martiniae* occurs from Kwaggaskop north of Vanrhynsdorp to the eastern side of the Pakhuis Pass near Clanwilliam (Map 79). At Clanwilliam, the plants grow singly or in small groups in shallow, stony soil overlaying sandstone ridges, close to a large colony of *L. violacea*. Plants often grow amongst bushes including *Euphorbia mauritanica* (Euphorbiaceae) and *Euryops speciosissimus* (Asteraceae). *L. martiniae* is encountered in three vegetation types: Knersvlakte Quartz Vygieveld, Leipoldtville Sand Fynbos and Cederberg Sandstone

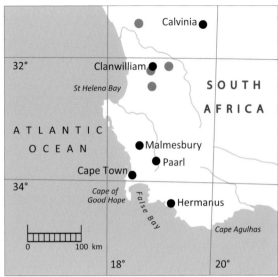

Map 79. Known distribution of *Lachenalia martiniae*.

Figure 204. *Lachenalia martiniae* on a sandstone ridge near Clanwilliam. Image: Graham Duncan.

Fynbos (Mucina & Rutherford, 2006). This species is threatened by agricultural expansion and road widening, and has a conservation status of Vulnerable (Duncan & Raimondo, 2009b).

NOTES. Pollinators are unknown for *L. martiniae*. The scapes detach from the bulbs shortly after capsule dehiscence and are blown away, scattering the seed widely.

79. LACHENALIA STAYNERI

Lachenalia stayneri W. F. Barker, *Journal of South African Botany* 45 (2): 214–216 (1979).
TYPE: South Africa, Western Cape, De Wet, north of Worcester, *F. J. Stayner s.n.* sub. NBG 88509 (NBG!, holo.!; iso!.).

ETYMOLOGY. *stayneri*: after Frank J. Stayner (1907–1981), Curator of the Karoo National Botanical Garden from 1959–1969.
DESCRIPTION. *Geophyte* 120–300 mm high. *Bulb* globose, 15–25 mm in diameter, solitary; tunic multilayered, outer layers dark brown, spongy; inner cataphyll translucent white, subterranean, adhering to leaf bases, apex obtuse. *Leaves* 2, lanceolate, 40–150 × 15–30 mm, prostrate, upper surface bright green to maroonish-green with scattered, small to very large wart-like pustules and prominent depressed longitudinal grooves, lower surface tinged with maroon; margins coriaceous, deep maroon; primary seedling leaf flat, prostrate. *Inflorescence* racemose, sturdy, many-flowered, sterile apex short; scape erect, 70–120 mm long, sturdy, light green, heavily marked with small to large brownish-maroon blotches; rachis light green with small brownish-maroon blotches, apex shading to light

electric blue; pedicels suberect, 3–7 mm long, white to light green; lower bracts ovate, becoming lanceolate above, 2–4 × 1–4 mm, white, plain or maroon-tinged. *Perianth* zygomorphic, oblong-campanulate, cernuous; tube cup-shaped, 1–2 mm long, light whitish-blue; outer tepals ovate, 5–7 × 4 mm, light whitish-blue, apical gibbosities dark reddish-brown; inner tepals obovate, 6–7 × 3 mm, white, median keels reddish-brown. *Stamens* well exserted, declinate; filaments white, 10–11 mm long. *Ovary* obovoid, 3 × 2 mm, bright green; style well exserted, straight, 8–9 mm long, white. *Capsule* obovoid, 6–7 × 5–6 mm. *Seed* globose, 1.1–1.2 × 1.3 mm, glossy, black; strophiole inflated, 0.8 mm long, smooth. *Chromosome number* 2n = 24 (Johnson & Brandham, 1997). Figure 205.

FLOWERING PERIOD. August to September, with a peak in mid-September.

HISTORY. *L. stayneri* was first collected in September 1947 by staff of the Karoo National Botanical Garden at Worcester, where numerous collections of this species have been made and where it still occurs in limited numbers. The type material was collected at De Wet, north of Worcester, in August 1972 by Frank Stayner; the species has also been recorded in August 1972 to the east of Worcester and in September 1971 at Robertson.

DISTINGUISHING CHARACTERS AND AFFINITIES. Duncan *et al.* (2005a) considered *L. stayneri* to be sister to *L. nervosa*. Both have paired, prostrate leaves that are usually covered in large pustules on the upper surface, with prominent longitudinal grooves, obovoid capsules and seeds with inflated strophioles. *L. nervosa* differs from *L. stayneri* in having ovate leaves and in its flowers, which are shortly pedicellate and widely campanulate with strongly recurved outer and inner tepals and recurved filaments. It also has smaller globose seeds (0.9 × 0.8–0.9) with slightly shorter strophioles (0.6–0.7 mm long).

Map 80. Known distribution of *Lachenalia stayneri*.

Figure 205. *Lachenalia stayneri* in Breede Shale Renosterveld near Worcester. Image: Patrick Fraser.

DISTRIBUTION, HABITAT AND CONSERVATION STATUS. *L. stayneri* is limited to the Worcester-Robertson Karoo, extending from De Wet just north of Worcester to east of Robertson (Map 80). At the Karoo Desert National Botanical Garden in Worcester, populations occur on a low-lying, gentle south-westerly slope in stony-clay soil, singly or in small groups, within the protection of low karroid scrub or in open aspects. Plants occur in Breede Shale Renosterveld vegetation (Mucina & Rutherford, 2006) and are often associated with *L. perryae*. The habitat of *L. stayneri* has been affected to a large extent by agricultural expansion, mainly for vineyards and deciduous fruit orchards, and the species has a conservation status of Endangered (Helme *et al.*, 2009).

NOTES. The scapes, flower buds and open flowers are heavily grazed by Cape hares (*Lepus capensis*) at the Karoo Desert National Botanical Garden, resulting in very poor seed production. The flowers are pollinated by honey bees (*Apis mellifera*). The scapes remain attached for a number of weeks following capsule dehiscence, dispersing the seed locally.

80. LACHENALIA JUNCIFOLIA

Lachenalia juncifolia Baker, *Journal of the Linnean Society* (*Botany*) 11: 409 (1871).
TYPE: South Africa, Western Cape, Caledon, 'Zwartberg und umgegend des Bades', *C. F. Ecklon & C. L. Zeyher* 51 (TCD!, holo.; SAM!, S! iso.).
SYNONYMY: *Lachenalia esterhuysenae* W. F. Barker, *Journal of South African Botany* 44 (4): 398–399 (1978). Type: South Africa, Western Cape, northern Cederberg, Sneeuwberg area above Bakleikraal, *E. E. Esterhuysen* 34149 (BOL!, holo.; K!, MO, NBG!, PRE!, S, iso.).

ETYMOLOGY. *juncifolia*: narrow leaves reminiscent of the genus *Juncus*.
DESCRIPTION. *Geophyte*, 60–200 mm high. *Bulb* subglobose, 8–20 mm in diameter, offset- and bulbil-forming; tunic multilayered, outer layers dark brown, spongy; inner cataphyll translucent white, subterranean, adhering to leaf bases, apex obtuse. *Leaves* 2, linear, 80–200 × 2–7 mm, spreading to suberect, bright green to yellowish-green; upper surface plain, lower surface with green or purplish-magenta bands in central portion becoming purplish-magenta below, canaliculate, or lower two-thirds slightly to deeply canaliculate, upper third terete; leaf bases loosely surrounding base of scape, 30–40 mm long, subterranean, white; primary seedling leaf terete, erect. *Inflorescence* racemose, many-flowered, sterile apex short; scape erect to suberect, 50–120 mm, sturdy, light green, lightly to heavily mottled with purplish-magenta; rachis light green, heavily mottled with purplish-magenta; pedicels spreading to suberect, white or light purple, 4–9 mm long; lower bracts ovate, becoming lanceolate above, 1–2 × 1–3 mm, white, or dark magenta with white margins. *Perianth* zygomorphic, oblong-campanulate, white, pink, whitish-magenta to dark magenta or purple, spreading to cernuous; tube cup-shaped, 1–3 mm long; outer tepals ovate, 5–6 × 2–4 mm, apical gibbosities dark brown, green or deep magenta, apices flat to slightly recurved; inner tepals obovate, 5–7 × 4 mm, apices flat to slightly recurved; median keels dull brown, deep magenta or purple. *Stamens* well exserted, more or less straight; filaments white, or lower half white, upper half light to deep purple, 8–12 mm long; pollen yellow, cream or brownish-magenta. *Ovary* ellipsoid, 2.0 × 1.5–2.0 mm, bright green; style well exserted, more or less straight, 8–12 mm long, white, or lower half white, upper half light magenta to purple. *Capsule* obovoid, 5–7 × 4–6 mm. *Seed* globose, 1.1–1.2 × 1.3 mm, glossy, black; strophiole rudimentary, 0.2 mm long, ridged. *Chromosome number* 2n = 22 (Johnson & Brandham, 1997; Spies *et al.*, 2002; Hamatani *et al.*, 2007; Spies *et al.*, 2008). Plate 26, Figure 206–208.
FLOWERING PERIOD. August to December, with a peak in mid-September.

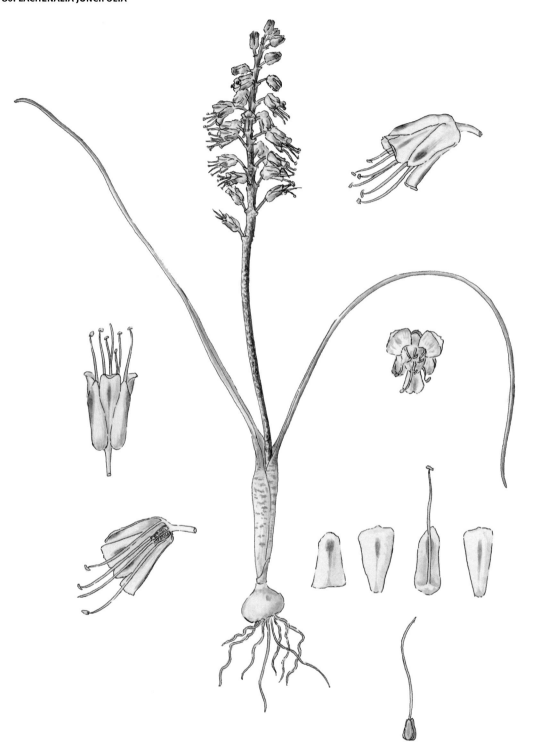

Plate 26. Watercolour painting of *Lachenalia juncifolia* from Porterville (*Bolus s.n.*, in BOL) courtesy of the Compton Herbarium, Kirstenbosch, South African National Biodiversity Institute. Artist: Winsome Barker.

Figure 206 (left). Purple form of *Lachenalia juncifolia* on sandy flats in the Tanqua Karoo. Image: Graham Duncan.
Figure 207 (right). White form of *Lachenalia juncifolia* in alluvial sand near Sutherland. Image: Graham Duncan.

HISTORY. *L. juncifolia* was described by J. G. Baker in the *Journal of the Linnean Society* (1871) from material collected by C. F. Ecklon and C. L. Zeyher near Caledon in September 1834. Isotype material was deposited in the South African Museum collection in the Compton Herbarium at Kirstenbosch and in the Swedish Museum of Natural History in Stockholm, but the whereabouts of the holotype remained unknown, even to Baker, who noted in his second monograph (1897a) that it was not in the Kew Herbarium; it was finally located in 2008 in the herbarium at Trinity College, Dublin, by Prof. John Parnell. *L. esterhuysenae* was described by W. F. Barker in the *Journal of South African Botany* from specimens first collected in the Pakhuis Pass east of Clanwilliam by Elsie Esterhuysen in September 1940 (Barker, 1978); the plant is simply a high altitude form of *L. juncifolia* and is here placed in synonymy. In 1989, Barker described a variety of *L. juncifolia* from the southern coastal part of the Western Cape as var. *campanulata*, which she distinguished on the basis of its 'succulent, semi-terete leaves with a very narrow channel above', and its flowers that are 'campanulate rather than oblong-campanulate'. The taxon is here upgraded to species level, but as the specific epithet *campanulata* has already been used for another species, *L. campanulata* Baker, it receives a new name, *L. magentea*.

80. LACHENALIA JUNCIFOLIA

DISTINGUISHING CHARACTERS AND AFFINITIES. *L. juncifolia* forms a distinctive but highly polymorphic species complex, consisting of numerous small to large, more or less continuously varying populations. Perianth colour varies from white to pink, whitish-magenta, dark magenta or purple, and filament colour from entirely white, to white in the lower half and purple above. Duncan *et al.* (2005a) placed *L. juncifolia* adjacent to *L. magentea*, with which it shares paired linear leaves, exserted, straight filaments and similar globose seeds with rudimentary strophioles. *L. magentea* differs in its larger, narrowly campanulate flowers with subequal tepals, and succulent, subterete leaves with a very narrow channel along the upper surface.

DISTRIBUTION, HABITAT AND CONSERVATION STATUS. *L. juncifolia* occurs from Calvinia south to the Roggeveld and Komsberg, the Cederberg and Koue Bokkeveld to Porterville, south-east to Ceres and Worcester, east to Touws River, south to Caledon and east to Herbertsdale (Map 81). It traverses 11 vegetation types: Hantam-, Tanqua-, Roggeveld- and Koedoesberge-Moordenaars Karoo, Cederberg- and Swartruggens Sandstone Fynbos, Ceres-, Western- and Central Rûens Shale Renosterveld, De Hoop Limestone Fynbos and Albertinia Sand Fynbos (Mucina & Rutherford, 2006). The plants occur in a variety of seasonally moist, mainly sandy habitats, across a range of altitudes. On Sneeuwkop in the northern Cederberg, a tall-growing form with white flowers (previously *L. esterhuysenae*) grows on rocky sandstone slopes in arid fynbos at 1,200 m; and at the base of the Komsberg Pass, another white-flowered form occurs in alluvial sand. At Op-die-Berg in the Swartruggens Mountains, a low-growing, mauve form grows on north-facing sandstone outcrops at 900 m in shallow cracks of rock sheets, in brownish-yellow sand; and in the southern Tanqua Karoo, a deep purple form grows on red sandy flats and lower hill slopes.

Across the range of *L. juncifolia*, the plants occur singly or in small to large groups, certain forms rapidly increasing by the formation of bulblets that are produced on stolons from the base of the bulb. Plant size diminishes and colony size increases towards the southern coastal parts of the species' range, for example in the De Hoop Nature Reserve, where these plants are frequently encountered close to sea level on lower hill slopes and seasonally moist flats in shallow sand overlaying limestone. In coastal areas, tepal colour varies in shades of pink, often with prominent blue perianth tubes.

Figure 208. Dwarf, pink form of *Lachenalia juncifolia* on limestone flats in the De Hoop Nature Reserve. Image: Cameron McMaster.

Although parts of this species' habitat have been affected by housing development in coastal parts of the southern Cape, it is not under immediate threat.

NOTES. Forms of *L. juncifolia* from the southern Cape flower in profusion following bush fires in the previous summer and autumn, but are not dependent on fire for flowering to occur. Pollination is effected by honey bees (*Apis mellifera*), but certain forms are at least partially self-fertile, regularly forming seed under isolated nursery conditions. The scapes remain attached to the bulbs for several weeks after capsule dehiscence and the seeds are dispersed locally by the shaking action of wind. Bulbs of populations on the Komsberg Pass remain dormant for one or more consecutive winter seasons of insufficient rain, suggesting strong local adaptation.

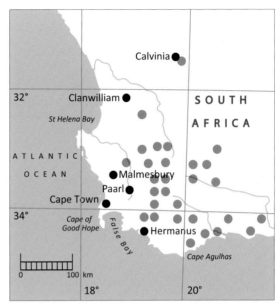

Map 81. Known distribution of *Lachenalia juncifolia*.

81. LACHENALIA MONILIFORMIS

Lachenalia moniliformis W. F. Barker, *Journal of South African Botany* 49 (4): 438–440 (1983).
TYPE: South Africa, Western Cape, Lemoenpoort, *P. L. Perry* 795 (NBG!, holo., iso.).

ETYMOLOGY. *moniliformis*: bead-like, descriptive of the raised green fleshy swellings along the upper two thirds of the leaf, reminiscent of a string of beads.

DESCRIPTION. *Dwarf geophyte*, 120–170 mm high. *Bulb* subglobose, 10–15 mm in diameter, offset- and bulbil-forming; tunic multilayered, outer layers dark brown, papery; inner cataphyll translucent white, loosely surrounding leaf bases, apex obtuse. *Leaves* 5–8, terete, 60–160 × 1–2 mm, rosulate, erect, bright green, lower third heavily banded with dark maroon and magenta, upper two-thirds with prominent circular, dark green raised fleshy bands; leaf bases loosely clasping, distinctly swollen, 20–30 mm long, white with short magenta bands; primary seedling leaf terete, erect. *Inflorescence* racemose, few- to many-flowered, moderately dense, lax, sterile apex short; scape erect to suberect, 50–90 mm long, light green, heavily marked with small dark maroon blotches, slender; rachis heavily marked with small dark maroon blotches in lower two thirds, shading to light electric blue in upper third; pedicels spreading to suberect, 5–6 mm long, light bluish-white; lower bracts ovate, swollen at base, becoming lanceolate above, 2–3 × 1–2 mm white. *Perianth* zygomorphic, oblong-campanulate; tube cup-shaped, 1 mm long, ice-blue; outer tepals ovate, 5–6 × 4 mm, light blue at base shading to white above, apices recurved, apical gibbosities reddish-brown; inner tepals obovate, 6–7 × 3 mm, white, apices recurved, apical marking reddish-brown. *Stamens* well exserted, declinate; filaments white, 9–11 mm long. *Ovary* obovoid, 2 × 2 mm, light green; style well exserted, declinate, 6–8 mm long, white. *Capsule* obovoid, 4–5 × 3–4 mm. *Seed* globose, 1.2–1.3 × 1.2–1.3 mm, glossy, black; strophiole rudimentary, 0.3 mm long, ridged. *Chromosome number* 2n = 22 (Spies *et al.*, 2008). Figures 33, 209.

81. LACHENALIA MONILIFORMIS

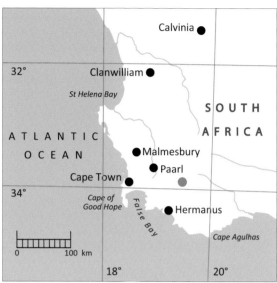

Map 82. Known distribution of *Lachenalia moniliformis*.

Figure 209. *Lachenalia moniliformis* on shale flats in the Worcester Valley. Image: Graham Duncan.

FLOWERING PERIOD. September to October, with a peak in late September.

HISTORY. *L. moniliformis* was discovered by the botanist Pauline Perry who collected a single plant on the Lemoenpoort Farm south-west of Worcester, in September 1978. Perry visited this locality again in the same year and collected seedlings for cultivation at the Karoo National Botanical Garden; the original bulb was then pressed to serve as the holotype (*Perry* 795, in NBG). In September 1979, a few more specimens were collected for preservation in the Compton Herbarium and the species was described in the *Journal of South African Botany* (Barker, 1983b). No further collections were made until late September 2008 when *L. moniliformis* was found in full flower at the type locality.

DISTINGUISHING CHARACTERS AND AFFINITIES. The most distinctive vegetative feature of *L. moniliformis* is its rosette of numerous terete leaves, their upper two-thirds with prominent dark green, raised fleshy swellings. *L. moniliformis* is the only member of the genus with this trait. The species forms bulblets on horizontal stolons that are produced from the base of the bulb.

Duncan *et al.* (2005a) suggested that *L. moniliformis* is allied with *L. hirta*. These two species share oblong-campanulate, cernuous perianths with very short perianth tubes, and inner tepals with recurved apices. *L. hirta* differs mainly in its solitary linear, canaliculate leaf covered with simple trichomes on the lower surface and margins. It also differs in its straight, included to shortly exserted stamens and smaller, matte black seeds (0.9–1.2 × 0.8–1.2 mm) that have shorter rudimentary strophioles (0.1 mm long).

Superficially, the flowers of *L. moniliformis* resemble those of *L. stayneri* and certain forms of *L. juncifolia*; however, *L. stayneri* is a much larger plant with two prostrate, broadly lanceolate leaves and *L. juncifolia* has paired linear, canaliculate leaves.

DISTRIBUTION, HABITAT AND CONSERVATION STATUS. *L. moniliformis* is restricted to Lemoenpoort Farm in the Worcester Valley (Map 82). It grows in Breede Shale Renosterveld vegetation (Mucina & Rutherford, 2006), singly or in small groups and occupies flats in dry, stony shale ground as well as in sandy depressions, in full sun or the light shade of low karroid scrub. *L. moniliformis* flowers simultaneously with *Gazania krebsiana* (Asteraceae), and *L. orchioides* subsp. *orchiodes* also occurs at this locality, flowering a month earlier. *L. moniliformis* has a conservation status of Critically Endangered because of vineyard expansion and overgrazing in the immediate surroundings of the only known population (Victor & Duncan, 2009h).

NOTES. The flowers are pollinated by honey bees (*Apis mellifera*). The scapes remain attached to the bulbs for several weeks following capsule dehiscence and the seeds are dispersed locally.

82. LACHENALIA POLYPHYLLA

Lachenalia polyphylla Baker, in Engler, H. G. A., *Botanische Jahrbücher* 15 (3): 7 (1892).
TYPE: South Africa, Western Cape, Tulbagh, *C. F. Ecklon & C. L. Zeyher* 50 (B!, holo.; K!, SAM!, iso.).

ETYMOLOGY. *polyphylla*: many-leafed.
DESCRIPTION. *Dwarf geophyte*, 60–200 mm high. *Bulb* globose, 8–15 mm in diameter, offset-forming, tunic multilayered, outer layers dark brown, papery; inner cataphyll translucent white, adhering to leaf bases, apex acute. *Leaves* 5–12, linear, 6–10 × 0.8–1.0 mm, terete, erect to suberect, rosulate, upper two-thirds olive-green to dark green, lower third dark maroon to magenta and covered with minute papillae; leaf bases loosely clasping base of scape, 10–20 mm long, subterranean, broadening abruptly and swollen, dark maroon to magenta in upper part, shading to white below, covered with minute papillae; primary seedling leaf terete, erect. *Inflorescence* racemose, few- to many-flowered, sterile apex short; scape erect to suberect, 100–180 mm long, light green, plain to heavily mottled with dull maroon or purplish-grey, slender; rachis heavily mottled with dull maroon to purplish-grey in lower two-thirds, shading to dull electric blue in upper third, slender; pedicels spreading to arcuate at anthesis, 4–7 mm, becoming suberect to erect in fruit, white to light purplish-white; lower bracts ovate, becoming narrow-lanceolate above, 1–2 × 0.5–2.0 mm, white to dull maroon. *Perianth* zygomorphic, oblong-campanulate, cernuous; tube cup-shaped, 2 mm long, light whitish-blue; outer tepals ovate, 5–6 × 3–4 mm, light whitish-blue, apical gibbosities dull rose-pink to reddish-brown, apices slightly recurved; inner tepals obovate, 5–6 × 3 mm, translucent white, median keels brown to reddish-brown, apices recurved. *Stamens* well exserted, recurved; filaments white, 7–9 mm long. *Ovary* obovoid, 2.0 × 1.5 mm, bright green; style well exserted, straight, 7–8 mm long, white, protruding well beyond stamens as ovary matures. *Capsule* obovoid, 4–5 × 4 mm, suberect to erect. *Seed* globose, 0.9 × 0.5 mm, matte, black, primary sculpturing rugose; strophiole rudimentary, 0.2 mm long, ridged. *Chromosome number* $2n = 22$ (Spies *et al.*, 2008). Plate 27, Figure 210.
FLOWERING PERIOD. September to October, with a peak in late September.
HISTORY. *L. polyphylla* was described by J. G. Baker (1892a) in volume 15 of Engler's *Botanische Jahrbücher*. The undated holotype sheet was collected by C. F. Ecklon and C. L. Zeyher at Tulbagh in the south-western Cape, from whence most collections of the species come. The holotype is housed in the Botanical Museum of Berlin-Dahlem, with isotype material in the Kew Herbarium and the

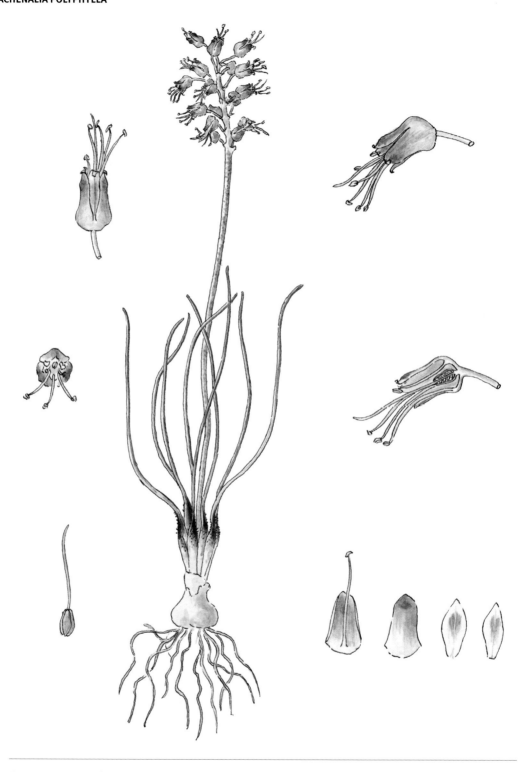

Plate 27. Watercolour painting of *Lachenalia polyphylla* from Tulbagh, courtesy of the Compton Herbarium, Kirstenbosch, South African National Biodiversity Institute. Artist: Winsome Barker.

South African Museum collection. *L. polyphylla* was not collected for a period of almost 20 years after 1972, and was considered possibly extinct in the wild until its rediscovery on Gouda Commonage west of Tulbagh (Duncan, 1992b).

DISTINGUISHING CHARACTERS AND AFFINITIES. In phylogenetic reconstructions, *L. polyphylla* is placed close to *L. multifolia* (Duncan *et al.*, 2005a), which shares similarly swollen, subterranean leaf bases but differs in its widely campanulate, white flowers with well exserted, narrowly spreading stamens, a shorter, sturdier scape (30–40 mm long) and intensely glaucous leaves that spiral in the upper half.

DISTRIBUTION, HABITAT AND CONSERVATION STATUS. *L. polyphylla* is limited to severely fragmented subpopulations between Piketberg and Gouda (Map 83). Much of its former habitat has been lost to agricultural expansion for winter cereal crops and the species is further compromised by invasive alien grasses, housing development and road construction. It is probably extinct at Tulbagh and is currently known mainly from locations around the town of Gouda to the south-west of Tulbagh and within the nearby Elandsberg Private Nature Reserve. The plants grow in Swartland- and Breede Shale Renosterveld vegetation (Mucina & Rutherford, 2006), in colonies on seasonally waterlogged flats in gravelly or stony clay soil, singly or in small groups. Until the late 1990s, *L. polyphylla* grew on Gouda Commonage in association with *L. contaminata* (which still survives) but it is probably extinct there as a result of alien grass infestation and indiscriminate mowing. *L. polyphylla* has a conservation status of Endangered (Raimondo & Ebrahim, 2009).

NOTES. The flowers of *L. polyphylla* are pollinated by honey bees (*Apis mellifera*). During the fruiting stage, the pedicels and capsules become re-orientated to a suberect or erect position. The scapes remain attached to the bulb for several weeks following capsule dehiscence and the seeds are dispersed by the shaking action of wind.

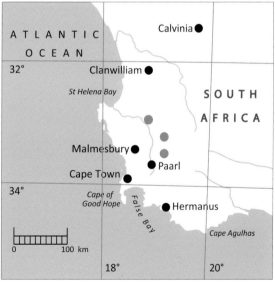

Map 83. Known distribution of *Lachenalia polyphylla*.

Figure 210. *Lachenalia polyphylla* on seasonally moist shale flats near Hermon. Image: Graham Duncan.

83. LACHENALIA HIRTA

Lachenalia hirta (Thunb.) Thunb., *Prodromus Plantarum Capensium*: 64 (1794). *Phormium hirtum* Thunb., *Nova Genera Plantarum*: 98 (1784).
TYPE: South Africa, Cape, precise locality unknown, C. P. *Thunberg* 8549 in Herbarium Thunberg (UPS!, holo., microfiche).
SYNONYMY: *Lachenalia hirta* (Thunb.) Thunb. var. *exserta* W. F. Barker, *South African Journal of Botany* 55 (6): 638 (1989). Type: South Africa, Western Cape, Piketberg district, 3 km south of 'The Rest', *T. M. Salter* 3867 (BOL!, holo.; BM!, K!, iso.).

ETYMOLOGY. *hirta*: hairy, descriptive of the rigid trichomes that develop along the leaf margins and lower leaf surfaces.

DESCRIPTION. *Geophyte*, 100–300 mm high. *Bulb* subglobose, 10–20 mm in diameter, offset-forming; tunic multilayered, outer layers dark brown, spongy; inner cataphyll subterranean, translucent white, loosely surrounding clasping leaf base, apex obtuse. *Leaf* solitary, linear, 100–200 × 5–20 mm, spreading to suberect, light to dark green, upper surface smooth, deeply canaliculate, bifacial, widening abruptly at base, lower surface marked with dark green transverse bands, covered with short to long rigid, simple trichomes, midrib prominent on lower surface; margins flat in upper part, undulate and crisped on leaf base, covered with short to long simple trichomes; leaf bases loosely surrounding base of scape, 20–60 mm long, with light to dark magenta bands; primary seedling leaf terete, suberect. *Inflorescence* racemose, few- to many-flowered, lax, sterile apex short; scape suberect, 70–120 mm long, light green, plain or densely to sparsely marked with dull maroon or purple blotches, slender, wiry; rachis light green or densely to sparsely marked with small dull maroon or purple blotches in lower half, shading to light blue in upper half, slender; pedicels suberect to spreading or slightly declinate, 10–15 mm, light green, blue or white; lower bracts ovate, becoming lanceolate above, 2–3 × 0.5–2.0 mm, light greenish- to bluish-white. *Perianth* zygomorphic, oblong-campanulate, cernuous, slightly sweet-scented; tube cup-shaped, 1 mm long, light blue to bluish-grey; outer tepals narrowly ovate, 6–7 × 3–4 mm, light blue shading to light yellow or greenish-yellow above, or uniformly bluish-grey or greenish-yellow, apices flat to slightly recurved, apical gibbosities brown to reddish-brown or green; inner tepals obovate, 7–8 × 4 mm, light blue to bluish-grey in lower half, shading to translucent white or greenish-yellow in upper half, or uniformly greenish-yellow, lower inner tepal slightly longer than upper inner tepals, apices slightly recurved; apical marking greenish- to reddish-brown; median keels greenish-blue. *Stamens* included to shortly or well exserted; filaments straight, 5–12 mm long, white. *Ovary* obovoid, 2–3 × 1–1.5 mm, bright green; style included to shortly or well exserted, straight, 6–7 mm long, white. *Capsule* obovoid, 6–8 × 5–7 mm. *Seed* globose, 0.9–1.2 × 0.8–1.2 mm, matte, black; strophiole rudimentary, 0.1 mm long, ridged. *Chromosome number* 2n = 22 (Hamatani *et al.*, 2004); 2n = 24 (de Wet, 1957); 2n = 22, 24 (Johnson & Brandham, 1997). Figures 211, 212.

FLOWERING PERIOD. August to September, with a peak from late August to early September.

HISTORY. Originally described as *Phormium hirtum* by Thunberg (1784) in his *Nova Genera Plantarum*, this taxon is the subject of the oldest colour illustration of any *Lachenalia* to which a definite date can be ascribed. It first appeared as a painting in the diary of Simon van der Stel's journey to Namaqualand in 1685/6. Leonardi Plukenett (1692) illustrated it in his *Leonardi Plukenetii Phytographia*, describing it under the phrase name *Hyacinthus Africanus Orchioides serpentarius, folio singularis, undato, piliscilliaribus fimbriato, floribus ex aureo punicatibus*. The added phrase *Codicis Comptoniani* indicated that the figure had been copied from the codex of Bishop Compton of London, the whereabouts of which is

Figure 211. Cultivated blue form of *Lachenalia hirta* from Citrusdal. Image: Graham Duncan.
Figure 212. Light yellow form of *Lachenalia hirta* on the Kamiesberg. Image: Tessa Oliver.

unknown. The figure appeared again in 1709 in James Petiver's *Gazophylacii Naturae & Artis, Herbarium Capense*, this time as an exact mirror image of Plukenett's figure. In 1794, the taxon finally became *L. hirta* in Thunberg's *Prodromus Plantarum Capensium*, a status that was recognised by Baker in both his revisions of the genus (1871, 1897a).

DISTINGUISHING CHARACTERS AND AFFINITIES. Duncan *et al.* (2005a) placed *L. hirta* adjacent *L. moniliformis* in a phylogenetic analysis. These two species share similar oblong-campanulate, cernuous flowers, but *L. moniliformis* has much shorter pedicels (5–6 mm long), declinate stamens and rosulate, terete linear leaves with peculiar bead-like thickenings.

Most forms of *L. hirta* have included to shortly exserted stamens. The var. *exserta* described by Barker (1989) has well-exserted stamens and is found mainly in the central part of the distribution range from Clanwilliam to Moorreesburg; however, the degree of stamen exsertion forms a continuum across the many geographical forms, making recognition of the var. *exserta* impractical. The outer tepal colour of most forms of this species, especially those occurring in sandy soils, varies in shades of light blue with greenish-yellow apices; forms growing in heavier clay soils tend to have uniformly bluish-grey outer tepals, and there is a greenish-yellow form from Riebeeck Kasteel, north-east of Malmesbury and a light yellow form from the Kamiesberg.

83. LACHENALIA HIRTA

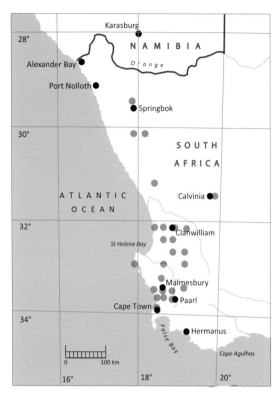

Map 84. Known distribution of *Lachenalia hirta*.

DISTRIBUTION, HABITAT AND CONSERVATION STATUS. *L. hirta* extends from Springbok in central Namaqualand south-east to Calvinia and further south to Kleinberg just north of Cape Town (Map 84). It is encountered in 12 vegetation types: Namaqualand Klipkoppe, Kamiesberg Mountains- and Agter-Sederberg Shrubland, Hantam Karoo, Graafwater-, Cederberg- and Olifants Sandstone Fynbos, Leipoldtville Sand Fynbos, Saldanha Granite Strandveld, Swartland Granite Renosterveld, and Swartland Shale- and Breede Shale Renosterveld (Mucina & Rutherford, 2006). The plants grow on rocky hillsides (Malmesbury), in deep sand (Cederberg) and on arid sandy clay flats (west of Kamieskroon). Populations of sparse, solitary individuals or small groups occur in open aspects or within the protection of low vegetation. The species occurs in association with *L. membranacea* and a mauve form of *L. pallida* at Wilgersbosdrift Farm north-east of Piketberg. *L. hirta* is not threatened.

NOTES. The flowers are pollinated by honey bees (*Apis mellifera*). During the fruiting stage, the pedicels and developing fruits become suberect. The scapes remain attached for several weeks following capsule dehiscence, and the seeds are dispersed locally.

84. LACHENALIA ATTENUATA

Lachenalia attenuata W. F. Barker ex G. D. Duncan, *Bothalia* 28 (2): 132–134 (1998).
TYPE: South Africa, Western Cape, Keisieberg, Montagu District, *G. J. Lewis* 2431 (SAM!, holo.).

ETYMOLOGY. *attenuata*: gradually tapering leaf.

DESCRIPTION. *Geophyte*, 65–220 mm high. *Bulb* subglobose, 8–10 mm in diameter, solitary; tunic multilayered, outer layers dark brown, hard; inner cataphyll translucent white, loosely surrounding leaf bases, apex obtuse. *Leaf* solitary, broadly linear, 35–140 × 7–10 mm, erect to suberect, light green, canaliculate, often proteranthous, upper surface plain, lower surface heavily marked with dark green transverse bands; leaf bases clasping, 10–20 mm long, white marked with purplish-magenta transverse bands; primary seedling leaf terete, erect. *Inflorescence* racemose, few- to many-flowered, lax, sterile apex short; scape erect to suberect, 35–200 mm long, light green, lower half with brownish-purple blotches, upper half yellowish-green or brownish-purple, slender; rachis light green, heavily mottled with brownish-purple, apex light blue; pedicels suberect, 1–4 mm long, white or brownish-purple; bracts ovate throughout, 1–3 × 1–3 mm. *Perianth* zygomorphic, oblong-campanulate, cernuous; tube cup-shaped, 2 mm long, light blue; outer tepals ovate, 6–7 × 3–4 mm, light blue, apices and apical gibbosities dull purplish-brown; inner tepals obovate, 6–7 × 4–5 mm, bases dull white, shading to greenish-yellow above, median keels brownish-green, upper inner tepals overlapping. *Stamens* declinate, included to shortly exserted; filaments 6–7 mm long, white; pollen ageing to black. *Ovary* obovoid, 3–4 × 3 mm, light green; style included, more or less straight, 7 mm long, white, protruding beyond perianth as ovary matures. *Capsule* obovoid, 6–7 × 5–6 mm. *Seed* ovoid, 1.3–1.4 × 1.2–1.3 mm, matte, black; strophiole 1.1 mm long, ridged. *Chromosome number* 2n = 14 (Spies *et al.*, 2009). Figure 213.

FLOWERING PERIOD. August to September, with a peak in late September.

HISTORY. The first collection of *L. attenuata* was made by Dr J. Muir in August 1933 near Riversdale in the southern Cape (*Muir* 4886, in K). In September 1946, Dr G. J. Lewis collected it on the Keisieberg in the Montagu District, and her material (*Lewis* 2431) forms the holotype, preserved in the South African Museum collection at Kirstenbosch. The manuscript name *L. attenuata* was appended to the Lewis collection by W. F. Barker and the species was formally described in *Bothalia* (Duncan, 1998a). Most recent collections of the species have been made in the Sutherland District.

DISTINGUISHING CHARACTERS AND AFFINITIES. Duncan *et al.* (2005a) considered *L. attenuata* to be allied to *L. whitehillensis* because these two species share oblong-campanulate, cernuous flowers with light blue perianth tubes and outer tepals, and usually solitary canaliculate leaves. *L. whitehillensis* differs mainly in its lanceolate leaf shape, longer (10–17 mm) well-exserted stamens, and seeds with shorter strophioles (0.5–0.6 mm).

DISTRIBUTION, HABITAT AND CONSERVATION STATUS. *L. attenuata* is known from three disjunct centres: the Roggeveld Plateau extending from the Gannaga Pass south of Middelpos to the Verlatekloof Pass south of Sutherland; a centre extending from the Hex River Pass to the Keisieberg in the Montagu District in the Little Karoo; and a third region in the Riversdale District in the southern Cape (Map

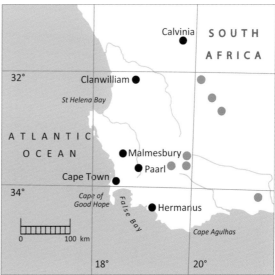

Map 85. Known distribution of *Lachenalia attenuata*.

Figure 213. *Lachenalia attenuata* in gravelly sand on the Roggeveld Plateau south of Middelpos.
Image: Graham Duncan.

85). At Agterkop Farm (Gannaga Pass), *L. attenuata* occurs on east-facing hill slopes in seasonally moist, gravelly sandy soil between granite boulders, and on the Gannaga Pass, it favours rock ledges of south-west-facing slopes. The plants occur in four vegetation types: Roggeveld Karoo, Breede Shale Fynbos, and Montagu- and Mossel Bay Shale Renosterveld (Mucina & Rutherford, 2006), singly or in small groups, amongst low annuals and grasses. *L. attenuata* is not threatened.

NOTES. In the wild, the leaves of *L. attenuata* are often proteranthous, senescing before flowering occurs in late spring. The flowers are pollinated by honey bees (*Apis mellifera*). The scapes remain attached for several weeks following capsule dehiscence and the seeds are dispersed locally by wind.

85. LACHENALIA WRIGHTII

Lachenalia wrightii Baker, *Journal of Botany (new series)* 7: 322 (1878). *Lachenalia unifolia* Jacq. var. *wrightii* (Baker) Baker, in Thiselton-Dyer (ed.), *Flora Capensis* 6: 431 (1897).
TYPE: South Africa, Western Cape, Simons Bay, *C. Wright* 219 (K!, holo.).

ETYMOLOGY. *wrightii*: after Charles Wright (1811–1886), 19th century botanical collector from Connecticut, USA.

DESCRIPTION. *Geophyte*, 100–260 mm high. *Bulb* subglobose, 10–15 mm in diameter, offset-forming; tunic multilayered, outer layers dark brown, spongy; inner cataphyll translucent white, adhering to leaf base, apex obtuse. *Leaf* solitary, linear, 50–160 × 2–12 mm, suberect, light to dark green, canaliculate, sometimes conduplicate, upper surface plain, lower surface banded with darker green in lower half; leaf base clasping, 55–70 mm long, white, upper part banded with brownish-maroon, subterranean part banded with magenta bands; primary seedling leaf terete, erect. *Inflorescence* racemose, few- to many-flowered, sterile apex short; scape erect to suberect, 80–160 mm long, light green, plain or lightly to heavily mottled with light to dark brownish-maroon; rachis light green, lower half with dull brown or purple mottling, shading to light blue, pink or white above; pedicels suberect at anthesis, 9–14 mm long, increasing in length towards inflorescence apex, white, light green or pinkish-purple, plain or mottled with brownish-maroon, usually becoming erect in fruit; lower bracts ovate, becoming lanceolate above, 1–3 × 1–3 mm, translucent white or dark brown with white margins. *Perianth* zygomorphic, oblong-campanulate, spreading to cernuous; tube cup-shaped, 3 mm long, white or light to deep blue; outer tepals ovate, 5–6 × 4–5 mm, lower half light to dark blue, upper half white, plain to heavily marked with purplish blotches, apical gibbosity large, light to dark brown; inner tepals obovate, 6–7 × 4–5 mm, white, apices slightly recurved, keels light blue. *Stamens* included to shortly exserted, straight; filaments white, 6–7 mm long. *Ovary* ellipsoid, 3 × 2 mm, light green; style included, 7–8 mm long, straight, white. *Capsule* ellipsoid, 6–8 × 4–5 mm, spreading to suberect. *Seed* globose, 0.9 × 0.9 mm, glossy, black; strophiole rudimentary, 0.1 mm long, ridged. *Chromosome number* unknown. Figure 214.

FLOWERING PERIOD. August to October, with a peak from mid-August to mid-September.

HISTORY. *L. wrightii* was first collected by the American Charles Wright in the late 19th century at Simons Town in the southern Cape Peninsula. Baker (1878) initially described it at species level but later, in his second monograph, re-assessed it as a variety of *L. unifolia* (Baker, 1897a). It is here returned to species status as it is morphologically distinct and breeds true in allopatric populations.

85. LACHENALIA WRIGHTII

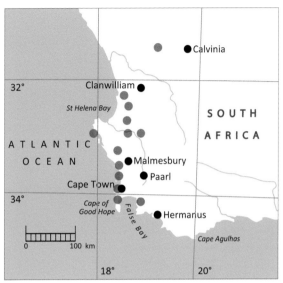

Map 86. Known distribution of *Lachenalia wrightii*.

Figure 214. *Lachenalia wrightii* on a granite outcrop near Vredenburg. Image: Dennis Tsang.

DISTINGUISHING CHARACTERS AND AFFINITIES. *L. wrightii* appears to be closely related to *L. unifolia*, which has a similar leaf and tepal colouring, and a similar subglobose bulb and globose seed. *L. unifolia* differs primarily in having tubular flowers with long perianth tubes (6 mm), oblong outer tepals, oblong-obovate inner tepals and longer filaments (9–15 mm).

DISTRIBUTION, HABITAT AND CONSERVATION STATUS. *L. wrightii* occurs mainly on the Cape west coast from just east of Elandsbaai to Mamre, and on the northern and southern Cape Peninsula, with outliers at Nieuwoudtville on the Bokkeveld Plateau and at Strand on the Cape Flats (Map 86). It occurs in nine vegetation types: Nieuwoudtville Roggeveld Dolerite Renosterveld, Leipoldtville-, Hopefield-, Atlantis- and Cape Flats Sand Fynbos, Swartland Granite- and Swartland Shale Renosterveld, Saldanha Granite Strandveld and Peninsula Granite Fynbos (Mucina & Rutherford, 2006). It is usually encountered on sandy flats, less frequently in humus-rich soil in seasonally moist shallow pans of granite outcrops. At a locality near Vredenburg, *L. wrightii* grows on a large granite outcrop in association with a mauve form of *L. pallida* that flowers later in the season. *L. wrightii* is not threatened.

NOTES. Honey bees (*Apis mellifera*) pollinate this species. The pedicels and ripening capsules usually become re-oriented into an erect position during the fruiting stage. Following capsule dehiscence, the scapes remain attached for several weeks and the seeds are dispersed locally.

1.3. Section ANGUSTAE

Angustae G. D. Duncan, *sect. nov.*, *a ceteris sectionibus perianthio anguste campanulato, tubo cupulato, tepalis interioribus expansis 45–50°, filamentis inclusis usque ad bene exsertis differt.* Type: *Lachenalia orthopetala* Jacq.

Perianth narrowly campanulate; tube cup-shaped, up to 3 mm long; outer tepals ovate; inner tepals obovate, radiating 45–50° from the longitudinal axis; filaments straight, narrowly spreading or declinate, included to well exserted.

9 species (9 taxa).

86. LACHENALIA ANGUINEA

Lachenalia anguinea Sweet, *The British Flower Garden* 2: t. 179 (1826). *Scillopsis anguinea* (Sweet) Lem., *L'Illustration Horticole* 3: 34 (1856).
TYPE: South Africa, Cape, precise locality unknown, specimen collected by W. Synnot, probably at Clanwilliam, figure in *The British Flower Garden* 2: t. 179 (1826) (lectotype, designated here).

ETYMOLOGY. *anguinea*: snake-like, with reference to the leaf shape and banding of the lower leaf surface.
DESCRIPTION. *Geophyte*, 100–600 mm high. *Bulb* subglobose, 15–20 mm in diameter, offset-forming, deep-seated; tunic multilayered, outer layers dark brown, spongy; inner cataphyll translucent white, loosely surrounding leaf bases, subterranean, apex obtuse. *Leaf* solitary, narrowly lanceolate, 300–450 × 15–30 mm, attenuate, spreading to suberect, light green or yellowish-green, upper surface plain, lower surface heavily marked with dark green transverse bands above ground level, shading to magenta bands below ground level, slightly to strongly canaliculate; leaf base loosely surrounding base of scape, 20–40 mm long, white with magenta bands; primary seedling leaf terete, erect. *Inflorescence* racemose, many-flowered, robust, sterile apex short; scape suberect, 140–300 mm long, light green, robust, uppermost portion inflated, plain or with brownish-magenta blotches; rachis light green throughout or light to deep pink above; pedicels spreading, suberect or arched, 10–25 mm long, white, lengthening markedly towards inflorescence apex; lower bracts ovate, 2–4 × 1–3 mm, green, becoming lanceolate above, translucent white. *Perianth* zygomorphic, narrowly campanulate, white, cream or greenish-cream, slightly spice-scented; tube cup-shaped, 2–3 mm long; outer tepals ovate, 5–6 × 5–6 mm, apical gibbosities light to bright green; inner tepals obovate, 6–7 × 2–3 mm, median keels light to deep green. *Stamens* well exserted, declinate; filaments white, 11–14 mm long. *Ovary* obovoid, 3–4 × 2–3 mm, bright green; style well exserted, declinate, 11–13 mm long, white. *Capsule* obovoid, 10–12 × 6–8 mm. *Seed* globose, 1.3–1.4 × 1.8 mm, glossy, black; strophiole rudimentary, 0.1–0.2 mm long, ridged. *Chromosome number* 2n = 30 + 2B (Johnson & Brandham, 1997). Figures 30, 215.
FLOWERING PERIOD. July to September, with a peak from late August to early September.
HISTORY. The English horticulturist, botanist and ornithologist Robert Sweet (1783–1835) described *L. anguinea* in 1826 in volume 2 of his *The British Flower Garden*. His description was accompanied by a coloured figure by Edwin Dalton Smith, drawn from a plant collected by the Irish soldier Captain Walter Synnot (1773–1851) at an unrecorded locality at the Cape, and cultivated in Britain by the English nurseryman and seedsman James Colville of King's Road, Chelsea. Synnot's collection is probably from the Clanwilliam district, where he settled in 1820 and became deputy magistrate in

THE GENUS LACHENALIA
86. LACHENALIA ANGUINEA

Figure 215. *Lachenalia anguinea* on deep red sand flats near Kleinzee, western Namaqualand, with *Ornithogalum xanthochlorum*. Image: Graham Duncan.

1821. He sent many bulbous plants to the London nurseryman Robert Sweet who credited him with the introduction of 'more new and rare bulbs from the Cape of Good Hope at one time than was ever done by any other individual' (Gunn & Codd, 1981). In 1856, Lemaire transferred the species to *Scillopsis* as *S. anguinea* (Sweet) Lem. in volume 3 of his *L'Illustration Horticole*, but it was returned to *Lachenalia* by Baker (1871) in his first monograph and this status was upheld in his final work (Baker, 1897a). Sölch & Roessler (1970) mistakenly listed one of several Dinter specimens amongst a mixed collection under Dinter's collection number 6666 as *L. anguinea* in their *Podromus einer Flora von Südwestafrika*; the specimen was later described as a new species, *L. nutans* (Duncan, 1998a).

Map 87. Known distribution of *Lachenalia anguinea*.

DISTINGUISHING CHARACTERS AND AFFINITIES. *L. anguinea* is allied to *L. nutans* (Duncan *et al.*, 2005a), with which it shares inflated scapes and well-exserted, declinate stamens, but *L. nutans* is a much smaller plant (up to 110 mm high) with a broadly lanceolate, glaucous leaf, with dark maroon margins. It has smaller, oblong-campanulate, strongly cernuous flowers, and smaller obovoid capsules (4–6 × 4 mm).
DISTRIBUTION, HABITAT AND CONSERVATION STATUS. The distribution of *L. anguinea* is in the dry sandy coastal plain extending from the western Richtersveld to Het Kruis, just north of the Piketberg Mountains (Map 87). The plants occur in a variety of vegetation types including Richtersveld Coastal Duneveld, Rooiberg Quartz Vygieveld, Namaqualand Strandveld, Namaqualand Klipkoppe Shrubland, Namaqualand Arid Grassland, Vanrhynsdorp Gannabosveld, Klawer Sandy Shrubland and Leipoldtville Sand Fynbos (Mucina & Rutherford, 2006). Typically, *L. anguinea* grows in large colonies in deep red sand in exposed situations or within the protection of low bushes. Near Kleinsee in western Namaqualand, *L. anguinea* occurs in association with the robust *Ornithogalum xanthochlorum* (Asparagaceae), and in the same area where deep sandy soil gives way to stony quartz outcrops, it grows adjacent to *L. framesii*, which is restricted to this substrate. *L. anguinea* is not threatened.
NOTES. *L. anguinea* is pollinated by honey bees (*Apis mellifera*). Its leaves are subject to heavy attack by the fungus *Uromyces lachenaliae* in winter, both in the wild and in cultivation. The infructescence detaches from the bulb shortly after the seed capsules have dehisced and is carried away by the wind, scattering the seeds widely. The bulbs are adapted to remain dormant for one or more years during periods of extended drought.

87. LACHENALIA LILIIFLORA

Lachenalia liliiflora Jacq. (sphalm. *liliflora*), *Collectanea* 5: 66–67 (1797). *Scillopsis liliiflora* (Jacq.) Lem., *L'Illustration Horticole* 3: 35 (1856).
TYPE: South Africa, Cape, collector and precise locality unknown, figure in N. J. Jacquin, *Icones Plantarum Rariorum* 2: 15, t. 387 (1793 or 1794) (neotype, designated here).

ETYMOLOGY. *liliiflora*: funnel-shaped flowers, as in certain *Lilium* species.
DESCRIPTION. *Geophyte*, 100–200 mm high. *Bulb* subglobose, 15–20 mm in diameter, offset-forming; tunic multilayered, outer layers light to dark brown; inner cataphyll translucent white, loosely surrounding leaf bases, apex obtuse. *Leaves* 2, lanceolate, 50–100 × 9–25 mm, suberect to spreading, dark green, coriaceous, slightly canaliculate, upper surface with depressed longitudinal grooves, smooth or sporadically to densely covered with small to large green pustules, lower surface plain or flushed with dull maroon, proteranthous; leaf margins flat to slightly undulate; leaf bases clasping, 15–20 mm long, white below, shading to light green above; primary seedling leaf flat, prostrate. *Inflorescence* racemose, dense, many-flowered, sterile apex short; scape erect to suberect, 40–120 mm long, sturdy, light green, plain or slightly to heavily suffused with dull greyish-purple; rachis colouring as for scape; pedicels suberect, 1–5 mm long, light to dark green or greyish-purple; lower bracts ovate, becoming lanceolate above, 1.5–4.0 × 0.5–4.0 mm, green, purplish-brown or greyish-purple, apices translucent white. *Perianth* zygomorphic, narrowly campanulate, suberect, white, sweet-scented; tube cup-shaped, 3 mm long, white; outer tepals ovate, 9–10 × 4–5 mm, apices slightly recurved, apical gibbosities dark brown, brownish-green or brownish-purple; inner tepals obovate, 13–15 × 5–6 mm, apices obtuse, apical marking brown, brownish-green or light to deep purple. *Stamens* included to shortly exserted, narrowly spreading; filaments white, 11–13 mm long. *Ovary* ellipsoid, 4–5 × 2–3 mm, yellowish-green; style included, straight, 10–11 mm long, white to light magenta, protruding beyond perianth

as ovary matures. *Capsule* ellipsoid, 9–10 × 4–5 mm. *Seed* ovoid, 0.9 × 0.8–0.9 mm, glossy, black; strophiole rudimentary, 0.2 mm long, ridged. *Chromosome number* 2n = 16 (Moffett, 1936; Fernandes & Neves, 1962; Johnson & Brandham, 1997; Hamatani *et al.*, 1998). Figures 216, 217.

FLOWERING PERIOD. September to October, with a peak in mid-October.

HISTORY. *L. liliiflora* was first illustrated by N. J. Jacquin on plate 387 in volume 2 of his *Icones Plantarum Rariorum*, produced in 1793 or 1794, and the description followed in volume 5 of *Collectanea* (Jacquin, 1797a). The plate is here designated as the neotype as it was not referred to in the original description, and the holotype perished in a fire in Vienna between 1939–1945. The specific epithet was originally spelt *liliflora* but the correct version is *liliiflora*, in keeping with the regulations governing syntax. The species was transferred to *Scillopsis* by Lemaire (1856) as *S. liliiflora* (Jacq.) Lem. in volume 3 of his *L'Illustration Horticole*, but was recognised by its original name by Baker in both of his monographs (1871, 1897a).

DISTINGUISHING CHARACTERS AND AFFINITIES. *L. liliiflora* resolved as sister to *L. orthopetala* in phylogenetic analysis (Duncan *et al.*, 2005a). These two species flower at the same time and share similar white, narrowly campanulate, suberect flowers with narrowly spreading stamens, but *L. orthopetala* differs in its narrower, shorter inner tepals (11–13 × 4 mm) with acute apices and brown, greenish-brown or dark magenta apical markings. *L. orthopetala* also differs in its 3–5 linear, deeply canaliculate leaves and larger seeds (1.0–1.1 × 0.9–1.0 mm). The leaves of both species are often proteranthous in the wild, being partially or completely shrivelled at flowering. The form of *L. liliiflora* that occurs on hill slopes within the Tygerberg Nature Reserve and the adjacent De Grendel Farm east of Cape Town has shorter, congested inflorescences when compared to the elongate racemes that are produced by typical forms from open flats.

Figure 216. *Lachenalia liliiflora* with elongate racemes on clay flats at Durbanville. Image: Graham Duncan.

87. LACHENALIA LILIIFLORA

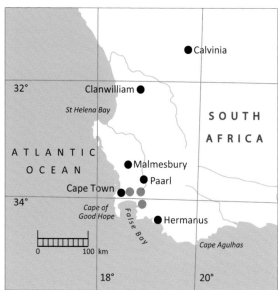

Map 88. Known distribution of *Lachenalia liliiflora*.

Figure 217. *Lachenalia liliiflora* with congested racemes on a stony clay slope at Tygerberg. Image: Adam Harrower.

A white form of *L. pallida* previously known as *L. unicolor* Jacq. var. *fragrans* (Jacq.) Baker from Schaapenberg on Vergelegen Estate at Somerset West, superficially resembles *L. liliiflora*. This taxon differs from *L. liliiflora* in having oblong-campanulate flowers with much shorter outer and inner tepals (4–5 and 7–8 mm, as opposed to 9–10 and 13–15 mm) and in its declinate stamens and style.

DISTRIBUTION, HABITAT AND CONSERVATION STATUS. Historically, *L. liliiflora* has always been restricted to a small area east of Cape Town that extends from Tygerberg to Paarl and Somerset West (Map 88). It still survives in the vicinity of Durbanville but many populations have been extirpated. The plants grow in Swartland Shale Renosterveld vegetation (Mucina & Rutherford, 2006), occurring as scattered individuals or in small groups on seasonally moist, stony clay flats that bake rock hard in summer, or on west- and north-east facing hillsides in open aspects and within the protection of low bushes. Near Durbanville, *L. liliiflora* grows within a few hundred metres of a population of *L. orthopetala*, but no natural hybrids have been recorded. *L. liliiflora* is protected within the Tygerberg Nature Reserve and De Grendel Farm, but elsewhere is severely threatened by agricultural expansion, invasive alien grasses, housing and industrial development. It has a conservation status of Endangered (Raimondo *et al.*, 2009).

NOTES. The sweet-scented flowers of *L. liliiflora* are pollinated mainly by honey bees (*Apis mellifera*) and are also visited by march flies (Diptera: Bibionidae) (Figure 69). The bulbs are occasionally eaten by Cape porcupines (*Hystrix africaeaustralis*). The scape remains attached for several weeks following capsule dehiscence and the seeds are dispersed locally.

88. LACHENALIA ORTHOPETALA

Lachenalia orthopetala Jacq., *Collectanea* 3: 240–241 (1790). *Scillopsis orthopetala* (Jacq.) Lem., *L'Illustration Horticole* 3: 34–35 (1856).

TYPE: South Africa, Cape, collector and precise locality unknown, figure in N. J. Jacquin, *Icones Plantarum Rariorum* 2: 6, t. 383 (1790) (neotype, designated here).

SYNONYMY: *Lachenalia ustulata* Banks ex Schult. & Schult. f., (name only), *Systema Vegetabilium* 7 (1): 601 (1829). Type: South Africa, Cape, precise locality unknown, F. Masson s.n. (BM!, holo.).

ETYMOLOGY. *orthopetala*: straight petals (tepals).

DESCRIPTION. *Geophyte*, 90–270 mm high. *Bulb* subglobose, 15–25 mm in diameter, offset-forming, yellow, tunic multilayered, outer layers dark brown, spongy; inner cataphyll translucent white, subterranean, adhering to leaf bases, apex obtuse. *Leaves* 3–5, linear, 100–220 × 3–10 mm, rosulate, erect to suberect or spreading, canaliculate, subulate, bright green, upper surface plain or with darker green or brown spots, proteranthous; leaf bases loosely surrounding base of scape, 20–40 mm long, white below, shading to light green tinged with dark magenta above; primary seedling leaf terete, erect. *Inflorescence* racemose, many-flowered, dense, sterile apex short; scape suberect, 50–180 mm long, suberect, greyish-maroon, plain or heavily marked with brownish-maroon blotches, slender, rigid; rachis greyish-maroon, apex shading to light green; pedicels suberect, 2–3 mm long, brownish-maroon, greyish-maroon or white; bracts cup-shaped throughout, 2–3 × 1–3 mm, green or white, apices occasionally greyish-maroon. *Perianth* zygomorphic, narrowly campanulate, suberect, white;

Figure 218 (left). *Lachenalia orthopetala* on seasonally moist clay flats at Durbanville. Image: Graham Duncan.
Figure 219 (right). Cultivated specimens of *Lachenalia orthopetala* from Durbanville. Image: Graham Duncan.

88. LACHENALIA ORTHOPETALA

tube cup-shaped, 3 mm long, white, median keels white, ageing to bright magenta; outer tepals narrowly ovate, 9–10 × 3–4 mm, straight, apices acute, apical gibbosities brown, greenish-brown or dark magenta; inner tepals narrowly obovate, straight, apices acute, 11–13 × 4 mm, apical marking brown, greenish-brown or dark magenta. *Stamens* included to shortly exserted, narrowly spreading; filaments white, 10–12 mm long. *Ovary* ellipsoid, 4 × 2 mm, yellowish-green; style included, straight, 11 mm long, white. *Capsule* ellipsoid, 8–10 × 4–5 mm. *Seed* ovoid, 1.0–1.1 × 0.9–1.0 mm, glossy, black; strophiole rudimentary, 0.2–0.3 mm long, ridged. *Chromosome number* 2n = 16 (Johnson & Brandham, 1997; Hamatani *et al.*, 2004; Spies *et al.*, 2008). Figures 218, 219.

FLOWERING PERIOD. September to October, with a peak in mid-October.

HISTORY. *L. orthopetala* was illustrated on plate 383 in volume 2 of his *Icones Plantarum Rariorum* in 1790 and described in volume 3 of N. J. Jacquin's *Collectanea* the same year. Because of the loss of the type specimen in a fire in Vienna between 1939–1945, the plate is designated here as the neotype as it was not referred to in the protologue of Jacquin's original description. Lemaire (1856) transferred the species to *Scillopsis* as *S. orthopetala* (Jacq.) Lem. but Baker (1871, 1897a) recognised its original name in both of his monographs.

Although Jacquin was the first to publish a description and figure of the species, the Scottish gardener and traveller Francis Masson probably discovered it. A Masson specimen collected at an unrecorded locality at the Cape, probably in the spring of 1773 or 1774, is preserved in the herbarium of Sir Joseph Banks and now forms part of the *Lachenalia* holdings in the British Museum (Natural History). Banks gave it the manuscript name *L. ustulata*, which was published by J. A. Schultes and J. H. Schultes (1829) under *L. orthopetala* in the seventh edition of Linnaeus's *Systema Vegetabilium*.

DISTINGUISHING CHARACTERS AND AFFINITIES. *L. orthopetala* resolved as sister to *L. liliiflora* in a phylogenetic analysis of unweighted morphological data (Duncan *et al.*, 2005a). *L. liliiflora* flowers at the same time of year and has similar white flowers. Diagnostic differences are discussed under *L. liliiflora*.

DISTRIBUTION, HABITAT AND CONSERVATION STATUS. *L. orthopetala* has a narrow range extending from Papkuilsvlei west of Citrusdal in the southern Olifants River Valley to Durbanville east of Cape Town (Map 89). The species is gregarious, growing in large colonies on seasonally inundated flats. It occurs in Swartland Shale Renosterveld vegetation (Mucina & Rutherford, 2006), in stony clay soils that bake rock hard in summer. Most of its habitat has been destroyed by wheat cultivation, and it is under continuing decline resulting from urban sprawl, invasive alien grasses and road construction east of Cape Town, where a few populations remain on degraded farmland and in roadside ditches. A small population on farmland in Durbanville grows in association with the Critically Endangered *Babiana secunda* (Iridaceae), in close proximity to a population of *L. liliiflora*, which occurs in drier, gently sloping terrain. *L. orthopetala* has a conservation status of Vulnerable (Raimondo & Duncan, 2009c).

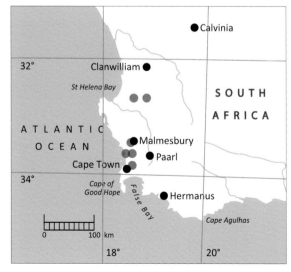

Map 89. Known distribution of *Lachenalia orthopetala*.

NOTES. The flowers of *L. orthopetala* are pollinated mainly by honey bees (*Apis mellifera*), and are also visited by two butterfly species, the Painted Lady (*Vanessa cardui*) and the exotic Cabbage White (*Pieris brassicae*). In the wild, the leaves are proteranthous, usually having senesced before the flowers open. The scape remains attached for several months after the capsules have dehisced and the seeds are dispersed locally.

89. LACHENALIA CONVALLARIOIDES

Lachenalia convallarioides Baker, *Journal of the Linnean Society* (*Botany*) 11: 407 (1871).
TYPE: South Africa, Eastern Cape, Kreli's Country, Caffraria, Mount Arthur Range, *J. H. Bowker* 444 (K!, holo.; TCD!, iso.).
SYNONYMY: *Lachenalia convallarioides* Baker var. *robusta* Baker, *Journal of the Linnean Society* 11: 407 (1871). Type: South Africa, Eastern Cape, Albany, *Williamson s.n.* (TCD!, holo.).

ETYMOLOGY. *convallarioides*: flower shape reminiscent of *Convallaria majalis*, 'Lily of the Valley'.
DESCRIPTION. *Dwarf geophyte*, 60–150 mm high. *Bulb* subglobose, 10–15 mm in diameter, offset-forming; tunic multilayered, outer layers light to dark brown, spongy; inner cataphyll translucent white, loosely surrounding leaf base, apex acute. *Leaf* 1 (2 leaves sometimes produced in cultivation), lanceolate, 100–200 × 6–15 mm, spreading to suberect, light green to yellowish-green, apex sometimes dull greenish-maroon, upper surface with distinct depressed longitudinal grooves, fleshy, slightly to moderately canaliculate, or conspicuously u-shaped in cross-section, tapering to a long attenuate apex, midrib absent or distinct on lower surface; leaf bases clasping, 5–15 mm long, lower half plain or with minute purple spots at base, shading to light green above; primary seedling leaf terete, erect. *Inflorescence* racemose, few- to many-flowered, lax, sterile apex short; scape erect to suberect, 40–100 mm long, light green, plain or with light brownish-purple tinge, slender; rachis light green, plain or with brownish-purple to dull maroon tinge; pedicels suberect, 3–5 mm long, white in lower half shading to brownish-maroon above; bracts cup-shaped throughout, 1–2 × 1–2 mm, white. *Perianth* zygomorphic, narrowly campanulate, translucent white to light pink, cernuous, lightly sweet-scented; tube cup-shaped, 1–2 mm long; outer tepals ovate, 5–6 × 3–4 mm, apices slightly recurved, apical gibbosities light- to dark brownish-magenta, brownish-green or light green, median keels light brownish-magenta, darkening as flowers mature; inner tepals obovate, 5–6 × 3–4 mm, apices recurved, median keels light brownish-magenta. *Stamens* included, straight; filaments white, 4 mm long; pollen ageing to black. *Ovary* obovoid, 1–2 × 1.0–1.5 mm, bright green or yellow; style included, straight, 3–4 mm long, white. *Capsule* obovoid, 4–5 × 3–4 mm. *Seed* globose, 0.9–1.2 × 0.9–1.0 mm, matte, black; strophiole rudimentary, 0.1–0.2 mm long, ridged. *Chromosome number* 2n = 30 (Johnson & Brandham, 1997). Plate 28, Figure 220.
FLOWERING PERIOD. August to November, with a peak in early October.
HISTORY. The earliest collection of *L. convallarioides* is that of Col. J. H. Bowker (1822–1900), a South African soldier and naturalist (after whom *L. bowkeri* was named), who found it near Cala, in 'Kreli's Country, Caffraria', in 1856. The territory known as 'Kreli's Country' is the area between the Great Kei and Bashee Rivers in the former Transkei (now the Eastern Cape), then under the control of Chief Kreli. The species has not been recorded from that locality again. *L. convallarioides* was described by Baker from Bowker's collection when he published his first monograph of the genus in the *Journal of the Linnean Society* in 1871, and the holotype sheet (*Bowker* 444) is housed in the Kew Herbarium. A sheet in the herbarium of Trinity College, Dublin collected by M. E. Barber (sister of

PLATE 28. Watercolour painting of *Lachenalia convallarioides* from Grahamstown (*Dold* 1018, in NBG) courtesy of The Editor, *Flowering Plants of Africa*, vol. 56, t. 2145 (1999). Artist: Fay Anderson.

89. LACHENALIA CONVALLARIOIDES

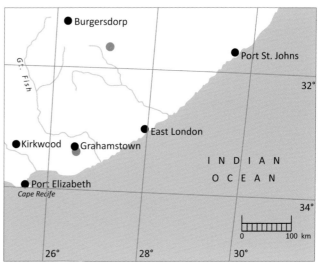

Map 90. Known distribution of *Lachenalia convallarioides*.

Figure 220. Cultivated specimen of *Lachenalia convallarioides* from Grahamstown. Image: Graham Duncan.

Bowker) at Zuurberg in Chief Kreli's territory has the same collecting number (444) and is thought to be the same collection as the holotype. Bowker is known to have collected specimens for his sister, who sent them to Harvey at Trinity College. Baker (1871) described a second, undated collection, gathered by T. Williamson in the Albany District, in the same volume of the *Journal of the Linnean Society* as *L. convallarioides* Baker var. *robusta* Baker. This name has since been placed in synonymy as it is merely a robust specimen with slightly larger dimensions (Dold & Phillipson, 1998).

L. convallarioides was in cultivation in England by the early 1900s if not before, according to an article in *The Garden*, in which the flowers are described as having a 'strong, heather-like odour' (Watson, 1904). Since its original publication, the species has only been recorded from the Albany District near Grahamstown. Three collections were made there in the late 1890s, one by Dr Selmar Schönland (*Schönland s.n.*, in SAM), another by C. A. Pym (*Pym s.n.*, in GRA) and a third by the nuns of the local Convent, in 1928 (specimen in GRA). L. Borman collected it again in 1942, but *L. convallarioides* was not recorded for just over 50 years until it was found again in the same area in 1994 by A. P. Dold, who made material available for cultivation at Kirstenbosch. A watercolour painting by Fay Anderson was subsequently published in *Flowering Plants of Africa* (Duncan & Anderson, 1999).

DISTINGUISHING CHARACTERS AND AFFINITIES. *L. convallarioides* is allied to *L. zeyheri* (Duncan *et al.*, 2005a), with which it shares narrowly campanulate white to light pink perianths with subequal tepals with rounded apices, and straight stamens; but *L. zeyheri* differs in having a congested raceme of larger, spreading to suberect flowers with longer filaments (5–6 mm). It also differs in having linear leaves that are deeply canaliculate in the lower half and terete above, and in its larger globose seeds (1.3–1.4 × 1.2–1.3 mm).

DISTRIBUTION, HABITAT AND CONSERVATION STATUS. *L. convallarioides* is known from only two locations in the Eastern Cape: one is an extant population at Grahamstown in the Albany District, the other a historical record to the north-east in the Mount Arthur Range near Cala, a disjunction of about 180 km (Map 90). Altitude for the species ranges from 17 m to 1,800 m above sea level. The plants grow in Suurberg Quartzite Fynbos at Grahamstown, and in Tarkastad Montane Shrubland vegetation near Cala (Mucina & Rutherford, 2006). At Grahamstown, they occur on south-facing rocky slopes in sandy soil, in full sun or within the partial shade of surrounding shrubs. The habitat at this location is potentially threatened by alien invasive plants and *L. convallarioides* has a conservation status of Critically Endangered (Victor & Dold, 2009).

NOTES. The narrowly campanulate, white flowers of *L. convallarioides* suggest pollination by honey bees (*Apis mellifera*). The scape remains attached for several months following capsule maturation and the seeds are dispersed locally.

90. LACHENALIA ZEYHERI

Lachenalia zeyheri Baker, *Journal of the Linnean Society* (*Botany*) 11: 407 (1871).
TYPE: South Africa, Western Cape, flats between Witzenberg and Schurfteberg, *C. L. Zeyher* 1694 (K!, holo., iso; SAM!, iso.).

ETYMOLOGY. *zeyheri*: after Carl L. Zeyher (1799–1858), 19th century plant collector.

DESCRIPTION. *Dwarf geophyte*, 25–200 mm high. *Bulb* subglobose, 10–13 mm in diameter, offset-forming; tunic multilayered, outer layers dark brown, spongy; inner cataphyll translucent white, adhering to leaf bases, apex obtuse. *Leaf* 1(–2), linear, 150–290 × 4–10 mm, bright to dark green, spreading to erect, deeply canaliculate in lower half, upper half terete; leaf bases clasping, 10–30 mm long, subterranean, white with light to dark magenta blotches; primary seedling leaf terete, erect. *Inflorescence* racemose, few- to many-flowered, dense; scape erect to suberect, 50–110 mm long, light green, lightly to heavily marked with dark maroon blotches; pedicels suberect, 3–5 mm long, white or light green; bracts cup-shaped throughout, 1.0–1.5 × 0.5–4.0 mm. *Perianth* zygomorphic, narrowly campanulate, spreading to suberect, white or light pink; tube cup-shaped, 2 mm long; white or light pink; outer tepals ovate, 4 × 3 mm, apices rounded, slightly recurved, apical gibbosities yellowish-green; inner tepals obovate, 5 × 3 mm, apices rounded, slightly recurved, median keels light yellowish-green. *Stamens* included to shortly exserted, straight; filaments white, 5–6 mm long. *Ovary* obovoid, 2.0 × 1.5 mm, light green; style included, straight, 5 mm long, white. *Capsule* obovoid, 3–4 × 2–3 mm. *Seed* globose, 1.3–1.4 × 1.2–1.3 mm, glossy, black; strophiole rudimentary, 0.1–0.2 mm long, ridged. *Chromosome number* 2n = 22 (Johnson & Brandham, 1997); 2n = 23 (Hamatani et al., 1998). Figures 39, 221.

FLOWERING PERIOD. September to October, with a peak in late September.

HISTORY. *L. zeyheri* was described by Baker (1871) in the *Journal of the Linnean Society*, from specimens collected by C. L. Zeyher at an unrecorded locality between the Schurfteberg and Witzenberg mountain ranges north of Tulbagh. No collection date appears on either the holotype or isotype sheets housed at Kew (*Zeyher* 1694), but they were probably gathered in late 1831 when Zeyher and his botanical partner C. F. Ecklon climbed the mountains around Tulbagh (Gunn & Codd, 1981). The species was included in Baker's second monograph in 1897 but more than a century passed before the plant was illustrated in colour for the first time, in *The Lachenalia Handbook* (Duncan, 1988a).

DISTINGUISHING CHARACTERS AND AFFINITIES. *L. zeyheri* is allied to the Eastern Cape endemic *L. convallarioides* (Duncan *et al.*, 2005a), and the relationship is discussed under that species. Superficially, *L. zeyheri* resembles certain forms of *L. contaminata*, but the latter differs mainly in having inner tepals that are distinctly longer than the outer tepals, and outer and inner tepals that have flat apices, maroon gibbosities and median keels. *L. contaminata* also differs from *L. zeyheri* in having a rosette of 3–11 narrower, linear leaves that are flat to slightly canaliculate in the lower half.

DISTRIBUTION, HABITAT AND CONSERVATION STATUS. *L. zeyheri* is limited to the mountain ranges extending from the northern Cederberg to just south of Worcester (Map 91). It occurs in six vegetation types including Cederberg-, Olifants-, Winterhoek- and Western Altimontane Sandstone Fynbos, Swartruggens Quartzite Fynbos and Ceres Shale Renosterveld (Mucina & Rutherford, 2006). The plants grow on exposed, seasonally waterlogged sandy-loam flats and along the edges of seasonal pools, and on moist sandy lower mountain slopes, often in large colonies sometimes numbering hundreds or thousands of individuals. *L. zeyheri* is frequently seen in the mountains north-west and north-east of Ceres. At a locality north-east of Op-die-Berg in the Swartruggens Mountains, it grows close to stands of *Gladiolus alatus* and *Moraea macronyx* (Iridaceae) and the shrub *Nylandtia spinosa* (Polygalaceae). Its numbers have been reduced to some extent by the expansion of citrus and deciduous fruit orchards, but it is not under immediate threat.

NOTES. The flowers are pollinated by honey bees (*Apis mellifera*) and are also visited by the Painted Lady butterfly (*Vanessa cardui*). Following capsule dehiscence, the scapes remain attached for several weeks and the seeds are dispersed locally by the shaking action of the wind.

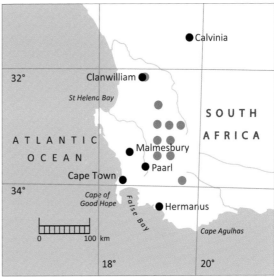

Map 91. Known distribution of *Lachenalia zeyheri*.

Figure 221. *Lachenalia zeyheri* on high-lying clay flats in the Swartruggens Mountains north-west of Ceres. Image: Graham Duncan.

91. LACHENALIA CONTAMINATA

Lachenalia contaminata Aiton, *Hortus Kewensis* 1: 460–461 (1789). *Scillopsis contaminata* (Aiton) Lem., *L'Illustration Horticole* 3: 34 (1856).

TYPE: South Africa, Cape, collector and precise locality unknown (cultivated specimen), (BM!, holo.).

SYNONYMY: *Lachenalia angustifolia* Jacq., *Collectanea* 5: 57–58 (1797). *Scillopsis angustifolia* (Jacq.) Lem., *L'Illustration Horticole* 3: 33–34 (1856). Type: South Africa, Cape, collector and precise locality unknown, figure in N. J. Jacquin, *Icones Plantarum Rariorum* 2: 16, t. 381 (1795) (neotype, designated here).

Lachenalia hyacinthoides Jacq., *Collectanea* 5: 58–59 (1797). *Scillopsis hyacinthoides* (Jacq.) Lem., *L'Illustration Horticole* 3: 35–36 (1856). Type: South Africa, Cape, collector and precise locality unknown, figure in N. J. Jacquin, *Icones Plantarum Rariorum* 2: 16, t. 382 (1795) (neotype, designated here).

Lachenalia albida Tratt., *Archiv der Gewachskunde* 2: t. 162 (1814). Type: South Africa, Cape, collector and precise locality unknown, figure in *Archiv der Gewachskunde* 2: t. 162 (1814) (lectotype, designated here).

Lachenalia fragrans Lodd., *Loddiges Botanical Cabinet*: t. 1140 (1826), illegitimate homonym, non *Lachenalia fragrans* Jacq., *Plantarum rariorum horti caesarei Schönbrunnensis* 1: t. 82 (1797), non *Lachenalia fragrans* Andrews, *The Botanist's Repository* 5: t. 302 (1803). Type: South Africa, Cape, collector and precise locality unknown, figure in *Loddiges Botanical Cabinet*: t. 1140 (1826) (lectotype, designated here).

Lachenalia hyacinthina Hort., *Möllers Deutsche Gartner-Zeitung* 21: 443 (1906). Type: South Africa, Cape, collector and precise locality unknown, figure in *Möllers Deutsche Gartner-Zeitung* 21: 445 (1906) (lectotype, designated here).

ETYMOLOGY. *contaminata*: descriptive of the maroon tepal markings or alternatively the spicy scent emitted by the flowers.

DESCRIPTION. *Geophyte*, 60–300 mm high. *Bulb* subglobose, 10–20 mm in diameter, offset-forming; tunic multilayered, outer layers dark brown, spongy; inner cataphyll translucent white, adhering to leaf bases, apex obtuse. *Leaves* 3–11, linear, 90–210 × 1–4 mm, rosulate, spreading to erect, light to dark green, lower half canaliculate, becoming terete above, upper surface plain or sporadically to heavily marked with purplish-maroon spots, lower surface plain, or lower half lightly to heavily shaded with purplish-maroon; leaf bases loosely surrounding base of scape, 20–35 mm long, subterranean, white, plain or with brownish-magenta blotches; primary seedling leaf terete, erect. *Inflorescence* racemose, many-flowered, dense, sterile apex short; scape erect to suberect, light green to dull purplish-maroon, marked with minute to relatively large, darker purplish-maroon blotches, 40–190 mm long; rachis light green, plain or lightly to heavily marked with purplish-maroon blotches; pedicels suberect, white, 1–2 mm long; bracts cup-shaped, green at base of inflorescence, shading to white above with purplish-maroon median keels, 1–3 × 1–4 mm. *Perianth* zygomorphic, narrowly campanulate, spreading to suberect, spice-scented; tube cup-shaped, white, 2–3 mm long; outer tepals ovate, white with purplish-maroon or brown gibbosities, median keels ageing to purplish-maroon, 6–7 × 3–4 mm; inner tepals narrowly obovate, spreading, white, median keels dark maroon or brown, ageing to purplish-maroon, 7–8 × 3–4 mm. *Stamens* included to well exserted, filaments more or less straight, white, 5–10 mm long; anthers maroon, pollen yellow. *Ovary* ellipsoid, 2–3 × 2 mm, light green, style included to well exserted, straight, white, 4–8 mm long. *Capsule* ellipsoid, 7–8 × 4–5 mm. *Seed* globose, glossy, black, 0.9–1.1 × 0.8–1.0 mm; strophiole rudimentary, 0.1–0.2 mm long, ridged. *Chromosome number* $2n = 14$

91. LACHENALIA CONTAMINATA

Figure 222 (left). *Lachenalia contaminata* on seasonally inundated clay flats near Wolseley. Image: Graham Duncan.
Figure 223 (right). Robust form of *Lachenalia contaminata* from near Worcester. Image: Graham Duncan.

(Crosby, 1986; Spies *et al.*, 2008); 2n = 14, 16 (Gouws, 1964); 2n = 16 (de Wet, 1957); 2n = 16+1B (Hamatani *et al.*, 2004); 2n = 16, 32, 16+1B (Johnson & Brandham, 1997). Figures 71, 222–224.

FLOWERING PERIOD. Early August to late October, with a peak in mid-September.

HISTORY. *L. contaminata* was first illustrated in watercolour in about 1700 for the Florilegium *The Flora Capensis of Jakob and Johann Philipp Breyne*. This illustration was one of three paintings of lachenalias from this Florilegium copied for the *Lachenalia* engraving used to illustrate the Breynes's *Prodromi Fasciculi Rariorum Plantarum* (1739), in which *L. contaminata* was described by the phrase name *Hyacinthus orchioides, aphyllus, serpentarius maior*. The species was in cultivation in England by at least the late 1760s and was introduced to Kew in 1774 by the Scottish plantsman Francis Masson (Aiton, 1789). It was listed as an unnamed variety (ε) of *Phormium orchioides* L. by Thunberg (1784) in his *Nova Genera Plantarum*, and finally described as *L. contaminata* by William Aiton (1789) in the first volume of *Hortus Kewensis*, from a cultivated specimen of unknown wild provenance that is preserved in the British Museum (Natural History). Various forms of *L. contaminata* were subsequently described and illustrated under different names, including Jacquin's *L. angustifolia* and *L. hyacinthoides*, which were reproduced in 1795 in the second volume of Jacquin's *Icones Plantarum Rariorum*. The well-known plate t. 178 reproduced in the second volume of N. J. Jacquin's *Hortus Botanicus Vindobonensis* (1772) was initially mistakenly identified as *Hyacinthus orchioides* L., but later placed with *L. angustifolia* (Jacquin, 1797a).

91. LACHENALIA CONTAMINATA

DISTINGUISHING CHARACTERS AND AFFINITIES. *L. contaminata* is regarded here as a distinctive but highly polymorphic species complex, occurring as geographically isolated, more or less continuously varying populations. The plants vary minimally within populations but variation between populations is marked. Forms with included stamens predominate in the western parts of its range, whereas those with shortly to well-exserted stamens occur mainly in the southern parts around Bredasdorp and Riviersonderend. An unusually robust, tall-growing form of up to 300 mm in height occurs at Riverlands near Malmesbury and at Rawsonville near Worcester, and is especially late-flowering (mid- to late October).

L. contaminata is allied to *L. bachmannii* (Duncan *et al.*, 2005a), sharing straight stamens, linear leaves, subglobose bulbs and globose seeds that have rudimentary strophioles. *L. bachmannii* differs in its taller stature (up to 420 mm high) and always has two deeply canaliculate or sometimes conduplicate leaves. It has shorter, broader outer tepals (5–6 × 4–5 mm), shorter filaments (4–5 mm long) and a larger seed (1.1–1.2 × 1.3–1.4 mm).

DISTRIBUTION, HABITAT AND CONSERVATION STATUS. *L. contaminata* is confined to the western, south-western and southern parts of the Western Cape, extending from Lamberts Bay to De Hoop Nature Reserve east of Bredasdorp (Map 92). It is fairly common across its range and is fairly abundant in the south-western parts around Darling, Worcester and Tulbagh. It occurs in 10 vegetation types: Leipoldtville-, Atlantis- and Cape Flats Sand Fynbos, Olifants Sandstone Fynbos, Swartland-, Breede-, Western Rûens- and Central Rûens Shale Renosterveld, Swartland Granite Renosterveld and De Hoop Limestone Fynbos (Mucina & Rutherford, 2006). *L. contaminata* is mostly encountered in dense colonies on exposed, seasonally inundated gravelly clay flats (Darling), but also grows on

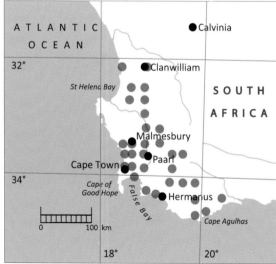

Map 92. Known distribution of *Lachenalia contaminata*.

Figure 224. Cultivated specimen of *Lachenalia contaminata* with well-exserted stamens from De Hoop Nature Reserve. Image: Graham Duncan/SANBI.

sandstone derived soils (Riviersonderend) and on limestone deposits (Bredasdorp). Population numbers have been much reduced by agricultural expansion, including increased winter cereal cultivation, trampling and overgrazing by livestock, and encroaching alien grasses. *L. contaminata* has a conservation status of Near Threatened (Helme & Raimondo, 2009c).

NOTES. The flowers of *L. contaminata* are pollinated mainly by honey bees (*Apis mellifera*) and are also visited by the generalist day-flying Heady Maiden moth (*Amata cerbera*) (Figure 71). Certain forms observed in cultivation at Kirstenbosch are, however, at least partially self fertile, regularly setting seed under controlled, insect-free nursery conditions. Following capsule dehiscence, the scapes remain attached for several months and the seeds are dispersed locally.

92. LACHENALIA BACHMANNII

Lachenalia bachmannii Baker (sphalm. *bachmanni*), in Engler, H. G. A., *Botanische Jahrbücher* 15 (3): 8 (1892).

TYPE: South Africa, Western Cape, near Hopefield, Malmesbury district, *F. E. Bachmann* 1232 (B!, holo.; Z!, iso.).

SYNONYMY: *Lachenalia brevipes* Baker, *Bulletin de L'Herbier Boissier* (2nd series) 1: 856 (1901). Type: South Africa, Western Cape, Hopefield district, *F. E. Bachmann* 1235 (Z!, holo.).

ETYMOLOGY. *bachmannii*: after Dr F. E. Bachmann (1856–ca.1916), German medical practitioner, naturalist and explorer.

DESCRIPTION. *Geophyte*, 140–420 mm high. *Bulb* subglobose, 17–20 mm in diameter, offset-forming; tunic multilayered, outer layers spongy, light to dark brown; inner cataphyll translucent white, loosely surrounding leaf bases, apex obtuse; *Leaves* 2, linear, 150–250 × 4–10 mm, erect to suberect or arcuate, bright green to yellowish-green, lower surface plain or suffused with dull maroon, deeply canaliculate to conduplicate, uppermost portion terete; leaf bases clasping, 50–200 mm long, white below, shading to light green above; primary seedling leaf terete, erect. *Inflorescence* racemose, many-flowered, dense, slender, sterile apex short; scape erect to suberect, 90–300 mm long, light green, plain or tinged with light brown; rachis light green, plain or tinged with light brown; pedicels suberect, 1–2 mm long, white; bracts cup-shaped throughout, 1–2 × 3–4 mm, translucent white. *Perianth* zygomorphic, narrowly campanulate, suberect, slightly spice-scented; tube cup-shaped, 3 mm long, white, or light blue at base shading to white above; outer tepals ovate, 5–6 × 4–5 mm, white, apical gibbosities dark brown to dull maroon; inner tepals obovate, 7–8 × 4–5 mm, white, keels dark brown to dull maroon, apices slightly recurved. *Stamens* included, more or less straight; filaments white, 4–5 mm long. *Ovary* ellipsoid, 2–3 × 1–2 mm, dark green; style included, more or less straight, 4–5 mm long, white. *Capsule* ellipsoid, 5–7 × 4–5 mm, suberect. *Seed* globose, 1.1–1.2 × 1.3–1.4 mm, glossy, black; strophiole rudimentary, 0.3 mm long, ridged. *Chromosome number* 2n = 16 (de Wet, 1957; Crosby, 1986; Johnson & Brandham, 1997; Hamatani *et al.*, 2004). Figures 38, 66, 225.

FLOWERING PERIOD. August to September, with a peak from early to mid-September.

HISTORY. *L. bachmannii* was described by J. G. Baker in volume 15 of Engler's *Botanische Jahrbücher*, from material collected by F. E. Bachmann in August 1885 near Hopefield (*Bachmann* 1232) in the Malmesbury District (Baker, 1892b). Baker subsequently described a second Bachmann collection (*Bachmann* 1235) from near Hopefield as *L. brevipes* in *Bulletin de L'Herbier Boissier*, stating that he considered it to be near to *L. bachmannii* (Baker, 1901a). Comparison of the holotype material has revealed it to be the same species, and *L. brevipes* is here placed in synonymy.

92. LACHENALIA BACHMANNII

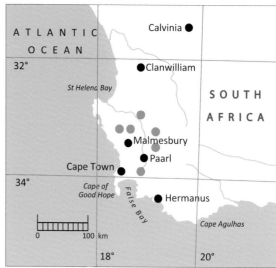

Map 93. Known distribution of *Lachenalia bachmannii*.

Figure 225. *Lachenalia bachmannii* in a seasonal pool at Saron near Tulbagh. Image: Graham Duncan.

DISTINGUISHING CHARACTERS AND AFFINITIES. *L. bachmannii* resolved adjacent to *L. contaminata* in a phylogenetic analysis of morphological data (Duncan *et al.*, 2005a) and the relationship is discussed under that species.

DISTRIBUTION, HABITAT AND CONSERVATION STATUS. *L. bachmannii* is native to the western and southwestern parts of the Western Cape from Piketberg to Hopefield, Moorreesburg, Saron, Gouda and Stellenbosch (Map 93). It is a seasonal aquatic, growing in standing pools, along their margins or in ditches in heavy clay, often in association with another aquatic geophyte, *Wurmbea stricta*. At a locality near Saron, it grows close to a population of *L. mediana* subsp. *rogersii* that favours less moist ground. *L. bachmannii* is confined to Swartland Shale Renosterveld vegetation (Mucina & Rutherford, 2006) and the plants occur in colonies numbering hundreds or sometimes thousands of individuals. The species is under extreme threat as a result of the draining of its fertile habitat for agricultural expansion, fertiliser runoff, trampling and overgrazing, and has a conservation status of Endangered (Helme & Raimondo, 2009a).

NOTES. The flowers of *L. bachmannii* are pollinated by honey bees (*Apis mellifera*) and solitary anthophorid bees (*Anthophora diversipes*), and are also visited by the Painted Lady butterfly (*Vanessa cardui*). The pollen is consumed by monkey beetles (*Pachycnema crassipes*) (Figure 66). The scapes remain attached for several months following capsule dehiscence and the seeds are dispersed locally.

93. LACHENALIA MAGENTEA

Lachenalia magentea G. D. Duncan, nom. nov. *Lachenalia juncifolia* Bak. var. *campanulata* W. F. Barker, *South African Journal of Botany* 55 (6): 640 (1989).

TYPE: South Africa, Western Cape, 19 km south of Riversdale on Blombos Road, *G. J. Lewis* 5628 (NBG!, holo.).

ETYMOLOGY. *magentea*: descriptive of the prominent magenta tepal keels and gibbosities.

DESCRIPTION. *Geophyte*, 60–150 mm high. *Bulb* subglobose, 10–15 mm in diameter, offset- and bulblet-forming; tunic multilayered, outer layers dark brown, spongy; inner cataphyll translucent white, subterranean, adhering to leaf bases, apex obtuse. *Leaves* 2, linear, 80–130 × 2–3 mm, spreading to suberect, succulent, subterete, bright green to yellowish-green; upper surface with very narrow channel, plain, lower surface with green or purplish-magenta bands in central portion becoming purplish-magenta below; leaf bases loosely surrounding base of scape, 20–30 mm long, subterranean, white; primary seedling leaf terete, erect. *Inflorescence* racemose, few- to many-flowered, sterile apex short; scape erect to suberect, 50–90 mm long, sturdy, light green, lightly to heavily mottled with purplish-magenta; rachis light green, heavily mottled with purplish-magenta; pedicels suberect, white, 5–9 mm long; lower bracts ovate, becoming lanceolate above, 1–2 × 1–3 mm, white, or dark magenta with white margins. *Perianth* zygomorphic, narrowly campanulate, cernuous; tube cup-shaped, 1–2 mm long, white to dull blue; outer tepals ovate, 5–6 × 3–4 mm, white, apices straight, apical gibbosities and surrounding zone light to deep magenta; inner tepals obovate, 5–7 × 4–5 mm,

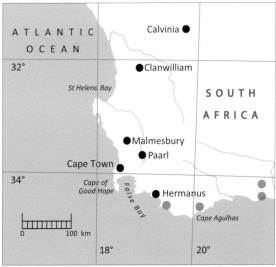

Map 94. Known distribution of *Lachenalia magentea*.

Figure 226. Cultivated specimens of *Lachenalia magentea* from Riversdale. Image: Graham Duncan/SANBI.

white, apices slightly recurved; keels light to deep magenta. *Stamens* shortly to well exserted, more or less straight; filaments white, 8–13 mm long; pollen yellow, ageing to dark brown. *Ovary* ellipsoid, 2 × 2 mm, bright green; style well exserted, often beyond the anthers, more or less straight, 8–12 mm long, white. *Capsule* obovoid, 5–7 × 4–6 mm. *Seed* globose, 1.1–1.2 × 1.3 mm, glossy, black; strophiole rudimentary, 0.2 mm long, ridged. *Chromosome number* unknown. Figure 226.

FLOWERING PERIOD. August to October, with a peak in September.

HISTORY. *L. magentea* was originally described as *L. juncifolia* Baker var. *campanulata* W. F. Barker in the *South African Journal of Botany* (Barker, 1989). Following a phylogenetic analysis of morphology (Duncan *et al.*, 2005a), it was considered expedient to elevate the taxon to specific rank, but the earlier specific epithet *campanulata* had already been used for another plant, *L. campanulata* Baker, and a new name became necessary.

DISTINGUISHING CHARACTERS AND AFFINITIES. *L. magentea* resolved adjacent to *L. juncifolia* in a phylogenetic analysis of morphology (Duncan *et al.*, 2005a), sharing paired linear leaves, exserted, straight filaments and similar globose seeds with rudimentary strophioles. *L. juncifolia* differs in its smaller, oblong-campanulate flowers, and in its canaliculate, less fleshy leaves.

DISTRIBUTION, HABITAT AND CONSERVATION STATUS. *L. magentea* is confined to the southern Cape coastal belt from Frikkies Bay west of Bredasdorp to just south of Riversdale (Map 94). It occurs in Agulhas Sand Fynbos, and Agulhas- and Canca Limestone Fynbos vegetation (Mucina & Rutherford, 2006), growing in colonies in shallow, seasonally moist sandy soil. *L. magentea* is not threatened.

NOTES. Honey bees (*Apis mellifera*) pollinate the flowers of *L. magentea*. The scapes remain attached to the bulbs for several weeks following capsule dehiscence and the seeds are dispersed locally by wind.

94. LACHENALIA KLIPRANDENSIS

Lachenalia kliprandensis W. F. Barker, *South African Journal of Botany* 53 (2): 168–169 (1987).
TYPE: South Africa, Western Cape, Kliprand, Bushmanland, *E. A. Schelpe s.n.* sub. *NBG* 129084 (NBG!, holo.).

ETYMOLOGY. *kliprandensis*: after the farm Kliprand in western Bushmanland.

DESCRIPTION. *Geophyte*, 85–200 mm high. *Bulb* globose, 15–20 mm in diameter, solitary; tunic multilayered, outer layers dark brown, spongy; inner cataphyll translucent white, adhering to leaf bases, apex obtuse. *Leaves* 2, ovate, 40–80 × 25–60 mm, prostrate, becoming spreading or suberect in cultivation, upper surface dark green to slightly glaucous, plain or with many irregularly scattered, dark green to light purple pustules; depressed longitudinal grooves prominent; leaf margins coriaceous, light purple; leaf bases clasping, 10–15 mm long, subterranean, light greenish-white below, shading to purple above; primary seedling leaf flat, prostrate. *Inflorescence* racemose, many-flowered, dense, sterile apex short; scape erect, 20–50 mm long, light green, apex sometimes inflated, plain or with irregularly scattered, light maroon blotches, or flushed with light maroon, subterranean, rarely shortly aerial; rachis light green with irregularly scattered maroon blotches or minute speckles, or flushed with light maroon, lower portion sometimes inflated; pedicels suberect, 1 mm long, light green or maroon; lower bracts ovate, becoming lanceolate above, 1–4 × 1–5 mm, light greenish-white. *Perianth* zygomorphic, narrowly campanulate, spreading to slightly cernuous, heavily spicy-sweet scented; tube cup-shaped, 3 mm long, white to greenish- or bluish-white; outer tepals ovate, 7–8 × 4–5 mm, white to light greenish-white, median keels and apical gibbosities greenish-brown; inner tepals obovate, 12–13 × 4–5 mm, translucent white, plain or apices light to bright magenta, keels brownish-magenta. *Stamens* included to shortly exserted, declinate; filaments white to light magenta, 10–13 mm long. *Ovary* ellipsoid, 3–4 × 2 mm, light bluish-green; style well exserted, declinate, 10–11 mm long, white to light magenta. *Capsule* ellipsoid, 10–12 × 5–7 mm. *Seed* ovoid, 0.9–1.0 × 0.8–0.9 mm, glossy, black; strophiole 0.5 mm long, ridged. *Chromosome number* 2n = 16 (Johnson & Brandham, 1997). Figure 227.

FLOWERING PERIOD. August to September, with a peak in late August.

HISTORY. *L. kliprandensis* was recorded for the first time by Prof. E. A. Schelpe (1924–1985) who collected a single dormant bulb (unusually with an attached infructescence, probably lodged within a bush) at Kliprand in western Bushmanland on 25th February 1982. When it flowered in the bulb nursery at Kirstenbosch in 1984, this plant was identified as an unknown species and the inflorescence and one leaf were pressed to serve as the type material. The plant was described three years later in the *South African Journal of Botany* (Barker, 1987) and the type collection continues to be grown in a pot at Kirstenbosch. The species was subsequently collected further north near Gamoep (*Duncan* 188, *Duncan* 380, both in NBG) but has since rarely been recorded.

DISTINGUISHING CHARACTERS AND AFFINITIES. *L. kliprandensis* is sister to *L. congesta* (Duncan et al., 2005a) sharing paired, prostrate, ovate leaves with subterranean scapes. The latter differs in its congested spike of much smaller, urceolate, yellowish-cream flowers with strongly recurved, undulate inner tepal apices and included, straight filaments. It differs further in its olive-green leaf upper surface and deep magenta lower surface, and it has much larger, globose seeds (1.8–1.9 × 1.8–1.9 mm) with rudimentary strophioles (0.1 mm long). In cultivation the scape of *L. kliprandensis* often elongates and its leaves become spreading or suberect.

Map 95. Known distribution of *Lachenalia kliprandensis*.

Figure 227. Cultivated specimen of *Lachenalia kliprandensis* from Gamoep, western Bushmanland. Image: Graham Duncan.

Forms of *L. kliprandensis* that have magenta-tipped inner tepals superficially resemble *L. carnosa*, which also has declinate stamens and ovate leaves, but the latter differs in its smaller, urceolate flowers with shorter, overlapping upper inner tepals (9–10 mm long), with the lower inner tepal longer and distinctly canaliculate, and in its included stamens, shorter style (7–8 mm long) and spreading to suberect leaves. The two species are allopatric, *L. carnosa* having a much wider range from northern Namaqualand to the southern Knersvlakte.

DISTRIBUTION, HABITAT AND CONSERVATION STATUS. *L. kliprandensis* is restricted to arid western Bushmanland extending from Gamoep south-east of Springbok to Kliprand east of Garies (Map 95). It occurs in Platbakkies Succulent- and Namaqualand Klipkoppe Shrubland vegetation (Mucina & Rutherford, 2006), as scattered individuals or in colonies on flats in deep red, gravelly sand, and on gentle slopes below quartz outcrops in stony red sand. *L. kliprandensis* is naturally rare but not under immediate threat, and has a conservation status of Rare (Victor & Duncan, 2009e).

NOTES. The large scented flowers of *L. kliprandensis* are pollinated by honey bees (*Apis mellifera*). The scape detaches shortly after capsule dehiscence, scattering the seeds over a wide area as it is blown about. During years of insufficient rainfall, the bulbs are adapted to remain dormant for one or more growing seasons.

1.4. Section LATAE

Latae G. D. Duncan, *sect. nov.*, *a ceteris sectionibus perianthio late campanulato, tubo cupulato, tepalis interioribus expansis 60–85°, filamentis inclusis usque ad bene exsertis differt.* Type: *Lachenalia comptonii* W. F. Barker

Perianth widely campanulate; tube cup-shaped, up to 3 mm long; outer tepals ovate; inner tepals obovate, oblong, linear or linear-oblong, radiating 60–85° from the longitudinal axis; filaments narrowly spreading, recurved or declinate, included to well exserted.

13 species (14 taxa).

95. LACHENALIA NERVOSA

Lachenalia nervosa Ker Gawl., *Curtis's Botanical Magazine* 36: t. 1497 (1812). *Scillopsis nervosa* (Ker Gawl.) Lem., *L'Illustration Horticole* 3: 34 (1856).

TYPE: South Africa, Cape, collector and precise locality unknown, figure in *Curtis's Botanical Magazine* 36: t. 1497 (1812) (lectotype, designated by Duncan & Linder Smith, 1999b).

SYNONYMY: *Lachenalia fragrans* Andrews, *The Botanist's Repository* 5: t. 302 (1803), illegitimate homonym, non *Lachenalia fragrans* Jacq., *Plantarum rariorum horti caesarei Schönbrunnensis* 1: t. 82 (1797), non *Lachenalia fragrans* Lodd., *Botanical Cabinet*: t. 1140 (1826). Type: South Africa, Cape, collector and precise locality unknown, figure in Andrews, *The Botanist's Repository* 5: t. 302 (1803) (lectotype, designated here).

Lachenalia latifolia Tratt., *Archiv der Gewachskunde* 2: t. 142 (1814), nom. superfl. Type: as for *L. fragrans* Andrews.

Lachenalia bowieana Baker, *Journal of the Linnean Society* 11: 410 (1871). Type: South Africa, Cape, precise locality unknown, unpublished painting of plant collected by Bowie, in Kew Herbarium, illustrated 10th June 1824 (K!, holo., designated here).

THE GENUS LACHENALIA
95. LACHENALIA NERVOSA

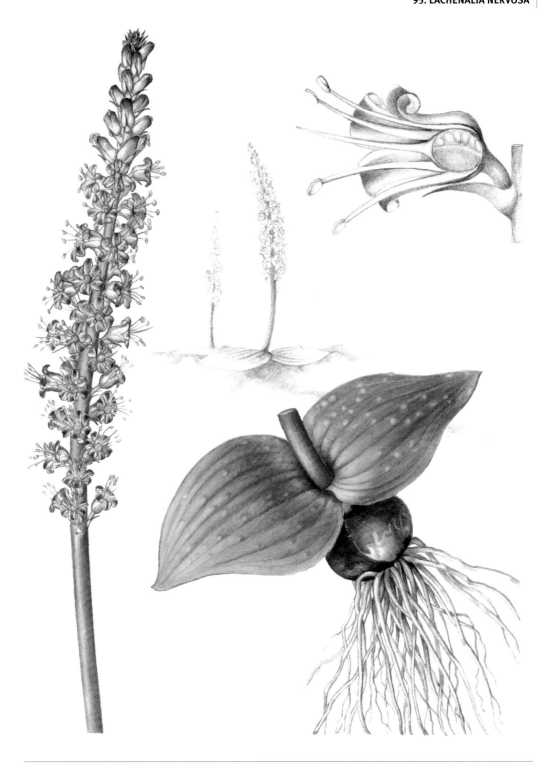

Plate 29. Watercolour painting of *Lachenalia nervosa* from Bredasdorp (*Thomas* 71, in NBG) courtesy of The Editor, *Flowering Plants of Africa*, vol. 56, t. 2144 (1999). Artist: Claire Linder Smith.

ETYMOLOGY. *nervosa*: veined, with reference to the prominent depressed longitudinal grooves between the veins of the upper leaf surface.

DESCRIPTION. *Geophyte*, 120–250 mm high. *Bulb* subglobose, 20–30 mm in diameter, solitary; tunic multilayered, outer layers dark brown, spongy; inner cataphyll translucent white, adhering to leaf bases, apex obtuse. *Leaves* 2, ovate, 70–120 × 25–55 mm, prostrate, upper surface bright green, yellowish-green or brownish-green, glabrous or covered with few to many small to large brown, green or yellowish-green pustules, depressed longitudinal grooves prominent, lower surface unmarked; leaf margins slightly coriaceous, maroon to brown; leaf bases clasping, 5–10 mm long, subterranean, white streaked with light magenta; primary seedling leaf flat, prostrate. *Inflorescence* racemose, many-flowered, dense, erect, sterile apex short; scape erect, 80–100 mm long, sturdy, light brownish-magenta to light purple; rachis light brownish-magenta to light greenish-brown; pedicels suberect at anthesis, becoming erect in fruit, 3–4 mm long, white, light whitish-magenta or brownish-green; lower bracts ovate, becoming lanceolate above, 3–4 × 1–5 mm, translucent white, plain or basal marking light to dark purplish-magenta. *Perianth* zygomorphic, widely campanulate, heavily spicy-sweet scented; tube cup-shaped, 2–3 mm long, white; outer tepals ovate, 5–8 × 3–4 mm, translucent white, median stripe and apical gibbosities purplish-magenta or light purplish-brown to brownish-pink, apices strongly recurved; inner tepals obovate, 6–7 × 3–4 mm, translucent white, median stripe broad, purplish-magenta, light purplish-brown or brownish-pink, apices strongly recurved. *Stamens* well exserted, recurved; filaments white to light whitish-magenta, 8–10 mm long. *Ovary* obovoid, 3–5 × 2–3 mm, light green; style well exserted, straight, white to light whitish-magenta, 7–8 mm long, protruding well beyond stamens during fruiting stage. *Capsule* obovoid, 7–8 × 5–6 mm, erect. *Seed* globose, 0.9 × 0.8–0.9 mm, glossy, black; strophiole inflated, 0.6–0.7 mm long, ridged. *Chromosome number* $2n = 16$ (Spies *et al.*, 2008); $2n = 24$ (Johnson & Brandham, 1997; Hamatani *et al.*, 2007). Plate 29, Figure 228.

FLOWERING PERIOD. September to November, with a peak in late October.

HISTORY. *L. nervosa* was introduced into cultivation at Kew in 1798 by Messrs Lee and Kennedy (Aiton, 1811) and described by the English botanist John Bellenden Ker Gawler (1764–1842) in *Curtis's Botanical Magazine* (Ker Gawler, 1812a). The plate accompanying Ker Gawler's brief text was illustrated from a plant sent to him by William Herbert, who had raised it from seed produced by a bulb imported from the Cape. The species had, however, already been illustrated nine years earlier, in the third volume of Andrews's *The Botanist's Repository* (1803), as *L. fragrans* Andrews, but the name is an illegitimate homonym as N. J. Jacquin had already published the name *L. fragrans* Jacq. six years earlier, in 1797, for a different species, in the first volume of his *Plantarum rariorum horti caesarei Schönbrunnensis*.

Ker Gawler (1811) published *L. lucida* on plate 1372 of *Curtis's Botanical Magazine*, from a specimen imported from the Cape by Messrs Chandler and Buckingham of Vauxhall Road, London. He considered it to be possibly synonymous with *L. fragrans* Andrews (now included under *L. nervosa* Ker Gawl.) but the figure of *L. lucida* depicts flowers with shortly exserted, declinate filaments whereas *L. nervosa* has well-exserted, strongly recurved filaments, and the name is more appropriately placed under *L. pallida* Aiton.

L. latifolia Tratt. was published by the Austrian botanist Leopold Trattinnick (1764–1849) in his *Archiv der Gewachskunde* (Trattinnick, 1814b), the black-and-white plate of which was copied from Andrews's plate 302 of *L. fragrans* in *The Botanist's Repository* (1803). In 1871, J. G. Baker described *L. bowieana* in the *Journal of the Linnean Society* from a painting in the Kew Herbarium. This painting was of a bulb grown at Kew that had been sent there from the Cape by the intrepid collector

James Bowie in 1823. A specimen collected by Bowie "near Samson's River in the district of George" is preserved in the British Museum (Natural History), but I have not been able to trace 'Samson's River' on any map. Bowie, renowned for writing misleading locality details onto his specimen labels in order to protect his business interests, might mischievously have 're-named' the Goliatsrivier (Goliath's River) near Cloete's Pass, north of Herbertsdale, an area in which the species could very well be expected to occur, as 'Samson's River' (Hugo Leggatt, pers. comm.).

Baker (1897a) recognised *L. latifolia*, *L. nervosa* and *L. bowieana* as separate species in his second revision. Under *L. latifolia* he cited a Zeyher specimen (*Zeyher* 4290, in K) from Swellendam, collected in November "among shrubs by the Buffeljagt's River", and under *L. nervosa* he cited a Burchell collection (*Burchell* 6438, in K) collected in November 1814 on dry hills on the east side of the Gouritz River near Mossel Bay. In the same work, Baker mistakenly listed a second Burchell specimen (*Burchell* 6346, in K), collected in November 1814 "Between Zout (Salt) River and Duyker River near Mossel Bay", under *L. unicolor*. In

Figure 228. *Lachenalia nervosa* on a shale slope near Napier. Image: Graham Duncan.

a fairly recent paper in *Flowering Plants of Africa*, a painting of *L. nervosa* by Claire Linder Smith was published depicting a specimen collected by Margaret Thomas near Bredasdorp in October 1983, and *L. latifolia* and *L. bowieana* were reduced to synonymy (Duncan & Linder Smith, 1999b).

DISTINGUISHING CHARACTERS AND AFFINITIES. Duncan *et al.* (2005a) considered *L. nervosa* to be sister to *L. stayneri*. The two species share prostrate leaves with usually pustulate upper surfaces, well-exserted stamens and globose seeds with inflated strophioles. *L. stayneri* differs in having broadly lanceolate leaves covered with much larger pustules, oblong-campanulate flowers borne on longer pedicels (up to 7 mm), declinate filaments, and larger seeds (1.1–1.2 × 1.3 mm).

DISTRIBUTION, HABITAT AND CONSERVATION STATUS. *L. nervosa* is confined to the southern Western Cape, extending from the Tradouw Pass north of Swellendam to Napier and Bredasdorp, and east to Groot-Brakrivier south-west of George (Map 96). It is fairly common around Swellendam and grows on seasonally damp sandy clay flats and in shale substrates on rocky east- and south-facing lower hill and mountain slopes. It occurs in four vegetation types: Central- and Eastern Rûens, Montagu- and Mossel Bay Shale Renosterveld (Mucina & Rutherford, 2006). The plants grow as scattered individuals or in small groups, and on a rocky outcrop at Napier, *L. nervosa* grows in association with *Eucomis regia* (Asparagaceae) within the protection of low bushes, close to a population of *L. pusilla*. *L. nervosa* is threatened by agricultural expansion, overgrazing and encroaching alien vegetation, and has a conservation status of Endangered (Helme & Raimondo, 2009e).

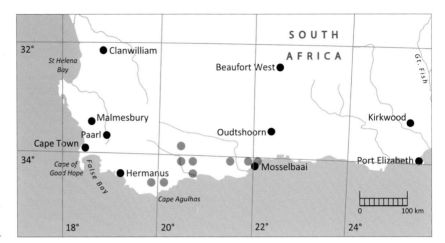

Map 96. Known distribution of *Lachenalia nervosa*.

NOTES. The widely campanulate flowers of *L. nervosa* emit a pervading, spicy-sweet scent reminiscent of a mixture of cloves and carnations. Pollinators have not been observed for this species but are likely to include bees. The pollen is consumed by monkey beetles (*Pachycnema crassipes*). The flowers are strongly self-fertile when grown in cultivation at Kirstenbosch in enclosed, insect-free conditions. During the fruiting stage, the developing capsules become erect. The scapes remain attached for several months after dehiscence and the seeds are dispersed locally.

96. LACHENALIA ANGELICA

Lachenalia angelica W. F. Barker, *South African Journal of Botany* 55 (6): 631–633 (1989).
TYPE: South Africa, Northern Cape, Road Hondeklipbaai to Springbok, *G. D. Duncan* 244 (NBG!, holo.).

ETYMOLOGY. *angelica*: angel-like, with reference to the white flowers.
DESCRIPTION. *Dwarf geophyte*, 60–95 mm high. *Bulb* globose, 5–10 mm in diameter, solitary; tunic multilayered, outer layers dark brown, spongy; inner cataphyll translucent white in subterranean portion, aerial portion light green, adhering to leaf base, apex acute. *Leaf* lanceolate, 20–35 × 10–15 mm, prostrate, becoming suberect at anthesis, upper surface glaucous, densely covered with minute stellate trichomes, lower surface deep maroon; margins slightly coriaceous, covered with minute stellate trichomes; leaf base 10–15 mm long, subterranean, suffused with deep maroon; primary seedling leaf flat, prostrate. *Inflorescence* racemose, few-flowered, sterile apex short; scape suberect, 25–50 mm long, slender, rigid, deep maroon; rachis deep maroon at base, white above; pedicels suberect, 2–5 mm long, white; lower bracts ovate, becoming lanceolate above, 2–3 × 1–2 mm, bases brownish-maroon, apices white. *Perianth* zygomorphic, widely campanulate, spreading to suberect, white, ageing to dull pink; tube cup-shaped, 2–3 mm long; outer tepals ovate, 5–6 × 3–4 mm, apices recurved, apical gibbosities light green; inner tepals obovate, 6–7 × 2–3 mm, apices recurved, median keels light yellow. *Stamens* well exserted, declinate; filaments white, 6–8 mm long. *Ovary* obovoid, 2 × 2 mm, bright green; style shortly exserted, declinate, 4–5 mm long, white. *Capsule* obovoid, 3–4 × 3–4 mm. *Seed* globose, 0.4 × 0.4 mm, secondary sculpturing reticulate; strophiole rudimentary, 0.1–0.2 mm long, ridged. *Chromosome number* unknown. Figures 48, 229, 230.

Figure 229. The diminutive *Lachenalia angelica* in red sand near Hondeklipbaai. Image: Graham Duncan.

Figure 230. Colony of *Lachenalia angelica* on a quartz hillside near Hondeklipbaai. Image: Graham Duncan.

96. LACHENALIA ANGELICA

FLOWERING PERIOD. September to October with a peak in mid-October.

HISTORY. *L. angelica* was discovered by the author east of Hondeklipbaai in western Namaqualand in August 1986. The plants were in a leafing state and their prostrate, glaucous, small solitary form resembled immature specimens of *L. trichophylla*, their upper surfaces covered in minute stellate trichomes. Some of these plants were cultivated in the Kirstenbosch Bulb Nursery, and when they flowered in October of 1986, they proved to be an undescribed species, which was published several years later in the *South African Journal of Botany* (Barker, 1989).

DISTINGUISHING CHARACTERS AND AFFINITIES. *L. angelica* resolved as sister to *L. polypodantha* in a phylogenetic analysis (Duncan et al., 2005a). These two species share similar lanceolate leaves that are covered with stellate trichomes on the upper surface, widely campanulate flowers that have exserted, declinate stamens, and small globose, matte black seeds that have reticulate secondary sculpturing. *L. polypodantha* differs in its usually dense, many-flowered raceme with the pedicels spreading at right angles, its longer, well-exserted stamens (7–12 mm) that have light to deep purple or ivory white filaments, its spreading leaves and its relatively larger seeds (0.9 × 0.8–0.9 mm).

DISTRIBUTION, HABITAT AND CONSERVATION STATUS. *L. angelica* is confined to a few populations in close proximity east of Hondeklipbaai in western Namaqualand (Map 97). The plants occur in Riethuis-Wallekraal Quartz Vygieveld vegetation (Mucina & Rutherford, 2006), in colonies containing thousands of individuals. They grow on seasonally moist flats and gentle west-facing, slightly drier slopes amongst quartz stones and pebbles, in light reddish-brown sand. Their leaves are proteranthous, having withered by the time flowering commences in late spring. At the type locality, the species occurs in association with large numbers of *Brunsvigia radula* (Amaryllidaceae) and occasional specimens of *Gladiolus equitans* (Iridaceae). The area is subject to regular, heavy morning fog, which provides considerable moisture for the plants in an arid environment. The populations are not under immediate threat but their close proximity to a road and the potential threat of mining places them at risk, affording the species a conservation status of Vulnerable (Victor et al., 2009a).

NOTES. The flowers are pollinated by honey bees (*Apis mellifera*), but in cultivation they are at least partially self-fertile, regularly producing seeds under enclosed, insect-proof conditions. As soon as capsule dehiscence has taken place, the base of the scape detaches and the seeds are dispersed as it is blown about. Occurring in an area of erratic rainfall, the bulbs are adapted to remain dormant for one or more growing seasons until favourable conditions return.

Map 97. Known distribution of *Lachenalia angelica*.

97. LACHENALIA CAMPANULATA

Lachenalia campanulata Baker, *Journal of Botany* 12: 6 (1874).
TYPE: South Africa, Eastern Cape, In lapidosis summii Montis Boschberg, 4,800 ft, *P. MacOwan* 1836 (K!, holo.; GRA!, iso.).
SYNONYMY: *Lachenalia rhodantha* Baker, in Thiselton-Dyer, W. T. (ed.), *Flora Capensis* 6: 430 (1897). Type: South Africa, Eastern Cape, Graaff Reinet, grassy slopes of the Oude Berg, 4,300 ft, *H. Bolus* 719 (K!, holo.).

ETYMOLOGY. *campanulata*: bell-shaped flowers.
DESCRIPTION. *Dwarf geophyte* 50–190 mm high. *Bulb* ovoid, 10–15 mm in diameter, strongly offset-forming; tunic multilayered, outer layers spongy, light brown; inner cataphyll translucent white, loosely surrounding leaf bases, apex acute. *Leaves* 2, rarely 1, linear, 50–180 × 2–10 mm, erect to suberect or spreading, light to dark green, deeply canaliculate to conduplicate, uppermost portion occasionally subterete, occasionally twisted, upper surface plain, with depressed longitudinal veins, lower surface plain or flushed with brownish-maroon near base; leaf bases loosely surrounding base of scape, 4–25 mm long, white in lower half, shading to light green above, with tiny magenta spots; primary seedling leaf terete, erect. *Inflorescence* racemose (up to three produced per season), few- to many-flowered, dense, sterile apex short; scape erect to suberect, 40–170 mm long, light green to light brownish-maroon, heavily marked with dark brownish-maroon or maroonish-magenta spots and blotches, slender; rachis light brownish-green with minute light to dark brownish-maroon

Figure 231 (left). Deep magenta form of *Lachenalia campanulata* on the Katberg Pass. Image: Graham Duncan.
Figure 232 (right). *Lachenalia campanulata* in grass tussocks on rocky mountain habitat of the Katberg Pass. Image: Hubert Kurzweil.

spots and blotches; pedicels suberect, 2–4 mm long, brownish-green or pinkish- to purplish-magenta; bracts cup-shaped throughout, 1.5–2.0 × 1.0–2.5 mm, brownish-magenta, apices white. *Perianth* zygomorphic, widely campanulate, spreading to suberect, lightly spice-scented; tube cup-shaped, light blue, 1 mm long; outer tepals ovate, 4–5 × 2.0–2.5 mm, white, or light pinkish-magenta in lower half, shading to bright magenta in upper half, apical gibbosities dark purplish-magenta; inner tepals linear-oblong, 4–5 × 1.5 mm, translucent white, median keels magenta or purplish, apices recurved. *Stamens* shortly exserted, recurved; filaments white, 4–5 mm long. *Ovary* obovoid, 2 × 2 mm, greenish-brown; style included, straight, 3 mm long, white, finally protruding beyond perianth as ovary enlarges. *Capsule* obovoid, 3.5–4.0 × 4.0–4.5 mm. *Seed* globose, 1.0–1.1 × 1.0–1.1 mm, glossy, black; strophiole rudimentary, 0.1–0.2 mm long, ridged. *Chromosome number* unknown. Figures 37, 231–233.

FLOWERING PERIOD. October to December, with a peak from early to mid-November.

HISTORY. *L. campanulata* was described by Baker (1874a) in volume 12 of the *Journal of Botany* from material collected in 1874, by the English botanist and teacher Prof. Peter MacOwan

Figure 233. Cultivated light pink form of *Lachenalia campanulata* from the Witteberge east of Lady Grey. Image: Graham Duncan.

(1830–1909), on the Boschberg near Somerset East (*MacOwan* 1836, in K). This material was in late flowering stage and had two filiform leaves, horizontal lowermost flowers, included stamens, and white tepals that were tinged with red. In November 1868, Harry Bolus collected the same species on the Oudeberg north-west of Graaff Reinet but Baker (1897a) considered it distinct and described it as *L. rhodantha*, mainly on account of its cernuous lowermost flowers, exserted stamens and reddish tepals. Baker was evidently unaware that tepal colour is variable for the species and that the stamens become drawn back into the perianth as they age. His observation that the flowers were cernuous was erroneous. In a revision of the *Lachenalia* species of the Eastern Cape, *L. rhodantha* was placed in synonymy under *L. campanulata* (Dold & Phillipson, 1998).

DISTINGUISHING CHARACTERS AND AFFINITIES. *L. campanulata* is sister to *L. macgregoriorum* (Duncan et al., 2005a), sharing similar widely campanulate, spreading to suberect flowers, recurved filaments, obovoid capsules and globose seeds. The latter differs in its much taller stature (160–300 mm high) with intensely glaucous, narrowly lanceolate leaves. It has maroonish-magenta flowers with longer, broader outer and inner tepals (6–8 × 3–4 mm as opposed to 5–6 × 3–4 mm for *L. campanulata*), longer, well-exserted, bright magenta filaments (10–11 mm long) and larger, matte black seeds (1.3–1.4 × 1.3–1.4 mm).

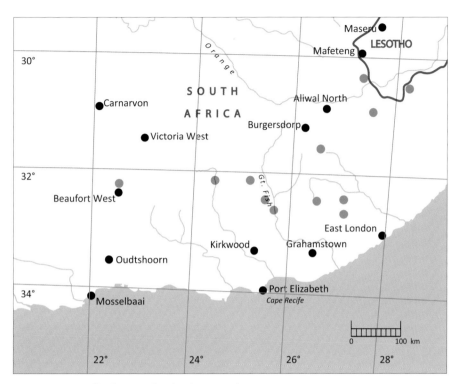

Map 98. Known distribution of *Lachenalia campanulata*.

DISTRIBUTION, HABITAT AND CONSERVATION STATUS. *L. campanulata* is native to the upper reaches of the Eastern Cape interior mountains, extending from the Oudeberg north-west of Graaff Reinet to Ongeluks Nek west of Matatiele, with a disjunct population in the Nuweveld Mountains in the Karoo National Park north of Beaufort West (Map 98). The species occurs in nine vegetation types: Karoo Escarpment, Amathole-, Tsomo-, East Griqualand- and Southern Drakensberg Highland Grassland, Camdebo Escarpment Thicket, Eastern Upper Karoo, and Senqu- and Tarkastad Montane Shrubland (Mucina & Rutherford, 2006). Altitudinal range varies from 1,220 m on the Boschberg at Somerset East to well over 2,000 m on the Sneeuberge west of Cradock and the Witteberge east of Lady Grey. The plants occur in large, dense colonies on summit plateaus, steep south-, south-west-, and east-facing rocky slopes, and at the base of cliffs. They grow in light shade or on rock ledges and flat, exposed rock sheets in shallow, damp black sandy soil, amongst short grass tussocks. *L. campanulata* is subject to sub-zero temperatures in winter in areas such as the Witteberge. Although the plants occur in a mainly summer rainfall zone, they are winter-growing and dormant in summer in the wild. *L. campanulata* is abundant and not under threat.

NOTES. The widely campanulate flowers of *L. campanulata* suggest pollination by honey bees (*Apis mellifera*). The scapes remain attached for several months after capsule dehiscence and the seeds are dispersed locally. On the Katberg Pass west of Cathcart, the flowers are heavily grazed by livestock and the area is subject to winter wild fires. *L. campanulata* is unusual within the genus in producing multiple inflorescences (up to three) per season, and in that its seeds germinate if sown directly upon ripening, not requiring an intervening dormant period (Rhoda McMaster, pers. comm.).

98. LACHENALIA COMPTONII

Lachenalia comptonii W. F. Barker, *South African Gardening and Country Life* 20: 14 (1930).
TYPE: South Africa, Western Cape, Karoo Garden at Whitehill, *R. H. Compton* 3533 (NBG!, holo.; BOL!, iso.).

ETYMOLOGY. *comptonii*: after Prof. R. H. Compton (1886–1979), second Director (from 1919 to1953) of the National Botanical Gardens at Kirstenbosch.

DESCRIPTION. *Geophyte*, 70–250 mm high. *Bulb* globose, 18–22 mm in diameter, solitary; tunic multilayered, outer layers dark brown, spongy; inner cataphyll translucent white, loosely surrounding leaf bases, apex acute. *Leaves* 2, lanceolate, 80–110 × 15–22 mm, spreading to suberect, light to dark green; upper surface trichomes simple, 3–12 mm long, sparse to dense; lower surface light green, flushed with maroon; margins flat, with long simple trichomes; leaf bases clasping, 20–40 mm long, white flushed with deep maroon; primary seedling leaf terete, erect. *Inflorescence* racemose, many-flowered, dense, sterile apex short; scape erect to suberect, 40–60 mm long, slender, deep maroon; rachis light maroon; pedicels suberect, 1 mm long, white; bracts cup-shaped throughout, 1–2 × 2–3 mm, translucent white. *Perianth* zygomorphic, widely campanulate, white, lightly spice-scented; tube cup-shaped, 2 mm long; outer tepals narrowly ovate, 6–7 × 3–4 mm, recurved, apical gibbosities

Figure 234 (left). *Lachenalia comptonii* on stony clay flats in the southern Tanqua Karoo. Image: Graham Duncan.
Figure 235 (right). Cultivated specimen of *Lachenalia comptonii* from Karoopoort. Image: Graham Duncan.

and median keels light green; inner tepals linear-oblong, 6–7 × 1–2 mm, recurved, median keels light green. *Stamens* well exserted, recurved; filaments bright purple or magenta in upper half, white below, 8–9 mm long. *Ovary* ellipsoid, 2–3 × 1–2 mm, light green; style well exserted, more or less straight, 8–9 mm long, bright purple or magenta in upper half, white below. *Capsule* ellipsoid, 5–7 × 3–4 mm. *Seed* globose, 0.9–1.0 × 0.8–0.9 mm, glossy, black; strophiole rudimentary, 0.1–0.2 mm long, ridged. *Chromosome number* 2n = 20 (Crosby, 1986; Johnson & Brandham, 1997). Figures 234, 235.

FLOWERING PERIOD. September to October, with a peak in late September.

HISTORY. The type material of *L. comptonii* was collected by Prof. R. H. Compton who discovered it on 24th September 1929

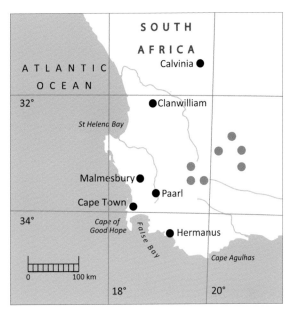

Map 99. Known distribution of *Lachenalia comptonii*.

within the original site of the Karoo National Botanical Garden at Whitehill, east of Matjiesfontein. The species was collected again by Compton in September 1931 on the Bonteberg east of Ceres (*Compton s.n.*, in K). *L. comptonii* was described in volume 20 of the periodical *South African Gardening and Country Life* (Barker, 1930a) and was illustrated in *The Flowering Plants of South Africa* (Barker, 1933a). Most subsequent collections have been made to the north-east of Ceres, and the species has been recorded as far north as the Roggeveld Mountains west of Sutherland.

DISTINGUISHING CHARACTERS AND AFFINITIES. Duncan *et al.* (2005a) placed *L. comptonii* as sister to *L. glaucophylla*, which shares spreading, widely campanulate flowers with exserted, recurved stamens and lanceolate leaves. *L. glaucophylla* differs in its much smaller cream flowers with shorter, narrower outer tepals (4 × 2 mm), shorter inner tepals (4 × 1 mm), obovoid capsules and a larger, ovoid, matte black seed (1.3–1.4 × 1.1–1.2 mm) with a longer, ridged strophiole (0.8 mm long). It also differs in its solitary, intensely glaucous, canaliculate leaf with a smooth upper surface and spotted lower surface.

DISTRIBUTION, HABITAT AND CONSERVATION STATUS. *L. comptonii* is confined to the arid Tanqua Karoo and southern Great Karoo, extending from the Roggeveld Mountains south-west of Sutherland to Karoopoort north of Ceres, to Whitehill east of Matjiesfontein (Map 99). The plants grow in Tanqua Karoo- and Ceres Shale Renosterveld vegetation (Mucina & Rutherford, 2006), in colonies on flats and south-east-facing lower hill slopes, in sandy or stony clay, exposed or within low bushes. Along the Ceres-Calvinia road, *L. comptonii* grows in association with *L. aurioliae*, which flowers much earlier in the season. The species is not threatened.

NOTES. The heavily spice-scented flowers of *L. comptonii* are pollinated by honey bees (*Apis mellifera*). The leaves are frequently proteranthous in the wild, desiccating before flowering commences. The scapes detach from the bulbs once capsule dehiscence has taken place and the infructescence is blown away, scattering seeds over a wide area.

99. LACHENALIA GLAUCOPHYLLA

Lachenalia glaucophylla W. F. Barker, *Journal of South African Botany* 44 (4): 409–412 (1978).
TYPE: South Africa, Northern Cape, 11 miles [17.6 km] from the junction of Williston-Calvinia and Klipwerf roads, from Downes, *M.L. Thomas s.n.* sub. *NBG* 105737 (NBG!, holo.).

ETYMOLOGY. *glaucophylla*: greenish-grey leaves.
DESCRIPTION. *Dwarf geophyte*, 70–110 mm high. *Bulb* globose, 10–13 mm in diameter, solitary, deep-seated; tunic multilayered, outer layers dark brown, spongy; inner cataphyll translucent white, adhering to leaf base, apex acute. *Leaf* solitary, narrowly lanceolate, 120–150 × 8–20 mm, distinctly curved, intensely glaucous, canaliculate or conduplicate, upper surface plain, lower surface heavily marked with green spots in upper portion, shading to dark purple spots below; leaf base clasping, 60–85 mm long, subterranean portion white with light magenta spots or blotches, shading to glaucous green with dark purple spots or blotches above. *Inflorescence* racemose, many-flowered, dense, sterile apex short; scape suberect, 25–70 mm long, slender, light green, marked with light to dark magenta blotches; rachis light green, marked with light to dark magenta blotches, 35–55 mm long; primary seedling leaf terete, erect; pedicels suberect, 1.5–3.0 mm long, white; bracts cup-shaped throughout, 0.5 × 0.5–1.0 mm, rudimentary, translucent white with minute, light to dark magenta speckles. *Perianth* zygomorphic, widely campanulate, suberect, cream, heavily spice-scented; tube cup-shaped, 1 mm long; outer tepals ovate, 4 × 2 mm, apical gibbosities light green or maroonish-brown, apices finally recurved; inner tepals linear-oblong, 4 × 1 mm, median keels light green or maroonish-

Map 100. Known distribution of *Lachenalia glaucophylla*.

Figure 236. *Lachenalia glaucophylla* in deep red sand near Middelpos. Image: Graham Duncan.

brown. *Stamens* well exserted, recurved; filaments white, 7–8 mm long, pollen ageing to black. *Ovary* obovoid, 1.5 × 2.0 mm, light green; style well exserted, more or less straight, 6.5 mm long, white. *Capsule* obovoid, 5–6 × 3–4 mm. *Seed* ovoid, 1.3–1.4 × 1.1–1.2 mm, glossy, black; strophiole 0.8 mm long, ridged. *Chromosome number* unknown. Figures 47, 67, 236.

FLOWERING PERIOD. October, with a peak in mid-October.

HISTORY. The first record of *L. glaucophylla* is of two young fruiting specimens collected by Dr Rudolf Marloth at Leliefontein in the Kamiesberg in October 1917. The species was described in the *Journal of South African Botany* (Barker, 1978). Most collections have been made in the Calvinia District, including the holotype material, which was collected by Margaret Thomas near Downes north of Calvinia in October 1974. A recent collection made by Gordon Summerfield north-west of Sutherland, in October 2001, is a significant range extension.

DISTINGUISHING CHARACTERS AND RELATIONSHIPS. *L. glaucophylla* is most closely related to *L. comptonii* (Duncan *et al.*, 2005a) and the relationship is discussed under that species.

DISTRIBUTION, HABITAT AND CONSERVATION STATUS. *L. glaucophylla* is centred in the Calvinia district on the western edge of the Great Karoo, with outlying populations to the north-west at Leliefontein in southern Namaqualand, and to the south-east at Danielskuil, south-west of Middelpos (Map 100). The plants occur in three vegetation types: Kamiesberg Granite Fynbos, Hantam Karoo and Koedoesberg-Moordenaars Karoo (Mucina & Rutherford, 2006), usually growing as scattered individuals or rarely in small groups of up to three plants, exposed or within the protection of low bushes. At Danielskuil, the bulbs occur in arid conditions in deep red sand and are buried down to 85 mm. *L. glaucophylla* is not under threat but is known from very few locations, and a conservation status of Rare is recommended.

NOTES. The flowers of *L. glaucophylla* are heavily spice-scented and pollinated mainly by honey bees (*Apis mellifera*). A blister beetle (*Lytta nitidula*) is a generalist pollinator that feeds on the nectar at the base of the perianth while its body becomes covered with pollen (Figure 67). The scapes of *L. glaucophylla* remain attached for several weeks following capsule dehiscence and the seeds are dispersed locally.

100. LACHENALIA MACGREGORIORUM

Lachenalia macgregoriorum W. F. Barker (sphalm. *macgregori*), *Journal of South African Botany* 45 (2): 199–201 (1979).

TYPE: South Africa, Northern Cape, Charlie's Hoek, Nieuwoudtville, *W. F. Barker* 9766 (NBG!, holo.; NBG!, PRE!, K!, iso.).

ETYMOLOGY. *macgregoriorum*: after the McGregor Family, formerly of Glen Lyon Farm, Nieuwoudtville, in recognition of their contribution to conservation on the Bokkeveld Plateau.

DESCRIPTION. *Geophyte*, 160–300 mm high. *Bulb* subglobose, 20–30 mm in diameter, offset-forming; tunic multilayered, outer layers light brown, spongy; inner cataphyll translucent white, adhering to leaf bases, apex obtuse. *Leaves* 2, narrowly lanceolate, 180–270 × 8–15 mm, proteranthous, erect to suberect, glaucous, upper surface plain, lower surface heavily marked with dull maroon blotches in lower third, shading to glaucous bands and blotches in central portion; leaf bases clasping, 30–45 mm long, white with light magenta spots and blotches; primary seedling leaf terete, erect. *Inflorescence* racemose, many-flowered, fairly dense, sterile apex short; scape erect to suberect, 120–200 mm long, sturdy, light green, heavily marked with light brownish- or purplish-maroon blotches and spots;

100. LACHENALIA MACGREGORIORUM

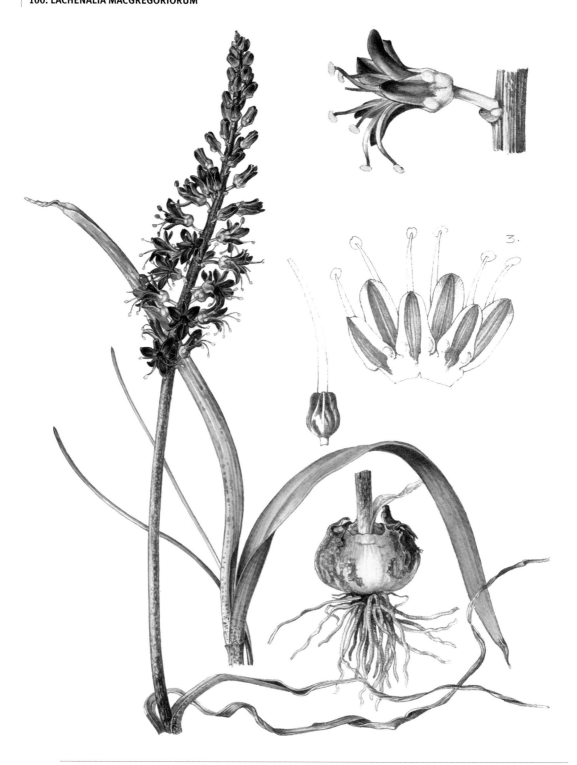

PLATE 30. Watercolour painting of *Lachenalia macgregoriorum* from Nieuwoudtville (*Barker* 9766, in NBG) courtesy of The Editor, *The Flowering Plants of Africa*, vol. 49, t. 1951 (1987). Artist: Ellaphie Ward-Hilhorst.

100. LACHENALIA MACGREGORIORUM

rachis heavily mottled with light to bright maroon spots; pedicels spreading to suberect, 2–5 mm long, lower half white, upper half pinkish-magenta; bracts cup-shaped throughout, 1–2 × 0.5–4.0 mm, light green. *Perianth* zygomorphic, widely campanulate, suberect to spreading; tube cup-shaped, 1–2 mm long, white; outer tepals ovate, 5–6 × 3–4 mm, white at base, shading to light greenish-white above, median keels broad, dark maroon-magenta, apical gibbosities dark purple-maroon; inner tepals linear-oblong, 6–8 × 3–4 mm, maroon-magenta, apical marking dark purple-maroon. *Stamens* well exserted, recurved; filaments bright magenta, 10–11 mm long. *Ovary* obovoid, 3 × 2 mm, light green to brownish-purple; style shortly exserted, straight, 9 mm long, bright magenta. *Capsule* obovoid, 7–8 × 4 mm. *Seed* globose, matte, black, 1.3–1.4 × 1.3–1.4 mm; strophiole rudimentary, 0.1 mm long, ridged. *Chromosome number* 2n = 22 (Spies *et al.*, 2008). Plate 30, Figure 237.

FLOWERING PERIOD. October to November, with a peak in early November.

HISTORY. *L. macgregoriorum* was discovered by W. F. Barker at Charlie's Hoek in the Municipal Wildflower Reserve just outside Nieuwoudtville on 5th November 1962, and is only known from this collection (*Barker* 9766, in NBG). The species was originally described as *L. macgregori* in honour of the MacGregor Family (Barker, 1979), but in order to convey W. F. Barker's stated intention, the specific epithet was emended to *macgregoriorum* and published in *The Flowering Plants of Africa*, accompanied by a watercolour plate by Ellaphie Ward-Hilhorst (Duncan, 1987). Plants of this collection were cultivated by Barker for many years and were subsequently donated to the bulb collection at Kirstenbosch, where they continue to be grown.

Map 101. Known distribution of *Lachenalia macgregoriorum*.

Figure 237. Cultivated specimens of *Lachenalia macgregoriorum* from Nieuwoudtville. Image: Graham Duncan.

DISTINGUISHING CHARACTERS AND AFFINITIES. *L. macgregoriorum* is sister to *L. campanulata* (Duncan et al., 2005a), which differs in its dwarf stature (usually less than 100 mm high) and linear, deeply canaliculate, light to dark green leaves. When compared with *L. macgregoriorum*, *L. campanulata* has shorter, narrower outer and inner tepals (4–5 × 2.0–2.5 mm and 4–5 × 1.5 mm, respectively), shortly exserted, white filaments and much smaller seeds (1.0–1.1 × 1.1–1.0 mm).

DISTRIBUTION, HABITAT AND CONSERVATION STATUS. *L. macgregoriorum* is only known from Charlie's Hoek, also known as Klipkoppies on the Bokkeveld Plateau immediately east of Nieuwoudtville (Map 101). It grows in Nieuwoudtville-Roggeveld Dolerite Renosterveld vegetation (Mucina & Rutherford, 2006) in clay soil under thorny bushes. The species has not been sighted for more than 45 years despite a number of searches, and is potentially threatened by alien annual grasses as well as by grazing and trampling by livestock. It has a conservation status of Vulnerable (Victor & Duncan, 2009f).

NOTES. The proteranthous leaves of *L. macgregoriorum* have withered by the time the flowers open in early summer. The widely campanulate perianths suggest pollination by honey bees (*Apis mellifera*). The infructescences remain attached for several months following capsule dehiscence and the seeds are dispersed locally by the shaking action of wind.

101. LACHENALIA GIESSII

Lachenalia giessii W. F. Barker, *Journal of South African Botany* 49 (4): 434–437 (1983).
TYPE: Namibia, Garub, west of Aus, Lüderitz District, *H. Merxmüller & J. W. Giess* 3041 (WIND, holo., missing; M!, PRE!, iso.).

ETYMOLOGY. *giessii*: after Johan Wilhelm Giess (1910–2000), prolific German collector of plants in Namibia.

DESCRIPTION. *Dwarf geophyte*, 35–160 mm high. *Bulb* globose, 10–30 mm in diameter, solitary; neck distinct, 15–50 mm long, tunic multilayered, outer layers rigid, dark brown or black, apices fasciculate; inner cataphyll translucent white, adhering to leaf bases, apex obtuse. *Leaves* (1–)2, linear to broadly lanceolate, 20–95 × 2–20 mm, spreading to suberect, glaucous to dark green, flat to canaliculate, flaccid; margins ciliolate; leaf bases loosely surrounding base of scape, 5–10 mm long, subterranean; primary seedling leaf terete, erect. *Inflorescence* racemose, few- to many-flowered, fairly dense, sterile apex short; scape erect to suberect, 20–80 mm long, light green, slender; rachis light green in lower half, shading to whitish-green in upper half; pedicels spreading, 2–12 mm long, white, elongating and often suberect in fruit; bracts cup-shaped throughout, 0.5–2.0 × 0.5–4.0 mm, light green to white. *Perianth* zygomorphic, widely campanulate, spreading to suberect, white, slightly spice-scented; tube cup-shaped, 1.5 mm long; outer tepals ovate, 5–9 × 4–5 mm, apices flat to slightly recurved, apical gibbosities green, brownish-green or reddish-purple; inner tepals linear-oblong, 6–9 × 3–4 mm, apices yellowish-green or reddish-purple, keels brown-magenta in upper half. *Stamens* shortly exserted, narrowly spreading; filaments white, 5–8 mm long; pollen ageing to black. *Ovary* globose, 2–3 × 2.0–2.5 mm, bright green; style shortly exserted, straight, 5–8 mm long, white. *Capsule* obovoid, 4–6 × 4–5 mm. *Seed* globose, 1.3–1.4 × 1.8 mm, glossy, black; strophiole rudimentary, 0.1 mm long, ridged. *Chromosome number* 2n = 32 (Spies et al., 2008). Figures 238–240.

FLOWERING PERIOD. July to September, with a peak in mid-August.

HISTORY. The first collection of *L. giessii* was made by Karen Regius south of Aus in south-western Namibia in September 1941 (*Regius s.n.*, in WIND). The species has subsequently been collected on numerous occasions in south-western Namibia, mainly by W. Giess, a former Curator of the

Windhoek Herbarium, after whom Barker (1983b) named the species. *L. giessii* was thought to be endemic to Namibia until recent collections were made at Klipbokkop in the eastern Richtersveld and Gamsberg in northern Bushmanland.

DISTINGUISHING CHARACTERS AND AFFINITIES. *L. giessii* is very variable with regard to its leaves, which vary from linear to broadly lanceolate, and from flat to slightly canaliculate. The species is sister to *L. pearsonii* (Duncan *et al.*, 2005), which differs in its bifacial leaf with a prominent abaxial midrib, shorter flower with narrower outer and inner tepals (4 × 2 mm and 4.5 × 1.0 mm, respectively), shorter filaments (4 mm long), shorter style (4 mm long) and much larger seed (2.1 × 1.0 mm).

DISTRIBUTION, HABITAT AND CONSERVATION STATUS. *L. giessii* extends from Zaris Farm (*Müller* 1349, in WIND), the most northerly record for the species and the genus (grid reference 2416 CD) in western Namibia, south to the Schakalberg (Jakalsberg) south of Rosh Pinah and south-east to Warmbad in southern Namibia. Its range extends into the Northern Cape, South Africa, occurring in the southern Richtersveld and east to the Gamsberg Mountains between Aggeneys and Pofadder (where a particularly dwarf form occurs (*Desmet* 3690 in NBG)) (Map 102). In western, south-western and

Figure 238 (top left). *Lachenalia giessii* with fleshy, broadly lanceolate leaves in stony red sand near Rosh Pinah, south-western Namibia. Image: Ted Oliver.

Figure 239 (bottom left). Cultivated specimens of *Lachenalia giessii* with narrowly lanceolate leaves from Namuskluft, south-western Namibia. Image: Graham Duncan.

Figure 240 (above). Cultivated specimen of *Lachenalia giessii* with linear leaves from Warmbad, southern Namibia. Image: Graham Duncan.

southern Namibia, it occurs in Semi-desert and Savanna Transition, Desert and Succulent Steppe, and Dwarf Shrub Savanna vegetation, respectively (Loots, 2005). In South Africa, the species occurs in two vegetation types: Stinkfonteinberge Eastern Apron Shrubland (Richtersveld) and Bushmanland Arid Grassland on the Gamsberg (Mucina & Rutherford, 2006). The plants occur in colonies, sometimes growing between small bushes or among long grasses. Habitats include stony soils on mountain and hill summits, loamy soil covered with phonolite boulders and aeolian sand on south- and west-facing rocky mountain slopes, dolomitic limestone and quartz ridges, flat, white quartz patches, and dry pans. *L. giessii* is locally common and not threatened.

NOTES. The widely campanulate flower shape of *L. giessii* suggests pollination by honey bees (*Apis mellifera*). During fruiting, the pedicels of certain forms elongate considerably. The scape detaches from the bulb shortly after capsule dehiscence and the seeds are dispersed as the infructescence is blown away.

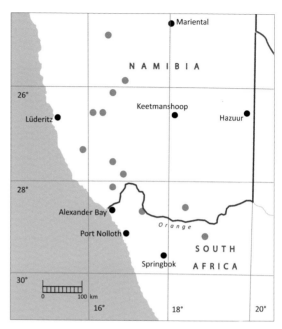

Map 102. Known distribution of *Lachenalia giessii*.

102. LACHENALIA NAMIBIENSIS

Lachenalia namibiensis W. F. Barker, *South African Journal of Botany* 53 (2): 170–172 (1987).
TYPE: Namibia, Lüderitz-Sud, 40 km north of Rosh Pinah, *P. Goldblatt* 7017 (NBG!, holo., MO, NBG!, iso.).

ETYMOLOGY. *namibiensis*: from Namibia.
DESCRIPTION. *Dwarf geophyte*, 35–75 mm high. *Bulb* ovoid, 10–15 mm diameter, offset-forming; neck distinct, 5–8 mm long; tunic multilayered, outer layers rigid, dark brown or black, apices fasciculate; inner cataphyll translucent white, adhering to leaf bases, apex obtuse. *Leaves* (1–)2, narrowly lanceolate, 30–60 × 3–12 mm, erect to suberect, dark green, bifacial, canaliculate, apex minutely cusped, abaxial midrib prominent; margins ciliolate; leaf bases clasping, 5–10 mm long, lower portion light green, upper portion light maroonish-brown; primary seedling leaf terete, erect. *Inflorescence* racemose, few- to many-flowered, moderately dense, sterile apex short; scape erect to suberect, 20–45 mm long, slender, light green with light brownish-magenta speckles; rachis light brownish-magenta; pedicels spreading to suberect, 2–7 mm long (often longer in cultivation), light green to light brownish-magenta; lower bracts ovate, becoming narrow-lanceolate above, 2–3 × 1–2 mm, light green. *Perianth* zygomorphic, widely campanulate, spreading to suberect, white; tube cup-shaped, 3 mm long; outer tepals narrowly ovate, 5–9 × 4–5 mm, apices recurved, apical gibbosities and keels green to purplish-brown; inner tepals linear-oblong, 7–9 × 2–3 mm, apices recurved, keels light to dark mauve or magenta. *Stamens* included, narrowly spreading; filaments white, 6–7 mm long. *Ovary* obovoid, 2 × 3

mm, light green; style included, straight, 7 mm long, becoming shortly exserted in fruit, white. *Capsule* ellipsoid, 7–9 × 5 mm. *Seed* globose, 1.2–1.3 × 1.3–1.4 mm, glossy, black; strophiole rudimentary, 0.1 mm long, ridged. *Chromosome number* 2n = 22 (Spies *et al.*, 2008). Figure 241.

FLOWERING PERIOD. August to September, with a peak in mid-August.

HISTORY. The Germans Hermann Merxmüller and Wilhelm Giess first collected *L. namibiensis*. They found it on the farm Witputs-Süd north of Rosh Pinah in southern Namibia in late August 1963, and their pressed material (*Merxmüller & Giess* 3205) is preserved in herbaria at Munich, Pretoria and Windhoek. The species was mistaken for *L. patula* in the account of *Lachenalia* for the *Prodromus einer flora von Südwestafrika* (Sölch & Roessler, 1970). Twenty years later, in August 1983, a collection made by John Lavranos at Namuskluft east of Rosh Pinah flowered in the bulb nursery at Kirstenbosch, but its identity was uncertain and pressed specimens were temporarily filed under the genus *Ornithogalum* in the Compton Herbarium. In late September 1983, Peter Goldblatt collected fruiting specimens 40 km north of Rosh Pinah. When these specimens flowered at Kirstenbosch in August the following year, they matched the previous two collections and it was realised that they represented an undescribed species, subsequently published in the *South African Journal of Botany* (Barker, 1987).

DISTINGUISHING CHARACTERS AND AFFINITIES. *L. namibiensis* is allied to *L. pearsonii* (Duncan *et al.*, 2005a), which differs in its much smaller flowers with shorter perianth tubes (1 mm long), shorter outer (4 × 2 mm) and inner tepals (4.5 × 1.0 mm), shorter filaments (4 mm long) and much larger seeds (2.1 × 2.0 mm). Both species are endemic to southern Namibia but are allopatric. *L. pearsonii* occurs on the Groot Karasberge in a summer rainfall area some distance to the north-east of *L. namibiensis*, and flowers from January to March.

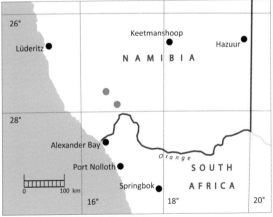

Map 103. Known distribution of *Lachenalia namibiensis*.

Figure 241. Cultivated specimen of *Lachenalia namibiensis* from Rosh Pinah, south-western Namibia. Image: Graham Duncan.

DISTRIBUTION, HABITAT AND CONSERVATION STATUS. *L. namibiensis* is confined to the arid south-western corner of Namibia in the Rosh Pinah and Zebrafontein districts (Map 103). Rainfall in this region is erratic and bulbs remain dormant for extended periods that often span several years. The plants grow in Desert and Succulent Steppe vegetation, in colonies on gravel or stony granitic flats, rocky outcrops or on mountain summits. At Zebrafontein north of Rosh Pinah, the species occurs in association with *L. giessii*. *L. namibiensis* is locally common and not under threat but has a conservation status of Rare (Loots, 2005).

NOTES. The widely campanulate flowers are adapted to pollination by honey bees (*Apis mellifera*). Shortly after capsule dehiscence, the base of the scape detaches and the infructescence is blown away, dispersing the seeds.

103. LACHENALIA PEARSONII

Lachenalia pearsonii (R. Glover) W. F. Barker, *Journal of South African Botany* 35 (5): 321–322 (1969). *Scilla pearsonii* R. Glover, *Annals of the Bolus Herbarium* 1: 105–106 (1915).
TYPE: Namibia, Groot Karasberge, Lord Hill, H. H. W. Pearson 7989 (BOL!, holo.; B, K!, iso).

ETYMOLOGY. *pearsonii*: after Prof. H. H. W. Pearson (1870–1916), first Director of the National Botanical Gardens of South Africa at Kirstenbosch, who discovered this species and made the first collection of plants.

DESCRIPTION. *Dwarf geophyte*, summer-growing, 50–180 mm high. *Bulb* globose, 10–20 mm in diameter, rarely offset-forming, neck distinct, 10–30 mm long; tunic multilayered, outer layers cartilaginous, dark brown or black, apex uppermost portion produced into a thick layer of long, flat, papery bristles; inner cataphyll tightly adhering to leaf bases, aerial portion bright green, subterranean portion translucent white, apex obtuse. *Leaves* 2, narrowly lanceolate, 55–100 × 5–9 mm; suberect, glaucous, canaliculate, abaxial midrib distinct; margins ciliolate; leaf bases clasping, 20–35 mm long, subterranean portion white, aerial portion glaucous; primary seedling leaf terete, erect. *Inflorescence* racemose, few- to many-flowered, sterile apex short; scape erect, 30–70 mm long, slender, yellowish-green; rachis white or light green, extremely slender; pedicels suberect in bud, perpendicular at anthesis, 4–7 mm long; light green; bracts cup-shaped throughout, 2 × 1 mm, white, base swollen. *Perianth* slightly zygomorphic, widely campanulate, white, slightly spice-scented; tube cup-shaped, 1 mm long; outer tepals narrowly ovate, 4 × 2 mm, apices recurved, apical gibbosities dull maroon; inner tepals linear, 4.5 × 1.0 mm, keels dull maroon, apices slightly recurved. *Stamens* shortly exserted, narrowly spreading; filaments white, 4 mm long. *Ovary* obovoid, 1.0 × 1.3 mm, light green; style shortly exserted, straight, 4 mm long. *Capsule* obovoid, 4.0 × 4–5 mm. *Seed* globose, 2.1 × 2.0 mm, glossy, black; strophiole rudimentary, 0.1–0.2 mm long, ridged. *Chromosome number* unknown. Figure 242.

FLOWERING PERIOD. Mid-January to early March.

HISTORY. *L. pearsonii* was discovered in flower on Lord Hill in the central Groot Karasberge north of Karasburg by Prof. Harold Pearson on 17th January 1913 during the Percy Sladen Memorial Expedition to the Groot Karasberge in southern Namibia. The expedition was funded by the Percy Sladen Memorial Trust, which was created following the death of the British naturalist Percy Sladen in 1900. Numerous plant species new to science were discovered during this Expedition, most of which were published in the first volume of *Annals of the Bolus Herbarium* in 1915. *L. pearsonii* was initially mistaken for a *Scilla* and was published as *S. pearsonii* R. Glover (Glover, 1915). It was later

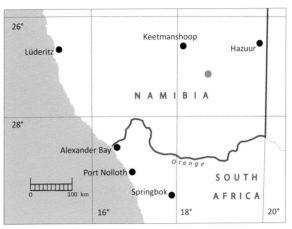

Map 104. Known distribution of *Lachenalia pearsonii*.

Figure 242. *Lachenalia pearsonii* amongst grasses in January in the central Groot Karasberge, southern Namibia. Image: Peter Bruyns.

transferred to *Lachenalia* because of the filament attachment at different levels on the perianth tube, and because it has two tepal whorls that differ in width and shape (Barker, 1969). Following its discovery, *L. pearsonii* has only been recollected twice, on both occasions by Dr Peter Bruyns, near the type locality, but at higher altitude. Flowering specimens were collected in January 1989 (*Bruyns 3554* in NBG), and in January 2000, vegetative specimens were collected that have flowered every subsequent year in the bulb nursery at Kirstenbosch.

DISTINGUISHING CHARACTERS AND AFFINITIES. *L. pearsonii* forms part of a small alliance of dwarf species that have widely campanulate, white flowers that are borne on long pedicels held more or less perpendicular to the rachis. The group includes *L. giessii*, *L. multifolia* and *L. namibiensis*. *L. pearsonii* and *L. glaucophylla* have the smallest flowers in the genus. Duncan *et al.* (2005a) consider *L. pearsonii* to be allied to *L. namibiensis*, these two species sharing widely campanulate flowers with narrowly spreading stamens, similar bifacial leaves with a distinct midrib on the lower surface, and ciliolate margins. The latter differs in its longer perianth tube (3 mm), much longer inner tepals (7–9 mm), included stamens, smaller, ovoid bulb (up to 15 mm in diameter) and shorter, lanceolate leaves (30–60 mm long). Floral symmetry is nearly actinomorphic in *L. pearsonii* and slightly less so in *L. namibiensis*. The globose seeds of *L. pearsonii* are large (2.1 × 2.0 mm) compared with those of *L. namibiensis* (1.2–1.3 × 1.3–1.4 mm).

DISTRIBUTION, HABITAT AND CONSERVATION STATUS. The species is currently only known from Lord Hill in the central Groot Karasberge of southern Namibia (Map 104). It occurs in Dwarf Shrub

Savanna vegetation (Loots, 2005) and is the only summer-growing member of the genus. It is a high altitude species, growing in large colonies from the middle slopes to just below the summit peak, between 1,800–2,200 m (Duncan, 1998b). The type material was recorded growing on sandstone of the middle slopes of the mountains, whereas recently collected material was found at higher altitude on vertical shale layers, amongst grasses and the dwarf succulent *Lithops karasmontana* (Mesembryanthemaceae). *L. pearsonii* is not threatened.

NOTES. *L. pearsonii* is the earliest-flowering *Lachenalia* species. Little is known of its ecology, but its widely campanulate, slightly scented white flowers are probably pollinated by bees. The scape remains attached to the bulb after the capsules have dehisced, and the seeds are dispersed by the shaking action of wind.

104. LACHENALIA MULTIFOLIA

Lachenalia multifolia W. F. Barker, *Journal of South African Botany* 44 (4): 394–396 (1978).
TYPE: South Africa, Western Cape, Karoo Poort, north of Ceres, *W. F. Barker* 3053 (NBG!, holo., iso.).

ETYMOLOGY. *multifolia*: many-leafed.

DESCRIPTION. *Dwarf geophyte*, 70–150 mm high. *Bulb* globose, 10–20 mm in diameter, tunic multilayered, outer layers light brown, papery; inner cataphyll translucent white, subterranean, adhering to clasping leaf bases, apex acute. *Leaves* 5–18, filiform, 150–170 × 1–2 mm, erect, terete, intensely glaucous, rosulate, uppermost portion weakly to strongly spirally twisted; leaf bases loosely surrounding scape, 20–40 × 10–20 mm, broad, swollen, subterranean, white; primary seedling leaf terete, erect. *Inflorescence* racemose, many-flowered, dense, sterile apex short; scape erect, 30–40 mm, dull purplish-brown or green; rachis light to dark green; bracts cup-shaped, 1–2 × 2–4 mm, white or green; pedicels perpendicular at anthesis, bending downwards in fruit, 5–10 mm long, white. *Perianth* zygomorphic, widely campanulate, white, strongly marzipan-scented; tube cup-shaped, 3 mm long; outer tepals ovate, 5–6 × 3–4 mm, apices slightly recurved, apical gibbosities bright green or light brown; inner tepals linear-oblong, 6–7 × 2 mm, apical marking bright green or light brown. *Stamens* well exserted, narrowly spreading; filaments white, 9–11 mm long. *Ovary* obovoid, 2–3 × 2 mm, bright green; style well exserted, straight, white, 8–10 mm long. *Capsule* obovoid, 4–5 × 4–5 mm, cernuous. *Seed* globose, 0.9 × 1.3 –1.4 mm, glossy, black; strophiole rudimentary, 0.1–0.2 mm long, ridged. *Chromosome number* unknown. Figures 243, 244.

FLOWERING PERIOD. September to October, with a peak in late September.

HISTORY. The first collection of *L. multifolia* was made north of Ceres in August 1925 (*Neilson s.n.*, in BOL) and it was subsequently found at several other locations including Karoo Poort, the Ceres Karoo

Figure 243. *Lachenalia multifolia* on sandy flats of the southern Tanqua Karoo. Image: Graham Duncan.

and east of Touwsrivier. W. F. Barker collected the type material at Karoo Poort on 26th September 1944, but the species languished undescribed for several decades until it finally appeared in the *Journal of South African Botany* (Barker 1978). It has since been recorded on several occasions, mainly in the southern Tanqua Karoo.

DISTINGUISHING CHARACTERS AND AFFINITIES. Duncan *et al.* (2005a) consider *L. multifolia* to be allied to *L. polyphylla*, which shares a rosette of filiform leaves with broad, swollen subterranean bases, exserted white stamens, and globose seeds with rudimentary strophioles. *L. polyphylla* differs in its dark green leaves with the maroon to magenta bases covered with minute papillae, its smaller, oblong-campanulate, whitish-blue perianths with pink or brown apical gibbosities to the outer tepals, its recurved stamens and its matte black seeds with rugose secondary sculpturing.

DISTRIBUTION, HABITAT AND CONSERVATION STATUS. *L. multifolia* is endemic to the arid Tanqua Karoo and the western edge of the Great Karoo, extending from just north of Calvinia, south to just east of Touwsrivier in the central Western Cape (Map 105). It is most commonly encountered in the Tanqua Karoo north of Karoo Poort, occurring on open flats and in rock pans and crevices of rocky quartz hillsides, in stony red or light yellowish-brown sand. The species occurs in four vegetation types: Hantam-, Tanqua- and Swartruggens Quartzite Karoo, and Western Little Karoo (Mucina & Rutherford, 2006). The plants occur in full sun or adjacent to low bushes, and grow singly or in small groups as part of larger colonies. At a locality in the southern Tanqua Karoo, *L. multifolia* grows in association with *L. comptonii*, a purple form of *L. juncifolia*, *L. suaveolens* and *L. violacea*. The species is not threatened.

NOTES. The heavily marzipan-scented flowers are pollinated by honey bees (*Apis mellifera*). During fruiting, the pedicels bend downwards and the scapes detach shortly after capsule dehiscence, dispersing the seeds over a wide area.

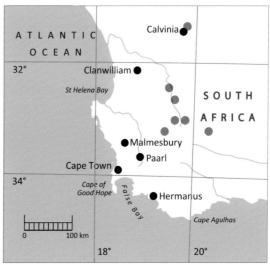

Map 105. Known distribution of *Lachenalia multifolia*.

Figure 244. Cultivated specimens of *Lachenalia multifolia* from the western Tanqua Karoo. Image: Graham Duncan/SANBI.

105. LACHENALIA PATULA

Lachenalia patula Jacq., *Collectanea* 4: 149–150 (1791). *Scillopsis patula* (Jacq.) Lem., *L'Illustration Horticole* 3: 35 (1856).
TYPE: South Africa, Cape, collector and precise locality unknown, figure in N. J. Jacquin, *Icones Plantarum Rariorum* 2: 12, t. 384 (1792 or 1793) (neotype, designated here).
SYNONYMY: *Lachenalia succulenta* Masson ex Baker, *Journal of Botany* 24: 336 (1886). Type: South Africa, Western Cape, Olifants River, drawing by Francis Masson (BM!, holo.).

ETYMOLOGY. *patula*: outspread, with reference to the orientation of the inner tepals.
DESCRIPTION. *Dwarf geophyte* 50–158 mm (rarely up to 230 mm) high. *Bulb* globose, 5–8 mm in diameter, offset-forming; tunic multilayered, outer layers rigid, dark brown to black; neck fasciculate; inner cataphyll translucent white, subterranean, adhering to leaf bases, apex obtuse. *Leaves* 2, broadly linear, 40–60 × 2–10 mm, subterete, erect to suberect, glabrous, light green, ageing to dull or bright maroonish-magenta, upper surface slightly to deeply canaliculate, apices apiculate; leaf bases clasping, 10–25 mm long, lower third white, shading to light maroon above; primary seedling leaf flat, prostrate. *Inflorescence* racemose, few- to many-flowered, sterile apex short; scape suberect, 30–70 mm long, slender to sturdy, dark maroon; rachis dark maroon in lower half, shading to light maroon or white above; pedicels suberect, 5–7 mm long, white to light maroon; bracts cup-shaped throughout, 0.5–1.0 × 1–3 mm, rudimentary, light to dark maroon at base, shading to white above. *Perianth* zygomorphic, widely campanulate, suberect, white to light pink; tube cup-shaped, 3 mm long; outer tepals ovate, 6–7 × 4–5 mm, apical gibbosities brownish-magenta or brownish-green, keels pinkish-magenta; inner tepals obovate, 8–9 × 7–8 mm, keels pinkish-magenta, apices slightly

Figure 245. A colony of *Lachenalia patula* amongst dolerite stones on the Knersvlakte near Vredendal. Image: Harry Hall.

THE GENUS LACHENALIA
105. LACHENALIA PATULA

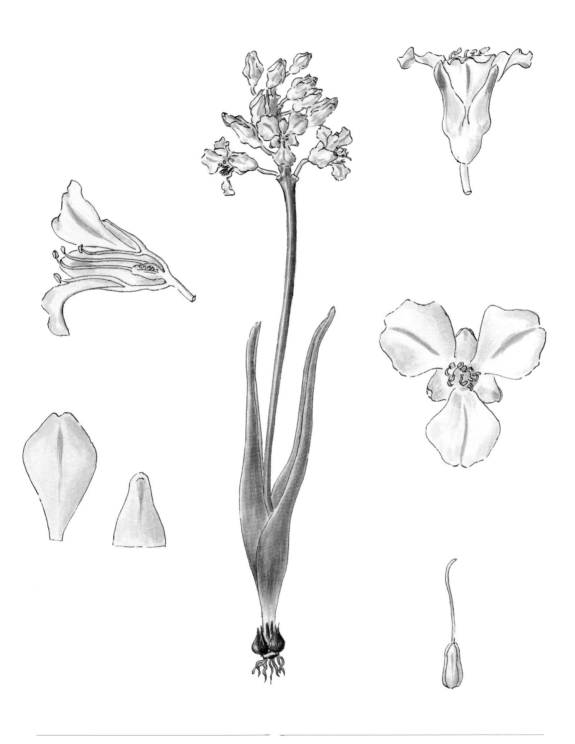

PLATE 31. Watercolour painting of *Lachenalia patula* from Vanrhynsdorp, courtesy of the Compton Herbarium, Kirstenbosch, South African National Biodiversity Institute. Artist: Winsome Barker.

recurved, rounded, margins undulate. *Stamens* included, declinate; filaments white, 5–9 mm long. *Ovary* ellipsoid, 3–4 × 2.0–2.5 mm, yellowish-green; style included, declinate, 6 mm long, white. *Capsule* ellipsoid, 7–8 × 4 mm. *Seed* globose, 0.5 × 0.5 mm; matte, black, primary sculpturing rugose; strophiole rudimentary, 0.2–0.3 mm long, ridged. *Chromosome number* 2n = 16 (Johnson & Brandham, 1997). Plate 31, Figure 245.

FLOWERING PERIOD. September to October, with a peak in late September.

HISTORY. *L. patula* was described by N. J. Jacquin (1791) in volume 4 of *Collectanea* and subsequently illustrated in the second volume of his *Icones Plantarum Rariorum* in 1792 or 1793. The type material was destroyed in a fire in Vienna between 1939–1945 and Barker (1989) designated the painting as the lectotype; strictly, this material is a neotype as it was not referred to by Jacquin in the protologue of his original description. The species was introduced into cultivation at Kew in 1795 (Aiton, 1811). *L. patula* was transferred to *Scillopsis* as *S. patula* (Jacq.) Lem. (Lemaire, 1856) but was returned to *Lachenalia* by Baker (1871) and this status was upheld in his second and final monograph of the genus (1897a).

In 1793, the Scottish plantsman Francis Masson completed a drawing of a *Lachenalia* species that was described almost a century later by Baker (1886) as *L. succulenta*. The drawing was made during one of two journeys of exploration to Namaqualand that Masson made, together with Thunberg, during Masson's second visit to the Cape between 1786–1795. The drawing, which is the holotype, was made from a specimen collected near the Olifants River, which forms the southernmost boundary of this species' range, probably in the vicinity of Klawer, and is preserved in the British Museum (Natural History). Baker (1897a) considered *L. succulenta* distinct from *L. patula* on account of its slightly shorter lower pedicels and purple scape, and because its outer tepals are (unusually) slightly longer than the inner ones, as opposed to the inner tepals being much longer than the outer ones in *L. patula*; the inner tepals do appear to be slightly shorter than the outer ones in Masson's drawing but this is probably a consequence of the angle at which the only fully open flower was depicted. The two names clearly represent the same species and the later *L. succulenta* was relegated to synonymy by Barker (1989), although she incorrectly cited Masson's drawing as an 'iconotype'.

L. patula was mistakenly included in Sölch and Roessler's (1970) *Prodromus einer Flora von Südwestafrika*, based on a collection made at Witputs-Süd (*Merxmüller & Giess 3205*, in B) in south-western Namibia, later described as *L. namibiensis* (Barker, 1987).

DISTINGUISHING CHARACTERS AND AFFINITIES. *L. patula* is one of the most distinctive members of the genus, having two erect, extremely fleshy, subterete, light green to maroonish-magenta leaves with apiculate apices. It has the smallest bulb within the genus and the second-smallest seed after *L. angelica*. Duncan *et al.* (2005a) consider *L. patula* to be allied to *L. angelica*, another dwarf, white-flowered species that shares long white pedicels, declinate stamens and extremely small globose seeds. *L. angelica*

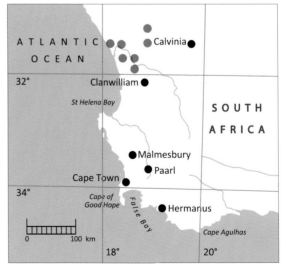

Map 106. Known distribution of *Lachenalia patula*.

differs in its single lanceolate, prostrate leaves, which are covered with minute stellate hairs on the upper surface, and in its much smaller, spreading flowers with strongly recurved outer and inner tepals, exserted stamens and obovoid capsules.

DISTRIBUTION, HABITAT AND CONSERVATION STATUS. Confined to the arid Knersvlakte, *L. patula* extends south-west and east of Nuwerus, south to Koekenaap, Vredendal, Vanrhynsdorp and Klawer (Map 106). The plants grow in colonies among low succulents, on black and white quartz, or among dolerite stones. Colonies occur in Knersvlakte Quartz- and Knersvlakte Shale Vygieveld vegetation (Mucina & Rutherford, 2006), on ridge summits and on southern and north-western aspects of hills. The species is not threatened.

NOTES. The leaves of *L. patula* are so fleshy that it is practically a geophytic succulent. Leaf colour is light green in winter but rapidly changes to maroonish-magenta as light intensity increases in early spring. Its large, widely campanulate white or light pink flowers suggest pollination by honey bees (*Apis mellifera*). The scapes detach from the bulbs soon after the capsules dehisce, the seeds being dispersed from the infructescence as it is blown away.

106. LACHENALIA POLYPODANTHA

Lachenalia polypodantha Schltr. ex W. F. Barker, *Journal of South African Botany* 45 (2): 212–214 (1979).

TYPE: South Africa, Northern Cape, 25 miles [40 km] north of O'kiep, *W.F. Barker* 9049 (NBG!, holo.).

ETYMOLOGY. *polypodantha*: many foot-like anthers.

DESCRIPTION. *Dwarf geophyte*, 55–150 mm high. *Bulb* globose, 5–20 mm in diameter, solitary, medium- to deep-seated up to 70 mm deep, tunic multilayered, outer layers spongy, light brown; inner cataphyll translucent white, adhering to leaf bases, subterranean, apex acute or obtuse. *Leaf* solitary, lanceolate to broadly lanceolate, 15–40 × 5–20 mm, spreading to arcuate, dark green or glaucous, fleshy, upper surface densely covered with short stellate trichomes, with shallow depressed longitudinal grooves; leaf base clasping, 20–50 mm long, white or glaucous, plain or flushed with minute maroon speckles; primary seedling leaf flat, prostrate. *Inflorescence* racemose, few- to many-flowered, dense, conical, sterile apex short; scape erect, 10–40 mm long, slender, maroonish-purple to light green or ivory white, plain or densely covered with minute maroon blotches; rachis light green, plain or densely covered with minute dull maroon blotches; pedicels perpendicular, 4–11 mm long, white or light green to glaucous; lower bracts ovate at base of inflorescence, becoming lanceolate above, 1–2 × 2–3 mm, light green. *Perianth* zygomorphic, widely campanulate, spreading, light bluish-white, ivory white or light violet; tube cup-shaped, 2–3 mm long; outer tepals ovate, 4–6 × 2.5–4.0 mm, apices slightly recurved, apical gibbosities light green to brown; keels dark violet, brownish-green or brown; inner tepals obovate, 4–7 × 2–3 mm, keels dark violet, greenish-violet, bright green or brown, apices recurved. *Stamens* well exserted, declinate; filaments uniformly ivory white, or lower third translucent white to light violet, upper two thirds light to dark violet, 7–12 mm long. *Ovary* obovoid, 2–4 × 2.5–3.0 mm, light bluish-green; style well exserted, declinate, 10–12 mm long, ivory or translucent white to light violet in lower third, upper two thirds light to dark violet, stigma white or dark magenta. *Capsule* obovoid, 4–5 × 3–4 mm. *Seed* globose, 0.7–0.9 × 0.7–0.9 mm, matte, black, secondary sculpturing reticulate; strophiole rudimentary, 0.2–0.3 mm long, ridged. *Chromosome number* unknown. Figures 246, 247.

106. LACHENALIA POLYPODANTHA

FLOWERING PERIOD. August to September, with a peak from late August to mid-September.

HISTORY. Rudolf Schlechter discovered *L. polypodantha* in fruit at Karoechas near Steinkopf in northern Namaqualand on 25th September 1896. He appended the manuscript name *L. polypodantha* Schltr. to his material and distributed it to seven local and foreign herbaria. More than 60 years later, W. F. Barker collected excellent flowering material north of O'kiep in central Namaqualand on 25th August 1959. The species remained undescribed for a further 20 years until it was finally published in the *Journal of South African Botany* (Barker, 1979a). *L. polypodantha* was not illustrated in *The Lachenalia Handbook* (Duncan, 1988a) as no living material was available for study at that time, but it has subsequently been collected on several occasions.

The subsp. *eburnea* is a very recent discovery and was collected for the first time by Gordon Summerfield in 2005 at Rietfontein Farm in western Bushmanland; it was at first thought to be merely a white-flowered form of *L. polypodantha*. In September 2008, Gordon Summerfield and I visited Rietfontein Farm where additional populations were found at peak flowering stage and the type material was collected. The plants were studied in detail and it was realised that it represented a new subspecies of *L. polypodantha*, based on morphological differences and a geographically disjunct distribution.

Figure 246 (left). *Lachenalia polypodantha* subsp. *polypodantha* on gravelly red sand flats near Springbok. Image: Graham Duncan.

Figure 247 (right). *Lachenalia polypodantha* subsp. *eburnea* in Bushmanland Sandy Grassland west of Gamoep. Image: Graham Duncan.

DISTINGUISHING CHARACTERS AND AFFINITIES. Phylogenetic analysis of morphology placed *L. polypodantha* as sister to *L. angelica* (Duncan *et al.*, 2005a). These two species share similar lanceolate leaves that are covered with short stellate trichomes on the upper surface, long pedicels, declinate, exserted stamens, and globose, matte black seeds with reticulate secondary sculpturing. *L. angelica* differs in having a much smaller, prostrate leaf which becomes suberect at anthesis, a rigid scape, few-flowered inflorescence, suberect pedicels, shorter, pure white filaments (6–8 mm long) and a much smaller seed (0.4 × 0.4 mm) with a shorter rudimentary strophiole (0.1–0.2 mm).

DISTRIBUTION AND HABITAT. *L. polypodantha* is endemic to the arid north-western and western part of the Northern Cape from Tatasberg in the northern Richtersveld to Springbok in central Namaqualand, and east to Pella and Rietfontein Farm in western Bushmanland (Map 107). Plants are solitary or occur in small to large groups. Populations usually grow on flats in deep red granitic sand or gravel, exposed or in low vegetation, or within grass tussocks (see under subspecies descriptions below for further details).

NOTES. The flowers of both subspecies of *L. polypodantha* are mainly pollinated by honey bees (*Apis mellifera*) but subsp. *eburnea* is also visited by a nectar-feeding blister beetle (*Lytta nitidula*). The scapes detach soon after the capsules have dehisced and the infructescences are blown away, dispersing the seeds over a wide area. The bulbs remain dormant in poor rainfall years; even during seasons of sufficient rainfall, a certain percentage of bulbs within a population remain dormant.

Key to the subspecies

1a Leaf dark green, spreading, upper surface flat, inner tepals light violet, filaments light to dark violet . . . . **a**. subsp. **polypodantha**
1b Leaf glaucous, arcuate, upper surface strongly canaliculate, inner tepals and filaments ivory white . . . . **b**. subsp. **eburnea**

a. subsp. polypodantha

DESCRIPTION. *Plants* 70–150 mm high. *Bulb* 10–20 mm in diameter. *Leaf* spreading, 15–40 × 5–20 mm, flat, dark green. *Scape* purple or light green, or with maroon blotches, 10–30 mm long; pedicels 4–9 mm long, white or light green. *Perianth* light bluish-white to light violet; tube 2–3 mm long; outer tepals 4–6 × 2.4–4.0 mm, apical gibbosities light to bright green, keels dark violet or brownish-green; inner tepals 6–7 × 2–3 mm, keels dark violet to greenish-violet. *Stamens* 7–12 mm long, filaments white or light violet in lower third, dark violet in upper two thirds. *Ovary* 2–3 × 2.5 mm; style 10–11 mm long, translucent white or light violet in lower third, dark violet in upper two thirds. *Seed* 0.9 × 0.8–0.9 mm. Figure 246.

FLOWERING PERIOD. August to September, with a peak from late August to mid-September.

DISTRIBUTION, HABITAT AND CONSERVATION STATUS. The subsp. *polypodantha* is the more widespread, occurring from the Tatasberg in northern Richtersveld, south-east to Springbok in central Namaqualand and east to Pella in western Bushmanland (Map 107). The plants occur in a variety of vegetation types including Southern Nababiepsberge Mountain Desert, Eastern Gariep Rocky Desert, Bushmanland Arid Grassland, Namaqualand Blomveld and Platbakkies Succulent Shrubland (Mucina & Rutherford, 2006). They grow as scattered individuals or in small groups within larger populations, exposed or adjacent to low bushy growth. The taxon has a conservation status of Rare (Victor & Duncan, 2009j).

b. subsp. eburnea G. D. Duncan, *subsp. nov.*, *a subsp. polypodantha foliis glaucis canaliculatis crassis, perianthio filamento et stylo eburneo, seminibus parvioribus globosis differt.*
TYPE: South Africa, Northern Cape, 2918 (Gamoep): Rietfontein farm, on flats in deep red sand, (–AD), *G. D. Duncan* 574 (NBG!, holo.)

ETYMOLOGY. *eburnea*: ivory, with reference to the tepal and stamen colour.
DESCRIPTION. *Plants* 55–110 mm high. *Bulb* 5–15 mm in diameter. *Leaf* arcuate, strongly canaliculate, glaucous, very succulent, apex acute, 25–40 × 9–17 mm. *Scape* 25–40 mm long, light green or ivory white, plain or with small dull maroon blotches; pedicels ivory white, 6–11 mm long. *Perianth* ivory; tube 3 mm long; outer tepals 4–5 × 2–3 mm, apical gibbosities bright green or brown; inner tepals 4–5 × 2 mm, keels bright green or brown. *Stamens* 8–9 mm long, filaments ivory. *Ovary* 3–4 × 2–3 mm, yellowish-green; style 11–12 mm long, ivory. *Seed* globose, 0.7–0.8 × 0.7–0.8 mm. Figure 247.
FLOWERING PERIOD. September.
DISTINGUISHING CHARACTERS AND AFFINITIES. The lanceolate leaf of subsp. *eburnea* differs from that of subsp. *polypodantha* in being arcuate, deeply canaliculate and glaucous, with an acute apex, and much fleshier. The perianths of both subspecies are similar but subsp. *eburnea* differs in being ivory, the outer tepals with bright green apical gibbosities, the inner tepals with green or brown keels, and the filaments are ivory, not light to dark violet as in subsp. *polypodantha*. The pedicels of subsp. *polypodantha* are somewhat thicker than those of subsp. *eburnea*.
DISTRIBUTION, HABITAT AND CONSERVATION STATUS. Subspecies *eburnea* is currently known only from Rietfontein Farm west of Gamoep in western Bushmanland (Figure 247). It occurs in Bushmanland Sandy Grassland vegetation (Mucina & Rutherford, 2006), and is solitary or grows in small groups within dense colonies, in deep red granitic sand. The subspecies is often associated with the indigenous tufted grass *Centropodia glauca* (gha-grass) (Poaceae), sometimes growing within the tussocks. At Rietfontein Farm, subsp. *eburnea* grows in association with *L. inconspicua* and *L. xerophila*. The taxon is not threatened but owing to its very limited known distribution, a conservation status of Rare is recommended.

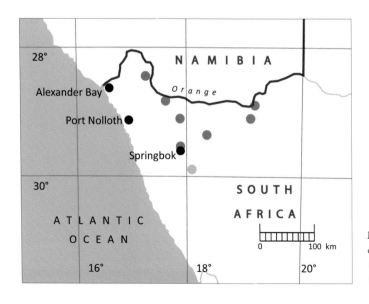

Map 107. Known distribution of *Lachenalia polypodantha*:
● = subsp. *polypodantha*;
○ = subsp. *eburnea*.

107. LACHENALIA NORDENSTAMII

Lachenalia nordenstamii W. F. Barker, *Journal of South African Botany* 49 (4): 428–432 (1983).
TYPE: South Africa, Northern Cape, Numees Mountains north of Hellskloof, Richtersveld, *R. B. Nordenstam* 1739 (NBG!, holo., flowering plant); mountain between Numees and Hellskloof, Richtersveld, *R. B. Nordenstam* 1762 (NBG!, holo., capsules and seeds).

ETYMOLOGY. *nordenstamii*: after Prof. Bertil Nordenstam (1936–), Swedish botanist and Emeritus Professor of the Swedish Museum of Natural History, Stockholm.

DESCRIPTION. *Dwarf geophyte*, 50–120 mm high. *Bulb* ovoid, 10–15 mm in diameter, solitary; neck distinct, slender, fibrous, 25–40 mm long; tunic multilayered, outer layers spongy, light brown; inner cataphyll adhering to leaf bases, apex obtuse. *Leaf* solitary, narrowly lanceolate, 20–40 × 5–8 mm, falcate, dark green to glaucous, canaliculate, distinctly bifacial, upper surface plain, lower surface blotched with purplish-brown in central and upper part, midrib prominent; leaf base clasping, 15–20 mm long, banded with maroon towards base. *Inflorescence* racemose, few-flowered, sterile apex short; scape suberect, 15–50 mm long, slender, glaucous, tinged with dull maroon; pedicels spreading to decurved, 2–3 mm long, glaucous; bracts minute, ovate at base of inflorescence, becoming lanceolate above, 0.8–1.0 × 0.5–0.6 mm. *Perianth* zygomorphic, widely campanulate, cernuous; tube cup-shaped, 1–2 mm long, brownish-yellow; outer tepals ovate, 4 × 2–3 mm, deep maroon, apical gibbosities dark brownish-maroon; inner tepals linear-oblong, 4–5 × 2 mm, deep maroonish-brown, apices recurved, keels maroon. *Stamens* well exserted, narrowly spreading; filaments stout, 7–10 mm long, upper half maroon, lower half greenish-yellow. *Ovary* ellipsoid, 2 × 1 mm, yellowish-green; style well exserted, more or less straight, 8–10 mm long, upper half maroon, lower half light greenish-yellow. *Capsule* obcordate, 5–6 × 10–12 mm, broadly winged, maroon. *Seed* oblong, 2.5 × 0.8–0.9 mm, matte, black; strophiole rudimentary, 0.3 mm long, ridged. *Chromosome number* unknown. Figure 248.

FLOWERING PERIOD. May to July, with a peak in late May.

HISTORY. Fruiting material of this rarely collected species was discovered by Prof. Bertil Nordenstam in the Numees Mountains north of Hellskloof, northern Richtersveld, on 3rd November 1962. He made vegetative collections the following day between Numees and Hellskloof, bulbs of which were cultivated at the Compton Herbarium at Kirstenbosch. This material flowered in May the following year to provide the holotype for the species. The plant was subsequently collected in the Aurus Mountains in 1977, near Rosh Pinah in south-western Namibia in 1990, and again in the Numees Mountains by Dr Graham Williamson in 1987. A bulb of this last collection was cultivated in the nursery at Kirstenbosch, flowered there in May 1990, and is illustrated in the accompanying photograph.

Figure 248. Cultivated specimen of *Lachenalia nordenstamii* from the northern Richtersveld. Image: Graham Duncan/SANBI.

107. LACHENALIA NORDENSTAMII

DISTINGUISHING CHARACTERS AND AFFINITIES. Phylogenetic analysis of morphology (Duncan et al., 2005a) placed *L. nordenstamii* as sister to the sympatric *L. buchubergensis*, which shares a short, few-flowered raceme, a solitary lanceolate leaf, a relatively large, ovoid bulb and large, broadly winged capsules that contain large oblong seeds with rudimentary, ridged strophioles. *L. buchubergensis* differs mainly in its broadly lanceolate, conduplicate, intensely glaucous leaf with undulate margins, and in its tubular grey flowers with shortly exserted, straight, narrow filaments.

DISTRIBUTION, HABITAT AND CONSERVATION STATUS. *L. nordenstamii* is confined to the arid far north-western corner of the Richtersveld in South Africa and the extreme south-western corner of Namibia, occurring between Numees and Hellskloof in South Africa, north-west to the Aurus Mountains in Namibia (Map 108). In South Africa, it occurs in Upper Annisvlakte Succulent Shrubland vegetation (Mucina & Rutherford, 2006) and in south-western Namibia it is endemic to Desert and Succulent Steppe Vegetation (Loots, 2005). it is associated with rocky terrain, growing on south- and west-facing dolomitic hills and mountain slopes. The plants occur as occasional solitary individuals or in small groups in sheltered rock crevices, in shallow sand (Duncan, 1999b). Like several other lachenalias from this arid area, including *L. buchubergensis*, *L. klinghardtiana* and *L. nutans*, *L. nordenstamii* flowers early, in late autumn and early winter. It has a conservation status of Rare in South Africa (Victor & Duncan, 2009i) and in Namibia (Loots, 2005).

NOTES. Pollination vectors for the species are as yet unknown. Like those of *L. buchubergensis* and *L. zebrina*, the scape of *L. buchubergensis* remains attached to the bulb following capsule dehiscence and the broadly winged, aerodynamic capsules detach from the rachis in gusts of wind and are blown away, dispersing the seeds.

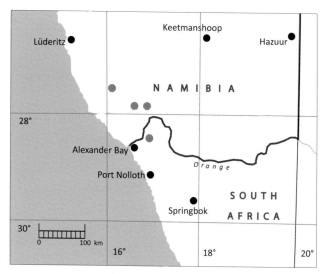

Map 108. Known distribution of *Lachenalia nordenstamii*.

1.5. Section URCEOLATAE

Urceolatae G. D. Duncan, sect. nov., *a ceteris sectionibus perianthio urceolato, tubo cupulato basim inflato, tepalis interioribus contracto ad orem tum expansis 25–40°, filamentis inclusis usque ad breviter exsertis differt.* Type: *Lachenalia verticillata* W. F. Barker

Perianth urceolate; tube cup-shaped, 1–3 mm long, slightly to strongly inflated at base; outer tepals ovate; inner tepals obovate, contracted at mouth, then radiating at an angle of 25–40° from the longitudinal axis, apices slightly to strongly recurved; filaments straight or declinate, included to shortly exserted.

19 species (19 taxa).

108. LACHENALIA MUTABILIS

Lachenalia mutabilis Lodd. ex Sweet, *The British Flower Garden* 5 (series 2), t. 129 (1832).
Orchiastrum mutabile (Lodd. ex Sweet) Lem., *L'Illustration Horticole* 2: 100 (1855).
TYPE: South Africa, Cape, collector and precise locality unknown, figure in *The British Flower Garden* 5 (series 2), t. 129 (1832) (lectotype, designated here).

ETYMOLOGY. *mutabilis*: changeable, descriptive of the changing colour of the flowers.

DESCRIPTION. *Geophyte*, 80–450 mm high. *Bulb* subglobose, 8–25 mm in diameter, offset- and bulbil-forming; tunic multilayered, outer layers spongy; inner cataphyll translucent white, plain or tinged light brown, adhering to leaf bases, apex obtuse. *Leaf* 1(–2), narrowly to broadly lanceolate, 70–300 × 10–35 mm, spreading to erect, canaliculate, light green or glaucous; margins entire to undulate and/or crisped, green or maroon, leathery; leaf bases clasping, 10–40 mm long, plain to faintly spotted or heavily banded with maroon; primary seedling leaf flat, prostrate. *Inflorescence* spicate, rarely racemose, many-flowered, dense, sterile apex long; scape erect, 40–130 mm long, slender to sturdy, light green, lower half plain or irregularly spotted or blotched, upper half shading to dull maroon, frequently strongly inflated; rachis light to dark brownish-maroon or brownish-green below, shading to light blue, mauve, intense electric blue or yellow above; pedicels usually absent, rarely 4–5 mm long; lower bracts ovate, becoming lanceolate above, 0.5–5.0 × 1.0–1.5 mm, white to light mauve. *Perianth* zygomorphic, urceolate, spreading to cernuous, tube cup-shaped, 1–3 mm long, light blue; outer tepals ovate, 5–8 × 4–5 mm, yellowish-green, greenish-brown, light to bright yellow or light to deep mauve, apical gibosities prominent, light to dark brown, maroonish-brown, purplish-maroon or dark yellow; inner tepals obovate, 7–9 × 3–6 mm, yellowish-green, translucent white or light to bright yellow, upper two tepals overlapping, lower inner tepal narrower, canaliculate and longer, apices slightly recurved, apical marking purplish-brown, reddish-brown or deep yellow, keels light to dark brown, green or yellow. *Stamens* declinate, included, rarely exserted up to 2 mm; filaments white, 6–9 mm long, rarely 11–13 mm long. *Ovary* ellipsoid, 3–4 × 2–3 mm, dark green; style included, rarely exserted, declinate, 5–7 mm long, rarely 11 mm long, white. *Capsule* ellipsoid, 6 × 4 mm. *Seed* globose, 1.2–1.4 × 1.3–1.4 mm, glossy, black; strophiole inflated, 0.6–0.7 mm long, smooth. *Chromosome number* $2n = 10$, 14 (Johnson & Brandham, 1997); $2n = 14$ (Hamatani *et al.*, 1998; Spies *et al.*, 2008); $2n = 12, 14, 24$ (Spies *et al.*, 2000); $2n = 14, 56$ (de Wet, 1957). Plate 32, Figures 249–251.

Figure 249. Robust, greenish form of *Lachenalia mutabilis* from gravelly slopes near Soebatsfontein. Image: Ivan van Niekerk.

380 THE GENUS LACHENALIA
108. LACHENALIA MUTABILIS

PLATE 32. Watercolour painting of *Lachenalia mutabilis* from Citrusdal, courtesy of the Compton Herbarium, Kirstenbosch, South African National Biodiversity Institute. Artist: Winsome Barker.

Figure 250 (left). Light yellow form of *Lachenalia mutabilis* with swollen scapes from sandy flats near Nieuwoudtville. Image: Graham Duncan.

Figure 251 (right). *Lachenalia mutabilis* with non-swollen scapes near Piketberg. Image: Graham Duncan.

FLOWERING PERIOD. July to September, with a peak from early to mid-August.

HISTORY. The name *L. mutabilis* was first published in April 1826 by the German nurseryman Conrad Loddiges (1738–1826) accompanying plate 1076 in volume 11 of his *Botanical Cabinet*, a collection of 2,000 botanical images produced in 20 volumes between 1817 and 1833 (Loddiges, 1826c). However, Loddiges's name is a *nomen nudum* as it was not accompanied by a description. A similar situation arose later that year (in 1826) and again in 1830 when the English nurseryman Robert Sweet (1783–1835) published the same name on page 420 of the first edition of *Hortus Brittanicus* and on page 529 of the second edition of that publication, both without descriptions. *L. mutabilis* was finally validated by Sweet (1832) in volume 5 of *The British Flower Garden*, which included a description and an illustration by E. D. Smith, designated here as the lectotype. In both of his monographs, Baker (1871, 1897a) mistakenly placed *L. mutabilis* under *L. orchioides* (L.) Aiton, citing a Zeyher collection from Riviersonderend (*Zeyher* 4289 in K, SAM); he mistakenly also cited the same specimen under *L. pustulata* Jacq. in his second monograph (Baker, 1897a). In the text accompanying plate 9433 of *Curtis's Botanical Magazine*, Hutchinson (1936) confirmed the status of *L. mutabilis* as a distinct species.

DISTINGUISHING CHARACTERS AND AFFINITIES. *L. mutabilis* is regarded here as a distinctive but highly polymorphic species complex, expressed as numerous geographically isolated, but more or less continuously varying, populations. Characters such as perianth orientation, tepal colouration, stamen length, seed size and chromosome number are stable within populations, whereas scape inflation is uniform within certain populations but variable in others. Different basic chromosome numbers (x = 5, 6 or 7) have been reported by various authors. Ornduff & Watters (1978) reported x = 5 for one collection, Johnson & Brandham (1997) reported 2n = 10 for six collections and 2n = 14 for two collections, and Spies *et al.* (2000) reported x = 6 and x = 7 for 35 specimens representing 16 populations. Variation in chromosome morphology occurs between different wild localities, and even between specimens from the same locality, although most specimens of *L. mutabilis* have four to eight very short chromosomes. The number of short chromosomes varies between different localities and between specimens collected at the same locality, and it is postulated that an aneuploid series might exist (Spies *et al.*, 2000). The difference in basic chromosome number for the species is not correlated to geographical location, as proximal populations sometimes have different basic chromosome numbers. Polyploidy is scarce in *L. mutabilis*; in a cytogenetic study of 35 specimens representing 16 populations, only 5% of specimens exhibited this phenomenon (Spies *et al.* 2000). Polymorphism of scape inflation is acceptable at the species level, but speciation is almost certainly taking place within the *L. mutabilis* complex as morphological divergence is being entrenched by breeding barriers at the chromosome level (Duncan, 2005).

L. mutabilis resolved as sister to *L. ventricosa* (Duncan *et al.*, 2005a). The two species share subglobose bulbs, lanceolate, canaliculate leaves and inflated strophioles, but *L. ventricosa* differs in having tubular flowers with longer, oblong outer tepals (10–11 mm), longer, oblong-obovate inner tepals (12–16 mm), much longer, well-exserted, straight stamens (15–23 mm long), straight styles (15–17 mm long) and much smaller, ovoid seeds (0.9–1.0 × 0.8–0.9 mm).

DISTRIBUTION, HABITAT AND CONSERVATION STATUS. *L. mutabilis* is widely distributed in the winter rainfall zone of the Northern and Western Cape, stretching from Anenous Pass west of Steinkopf in the southern Richtersveld to De Hoop Nature Reserve east of Bredasdorp in the southern Cape (Map 109). It occurs as scattered individuals, in small groups or large colonies numbering hundreds of plants, among low bushes or succulent groundcover, and is usually encountered in open aspects in deep white, yellowish-brown or red sand on flats, or in gravelly clay on north-west- and south-facing stony hillsides. *L. mutabilis* also grows in shallow, humus-rich soil in depressions of granite outcrops and sandy alluvial soil in river beds. The species occurs in numerous vegetation types including Anenous Plateau-, Namaqualand Klipkoppe- and Kamiesberg Mountains Shrubland, Namaqualand Coastal Duneveld, Namaqualand Heuweltjieveld, Namaqualand Blomveld, Vanrhynsdorp

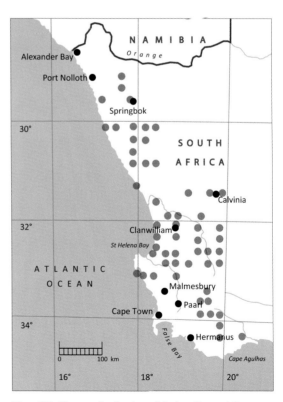

Map 109. Known distribution of *Lachenalia mutabilis*.

Gannabosveld, Lamberts Bay Strandveld, Leipoldtville- and Hopefield Sand Fynbos, Nieuwoudtville Shale Renosterveld, Hantam Karoo, Bokkeveld-, Piketberg-, Olifants-, Cederberg- and Hawequas Sandstone Fynbos, Saldanha Flats Strandveld, Langebaan Dune Strandveld, Breede Alluvium Fynbos and Western-, Central- and Eastern Rûens Shale Renosterveld (Mucina & Rutherford, 2006).

In the arid northern parts of its range in the southern Richtersveld and Namaqualand, the flowers of *L. mutabilis* are predominantly spreading, with dull yellowish-green tepals and brownish-green or light blue rachis apices; towards the central and southern parts from Vanrhynsdorp southwards, the flowers tend to be slightly to strongly cernuous with brighter yellowish-green, light to bright yellow or light to deep mauve outer tepals. Forms in the Citrusdal and Clanwilliam areas have narrower, longer racemes with intense, electric blue rachis apices. A form with inflated scapes from Sewefontein in western Namaqualand and another from the Gifberg south of Vanrhynsdorp are exceptional in having fertile flowers with distinct pedicels 4 mm long and exserted stamens. A particularly attractive bright yellow form occurs just north of Calvinia and in the southern Tanqua Karoo. *L. mutabilis* is especially abundant in the Citrusdal and Clanwilliam areas and the species is not threatened.

NOTES. The flowers of *L. mutabilis* are efficiently pollinated by honey bees (*Apis mellifera*) and the copious production of seed frequently depletes the bulbs severely. The scape detaches once the capsules are mature and the seeds are dispersed from the infructescence as it is blown about. Several instances of strongly fasciated inflorescences have been recorded from the Olifants River Valley. *L. mutabilis* is potentially invasive and has become naturalised in western Australia.

109. LACHENALIA AURIOLIAE

Lachenalia aurioliae G. D. Duncan, *Bothalia* 27 (1): 7–9 (1997).
TYPE: South Africa, Western Cape, hillside facing Hesperus Old Age Home, Beaufort West, *A. U. Batten* 468 (NBG!, holo.; PRE!, iso.).

ETYMOLOGY. *aurioliae*: after Dr Auriol Batten (1918–), botanical artist and author, who collected the type material in 1984.

DESCRIPTION. *Geophyte*, 45–130 mm high. *Bulb* ovoid, 15–25 mm in diameter, offset- and bulbil-forming; neck distinct, 10–15 mm long; tunic multilayered, outer layers light to dark brown, cartilaginous, apical fascicles 20–40 mm long; inner cataphyll translucent white, adhering to leaf bases, apex obtuse. *Leaves* 1–2, broadly lanceolate to ovate, 50–110 × 17–45 mm, spreading to suberect, yellowish-green or glaucous, flat to slightly canaliculate, apices slightly recurved, upper surface with faint depressed longitudinal grooves, plain or faintly spotted with dull green or purple, lower surface with darker green blotches and transverse bands, midrib distinct on lower surface; leaf bases clasping, 20–30 mm long, spotted or banded with brownish-magenta; primary seedling leaf terete, erect. *Inflorescence* spicate or racemose, moderately dense, sterile apex short; scape erect to suberect, 45–140 mm long, fairly sturdy, sometimes slightly inflated above, lower half light green with distinct, irregularly scattered maroon blotches, upper half densely mottled with minute brownish-magenta spots; rachis light green in lower half, shading to greenish-brown above, mottled with maroon spots; pedicels absent or suberect, up to 1 mm long, green or white; lower bracts broadly ovate, becoming ovate above, 1–4 × 1–3 mm, greenish-white, plain or with light brownish-magenta apices. *Perianth* zygomorphic, urceolate, arranged in 3-flowered whorls, spreading to cernuous; tube cup-shaped, 2–3 mm long, light blue or light yellowish-white with light blue spots; outer tepals ovate, 5–9 × 4–6 mm, light blue to light grey, ageing to dull red, apical gibbosities dull reddish-brown to purplish-brown, median keels minutely blue-spotted; inner tepals obovate, 6–10 × 3–6 mm, translucent white to yellowish-green, slightly recurved, keels dark blue, brownish-blue, purple or green. *Stamens* included to shortly exserted, declinate; filaments white, 6–10 mm long. *Ovary* ellipsoid, 2–4 × 2–4 mm, light green; style included, declinate, 5–8 mm long, white, protruding beyond stamens as ovary matures. *Capsule* ellipsoid, 6–10 × 5–7 mm. *Seed* ovoid, 1.1–1.4 × 1.3–1.4 mm, glossy, black; strophiole 0.6–1.1 mm long, ridged. *Chromosome number* unknown. Figures 252, 253.

FLOWERING PERIOD. June to August, with a peak in July.

HISTORY. *L. aurioliae* was collected for the first time by Rudolf Marloth in August 1922 at Droogekloof Farm in the Prince Albert District of the southern Great Karoo. It was subsequently recorded at Leeuwkloof Farm in the Nuweveld Mountains north of Beaufort West in 1935, and many collections have been made in the vicinity of this town. The plant remained undescribed for more than 70 years until its publication in *Bothalia* (Duncan, 1997). Within the past decade, its known range has been extended considerably to the west with a number of collections from the Little and Tanqua Karoos.

DISTINGUISHING CHARACTERS AND AFFINITIES. *L. aurioliae* is a very variable, inconspicuous plant. The bulb is surrounded by hard, dark brown outer scales with distinctive fasciculate apices that are especially long in forms from the Tanqua Karoo. The broadly lanceolate to ovate, glaucous leaves are heavily spotted or banded with magenta on the leaf bases and the spicate or racemose inflorescence consists of urceolate, spreading to cernuous flowers arranged in distinct three-flowered whorls. Outer tepal colour varies from light blue to light grey with prominent brown apical gibbosities, and the inner tepals from translucent white to yellowish-green, with slightly recurved apices. The stamens are declinate, and included to shortly exserted.

Figure 252 (left). *Lachenalia aurioliae* on stony clay flats of the Tanqua Karoo. Image: Graham Duncan.
Figure 253 (right). Cultivated specimen of *Lachenalia aurioliae* from Beaufort West. Image: Graham Duncan.

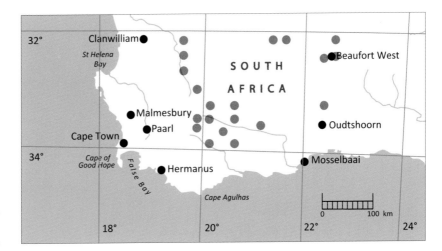

Map 110. Known distribution of *Lachenalia aurioliae*.

In a phylogenetic analysis of morphological data (Duncan *et al.* 2005), *L. aurioliae* appears to be related most closely to *L. karooica*, with which it shares ovoid bulbs and seeds. *L. karooica* differs in having oblong-campanulate flowers with strongly recurved inner tepals, well-exserted, recurved filaments, spirally arranged flowers, seeds that have reticulate secondary sculpturing and bulbs that lack fasciculate outer layers. By comparison with *L. aurioliae*, *L. karooica* has a more northerly and easterly distribution in the Great Karoo, extending from Prieska to Cradock, with an outlier at Karoopoort in the southern Tanqua Karoo.

DISTRIBUTION, HABITAT AND CONSERVATION STATUS. *L. aurioliae* is a species of arid habitats occurring from the Tanqua Karoo National Park in the Northern Cape, south to Montagu in the western Little Karoo and north-east to Beaufort West in the southern Great Karoo (Map 110). The plants occur across numerous vegetation types including Tanqua-, Swartruggens Quartzite-, Gamka-, Western Little-, and Western Upper Karoo, and Matjiesfontein Shale Renosterveld (Mucina & Rutherford, 2006). They grow as scattered individuals or in colonies, mostly on arid, stony shale flats, less frequently on lower hill slopes and in dry river beds. Forms from Beaufort West flower up to a month earlier than those further west in the Tanqua Karoo. The bulbs are adapted to remain dormant during extended droughts that can span several years. *L. aurioliae* is not threatened.

NOTES. *L. aurioliae* is pollinated by honey bees (*Apis mellifera*). The developing flower buds and leaves are heavily grazed by Cape hares (*Lepus capensis*) in the Tanqua Karoo. The base of the scape detaches from the bulb once capsule dehiscence has occurred, and the seeds are dispersed from the ripe infructescence as it is blown about.

110. LACHENALIA VERTICILLATA

Lachenalia verticillata W. F. Barker, *Journal of South African Botany* 44 (4): 405–407 (1978).
TYPE: South Africa, Northern Cape, 3 miles [4.8 km] south-east of Steinkopf, *T. M. Salter* 3746 (BOL!, holo.; BM!, K!, iso.).

ETYMOLOGY. *verticillata*: flowers arranged in three-flowered whorls.
DESCRIPTION. *Geophyte*, 100–250 mm high. *Bulb* globose, 10–20 mm in diameter, solitary; tunic multilayered, outer layers light brown, spongy; inner cataphyll translucent white, upper portion

Plate 33. Watercolour painting of *Lachenalia verticillata* from Steinkopf (*Müller-Doblies* 88125, in NBG) courtesy of The Editor, *Flowering Plants of Africa*, vol. 59, t. 2202 (2005). Artist: Fay Anderson.

tinged with green or dull red, apex acute. *Leaf* solitary, lanceolate, 70–90 × 12–20 mm, suberect to spreading or arcuate, canaliculate, intensely glaucous, upper surface unmarked or sporadically spotted with dull green, lower surface heavily blotched with dark green or purplish in upper portion, shading to purplish-magenta below; margin thickened, green or maroon, undulate or flat; leaf base clasping, 30–50 mm long, glaucous in upper portion, white below, heavily banded with purplish-magenta; primary seedling leaf terete, erect. *Inflorescence* spicate, many-flowered, erect, sterile apex short; scape erect to suberect, 60–120 mm long, slender, heavily blotched with dull purplish-magenta, blotches decreasing in size towards apex; rachis heavily mottled with dull magenta; lower bracts ovate, becoming lanceolate above, 0.5–1.0 × 2 mm, greenish-white. *Perianth* zygomorphic, urceolate, spreading, arranged in distinct 3-flowered whorls; tube cup-shaped, 3 mm long, dull blue; outer tepals ovate, 5–6 × 4 mm, yellowish-green, apices recurved, apical gibbosities dark green; inner tepals obovate, 8–10 × 3–4 mm, greenish-white, keels green, apices purple, strongly recurved, upper two tepals overlapping, lower inner tepal 2 mm longer than upper inner tepals, canaliculate. *Stamens* included, declinate; filaments white, 5–6 mm long. *Ovary* ellipsoid, 3 × 2 mm, light yellowish-green; style included, declinate, 5–6 mm long, white. *Capsule* ellipsoid, 7–8 × 5–6 mm. *Seed* globose, 1.1–1.2 × 1.3 mm, glossy, black; strophiole 0.6–0.7 mm long, ridged. *Chromosome number* 2n = 16 (Crosby, 1986). Plate 33, Figure 254.

Figure 254. Cultivated specimen of *Lachenalia verticillata* from Steinkopf. Image: Graham Duncan.

FLOWERING PERIOD. August to October, with a peak in mid-September.

HISTORY. This rare and beautiful species was recorded for the first time by the eminent Cape botanist Harry Bolus (1834–1911), who found it near O'kiep in central Namaqualand in September 1883. Half a century passed before the next collection was made by the English naval officer Paymaster-Captain T. M. Salter, who found it south-east of Steinkopf in the southern Richtersveld, in September 1933. A further 45 years elapsed before the species was finally published in the *Journal of South African Botany*, with the Salter collection (*Salter* 3746) selected as the type material (Barker 1978). *Lachenalia verticillata* was subsequently collected near Van Zylsrus in central Richtersveld in 1977, at Vaalputs in western Bushmanland in 1983, and south of Steinkopf in 1988. A watercolour painting by Fay Anderson of the latter collection (*Müller-Doblies* 88125, in NBG) was published as plate 2202 of *Flowering Plants of Africa* (Duncan et al., 2005b).

DISTINGUISHING CHARACTERS AND AFFINITIES. The species is sister to *L. concordiana* (Duncan et al., 2005a), with which it shares perianths that are arranged in 3-flowered whorls and included, declinate stamens. *L. concordiana* differs in its light greenish-yellow, oblong-campanulate flowers, much narrower, bifacial, dark green leaf (8–10 mm wide) with prominent dark green blotches and a distinct midrib on the lower surface. It also has a much larger, ovoid seed (1.3–1.4 × 0.8–0.9 mm) with a slightly longer strophiole (0.8 mm long).

DISTRIBUTION, HABITAT AND CONSERVATION STATUS. *L. verticillata* is confined to the arid far north-western part of the Northern Cape, extending from the upper eastern slopes and summit ridge of the Stinkfonteinberge south-west of Van Zylsrus in the central Richtersveld, south to Springbok in central Namaqualand, and south-east to Vaalputs in western Bushmanland (Map 111). The plants occur in Central Richtersveld Mountain Shrubland, Namaqualand Blomveld and Platbakkies Succulent Shrubland vegetation types (Mucina & Rutherford, 2006). They grow singly or in small groups among low succulent vegetation, in stony, well drained red sand. *L. verticillata* has a conservation status of Rare (Victor & Duncan, 2009l).

NOTES. Pollinators of this species are unknown but the urceolate flowers and extended lower inner tepal suggests that they are likely to include bees. Shortly after capsule dehiscence, the scape detaches from the bulb and the infructescence is blown away, dispersing seeds over a wide area. The bulbs are adapted to remain dormant during periods of insufficient rainfall.

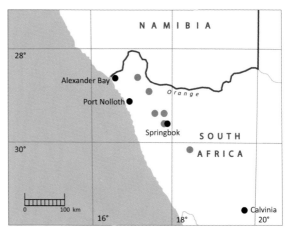

Map 111. Known distribution of *Lachenalia verticillata*.

111. LACHENALIA PEERSII

Lachenalia peersii Marloth ex W. F. Barker, *Journal of South African Botany* 44 (4): 391–394 (1978). TYPE: South Africa, Western Cape, Hermanus, *V. S. Peers s.n.* sub. Herb. Marloth 7263 (PRE!, holo.); *V. S. Peers s.n.* sub. BOL 16360 (BOL!, iso.).

ETYMOLOGY. *peersii*: after Victor Stanley Peers (1874–1940), amateur archaeologist, conservationist and cultivator of South African plants.

DESCRIPTION. *Geophyte*, 120–350 mm high. *Bulb* subglobose, 15–25 mm in diameter, offset-forming; tunic multilayered, outer layers light brown, spongy; inner cataphyll translucent white, loosely surrounding leaf bases, apex obtuse. *Leaves* (1–)2, lanceolate, 90–300 × 10–25 mm, spreading, coriaceous, upper surface light green to maroon or brownish-maroon with depressed longitudinal grooves, lower surface flushed with maroon, midrib absent to prominent; leaf margins flat, occasionally slightly undulate; leaf bases loosely clasping, 20–30 mm long; upper portion dark maroon, subterranean portion light maroonish-white; primary seedling leaf flat, prostrate. *Inflorescence* racemose, few- to many-flowered, lax, sterile apex short to long; scape erect to suberect, 50–290 mm long, slender, light green to dark brownish-maroon; rachis colour as for scape, plain or with minute brown speckles; pedicels erect to suberect, 4 mm long, white, green or brownish-magenta; lower bracts ovate, becoming lanceolate above, 2–6 × 1–4 mm, white or light to dark magenta. *Perianth* zygomorphic, urceolate, spreading to cernuous, white to cream, ageing to magenta, strongly carnation-scented; tube cup-shaped, 2–3 mm long; outer tepals ovate, 6–7 × 4–5 mm, apical gibbosities dark brown to greenish-brown; inner tepals obovate, 7–8 × 4–5 mm, plain or with light brown marking in upper third, apices strongly recurved. *Stamens* included to shortly exserted, straight; filaments white, 7–8 mm long. *Ovary* obovoid, 3 × 2 mm, light green; style included, straight, 4 mm long, white. *Capsule* obovoid, 7–8 × 6 mm.

PLATE 34. Watercolour painting of *Lachenalia peersii* from Betty's Bay (*Duncan* 280, in NBG) courtesy of The Editor, *Curtis's Botanical Magazine* vol. 20, t. 476 (2003). Artist: Vicki Thomas.

Seed globose, 0.9–1.0 × 0.8–0.9 mm, glossy, black; strophiole inflated, 0.6–0.7 mm long, smooth. *Chromosome number* 2n = 14 (Johnson & Brandham, 1997; Hamatani *et al.*, 2004). Plate 34, Figure 255.
FLOWERING PERIOD. October to November, with a peak in late October.
HISTORY. The first collection of *L. peersii* was made in 1912 at Hermanus by the Australian Victor Stanley Peers, whose name is usually associated with the cave in the Fish Hoek Valley in the southern Cape Peninsula, known as 'Peers's Cave'. Peers grew plants in his garden in Wynberg, Cape Town, and in 1915 presented flowering specimens to Dr Rudolf Marloth who added them to his herbarium and appended the manuscript name *L. peersii* after its collector (Barker, 1980b). Two years later, a black-and-white photograph of a pot of flowering specimens taken by Peers was published in the periodical *South African Gardening and Country Life*, accompanied by the caption '*Lachenalia peersii* a new variety found at Hermanus and recently named'. Additional flowering specimens from Peers's garden were preserved in November 1919 and added to the Bolus Herbarium collections at the University of Cape Town. The annotation 'Lachenalia peersii Marl. Typus in Litt.' was made to the sheets by Dr Louisa Bolus, who evidently assumed it to have been validly published. However, as no taxonomic description of the species could be found in the literature, Marloth's manuscript name was finally validated in the *Journal of South African Botany* (Barker, 1978).
DISTINGUISHING CHARACTERS AND AFFINITIES. *L. peersii* is allied to *L. lactosa* (Duncan *et al.*, 2005a) with which it shares straight stamens and globose seeds that have inflated strophioles. *L. lactosa* differs in having much smaller, oblong-campanulate flowers with shorter, light blue perianth tubes (0.5–1.0 mm long), shorter outer (3–5 × 5 mm) and inner (5–7 × 3 mm) tepals, heavily blotched scapes, and pedicels that re-orientate the capsules into an erect or suberect position.

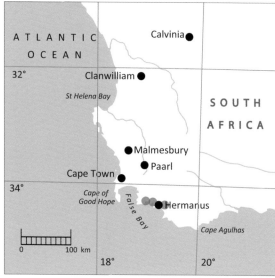

Map 112. Known distribution of *Lachenalia peersii*.

Figure 255. *Lachenalia peersii* on a fynbos slope at Hermanus. Image: Graham Duncan.

DISTRIBUTION, HABITAT AND CONSERVATION STATUS. *L. peersii* is restricted to a short stretch of southern Cape coastline extending from the mouth of the Palmiet River in the west, to Hermanus in the east (Map 112). It occurs in Kogelberg Sandstone Fynbos vegetation (Mucina & Rutherford, 2006) and is locally plentiful in stony, sandy soil in light shade of forest on lower south-facing mountain slopes, on coastal sandy plains, or in cracks of rocky sandstone outcrops close to the high water mark, in full sun. At Hermanus, it grows in association with the pyrophyte *L. montana*, flowering simultaneously in October and November (Duncan, 2003c). The species is threatened by coastal housing developments and has a conservation status of Vulnerable (Duncan & Raimondo, 2009d).

NOTES. The strongly carnation-scented blooms appear in profusion when there have been bush fires in the previous summer, but unlike *L. montana*, *L. peersii* does not depend on fire for flowering to occur. The pollinators of this species are unknown. The scapes remain attached for several weeks following capsule dehiscence and the seeds are dispersed by the shaking action of wind.

112. LACHENALIA CERNUA

Lachenalia cernua G. D. Duncan, *Bothalia* 36 (2): 150–152 (2006).
TYPE: South Africa, Western Cape, Palmiet Valley Farm, hillside behind homestead, in open aspects and semi-shade of sandstone boulders, *G. D. Duncan* 470 (NBG!, holo.; PRE!, iso.).

ETYMOLOGY. *cernua*: slightly drooping, with reference to the orientation of the flowers in the early to mid-flowering stage.

DESCRIPTION. *Geophyte*, 150–270 mm high. *Bulb* globose, 15–25 mm in diameter, offset- and bulblet-forming; tunic multilayered, outer layers spongy, dark brown; inner cataphyll translucent white, loosely surrounding leaf bases, apex obtuse. *Leaves* 1–2, lanceolate, 110–260 × 10–28 mm, spreading, slightly canaliculate, light to dark green or maroon, upper surface sporadically to heavily marked with dark maroon blotches; leaf bases clasping, 15–85 mm long, subterranean, white, sometimes forming marginal bulblets; primary seedling leaf flat, prostrate. *Inflorescence* racemose, few- to many-flowered, sterile apex short; scape erect to suberect, 60–130 mm long, light green, sporadically to heavily marked with maroon blotches; rachis light green, plain to heavily blotched with maroon; pedicels suberect, 2 mm long, white; lower bracts ovate, becoming lanceolate above, 1–5 × 1–7 mm, translucent white. *Perianth* zygomorphic, urceolate, suberect in bud, cernuous at early to mid-flowering, becoming spreading at late flowering and fruiting, creamy-white; tube cup-shaped, 3 mm long, upper part sometimes tinged with dull blue; outer tepals ovate, 6–8 × 5 mm, base sometimes tinged with dull blue, apical gibbosities yellowish-green; inner tepals obovate, 11–12 × 5 mm, apices slightly recurved, keels creamy light green. *Stamens* shortly exserted, more or less straight; filaments white, 11–13 mm long. *Ovary* ellipsoid, 4 × 3 mm, bright green; style included to shortly exserted, straight, 12 mm long, white. *Capsule* ellipsoid, 9–10 × 6 mm, olive green. *Seed* globose, 1.2 × 1.1 mm, glossy, black; strophiole inflated, 1 mm long, smooth. *Chromosome number* unknown. Figure 256.

FLOWERING PERIOD. September to October with a peak in late September.

HISTORY. The earliest known record of *L. cernua* is a sheet in the National Herbarium, Pretoria, collected by the amateur botanist Dr F. Z. van der Merwe in October 1937 (*van der Merwe s.n. sub.* PRE35699) on a hillside above the spa baths at Goudini in the Worcester Valley. Two further collections in NBG were made at this locality by J. W. Loubser in October 1971 and September 1972 (*Loubser* 2181, *Loubser s.n.*). In early October 2000, I collected flowering specimens on a hillside just south of Wolseley (*Duncan* 428 in NBG). In July 2001, leafing specimens of an unidentified *Lachenalia* were collected

112. LACHENALIA CERNUA

by Adam Harrower at the naval base at Klavervlei near Simonstown in the southern Cape Peninsula (*Harrower* 104 in NBG). When one of these flowered in the Kirstenbosch Nursery in October of the same year, the flowers matched those of the Goudini and Wolseley collections. The discovery of the Simonstown population is remarkable for a genus previously thought to have been extensively documented in the southern Cape Peninsula. Together with *L. patentissima*, *L. cernua* is one of two new *Lachenalia* species from the Cape Peninsula to be found since the publication of the endemic Peninsula species *L. capensis*, and *L. variegata*, which also grows on the Cape Peninsula (Barker, 1949). The restricted environment of the naval base no doubt accounts for *L. cernua* having remained undetected there for decades. No other populations are known to occur on the Peninsula. In late September 2002, the Wolseley locality was visited again, and the type collection was made (*Duncan* 470 in NBG).

DISTINGUISHING CHARACTERS AND AFFINITIES. *L. cernua* is sister to *L. variegata* (Duncan et al., 2005a), having a similar many-flowered raceme of spreading to cernuous flowers, included to shortly exserted stamens, a usually solitary lanceolate leaf, and a globose seed with an inflated strophiole. *L. variegata* differs in being a sturdier plant with oblong-campanulate, yellowish-green flowers, variegated outer tepals, declinate stamens and a leathery leaf with thickened margins.

DISTRIBUTION, HABITAT AND CONSERVATION STATUS. *L. cernua* has a disjunct distribution in the south-western Cape and is currently known from only three populations, one from the western end of the Worcester Valley at Goudini, another to the north-west of Goudini near Wolseley, and the third near Simonstown in the southern Cape Peninsula (Map 113). The population at Goudini is the closest spatially to the Simonstown population, a disjunction of more than 100 km. *L. cernua* occurs in Peninsula- and Hawequas Sandstone Fynbos vegetation (Mucina & Rutherford, 2006). The

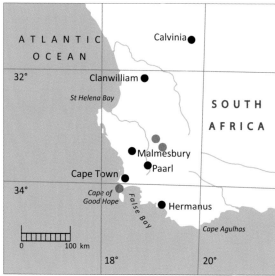

Map 113. Known distribution of *Lachenalia cernua*.

Figure 256. Cultivated specimen of *Lachenalia cernua* from sandstone slopes at Wolseley. Image: Graham Duncan.

Wolseley population occurs on an east-facing hill slope in semi-shade of large sandstone boulders, and at a slightly lower altitude in full sun. Plants growing in semi-shade tend to occur in small groups and flower erratically, whereas those in full sun usually occur singly and flower more reliably. At this locality, *L. cernua* grows in association with *L. lutea*. At Klavervlei near Simonstown, plants grow under similar conditions, mainly between large sandstone boulders on shaded, east- and south-east-facing ridges at an altitude of 363 m. The community includes *Moraea ramosissima* (Iridaceae), which flowers at the same time of year. Other notable companion species include *Protea cynaroides* (Proteaceae) and *Watsonia tabularis* (Iridaceae). A conservation status of Rare is recommended.

NOTES. The flowers are pollinated by honey bees (*Apis mellifera*) and although these plants flower prolifically following bush fires, they are not dependent on fire for flowering to occur. The scapes remain attached for several weeks following capsule dehiscence and the seeds are dispersed locally.

113. LACHENALIA ELEGANS

Lachenalia elegans W. F. Barker, *The Flowering Plants of South Africa* 13: t. 508 (1933).
TYPE: South Africa, Northern Cape, Nieuwoudtville, *H. Buhr s.n.* sub. BOL 19022 (BOL!, holo.).

ETYMOLOGY. *elegans*: elegant inflorescences.
DESCRIPTION. *Geophyte*, 150–250 mm high. *Bulb* globose, 15–20 mm in diameter, offset-forming; tunic multilayered, outer layers dark brown, spongy; inner cataphyll translucent white, loosely surrounding leaf bases, apex acute. *Leaves* 1–2, lanceolate, 60–115 × 9–25 mm, suberect, light green with depressed longitudinal grooves, canaliculate, upper surface plain; leaf margin thickened, purplish-maroon; leaf bases clasping, 5–35 mm long, white in lower portion, light to dark green or maroon above; primary seedling leaf flat, prostrate. *Inflorescence* spicate, few- to many-flowered, erect, slender, sterile apex long, pedicellate; scape erect, 60–110 mm long, maroon; rachis apex blue to brownish-mauve, 50–150 mm long, apical pedicels 1–2 mm long, light blue; lower bracts ovate, becoming lanceolate above, 0.5–1.0 × 1–3 mm, green to white. *Perianth* zygomorphic, urceolate, strongly suberect, heavily sweet-scented, ageing to dull red; tube cup-shaped, bright blue, 3 mm long; outer tepals ovate, 4–5 × 4–5 mm, bright blue, apical gibbosities purple or brown; inner tepals obovate, 6–7 × 4–5 mm, lower half bright blue, upper half translucent white, keels broad, deep magenta, apical margins recurved, undulate, white. *Stamens* included, straight; filaments white, 4–5 mm long. *Ovary* obovoid, 3–4 × 2 mm, olive green; style included, straight, 4–5 mm long, white. *Capsule* obovoid, 4–6 × 3–4 mm, yellowish-green. *Seed* globose, 1.3–1.4 × 1.4 mm, glossy, black; strophiole smooth, inflated, 0.8 mm long. *Chromosome number* unknown. Figure 257.
FLOWERING PERIOD. October to November, with a peak in mid-October.
HISTORY. *L. elegans* was described and illustrated by W. F. Barker in *The Flowering Plants of South Africa* from material collected by H. Buhr at Nieuwoudtville in October 1930 (Barker, 1933b). After numerous additional collections resembling *L. elegans* were made in inland parts of the north-western and south-western Cape in the ensuing 50 years, Barker described three additional taxa as varieties within *L. elegans*: var. *flava*, var. *membranacea* and var. *suaveolens* (Barker, 1989).
DISTINGUISHING CHARACTERS AND AFFINITIES. In a phylogenetic analysis of morphology (Duncan *et al.*, 2005a), var. *flava*, var. *membranacea* and var. *suaveolens* segregated from *L. elegans*, and the latter resolved as sister to *L. sessiliflora*. The three 'varieties' show a clear geographic pattern and are evolving in isolation, forming distinct evolutionary units, and are here upgraded to species level. *L. elegans* and *L. sessiliflora* share strongly suberect flowers with included stamens. *L. sessiliflora* differs in its

113. LACHENALIA ELEGANS

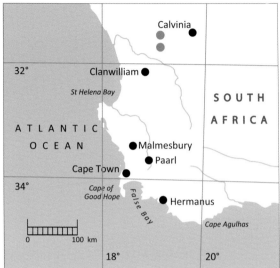

Map 114. Known distribution of *Lachenalia elegans*.

Figure 257. Cultivated specimens of *Lachenalia elegans* from Nieuwoudtville. Image: Graham Duncan.

much longer, narrower, linear-oblong inner tepals (13–15 × 2–3 mm) with flat margins, purplish- or pinkish-brown outer tepals, much longer, recurved filaments (8–10 mm), much smaller globose seed (0.4 × 0.4–0.5 mm), which has a shorter strophiole (0.5 mm long), and narrowly lanceolate leaves.

DISTRIBUTION, HABITAT AND CONSERVATION STATUS. *L. elegans* is known from only two locations in the Nieuwoudtville district (Map 114). It occurs in Bokkeveld Sandstone Fynbos vegetation (Mucina & Rutherford, 2006), growing in colonies in sandy, seasonally marshy ground and at the edges of pools, amongst low bushes, in full sun. Its habitat is under threat from agricultural activity and a conservation status of Endangered is recommended.

NOTES. *L. elegans* has strongly sweet-scented flowers and is pollinated by honey bees (*Apis mellifera*). The scapes remain attached for several weeks following capsule dehiscence and the seeds are dispersed locally by the shaking action of wind. The flowers are at least partially self-fertile, regularly producing seeds under enclosed, insect-proof conditions in cultivation.

114. LACHENALIA KAROOPOORTENSIS

Lachenalia karoopoortensis G. D. Duncan, nom. nov. *Lachenalia elegans* W. F. Barker var. *flava* W. F. Barker, *South African Journal of Botany* 55 (6): 634–635 (1989).
TYPE: South Africa, Western Cape, Karoo Poort, north of Ceres, *R. H. Compton s.n.*, Hort. NBG 333/41 (NBG!, holo.).

ETYMOLOGY. *karoopoortensis*: after Karoo Poort at the entrance to the Tanqua Karoo, where the type collection was made.

THE GENUS LACHENALIA 395
114. LACHENALIA KAROOPOORTENSIS

PLATE 35. Watercolour painting of *Lachenalia karoopoortensis* from near Ceres (*Dymond s.n.*, in BOL) courtesy of the Compton Herbarium, Kirstenbosch, South African National Biodiversity Institute. Artist: Winsome Barker.

114. LACHENALIA KAROOPOORTENSIS

DESCRIPTION. *Geophyte*, 130–250 mm high. *Bulb* globose, 15–20 mm in diameter, solitary, rarely offset-forming; tunic multilayered, outer layers dark brown, spongy; inner cataphyll translucent white, loosely surrounding leaf bases, apex acute. *Leaf* solitary, rarely 2, broadly lanceolate, 60–100 × 20–30 mm, glaucous, spreading to suberect, upper and lower surface with dark green blotches; margins cartilaginous, crisped, maroon; primary seedling leaf flat, prostrate. *Inflorescence* spicate, few- to many-flowered; scape erect, 60–100 mm long, light green or maroon, with light to dark maroon spots or blotches; rachis lower half green or maroon, upper half greenish-yellow, plain or maroon-blotched or speckled; lower bracts ovate, becoming lanceolate above, 0.5–3.0 × 1.0 mm, green to white. *Perianth* urceolate, spreading, sweet-scented; tube cup-shaped, 2–3 mm long, bright yellow; outer tepals ovate, 5–7 × 4–5 mm, bright yellow; apical gibbosities green or deep maroon; inner tepals obovate, 5–7 × 3–4 mm, protruding up to 1 mm, bright yellow, subapical markings broad, crescent-shaped, deep maroon, margins bright white, narrow. *Stamens* included, straight; filaments 4 mm long, white. *Ovary* ellipsoid, 3–4 × 2–3 mm, yellowish-green; style included, straight, 2 mm long, white. *Capsule* ellipsoid, 6–7 × 4–5 mm. *Seed* globose, 1.3–1.4 × 1.3–1.4 mm, glossy, black; strophiole smooth, inflated, 0.7–0.8 mm long. *Chromosome number* 2n = 42 (Johnson & Brandham, 1997). Plate 35.

FLOWERING PERIOD. Early July to mid-August.

HISTORY. *L. karoopoortensis* was first collected by Prof. R. H. Compton at Karoo Poort in 1941 and published as *L. elegans* W. F. Barker var. *flava* W. F. Barker in the *South African Journal of Botany*, accompanied by a painting by W. F. Barker of a wild specimen, which was reproduced in monochrome (Barker, 1989). In 2001, a painting by Joanna Langhorne of cultivated specimens was published in *Curtis's Botanical Magazine* (Duncan, 2001). Following a phylogenetic analysis (Duncan et al., 2005a), it became clear that the taxon merited upgrading to specific rank. The specific epithet *flava* had already been published for another species (*L. flava* Andrews) and a new name was therefore required.

DISTINGUISHING CHARACTERS AND AFFINITIES. *L. karoopoortensis* is allied to *L. membranacea* (Duncan et al., 2005a). *L. membranacea* differs in its light yellow to greenish-yellow or rarely light blue outer tepals, and longer inner tepals (protruding 2 mm or more), which lack maroon subapical markings and have broad, membranous, translucent white margins. *L. membranacea* has a much wider distribution than *L. karoopoortensis*, occurring in discrete populations on rocky mountain slopes, ridges and on rock sheets from the Bokkeveld Plateau to Piketberg, with isolated records in central Namaqualand. It usually flowers later (late July to late September), with a peak in late August.

DISTRIBUTION, HABITAT AND CONSERVATION STATUS. *L. karoopoortensis* is restricted to the south-western Cape, occurring from the Elandskloof Mountains west of Wolseley to Karoo Poort in the southern Tanqua Karoo (Map 115). It grows in Tanqua Karoo and Hawequas Sandstone Fynbos vegetation (Mucina & Rutherford, 2006), on sandy-loamy, stony flats and hillsides, in small colonies amongst low scrub. *L. karoopoortensis* is not threatened.

NOTES. *L. karoopoortensis* is pollinated by honey bees (*Apis mellifera*). Following capsule dehiscence the scapes remain attached for several weeks and the seeds are dispersed locally.

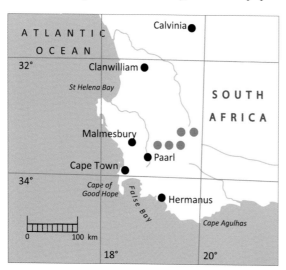

Map 115. Known distribution of *Lachenalia karoopoortensis*.

115. LACHENALIA MEMBRANACEA

Lachenalia membranacea (W. F. Barker) G. D. Duncan, stat. nov. *Lachenalia elegans* W. F. Barker var. *membranacea* W. F. Barker, *South African Journal of Botany* 55 (6): 636–637 (1989).
TYPE: South Africa, Northern Cape, Nieuwoudtville, *H. Buhr s.n.* sub. hort. NBG 646/29 (BOL!, holo.).

ETYMOLOGY. *membranacea*: descriptive of the broad membranous, translucent white margins of the inner tepals.

DESCRIPTION. *Geophyte*, 110–250 mm high. *Bulb* globose, 15–20 mm in diameter, offset-forming; tunic multilayered, outer layers dark brown, spongy; inner cataphyll translucent white, loosely surrounding leaf bases, apex acute. *Leaves* 1–2, lanceolate or ovate, 80–120 × 15–45 mm, spreading or rarely suberect, flat to slightly canaliculate, bright green to glaucous, upper surface heavily marked with large, darker green or purplish-maroon spots, rarely plain; leaf margins thickened, purplish-maroon; leaf bases clasping, 5–35 mm long, lower portion white, light to dark green or maroon above; primary seedling leaf flat, prostrate. *Inflorescence* spicate, many-flowered, dense, sterile apex 15–50 mm long, vestigial flowers borne on white, light to deep pink, blue or mauve pedicels 1–15 mm long; scape erect to suberect, 80–110 mm long, light green to pinkish-maroon, plain or with purplish-brown blotches; rachis light green to brown below, shading to white, light to deep pink, blue or mauve

Figure 258 (left). *Lachenalia membranacea* on a sandy slope north-west of Piketberg. Image: Graham Duncan.
Figure 259 (right). Cultivated specimens of *Lachenalia membranacea* from sandstone slopes of the Gifberg. Image: Graham Duncan.

above; lower bracts ovate, becoming lanceolate above, 1–6 × 1–4 mm, green to white. *Perianth* broadly urceolate, spreading, sweet-scented; buds light to bright pink or blue, tepals ageing to dull red or pink; tube cup-shaped, 2–3 mm long, white, light yellow to greenish-yellow or rarely light blue; outer tepals ovate, 5–6 × 4–5 mm, light yellow to greenish-yellow or light blue; apical gibbosities light to bright green, brownish-green or dark brown; inner tepals obovate, 6–7 × 4 mm, light greenish-yellow or translucent white; keels light green or brownish-purple, apices recurved, undulate, with broad, membranous, translucent margins. *Stamens* included, straight; filaments 5–6 mm long, white; anthers maroon prior to anthesis. *Ovary* ellipsoid, 2–3 × 2 mm, light green; style included, straight, 3–4 mm long, white. *Capsule* ellipsoid, 4–6 × 3–4 mm. *Seed* globose, 1.3–1.4 × 1.3 mm, glossy, black; strophiole smooth, inflated, 0.8 mm long. *Chromosome number* 2n = 14, 28 (Ornduff & Watters, 1978; Johnson & Brandham, 1997). Figures 258, 259.

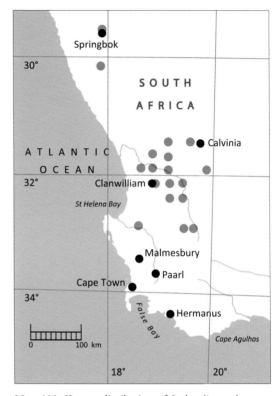

Map 116. Known distribution of *Lachenalia membranacea*.

FLOWERING PERIOD. Late July to September, with a peak in late August.

HISTORY. *L. membranacea* was first collected by the German traveller Rudolf Schlechter on 12th August 1897 on the Pakhuis Mountains near Clanwilliam. Schlechter distributed this material (*Schlechter* 10816) to ten local and foreign herbaria, but almost a century passed before the taxon was finally published, as *L. elegans* var. *membranacea* (Barker, 1989). The taxon is here upgraded to species level.

DISTINGUISHING CHARACTERS AND AFFINITIES. *L. membranacea* is allied to *L. karoopoortensis* and *L. suaveolens* (Duncan et al., 2005a). *L. karoopoortensis* differs in having a bright yellow perianth with shorter inner tepals (protruding up to 1 mm) that have crescent-shaped, deep maroon subapical markings and narrow, bright white margins, and in usually having solitary glaucous leaves with crisped lower margins. *L. karoopoortensis* commences flowering earlier than *L. membranacea*, in early July. *L. suaveolens* differs from *L. membranacea* in its light to deep pink, dull maroon or deep purple outer tepals, which have deep purple or maroon apical gibbosities, and in its pink, maroon or deep purple inner tepals, which have deep purple or maroon keels and conspicuous narrow, bright, solid white margins. It has a slightly shorter style (2 mm long), the flowers are heavily carnation-scented and its leaves are usually suberect. *L. suaveolens* occurs in clay or sandy soils and commences flowering later than *L. membranacea*, in early September.

DISTRIBUTION, HABITAT AND CONSERVATION STATUS. *L. membranacea* extends from the Bokkeveld Plateau at Nieuwoudtville to Van Rhyn's Pass and along the western mountain ranges to the eastern Cederberg and Wupperthal, south-west to the northern Piketberg, and to the east slopes of the Swartruggens Mountains. There are two disjunct collections from Namaqualand, one from Springbok (*Herre s.n.* in NBG) and another from Skilpad in the Namaqua National Park (specimen in BLFU) (Map 116).

The plants traverse five vegetation types: Bokkeveld- and Cederberg Sandstone Fynbos, Leipoldtville Sand Fynbos, Swartruggens Quartzite Karoo and Namaqualand Blomveld (Mucina & Rutherford, 2006). They occur in sandy soils on rocky sandstone mountain slopes, along ridges, in shallow soil on rock sheets, and on flats. They are encountered in small to large colonies, rarely as solitary individuals. *L. membranacea* is sympatric with *L. elegans* and *L. suaveolens* on the Bokkeveld Plateau, and with *L. suaveolens* on the Cederberg, but always occurs in discrete populations. At Wilgerbosdrift Farm north-west of Piketberg, it occurs in mixed populations with *L. hirta* and a deep mauve form of *L. pallida*. *L. membranacea* is not threatened.

NOTES. *L. membranacea* is pollinated by honey bees (*Apis mellifera*) and by the solitary bee *Anthophora diversipes*. Following capsule dehiscence, the scapes remain attached to the bulbs for several weeks and the seeds are dispersed locally.

116. LACHENALIA SUAVEOLENS

Lachenalia suaveolens (W. F. Barker) G. D. Duncan, stat. nov. *Lachenalia elegans* W. F. Barker var. *suaveolens* W. F. Barker, *South African Journal of Botany* 55 (6): 635–636 (1989).
TYPE: South Africa, Northern Cape, Nieuwoudtville, *H. Buhr s.n.* sub. NBG 1584/30 (BOL!, holo.).

ETYMOLOGY. *suaveolens*: sweet-smelling flowers.
DESCRIPTION. *Geophyte*, 160–400 mm high. *Bulb* globose, 15–20 mm in diameter, offset-forming; tunic multilayered, outer layers dark brown, spongy; inner cataphyll translucent white, loosely surrounding leaf bases, apex acute. *Leaves* 1–2, lanceolate or ovate, 80–150 × 15–50 mm, suberect to erect, rarely spreading, upper surface plain or densely marked with purplish-brown spots; leaf bases clasping, 12–35 mm long, white in lower portion, light to dark green or maroon above; primary seedling leaf flat, prostrate. *Inflorescence* spicate, many-flowered, dense, sterile apex 30–45 mm long, rachis light to bright pink, blue or brownish-mauve; scape erect, 80–200 mm long, rarely inflated above, dull pinkish-maroon, or light green with purplish-blotches; lower bracts ovate, becoming lanceolate above, 1–3 × 1–4 mm, green to white. *Perianth* urceolate, spreading, heavily carnation-scented; tube cup-shaped, 2–3 mm long, light to deep pink or deep purple; outer tepals ovate, 4–5 × 4–5 mm, light to deep pink, dull maroon or deep purple, apical gibbosities deep maroon or purple; inner tepals obovate, 6–7 × 2–3 mm, dull pink, deep maroon or deep purple, keels deep maroon or purple, lower inner tepal often strongly canaliculate, apices recurved,

Figure 260. Deep purple form of *Lachenalia suaveolens* in renosterveld south of Nieuwoudtville.
Image: Graham Duncan.

116. LACHENALIA SUAVEOLENS

with narrow, bright white margins. *Stamens* included, straight; filaments 4–5 mm long, white. *Ovary* ellipsoid, 2–3 × 2–3 mm, yellowish-green; style included, straight, 2 mm long, white. *Capsule* ellipsoid, 4–6 × 3–4 mm. *Seed* globose, 1.3–1.4 × 1.3 mm, glossy, black; strophiole smooth, inflated, 0.8 mm long. *Chromosome number* 2n = 14 (Johnson & Brandham, 1997). Figures 261, 262.

FLOWERING PERIOD. Early September to October, with a peak in mid September.

HISTORY. *L. suaveolens* was first recorded by H. Buhr who collected it on 17th September 1929 at Nieuwoudtville. The plant was published 60 years later as *L. elegans* var. *suaveolens*, in the *South African Journal of Botany*, the description being accompanied by a watercolour painting by W. F. Barker that was reproduced in monochrome (Barker, 1989). The results of a morphological phylogenetic analysis (Duncan *et al.*, 2005a), indicated that this taxon is deserving of specific rank.

DISTINGUISHING CHARACTERS AND AFFINITIES. *L. suaveolens* is allied to *L. membranacea* (Duncan *et al.*, 2005a). *L. membranacea* differs in its light yellow to greenish-yellow or rarely light blue outer tepals and light greenish-yellow or white inner tepals with broad membranous, translucent margins, and in its longer style (3–4 mm long) and spreading leaves. *L. membranacea* has a sweet scent that differs from that of *L. suaveolens*, and *L. membranacea* commences flowering earlier than *L. suaveolens*, in late July.

DISTRIBUTION, HABITAT AND CONSERVATION STATUS. *L. suaveolens* is distributed from the Bokkeveld Plateau to the southern Tanqua Karoo. It is common within the town of Nieuwoudtville and in surrounding areas, at Lokenburg to the south and in the Cederberg, and is occasional on flats of the southern Tanqua Karoo

Figure 261 (left). Robust pink form of *Lachenalia suaveolens* in the Nieuwoudtville Wild Flower Reserve. Image: Graham Duncan.

Figure 262 (right). Dwarf form of *Lachenalia suaveolens* on sandy-loamy flats of the southern Tanqua Karoo. Image: Graham Duncan.

(Map 117). *L. suaveolens* occurs in Nieuwoudtville Shale Renosterveld, Cederberg Sandstone Fynbos and Tanqua Karoo vegetation (Mucina & Rutherford, 2006). The plants usually grow in stony clay on the Bokkeveld Plateau, on stony sandstone slopes and plateaus of the Cederberg, and in sandy-loamy soils of the southern Tanqua Basin. The species' distribution overlaps with those of *L. elegans* and *L. membranacea* on the Bokkeveld Plateau, with that of *L. membranacea* on the Cederberg, and with that of *L. karoopoortensis* in the southern Tanqua Karoo, but *L. suaveolens* always occurs in discrete populations. Its numbers have been much reduced by agricultural extension around Nieuwoudtville, but *L. suaveolens* is not generally under threat.

NOTES. Honey bees (*Apis mellifera*) pollinate the strongly carnation-scented flowers of *L. suaveolens*. Following capsule dehiscence, the scapes remain attached for several weeks and the seeds are dispersed locally.

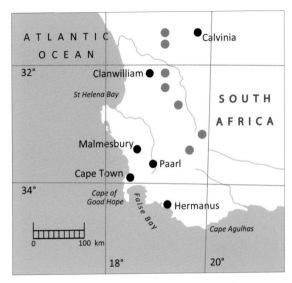

Map 117. Known distribution of *Lachenalia suaveolens*.

117. LACHENALIA FISTULOSA

Lachenalia fistulosa Baker, *The Gardener's Chronicle* (series 2) 21: 668 (1884).

TYPE: South Africa, Cape, precise locality unknown, *ex hort T. S. Ware*, April 1884 (K!, holo.).

SYNONYMY. *Lachenalia convallariodora* Stapf, *Curtis's Botanical Magazine* 168, t. 8955 (1923). Type: South Africa, Cape, collector and precise locality unknown, *ex hort* Kew Gardens, 12th May 1922 (K!, holo.).

ETYMOLOGY. *fistulosa*: hollow, formed by the outer and inner tepals.

DESCRIPTION. *Geophyte*, 75–300 mm high. *Bulb* subglobose, 20–25 mm in diameter, offset-forming; tunic multilayered, outer layers dark brown, spongy; inner cataphyll translucent white, loosely surrounding leaf bases, apex obtuse. *Leaves* 1–2, lanceolate, 45–120 × 10–35 mm, spreading or suberect, bright green, upper surface plain or marked with green, dark brown, purplish-brown or magenta spots and depressed longitudinal grooves, leaf margin entire to slightly undulate; leaf bases clasping, 10–30 mm long, lower half white, shading to light green above; primary seedling leaf flat, prostrate. *Inflorescence* spicate, many-flowered, dense; sterile apex short; scape erect to suberect, 40–120 mm long, slender to sturdy, light green, sometimes marked with few to many brownish-magenta spots or blotches; rachis light green or maroon, plain or heavily marked with maroon blotches; lower bracts ovate, becoming lanceolate above, 1–4 × 1–3 mm, green to translucent white. *Perianth* zygomorphic, urceolate, suberect, white, cream, yellow, light blue, lilac or violet, ageing to dull red, heavily sweet-scented; tube cup-shaped, 2–3 mm long; outer tepals ovate, 7–8 × 3–5 mm, plain or with minute maroon or blue speckles, apical gibbosities brown or green; inner tepals obovate, 8–9 × 3–4 mm, apices recurved, upper two tepals overlapping. *Stamens* included, straight; filaments white, 5–7 mm long. *Ovary* ellipsoid, 2–3 × 2–3 mm; yellowish-green to bright green; style included,

117. LACHENALIA FISTULOSA

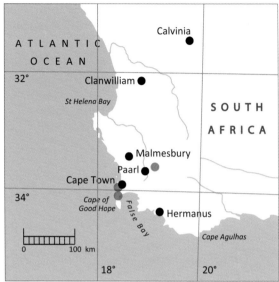

Map 118. Known distribution of *Lachenalia fistulosa*.

Figure 263. *Lachenalia fistulosa* in renosterveld after a fire on Lion's Head, northern Cape Peninsula. Image: Graham Duncan.

straight, 2–4 mm long, white. *Capsule* ovoid, 5–6 × 3–4 mm. *Seed* globose, 0.9–1.0 × 0.8–0.9 mm, glossy, black; strophiole inflated, 0.6–0.7 mm long, smooth. *Chromosome number* $2n = 14$ (Johnson & Brandham, 1997; Hamatani *et al.*, 2004); $2n = 28$ (Spies *et al.*, 2008). Figure 263.

FLOWERING PERIOD. September to November, with a peak in late September.

HISTORY. Plants of *L. fistulosa* were probably first collected and imported into England in the early 1880s. The species was described by J. G. Baker in May 1884 in the second series of *The Gardener's Chronicle*, from a single living plant in the collection of Mr T. S. Ware of Tottenham, England; the holotype is housed in the Kew Herbarium. The species was illustrated many years later under the name *L. convallariodora*, on plate 8955 of *Curtis's Botanical Magazine* (Stapf, 1923); this taxon is conspecific with Baker's *L. fistulosa* and the name *L. convallariodora* was reduced to synonymy by W. F. Barker (1989).

DISTINGUISHING CHARACTERS AND AFFINITIES. *L. fistulosa* resolved as sister to the Cape Peninsula endemic, *L. capensis* (Duncan *et al.*, 2005a). These two species share similar lanceolate leaves, heavily sweet-scented flowers, straight stamens and globose seeds with inflated strophioles. *L. capensis* differs mainly in its tubular flowers with straight inner tepals, in having the lower inner tepal shorter than the upper inner tepals, and in its slightly longer strophiole (0.7–0.8 mm).

DISTRIBUTION, HABITAT AND CONSERVATION STATUS. *L. fistulosa* is restricted to the northern and southern Cape Peninsula, with a disjunct, late-flowering population from Miaspoort in the Du Toitskloof Mountains north-east of Cape Town (Map 118). It occurs in Peninsula Shale Renosterveld, Peninsula Granite Fynbos and Hawequas Sandstone Fynbos vegetation (Mucina & Rutherford,

2006) and is locally common. It grows in a variety of aspects, mainly on south- and north-facing rocky slopes in heavy, stony clay or granitic soils, rarely on sandstone. The plants occur singly or in large colonies numbering hundreds or even thousands of individuals. Wide variation in tepal colouration is common within populations. Although sometimes seen in full sun, *L. fistulosa* is often heavily shaded under exotic *Pinus pinea* trees. On the southern slopes of Lion's Head in the northern Cape Peninsula, it grows in association with *L. orchioides* subsp. *orchioides*, which flowers earlier, and alongside the white-flowered *Aristea spiralis* and cream-coloured *Moraea tricuspidata* (Iridaceae). *L. fistulosa* is threatened by housing development on the Cape Peninsula and a conservation status of Vulnerable is recommended.

NOTES. *L. fistulosa* flowers profusely in the first spring season following summer fires, but flowering decreases rapidly as the surrounding bush grows up. In ensuing winter seasons, some bulbs produce leaves but eventually most individuals remain dormant until the next fire occurs. Under cultivation, the plants grow and flower annually. The flowers are pollinated by honey bees (*Apis mellifera*). After fires, the large black, hairy monkey beetle (*Peritrichia cinerea*) is sometimes seen clambering over the flowers near Lion's Head, but this species plays no role in pollination. The scape remains attached for a number of weeks following capsule dehiscence and the seeds are dispersed locally by the shaking action of wind.

118. LACHENALIA SESSILIFLORA

Lachenalia sessiliflora Andrews, *The Botanist's Repository* 7: t. 460 (1807).
TYPE: South Africa, Cape, collector and precise locality unknown, figure in *The Botanist's Repository* 7: t. 460 (1807) (lectotype, designated here).
SYNONYMY. *Lachenalia muirii* W. F. Barker, *South African Gardening and Country Life* 20: 14 (1930). Type: South Africa, Western Cape, Government Reserve opposite Stilbaai, *J. Muir* 1875 (BOL!, holo.).

ETYMOLOGY. *sessiliflora*: stalkless flowers.
DESCRIPTION. *Geophyte*, 85–400 mm high. *Bulb* globose, 10–15 mm in diameter, yellowish-white, offset-forming; tunic multilayered, outer layers dark brown, spongy; inner cataphyll translucent white, loosely surrounding leaf bases, apex obtuse. *Leaves* 2, narrowly lanceolate, 100–170 × 2–10 mm, erect to suberect, yellowish-green to olive green, slightly canaliculate, coriaceous, upper surface plain or marked with light to dark green blotches, lower surface plain or flushed with light to dark maroon in lower third; margins coriaceous; leaf bases loosely surrounding scape, 20–40 mm long, white, subterranean; primary seedling leaf flat, prostrate. *Inflorescence* spicate, few- to many-flowered, fairly dense, lax to sturdy, sterile apex short; scape erect to suberect, 60–300 mm long, dull purplish- to maroonish-brown, often with sporadic purplish blotches, slender to sturdy; rachis dull purplish-brown to maroonish-brown in lower third, shading to light purplish-grey or brown in upper two thirds; bracts ovate, 1–3 × 2–3 mm, light green in lower half shading to translucent white in upper half. *Perianth* zygomorphic, urceolate, suberect; tube cup-shaped, 2–3 mm long, light blue; outer tepals ovate, 7–8 × 3–4 mm, light purplish- to pinkish-brown, apical gibbosities deep purplish-brown, large; inner tepals linear-oblong, 13–15 × 2–3 mm, apices obtuse, recurved, lower tepal 1–2 mm longer than upper tepals, translucent white, keels deep maroon to magenta, conspicuous. *Stamens* included, recurved; filaments white, 8–10 mm long. *Ovary* ellipsoid, 2.5 × 1.5 mm, dark olive-green; style included, straight, 2–4 mm long, white. *Capsule* ellipsoid, 4–7 × 3–5 mm. *Seed*

Plate 36. Watercolour painting of *Lachenalia sessiliflora* from Bredasdorp (*Hilton-Taylor s.n.*, in NBG), courtesy of Fay Anderson. Artist: Fay Anderson.

globose, 0.4 × 0.4–0.5 mm, glossy, black; strophiole inflated, 0.5 mm long, smooth. *Chromosome number* $2n = 14$ (Johnson & Brandham, 1997, Hamatani *et al.*, 2007). Plate 36, Figure 264.

FLOWERING PERIOD. October to early January, with a peak from mid- to late November.

HISTORY. *L. sessiliflora* was described and illustrated in volume 7 of Andrews's *The Botanist's Repository* (Andrews, 1807b). The depicted specimen was painted in 1803 from a cultivated plant from the collection of G. Hibbert, but no information is known of its wild provenance. The tepal colour in the plate is dull reddish-pink but the accompanying description states it to be 'of purple colour', having 'wedge-shaped' outer tepals 'with the inner ones twice their length, narrow and appearing cut-off at the end'. Baker (1871) erroneously placed this species under *L. glaucina* Jacq. and inadvertently misspelt it *L. sessilifolia* Andrews, a mistake rectified in his later monograph (Baker 1897a). Dr J. Muir collected plants at Melkhoutfontein near Stilbaai south of Riversdale; his specimens form the holotype of the species later named *L. muirii* in the periodical *South African Gardening and Country Life* (Barker, 1930a). Andrews's (1807) description, drawing attention to the two distinctly different filament series lengths (upper filaments 2 mm shorter than the lower filaments) and the accompanying plate showing the slender, erect scape, with the flowers crowded together at the top of the rachis, clearly match the later name *L. muirii* W. F. Barker, and the latter is placed in synonymy.

Figure 264. *Lachenalia sessiliflora* on sandy flats in the De Hoop Nature Reserve. Image: Graham Duncan.

DISTINGUISHING CHARACTERS AND AFFINITIES. *L. sessiliflora* is sister to *L. elegans* (Duncan *et al.*, 2005a), the two species sharing sessile, suberect perianths with leathery leaves. *L. elegans* differs in its obovate, much shorter and broader inner tepals (6–7 × 4–5 mm) that have undulate margins, its bright blue and rose-pink outer tepals, its much shorter, straight stamens (4–5 mm), and its much larger globose seeds (1.3–1.4 × 1.4 mm), which have a longer inflated strophiole (0.8 mm long).

DISTRIBUTION, HABITAT AND CONSERVATION STATUS. *L. sessiliflora* is limited to a narrow coastal stretch of the southern Cape, extending from west of Bredasdorp to east of Stilbaai (Map 119). It occurs in De Hoop- and Canca Limestone Fynbos vegetation (Mucina & Rutherford, 2006), growing on top of low limestone ridges and hills, and in stony sand on limestone flats. The plants grow singly or in small groups as part of larger colonies, in full sun or within the light shade of restios and low bushy cover, often in small pockets or cracks within limestone outcrops. The species is locally common. On flats in the De Hoop Nature Reserve east of Bredasdorp, it grows in association with *L. contaminata* and *L. juncifolia*, both of which flower much earlier in the season. *L. sessiliflora* is threatened by coastal housing development, invasive alien plants and crop cultivation, and has a conservation status of Vulnerable (Raimondo, 2009).

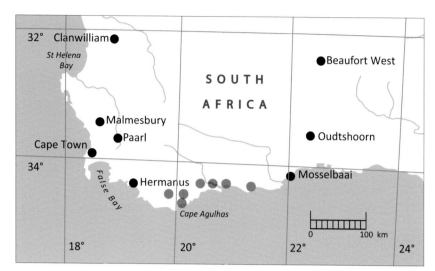

Map 119. Known distribution of *Lachenalia sessiliflora*.

NOTES. *L. sessiliflora* is stimulated to flower following bush fires in the previous summer, but is not dependent on fire for flowering to occur. Pollination agents have not yet been observed but are thought to be solitary bees with long proboscides that could brush against the very short style. *L. sessiliflora* has a long flowering period, which sometimes extends into early January. Even within populations, the flowering period is protracted, with some individuals flowering several weeks earlier than others. The leaves of *L. sessiliflora* are often proteranthous in the wild. The scape remains attached for several weeks following capsule dehiscence and the seeds are dispersed locally.

119. LACHENALIA CARNOSA

Lachenalia carnosa Baker, *Journal of the Linnean Society (Botany)* 11: 407 (1871).
TYPE: South Africa, Northern Cape, Little Namaqualand between Uitkomst and Geelbekskraal, *J. F. Drège* 2689a (K!, holo.; S!, iso.).
SYNONYMY: *Lachenalia ovatifolia* L. Guthrie, *Annals of the Bolus Herbarium* 4: 119 (1927). Type: South Africa, Northern Cape, Little Namaqualand, *M. H. Giffen s.n.* sub. BOL 140448 (BOL!, holo.).

ETYMOLOGY. *carnosa*: fleshy, alluding to the leaves.
DESCRIPTION. *Geophyte*, 80–250 mm high. *Bulb* globose, 10–20 mm in diameter, solitary; tunic multilayered, outer layers dark brown, spongy; inner cataphyll translucent white, adhering to leaf bases, apex obtuse. *Leaves* 2, very rarely 1, broadly ovate, 25–90 × 25–140 mm, spreading to suberect, upper surface light to bright green or glaucous, plain or with small to large green or brown pustules and/or spots, lower leaf surface light green, flushed with light to dark maroon, depressed longitudinal grooves prominent; leaf bases clasping, 10–20 mm long, light green flushed with maroon, subterranean; margins coriaceous, dark maroon; primary seedling leaf flat, prostrate. *Inflorescence* spicate, many-flowered, fairly dense, sturdy; sterile apex short; scape erect to suberect, 30–90 mm long, light green, plain or with maroon speckles; rachis light green, plain or mottled with maroon, apex light electric blue; lower bracts broadly ovate, becoming lanceolate above, 1–6 × 2–4 mm, magenta. *Perianth* zygomorphic, urceolate, spreading to suberect; tube cup-shaped, 3 mm long,

light blue to almost white; outer tepals ovate, 8–9 × 4–5 mm, light blue at base, often fading to white above, apical gibbosities magenta, brownish-mauve or greenish-brown; inner tepals obovate, 9–10 × 5–6 mm, translucent white, apices mauve to magenta, recurved, upper two tepals overlapping, lower tepal longer and narrower, caniculate. *Stamens* included, declinate; filaments white, 8–9 mm long. *Ovary* ellipsoid, 3–4 × 2 mm, light yellowish-green; style included, declinate, 7–8 mm long, white. *Capsule* ellipsoid, 10 × 6 mm. *Seed* ovoid, 0.9–1.0 × 0.8–0.9 mm, glossy, black; strophiole rudimentary, 0.1–0.2 mm long, ridged. *Chromosome number* 2n = 16 (Johnson & Brandham, 1997; Hamatani *et al.*, 1998). Figures 265, 266.

FLOWERING PERIOD. August to September, with a peak in late August.

HISTORY. The German horticulturist and botanical collector J. F. Drège (1794–1881) found *L. carnosa* in 'Little Namaqualand', between Uitkomst and Geelbekskraal on the Kamiesberg, probably in 1837. The species was described in the *Journal of the Linnean Society* (Botany) (Baker, 1871). *L. ovatifolia* was described by the Cape Town botanist Louise Guthrie in 1927, from material collected on 23rd August 1926 by M. H. Giffen at an undisclosed locality in Namaqualand, and illustrated in watercolour by W. F. Barker in *The Flowering Plants of South Africa* (Barker, 1931b). The holotype in the Bolus Herbarium at the University of Cape Town matches Drège's specimens and hence *L. ovatifolia* was placed in synonymy with *L. carnosa* (Barker, 1989).

Figure 265 (left). *Lachenalia carnosa* with suberect leaves on a granite outcrop near Springbok. Image: Graham Duncan.

Figure 266 (right). *Lachenalia carnosa* with spreading leaves on granitic gravel flats near Kamieskroon. Image: Ivan van Niekerk.

DISTINGUISHING CHARACTERS AND AFFINITIES. *L. carnosa* is allied to *L. congesta* and *L. kliprandensis* (Duncan *et al.*, 2005a). *L. congesta* differs in its smaller urceolate, yellowish-cream perianth, in its inner tepals, which have undulate margins, and in its much shorter, straight filaments (4–5 mm long) and style (2–3 mm long). It also has much larger, globose seeds (1.8–1.9 × 1.8–1.9 mm) and prostrate leaves. *L. kliprandensis* superficially resembles *L. carnosa*, but the former differs in having much larger, narrowly campanulate perianths with longer inner tepals (12–13 mm long), longer, shortly exserted filaments (10–13 mm long), a longer style (10–11 mm long) and prostrate leaves. *L. kliprandensis* is restricted to western Bushmanland.

DISTRIBUTION, HABITAT AND CONSERVATION STATUS. *L. carnosa* is endemic to the north-western Northern and Western Cape, extending from Kosies in northern Namaqualand to Klawer in the southern Knersvlakte (Map 120). It is common in central and western Namaqualand, and the Kamiesberg, and is mostly associated with granite hill and mountain slopes that have gravelly-clay or sandy soil, being found less frequently on dunes and quartz flats. The plants grow in large colonies among succulents and low shrubs, in full sun or light shade, in rock crevices and shallow pockets of granite boulders, or in seepage areas with a south-westerly or south-easterly aspect. *L. carnosa* favours three vegetation types: Namaqualand Blomveld, Kamiesberg Granite Fynbos and Knersvlakte Quartz Vygieveld (Mucina & Rutherford, 2006). The species is not threatened.

NOTES. The leaves of *L. carnosa* vary markedly in orientation from spreading to suberect, even within the same population. The species is pollinated by honey bees (*Apis mellifera*). Following capsule dehiscence, the scape detaches and the seeds are dispersed as the infructescence is blown about.

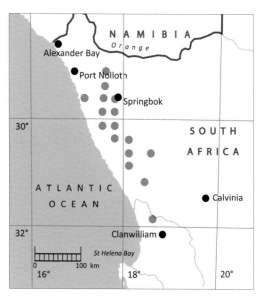

Map 120. Known distribution of *Lachenalia carnosa*.

120. LACHENALIA CONGESTA

Lachenalia congesta W. F. Barker, *Journal of South African Botany* 44 (4): 399–402 (1978).
TYPE: South Africa, Northern Cape, Sutherland, *F. J. Stayner s.n.* sub. NBG 93575 (NBG!, holo.).

ETYMOLOGY. *congesta*: congested inflorescences.

DESCRIPTION. *Geophyte*, 50–140 mm high. *Bulb* solitary, globose, 15–25 mm in diameter, neck distinct, 20–40 mm long; tunic multilayered, outer layers light brown, spongy; inner cataphyll translucent greenish-white, adhering to leaf bases, apex obtuse. *Leaves* 2, ovate, 35–80 × 15–55 mm, prostrate, leathery, apices sometimes notched, upper surface olive-green, lower surface flushed with maroon; margins coriaceous, maroon; leaf bases clasping, 20–40 mm long, subterranean, white; primary seedling leaf flat, prostrate. *Inflorescence* spicate, many-flowered, dense, sterile apex short; scape erect, subterranean, rarely slightly aerial, white; rachis light green; lower bracts ovate, becoming lanceolate above, 3–4 × 2–5 mm, translucent white to light green. *Perianth* zygomorphic, urceolate, suberect,

120. LACHENALIA CONGESTA

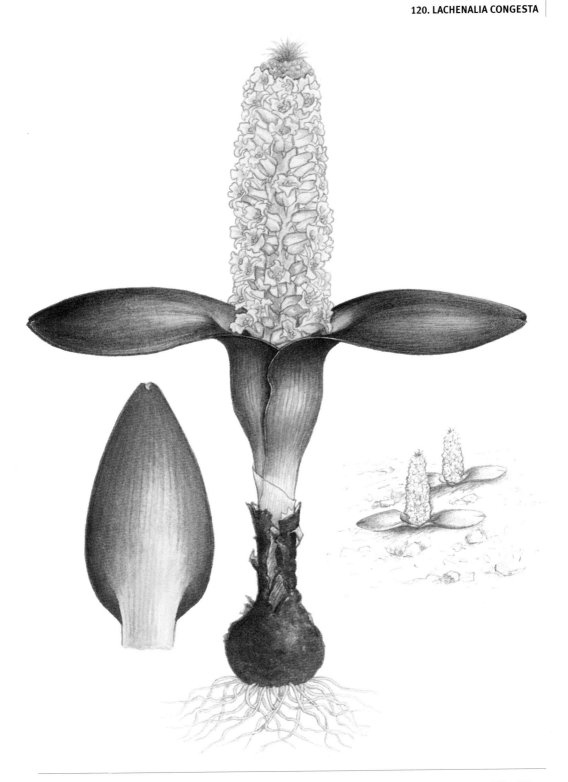

Plate 37. Watercolour painting of *Lachenalia congesta* from Sutherland (*Duncan* 530, in NBG) courtesy of The Editor, *Curtis's Botanical Magazine* vol. 26, t. 649 (2009). Artist: Marieta Visagie.

yellowish-cream, heavily spicy sweet-scented; tube cup-shaped, 3 mm long, light blue at base; outer tepals narrowly ovate, 6–8 × 4–5 mm, apical gibbosities light to bright green, apices recurved, margins undulate; inner tepals obovate, 9–10 × 3–5 mm, keels light to bright green, upper two tepals slightly overlapping, lower inner tepal slightly shorter. *Stamens* included, straight; filaments white, 4–5 mm long. *Ovary* ellipsoid, 3–4 × 2 mm, bright green; style included, straight, 2–3 mm long, white. *Capsule* ellipsoid, 7–13 × 5–6 mm. *Seed* globose, 1.8–1.9 × 1.8–1.9 mm, glossy, black; strophiole rudimentary, 0.1 mm long, ridged. *Chromosome number* 2n = 26, 28 (Johnson & Brandham, 1997). Plate 37, Figure 267.

FLOWERING PERIOD. May to August, with a peak in late July.

HISTORY. The first known collection of *L. congesta* was made by Percival Ross Frames (1863–1947), a South African solicitor, collector and cultivator of succulent plants. He found flowering specimens west of Middelpos in early August 1933, and brought them to the Bolus Herbarium. There, the specimens were painted in watercolour by Miss G. J. Lewis, the botanist in charge of the South African Museum Herbarium (SAM). In 1953, the taxon was collected near Sutherland by J. P. H. Acocks, and in 1955, a further collection was made in the same area by O. A. Leistner. A Sutherland collection made in 1970 by Frank Stayner, then Curator of the Karoo National Botanical Garden at Worcester, was selected as the holotype (Barker, 1978). The description is accompanied by Lewis's painting, reproduced in monochrome.

DISTINGUISHING CHARACTERS AND AFFINITIES. *L. congesta* is sister to *L. kliprandensis* (Duncan et al., 2005a), these two species sharing prostrate, ovate leaves with a usually subterranean scape. *L. kliprandensis* differs in its much larger, narrowly campanulate, spreading white flowers with plain or magenta inner tepal apices, strongly declinate, shortly exserted stamens, and much smaller, ovoid seed (0.9–1.0 × 0.8–0.9 mm), which has a much longer strophiole (0.5 mm long).

DISTRIBUTION, HABITAT AND CONSERVATION STATUS. This unusual species occurs at relatively high altitude and is native to the Roggeveld Escarpment in the western Great Karoo, extending from just west of Middelpos, south-east to the Sutherland district and the Komsberg, at altitudes up to

Figure 267. *Lachenalia congesta* on shale gravel flats near Williston. Image: Graham Duncan.

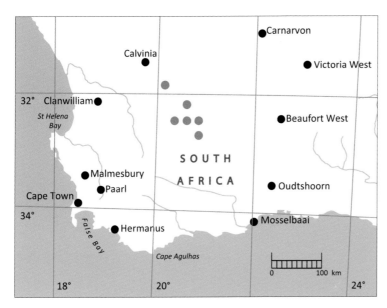

Map 121. Known distribution of *Lachenalia congesta*.

1,600 m (Map 121). *L. congesta* is limited to Roggeveld Shale Renosterveld vegetation (Mucina & Rutherford, 2006), occurring in small to large colonies on seasonally moist flats, steeply rolling east-facing mountain slopes and summit ridges in shale gravel or heavy gravelly clay. The plants grow in full sun, and at Sutherland they are frequently subject to sub-zero winter temperatures (down to as low as -15°C). They can be covered by up to 30 cm of snow for up to three days during the flowering period. The only visible damage following such heavy snow is bending of the scapes and slight bruising of the flowers (Duncan, 2009). *L. congesta* also experiences severe droughts that can span several years, during which the deep-seated bulbs remain dormant. Although several populations are subject to overgrazing, *L. congesta* is not under immediate threat.

NOTES. The flowers of *L. congesta* are strongly spicy honey-scented and are pollinated by honey bees (*Apis mellifera*). The flowering phenology of *L. congesta* is unusual in that flowering often extends over two months. Early-flowering individuals commence flowering in late June, thereafter, a continuous and increasing number of individuals come into flower until the peak flowering period is reached in late July, whereafter individuals flower sporadically until the end of August. Not all mature plants flower every year, despite having healthy leaves and growing in conditions that are seemingly ideal for flowering. This phenomenon is possibly a survival strategy that ensures that the bulbs of at least some individuals are in adequate condition to flower and produce seed when sufficient rains next fall, while those that have flowered during the current season have time to replenish their resources (Duncan, 2009).

At a locality south of Sutherland, a large *L. congesta* colony grows interspersed with *Colchicum latifolium* (Colchicaceae) and sporadic plants of *Massonia depressa* (Asparagaceae). The foliage of the first two species is remarkably similar and their identity can only be ascertained upon examination of the lower leaf surfaces, which reveals the magenta flush (resulting from the presence of anthocyanin pigment in the vacuoles of the subepidermal cell layer) that characterises *L. congesta* (Duncan, 2009). The scape breaks off at its base once the capsules have dehisced and the seeds are dispersed as the infructescence is blown about against bushes and stones. The developing inflorescences are heavily grazed by Cape hares (*Lepus capensis*).

121. LACHENALIA FRAMESII

Lachenalia framesii W. F. Barker, *Botaniska Notiser* 119 (Fasc. 2): 204–205, 207 (1966).
TYPE: South Africa, Western Cape, 20 miles [32 km] north of Vanrhynsdorp, *P. Ross-Frames s.n.* sub. BOL19614 (BOL!, holo.).

ETYMOLOGY. *framesii*: in honour of Percival Ross-Frames (1863–1947) of Kenilworth, Cape Town.
DESCRIPTION. *Dwarf geophyte*, 80–150 mm high. *Bulb* subglobose, 7–15 mm in diameter, offset-forming; tunic multilayered, outer layers dark brown, spongy; inner cataphyll translucent white, adhering to leaf bases, apex obtuse. *Leaves* 1–2, lanceolate, 40–120 × 7–20 mm, suberect to spreading, light to dark green, unmarked, canaliculate, upper surface with depressed longitudinal grooves, midrib distinct on lower surface; margins flat to undulate or crisped; leaf bases clasping, 20–40 mm long, yellowish-green, sometimes flushed with maroon; primary seedling leaf terete, erect. *Inflorescence* spicate, few- to many-flowered, fairly dense, sterile apex short to long; scape erect to suberect, 50–100 mm long, slender, light to bright green or flushed with minute dull maroon speckles; rachis light green, sterile apex electric blue; lower bracts ovate, becoming lanceolate above, 1 × 1–2 mm, light greenish-white to dull maroon. *Perianth* zygomorphic, urceolate, suberect; tube cup-shaped, 3 mm long, bright greenish-yellow; outer tepals ovate, 5–6 × 3–4 mm, light to bright greenish-yellow, apical gibbosities light to

Figure 268 (left). *Lachenalia framesii* amongst quartz stones near Komaggas, western Namaqualand. Image: Graham Duncan.
Figure 269 (right). Cultivated specimens of *Lachenalia framesii* from Vanrhynsdorp. Image: Graham Duncan.

dark green or dark greenish-yellow; inner tepals obovate, 8–9 × 4–5 mm, translucent white, apices recurved, light to bright magenta, keels light green, upper two tepals overlapping, lower inner tepal slightly longer. *Stamens* included, declinate; filaments white, 5–8 mm long. *Ovary* ellipsoid, 3–4 × 2–3 mm, bluish-green; style included, declinate, 4–6 mm long, white. *Capsule* ellipsoid, 6–7 × 4–5 mm. *Seed* ovoid, 0.9 × 0.8–0.9 mm, glossy, black, secondary sculpturing reticulate; strophiole rudimentary, 0.1–0.2 mm long, ridged. *Chromosome number* 2n = 16 (Spies *et al.*, 2008). Figures 31, 268, 269.

FLOWERING PERIOD. July to August, with a peak in early August.

HISTORY. Percival Ross-Frames, a South African collector and cultivator of succulent plants, discovered this species north of Vanrhynsdorp in 1931. The heavily undulate, crisped leaf margins of his plants led W. F. Barker to believe it to be *L. undulata*, which had been described by Baker (1886) on the basis of a drawing made by Francis Masson in 1793, housed in the British Museum (Natural History). Barker had unfortunately not seen Masson's drawing and her painting of a specimen from the Ross-Frames collection was subsequently published in *The Flowering Plants of South Africa* under the name *L. undulata*, with a description in English (Barker, 1931c). Upon seeing Masson's drawing some years later, her error became evident and *L. framesii* was described anew in *Botaniska Notiser*, illustrated with the same painting, which was reproduced in monochrome (Barker, 1966).

DISTINGUISHING CHARACTERS AND AFFINITIES. *L. framesii* appears most closely allied to *L. valeriae* (Duncan *et al.*, 2005a), with which it shares greenish-yellow perianths with recurved inner tepals, with the lower inner tepal longer and canaliculate, with included, declinate stamens and ovoid seeds. *L. valeriae* differs in its more robust nature (up to 350 mm high) and in having longer outer (7–8 mm) and inner (9–13 mm) tepals, with bright magenta restricted to the lower inner tepal. Its ovoid seeds are also much larger (1.4 × 1.3 mm). When compared with *L. framesii*, *L. valeriae* has leaves that are longer and broader (120–170 × 20–30 mm), and usually densely covered with small dome-shaped pustules on the upper surface. The distribution of *L. framesii* extends close to that of *L. valeriae* in the Komaggas Flower Reserve, but the species are allopatric (Duncan & Edwards, 2002).

DISTRIBUTION, HABITAT AND CONSERVATION STATUS. *L. framesii* is endemic to the north-western Northern and Western Cape, stretching from Komaggas on the Namaqualand coastal plain, inland to Vleikraal Farm east of Klawer, to Gannabos near Loeriesfontein (Map 122). It is common around Komaggas and on the Knersvlakte around Vanrhynsdorp, occurring in dense colonies numbering hundreds or sometimes thousands of individuals. *L. framesii* is always associated with stony quartzitic patches on rolling plains, flats, hillsides and ridges. It occurs in several vegetation types including Riethuis-Wallekraal- and Knersvlakte Quartz Vygieveld, and Hantam Karoo (Mucina & Rutherford, 2006). Near Komaggas, *L. framesii* grows in association with *Brunsvigia radula* and *Strumaria truncata* (Amaryllidaceae) on quartz flats adjacent to a large colony of *L. anguinea*, the latter confined to deep red sand. Further south near Soebatsfontein, *L. framesii* grows in association with the dwarf irid *Moraea serpentina*. *L. framesii* is not threatened.

Map 122. Known distribution of *Lachenalia framesii*.

NOTES. *L. framesii* is pollinated by honey bees (*Apis mellifera*). The base of the scape detaches from the bulb once the capsules have dehisced and the seeds are dispersed as the infructescence is blown away.

122. LACHENALIA VALERIAE

Lachenalia valeriae G. D. Duncan, *Bothalia* 32 (2): 190–192 (2002).
TYPE: South Africa, Northern Cape, north-western Namaqualand, Molyneux Nature Reserve, Kleinsee, on west-facing granite slopes in brownish-red sand, *G. D. Duncan* 444 (NBG!, holo.).

ETYMOLOGY. *valeriae*: after Valerie Fay Anderson (1931–), highly accomplished botanical artist and illustrator of numerous southern African botanical monographs.

DESCRIPTION. *Geophyte*, 100–350 mm high. *Bulb* subglobose, 15–20 mm in diameter, solitary or occasionally clump-forming; tunic multilayered, outer layers dark brown, spongy; inner cataphyll translucent white, loosely surrounding leaf bases, apex obtuse. *Leaves* 2, narrowly to broadly lanceolate, 120–170 × 20–30 mm, spreading to suberect, canaliculate, upper surface dark green, usually densely covered with minute, dark green pustules, depressed longitudinal grooves distinct, lower surface plain or lightly to heavily flushed with maroonish-magenta; leaf bases clasping, 20–50 mm long, upper portion yellowish-green or flushed with maroonish-magenta, subterranean portion white; primary seedling leaf terete, suberect. *Inflorescence* spicate, many-flowered, fairly dense, sterile apex short to long; scape erect to suberect, 70– 230 mm long, lower half light green with minute brownish-purple speckles, upper half heavily mottled with brownish-purple, uppermost portion

Figure 270. Greenish yellow form of *Lachenalia valeriae* in red sand near Kleinsee. Image: Graham Duncan.

THE GENUS LACHENALIA 415
122. LACHENALIA VALERIAE

Plate 38. Watercolour painting of *Lachenalia valeriae* from Kleinsee (*Duncan* 444, in NBG) courtesy of Fay Anderson. Artist: Fay Anderson.

sometimes inflated; rachis light purplish-brown in lower portion, shading to electric blue in upper portion; lower bracts ovate, becoming lanceolate above, 1–2 × 1–4 mm. *Perianth* zygomorphic, urceolate, slightly suberect; tube cup-shaped, 2–3 mm long, white to cream, sometimes blue-tinged; outer tepals ovate, 7–8 × 4–5 mm, white, or cream at base shading to greenish-yellow above, apical gibbosities and keels bright green; inner tepals obovate, 9–13 × 3–5 mm, white to cream, slightly recurved, upper inner tepals overlapping, keels bright green; lower inner tepal deeply canaliculate, lower half white to cream, upper half light to bright magenta, median keel light greenish-yellow. *Stamens* included, declinate; filaments 7 mm long, white; anthers dull maroon, pollen yellow. *Ovary* ellipsoid, 3–4 × 2 mm, light green; style included, declinate, 6–7 mm long, white. *Capsule* ellipsoid, 8 × 5 mm. *Seed* ovoid, 1.4 × 1.3 mm, glossy, black; strophiole rudimentary, 0.2–0.3 mm long, ridged. *Chromosome number* 2n = 16 (Spies *et al.*, 2008). Plate 38, Figures 270, 271.

FLOWERING PERIOD. July to August, with a peak in late July.

HISTORY. Walter Wisura, a retired Kirstenbosch horticulturist in charge of succulent plants, first collected *L. valeriae* in July 1970. He found vegetative material in sand dunes, a short distance east of the mouth of the Holgat River and above the Holgat River Canyon, north of the coastal town of Port Nolloth in the southern Richtersveld. The bulbs were cultivated in the Kirstenbosch Nursery and flowered there the same year. W. F. Barker recognised these plants as an undescribed taxon and preserved a few specimens for the Compton Herbarium collection. During the summer of 1993, dormant bulbs of an unidentified *Lachenalia* were gathered by John Lavranos west of Komaggas in western Namaqualand, also in sand dunes. When one of these bulbs flowered a few years later at Kirstenbosch, it matched the previous collections and was painted by Fay Anderson (Duncan,

Figure 271. White form of *Lachenalia valeriae* between shale rock slabs near Nababeep. Image: Barbara Kotze.

2003b). A subsequent collection of vegetative specimens was made by Ernst van Jaarsveld in the Molyneux Nature Reserve at Kleinsee, and the type collection was made there in August 2001 (*Duncan* 444, in NBG). *L. valeriae* has since been recorded at several other localities in central and southern Richtersveld, and in Namaqualand.

DISTINGUISHING CHARACTERS AND AFFINITIES. *L. valeriae* is closely allied to *L. framesii* (Duncan et al., 2005a). *L. framesii* differs in its dwarf habit (usually shorter than 100 mm) and in its shorter outer (5–6 mm) and inner tepals (8–9 mm), both with light to bright magenta apices. It also differs vegetatively in its shorter, narrower canaliculate leaves (40–120 × 7–20 mm) and much smaller seeds (0.9 × 0.8–0.9 mm).

DISTRIBUTION, HABITAT AND CONSERVATION STATUS. Native to the north-western Northern Cape, *L. valeriae* extends from the Goariepvlakte in central Richtersveld to the Namaqua National Park west of Kamieskroon (Map 123). It is encountered in Northern Richtersveld Yellow Duneveld, Namaqualand Sand Fynbos, Namaqualand Strandveld and Namaqualand Klipkoppe Shrubland vegetation (Mucina & Rutherford, 2006). The plants grow in colonies, usually on exposed, deep red sandy coastal plains, in sandy pockets of east- and west-facing slopes of granite outcrops (Duncan 2003e). Less frequently, they occur in horizontal crevices of shale rock slabs, such as at Nababeep north of Springbok. At its type locality, near the mouth of the Buffels River in the Molyneux Nature Reserve at Kleinsee, *L. valeriae* grows on west-facing slopes in rock cracks, in partial shade provided by low succulent vegetation including *Aloe framesii* (Xanthorrhoeaceae) and *Pelargonium fulgidum* (Geraniaceae) (Duncan & Edwards, 2002). *L. valeriae* is not threatened.

NOTES. The species is pollinated by honey bees (*Apis mellifera*). The foliage and partially open inflorescences are grazed by angulate tortoises (*Chersina angulata*) and steenbuck (*Raphicerus campestris*) (Duncan, 2003b). The scapes detach from the bulbs once the capsules have dehisced, and the infructescences are blown away, dispersing the seeds.

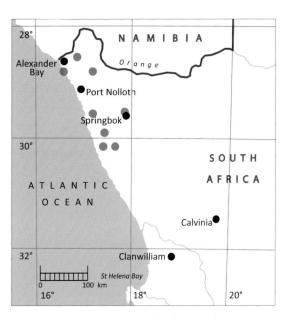

Map 123. Known distribution of *Lachenalia valeriae*.

123. LACHENALIA KRUGERI

Lachenalia krugeri G. D. Duncan, sp. nov.
A L. framesii foliis late lanceolatis glaucis cum marginibus coriaceis, floribus 3 in verticillos, tepalis interioribus non-recurvatis, seminique multo grandiore ovoideo cum strophiolo grande plano porcato differt.
TYPE: South Africa, Northern Cape, 2917 (Springbok): Kleinsee Namaqua Diamond Mine, in open aspects among low bushes on deep red sandy flats (–CA), *P. Kruger s.n.* (NBG!, holo.); *G. D. Duncan* 445 (NBG!, para.).

123. LACHENALIA KRUGERI

ETYMOLOGY. *krugeri*: after Paul Kruger (1967–), formerly Environmental Officer for Namaqualand Mines at Kleinsee, who discovered this species in 2001, and who greatly assisted the author in locating lachenalias in western Namaqualand.

DESCRIPTION. *Geophyte*, 110–300 mm high. *Bulb* globose, 15–20 mm in diameter, solitary; tunic multilayered, outer layers light brown, spongy; inner cataphyll translucent white, adhering to leaf base, apex acute. *Leaf* solitary, lanceolate to broadly lanceolate, 60–210 × 25–30 mm, spreading to suberect, glaucous, slightly to deeply canaliculate, leathery, upper surface plain or with sporadic brownish-maroon blotches, depressed longitudinal grooves prominent, lower surface with bright magenta blotches in lower part; margins heavily undulate and crisped, thickened, coriaceous, deep maroon; leaf base clasping, 30–40 mm long, aerial portion green with dark maroon bands, subterranean portion translucent white with bright magenta bands; primary seedling leaf terete, erect. *Inflorescence* spicate, flowers arranged in 3-flowered whorls, sterile apex short; scape erect, 50–150 mm long, yellowish-green, upper half plain, lower half mottled with light purple; rachis yellowish-green, narrowing towards apex; lower bracts ovate, becoming lanceolate above, 1–2 × 1–3 mm, translucent white. *Perianth* zygomorphic, urceolate, spreading to slightly cernuous; tube cup-shaped, 1–2 mm long, yellowish-green marked with dull blue; outer tepals ovate, 5–7 × 4–5 mm, yellowish-green, apices slightly flared, apical gibbosity brown or bright green; inner tepals obovate, 7–8 × 3–5 mm, translucent white, central and lower portion bright green, margins deep purple, apices slightly recurved, upper two tepals overlapping, lower tepal longer and narrower, canaliculate. *Stamens* included, declinate; filaments white, 7–9 mm long, pollen yellow at anthesis, ageing to black. *Ovary* ellipsoid, 2 × 2 mm, dark green; style included, declinate, 6 mm long, white. *Capsule* ellipsoid, 6–7 × 4–5 mm. *Seed* ovoid, 1.2 × 1.5 mm, glossy, black; strophiole flat, 0.5–0.6 mm long, smooth. *Chromosome number* unknown. Figures 272, 273.

FLOWERING PERIOD. June to July with a peak in mid-June.

HISTORY. This interesting, very seldom collected species was discovered by Paul Kruger in 2001 within the restricted mining area of the Namaqua Diamond Mine at Kleinsee in western Namaqualand. A specimen was collected there in fruit in early August that year (*Duncan* 445, in NBG). Additional specimens were collected on Dreyerspan Farm just north of Kleinsee in 2001 (*Duncan* 446, in NBG); these flowered in the Kirstenbosch Bulb Nursery in 2002 and were confirmed as undescribed. In July 2008, Paul Kruger found additional plants in flower at Kleinsee (*Kruger s.n.*, in NBG), one of which forms part of the type material. In September 2006, a single specimen was collected by Adam Harrower some distance to the north, near Anenous Pass west of Steinkopf in the southern Richtersveld (*Harrower* 3845, in NBG).

DISTINGUISHING CHARACTERS AND AFFINITIES. *L. krugeri* has not yet been phylogenetically analysed but appears to be closely allied to *L. framesii*, which has similar greenish-yellow perianths, included, declinate stamens and ovoid seeds. *L. framesii* differs in its bright green leaves, which have non-thickened green margins, and unmarked upper

Figure 272. *Lachenalia krugeri* in deep red sand at Kleinsee. Image: Paul Kruger.

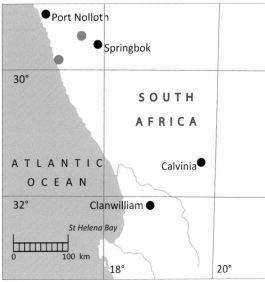

Map 124. Known distribution of *Lachenalia krugeri*.

Figure 273. Cultivated specimens of *Lachenalia krugeri* from Dreyerspan north of Kleinsee. Image: Graham Duncan.

leaf surfaces, in its smaller, spirally arranged flowers, the inner tepals with prominent magenta apices, and in having much smaller (0.9 × 0.8–0.9 mm) seeds that have a rudimentary strophiole of 0.1–0.2 mm in length.

DISTRIBUTION, HABITAT AND CONSERVATION STATUS. *L. krugeri* is currently known from three locations in the north-western Northern Cape, one at Kleinsee, another to the north of this town, and the third near Anenous Pass west of Steinkopf (Map 124). The plants occur in Namaqualand Strandveld and Anenous Plateau Shrubland vegetation (Mucina & Rutherford, 2006). They grow as scattered, solitary individuals in deep red sand, exposed or within the protection of low bushes. *L. krugeri* has been affected by mining activity around Kleinsee and a conservation status of Vulnerable is recommended.

NOTES. *L. krugeri* is pollinated by honey bees (*Apis mellifera*). The scapes detach from the bulbs shortly after capsule dehiscence and the infructescences are blown away, scattering the seeds.

124. LACHENALIA AMELIAE

Lachenalia ameliae W. F. Barker, *Journal of South African Botany* 49 (4): 441–444 (1983).
TYPE: South Africa, Western Cape, south of Bloutoring Station, Montagu district, *C. E. Malan* 90 (NBG!, holo.; BOL!, PRE!, iso.).

ETYMOLOGY. *ameliae*: after A. Amelia Mauve, née Obermeyer (1907–2001), 20th century author of numerous botanical publications.

DESCRIPTION. *Dwarf geophyte*, 40–115 mm high. *Bulb* globose, 8–15 mm in diameter, neck distinct, 10–15 mm long; solitary; tunic multilayered, outer layers dark brown to black, cartilaginous; inner

124. LACHENALIA AMELIAE

cataphyll translucent white, adhering to leaf bases, apex obtuse. *Leaves* 1–2, lanceolate to broadly lanceolate, 30–60 × 5–20 mm, spreading, upper surface light to dark green, covered with short to long simple trichomes 1–7 mm long, or trichomes restricted to the leaf margin; lower leaf surface green or dark maroon; leaf bases clasping, 10–15 mm long, subterranean, upper portion light green or white with small to large magenta spots or blotches, lower portion white; primary seedling leaf terete, erect. *Inflorescence* spicate, few-flowered, fairly dense, sterile apex short; scape erect to suberect, 10–60 mm long, sturdy, light green, lightly to heavily mottled with dark maroon or purple; rachis light green mottled with dark maroon; bracts ovate, 0.5–1.0 × 1–2 mm. *Perianth* zygomorphic, urceolate, suberect, light yellow to greenish-yellow; tube cup-shaped, 2–3 mm long; outer tepals narrowly ovate, 6–7 × 4 mm, apical gibbosities dark green or light magenta, keels light to dark magenta, apices recurved; inner tepals obovate, 9–10 × 3–4 mm, upper two tepals overlapping, lower tepal slightly longer, keels light green or magenta, apices recurved, greenish-yellow or dark magenta. *Stamens* included to shortly exserted, declinate; filaments 7–9 mm long, white. *Ovary* ellipsoid, 4 × 3 mm; light green; style included, declinate, 5–6 mm long, white, exserted as ovary matures. *Capsule* ellipsoid, 6–7 × 4–5 mm. *Seed* ovoid, 1.0–1.1 × 1.2–1.3 mm, matte, black; strophiole 0.5–0.6 mm long, ridged. *Chromosome number* 2n = 18 (Johnson & Brandham, 1997). Figures 274, 275.

FLOWERING PERIOD. August to September, with a peak in early September.

HISTORY. The earliest known record of *L. ameliae* is that of Dr Rudolf Marloth who found it in September 1921 at Zwartkoppies near Spes Bona, south-east of Wupperthal in the Western Cape.

Figure 274 (left). Magenta-tipped form of *Lachenalia ameliae* on stony clay flats in the Tankwa Karoo National Park. Image: Marinda Koekemoer.

Figure 275 (right). Cultivated, yellow form of *Lachenalia ameliae* from Montagu. Image: Graham Duncan/SANBI.

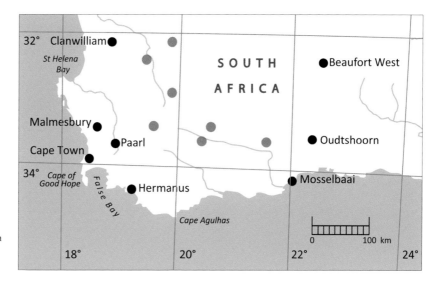

Map 125. Known distribution of *Lachenalia ameliae*.

That specimen is housed in the National Herbarium (PRE). Two further collections were made in 1935, one at Gansfontein Farm in the Tanqua Karoo, the other just east of Montagu in the Little Karoo. In September 1974, a population was found by Amelia Mauve, Inge Oliver and Christien Malan north-east of Montagu. In 1979, W. F. Barker and C. Malan revisited this site and made the type collection (Barker, 1983b). *L. ameliae* has since been collected infrequently.

DISTINGUISHING CHARACTERS AND AFFINITIES. *L. ameliae* is allied to *L. mathewsii* (Duncan *et al.*, 2005a), with which it shares a yellow perianth with declinate stamens and ovoid seeds, but *L. mathewsii* differs in having a many-flowered raceme of oblong-campanulate flowers and linear, attenuate, glaucous leaves. The inner tepal apices of *L. ameliae* in populations in the species' western range in the Tanqua Karoo are dark magenta, whereas those of forms from Montagu and further east are light greenish-yellow. The upper leaf surface and/or margins are covered with simple trichomes, the density and length of which vary considerably within, and between, individuals of the same population. Trichome length varies from 1–7 mm: some individuals have both short and long trichomes whereas others have only short ones. Plants from the Tanqua Karoo usually have very short trichomes whereas those from Montagu have both short and long trichomes.

DISTRIBUTION, HABITAT AND CONSERVATION STATUS. *L. ameliae* has a fairly limited known distribution, extending from the northern Tanqua Karoo to Karoo Poort north-east of Ceres, south-east to Montagu. Disjunct records near the Anysberg and Calitzdorp suggest that further populations are likely to be located in the central Little Karoo (Map 125). The plants occur in Tanqua- and Eastern Little Karoo, Ceres Shale Renosterveld and Montagu Shale Fynbos vegetation (Mucina & Rutherford, 2006). They grow as scattered individuals on red, sandy or stony clay flats or on low mounds, exposed or within the protection of low scrub. *L. ameliae* has a conservation status of Near Threatened because parts of its range are subject to agricultural expansion, overgrazing and trampling by livestock (Vlok *et al.*, 2009).

NOTES. Occurring in arid terrain with erratic rainfall patterns, the bulbs of *L. ameliae* are adapted to remain dormant for one or more growing seasons. The urceolate flowers suggest pollination by honey bees (*Apis mellifera*). The scape detaches from the bulb at its base in early summer and the seeds are dispersed by wind as the infructescence is blown about.

125. LACHENALIA NAMAQUENSIS

Lachenalia namaquensis Schltr. ex W. F. Barker, *Journal of South African Botany* 44 (4): 402–405 (1978).
TYPE: South Africa, Northern Cape, 6 miles [9.6 km] south of Steinkopf, *W. F. Barker* 9020 (NBG!, holo.).

ETYMOLOGY. *namaquensis*: of Namaqualand.

DESCRIPTION. *Geophyte*, 80–230 mm high. *Bulb* globose, 10–12 mm in diameter, neck distinct, 5–10 mm long, offset- and bulbil-forming; tunic multilayered, outer layers dark brown, spongy; inner cataphyll translucent white, adhering to leaf bases, apex obtuse. *Leaves* 1–2, linear, 80–140 × 5–15 mm, erect to suberect or falcate, light to dark green, lower two thirds canaliculate, upper third subterete, attenuate; leaf bases clasping, 20–40 mm long, subterranean portion white or light purplish-magenta, aerial portion light green; primary seedling leaf terete, erect. *Inflorescence* spicate, few- to many-flowered, lax, sterile apex short; scape erect to suberect, 60–120 mm long, slender, light green; rachis light green in lower half, shading to greenish-pink or greenish-blue in upper half, apex pink or electric blue; bracts ovate, 0.5–1.0 × 1–3 mm, green or white. *Perianth* zygomorphic, urceolate, suberect; tube cup-shaped, 3 mm long, light to bright blue or pinkish-magenta; outer tepals ovate, 5–7 × 4–5 mm, light to dark pinkish-magenta, apical gibbosities light green or purplish-magenta; inner tepals obovate, 8–9 × 4–5 mm, apices recurved, translucent light to dark pinkish-white, keels light to dark magenta, upper two tepals overlapping, lower tepal narrower and slightly longer, canaliculate, deep magenta. *Stamens* included, declinate; filaments light magenta in lower half, shading to white above, 4–7 mm long. *Ovary* ellipsoid, 3–4 × 3 mm, light green; style included, declinate, 5–6 mm long, white. *Capsule* ellipsoid, 6–8 × 4–5 mm. *Seed* ovoid, 0.9–1.0 × 0.9 mm, glossy, black, secondary sculpturing reticulate; strophiole 0.5 mm long, ridged. *Chromosome number* $2n = 16$ (Crosby, 1986; Johnson & Brandham, 1997; Spies *et al.*, 2008). Plate 39, Figure 276.

FLOWERING PERIOD. August to October, with a peak in late September.

HISTORY. Rudolf Schlechter first collected this species near Steinkopf in northern Namaqualand on 23rd September 1897, and distributed pressed material to eight local and European herbaria under his manuscript name *L. namaquensis* Schltr.. The species was finally described 80 years later in the *Journal of South African Botany* (Barker, 1978) with the type material collected by Barker south of Steinkopf in August 1959. *L. namaquensis* has been collected many times in the vicinity of this town and around Springbok. Annotations

Figure 276. *Lachenalia namaquensis* on a gravelly clay hillside west of Springbok. Image: Graham Duncan.

THE GENUS LACHENALIA
125. LACHENALIA NAMAQUENSIS

Plate 39. Watercolour painting of *Lachenalia namaquensis* from Steinkopf, courtesy of the Compton Herbarium, Kirstenbosch, South African National Biodiversity Institute. Artist: Winsome Barker.

to pressed material in several herbaria indicate that Barker had originally intended to name the species *L. pillansii* W. F. Barker for Neville Pillans, who collected it at Klipfontein near Steinkopf in October 1926, but she was probably unaware of the existence of Schlechter's material and manuscript name when she made these annotations.

DISTINGUISHING CHARACTERS AND AFFINITIES. *L. namaquensis* is sister to *L. mathewsii* (Duncan *et al.*, 2005a), which shares similar linear, canaliculate leaves that taper to a long attenuate apex. *L. mathewsii* differs in having a raceme of spreading, oblong- campanulate, bright yellow flowers with bright green apical gibbosities and keels. It also differs in its shortly exserted filaments and in its seed, which has a rudimentary strophiole of 0.2 mm in length.

DISTRIBUTION, HABITAT AND CONSERVATION STATUS. *L. namaquensis* is confined to the north-western Northern Cape, extending from Stinkfontein in the southern Richtersveld to a locality south-west of Garies in southern Namaqualand (Map 126). The species traverses Namaqualand Shale Shrubland and Namaqualand Heuweltjieveld vegetation (Mucina & Rutherford, 2006). It is plentiful around Steinkopf and to the west of Springbok, forming colonies in pockets of gravelly red clay on granite boulders or in stony white sandy soil on quartz outcrops. It is usually encountered towards the bottom of east- and south-facing slopes among low-growing succulents. At a locality west of Springbok, it grows in association with robust specimens of *L. glauca*. *L. namaquensis* is not threatened.

NOTES. The flowers of *L. namaquensis* are pollinated by honey bees (*Apis mellifera*). The scapes remain attached to the bulbs for several weeks following capsule dehiscence and the seeds are dispersed by the shaking action of wind. *L. namaquensis* is somewhat unusual in producing numerous long stolons that originate from the base of the bulb, each terminating in a bulblet just below ground level.

Map 126. Known distribution of *Lachenalia namaquensis*.

126. LACHENALIA BOLUSII

Lachenalia bolusii W. F. Barker, *Journal of South African Botany* 45 (2): 193–196 (1979).
TYPE: South Africa, Northern Cape, Klipfontein, *H. Bolus* 6592 (BOL!, holo.; K! iso.).

ETYMOLOGY. *bolusii*: after Dr Harry Bolus (1834–1911), English plant collector and author.

DESCRIPTION. *Geophyte* 85–340 mm high. *Bulb* globose, 10–20 mm in diameter, neck distinct, 5–10 mm long, offset-forming, deep-seated; tunic multilayered, outer layers dark brown, spongy; inner cataphyll translucent white, adhering to leaf bases, apex obtuse. *Leaf* solitary, narrowly to broadly lanceolate, 30–140 × 9–28 mm, suberect to spreading, light to dark green, upper surface with depressed longitudinal grooves, flat to slightly canaliculate, lower surface plain or with purplish-maroon bands in lower third; leaf margins slightly to heavily undulate and crisped, smooth or papillate near base;

leaf base tightly clasping, 20–120 mm long, subterranean portion white, aerial portion heavily banded with purplish-maroon or magenta; primary seedling leaf terete, erect. *Inflorescence* racemose, few- to many-flowered, lax, sterile apex short; scape erect to suberect, 60–220 mm long, slender, light green, plain or heavily marked with minute purple speckles; rachis light green, plain or heavily flushed dull purplish-brown in upper half; pedicels suberect, 4–8 mm long, light green, whitish-magenta or brownish-green; lower bracts ovate, becoming lanceolate above, 1–2 × 1–3 mm, white. *Perianth* zygomorphic, urceolate, cernuous; tube 2 mm long, light blue; outer tepals ovate, 6–7 × 5 mm, light blue, apical gibbosities light purple or brown; inner tepals obovate, 7–9 × 5–6 mm, greenish-white; apical marking purplish-brown, keels light blue, apices slightly recurved. *Stamens* included, straight; filaments white, 7–8 mm long. *Ovary* obovoid, 3–4 × 3 mm; style included, straight, 6 mm long, white. *Capsule* obovoid, 7–8 × 6 mm. *Seed* globose, matte, black, 1.2–1.3 × 1.3–1.4 mm; strophiole inflated, 0.6–0.7 mm long, smooth. *Chromosome number* 2n = 18 (Spies *et al.*, 2009). Figure 277.

FLOWERING PERIOD. July to September, with a peak in late August.

HISTORY. Dr Harry Bolus was the first to collect plants of this species, at Klipfontein north-west of Steinkopf in northern Namaqualand in August 1883. This collection forms the type material. In 1896, Rudolf Schlechter found plants of this taxon on the Biedouwberg near Wupperthal in the Western Cape, and it has since been recorded at numerous other localities in the Richtersveld, Namaqualand, and on the Pakhuis Pass east of Clanwilliam. The species remained unpublished for almost a century after its discovery until finally appearing in the *Journal of South African Botany* (Barker, 1979a).

DISTINGUISHING CHARACTERS AND AFFINITIES. The heavily banded leaf bases and perianth colour of certain forms of *L. hirta* and *L. unifolia* strongly resemble those of *L. bolusii*. Phylogenetic analysis of morphology suggested, however, that *L. bolusii* is more closely allied with *L. elegans* (Duncan *et*

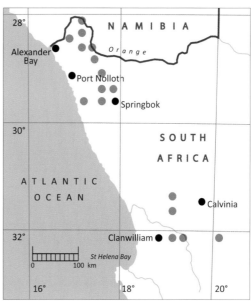

Map 127. Known distribution of *Lachenalia bolusii*.

Figure 277. Cultivated specimens of *Lachenalia bolusii* from the southern Richtersveld. Image: Graham Duncan.

al., 2005a), these two species sharing straight stamens, obovate capsules and globose seeds. *L. elegans* differs in its sessile, urceolate, suberect, heavily sweet-scented flowers, shorter stamens (4–5 mm long) and leathery leaf with flat margins.

DISTRIBUTION, HABITAT AND CONSERVATION STATUS. *L. bolusii* has a disjunct distribution and is native mainly to the north-western and western parts of the Northern and Western Cape. It extends from Helskloof in the northern Richtersveld to Springbok, and from Calvinia to the Biedouw Valley east of Clanwilliam, with an outlier at Soekop Farm south of Middelpos in the Roggeveld Mountains (Map 127). It occurs in eight vegetation types: Western Gariep Hills Desert, Noms- and Kahams Mountain Desert, Central Richtersveld Mountain, Namaqualand Shale- and Agter-Sederberg Shrubland, Bokkeveld Sandstone Fynbos and Roggeveld Karoo (Mucina & Rutherford, 2006). The plants grow as scattered individuals within the protection of bushes on steep, rocky, south-facing granite slopes in heavy clay, less frequently on dry open flats or in pockets of sandy soil on flat rocks. In the arid northern parts of its range, the bulbs of *L. bolusii* are deep-seated and adapted to remain dormant through prolonged droughts. *L. bolusii* is not threatened.

NOTES. *L. bolusii* is pollinated by honey bees (*Apis mellifera*). Following capsule dehiscence, the scapes remain attached to the bulbs and the seeds are dispersed locally by the shaking action of wind.

2. Subgenus POLYXENA

Subgenus ***Polyxena*** (Kunth) G. D. Duncan, stat. nov. *Polyxena* Kunth, *Enumeratio Plantarum* 4: 294 (1843).
TYPE: *Polyxena pygmaea* (Jacq.) Kunth (= *Lachenalia pygmaea* (Jacq.) G. D. Duncan).

Perianth actinomorphic, arising radially from the pedicel.

7 species (8 taxa).

127. LACHENALIA PYGMAEA

Lachenalia pygmaea (Jacq.) G. D. Duncan, comb. nov. *Polyanthes pygmaea* Jacq., *Collectanea* 5: 56–57 (1797); *Polyxena pygmaea* (Jacq.) Kunth, *Enumeratio Plantarum* 4: 294 (1843).
TYPE: South Africa, Cape, collector and precise locality unknown, figure in N. J. Jacquin, *Icones Plantarum Rariorum* 2: 16, t. 380 (1795) (neotype, designated here).
EPITYPE: South Africa, Western Cape, Melkkamer near De Hoop, sandy flat with calcareous outcrops, *Müller-Doblies 84021r* (PRE!, epitype; B, BOL!, BR, BTU, G, GRA!, K, M, MO, NBG!, S, Z, iso-epitypes, designated by Müller-Doblies & Müller-Doblies, 1997).
SYNONYMY. *Massonia violacea* Andrews, *The Botanist's Repository* 1: t. 46 (1797). Type: South Africa, Cape, collector and precise locality unknown, cultivated by G. Hibbert, figure in *The Botanist's Repository* 1: t. 46 (1797) (lectotype, designated here).
Hyacinthus bifolius Boutelou ex Cav., *Anales de Ciencias Naturales* 5 (14): t. 41, Figure 1 (1802). Type: South Africa, Cape, collector and precise locality unknown, figure in *Anales de Ciencias Naturales* 5: t. 41, Figure 1 (1802) (lectotype, designated here).

ETYMOLOGY. *pygmaea*: descriptive of the dwarf stature of the plant.
DESCRIPTION. *Dwarf geophyte*, 20–40 mm high. *Bulb* subglobose, 12–25 mm in diameter, neck distinct,

127. LACHENALIA PYGMAEA

Figure 278. Mauve form of *Lachenalia pygmaea* in granitic gravel on the Kamiesberg. Image: Lita Cole.

5–10 mm long, offset-forming; tunic multilayered, outer layers dark brown, spongy; inner cataphyll translucent white, adhering to leaf bases, apex obtuse. *Leaves* 2, broadly lanceolate, 45–80 × 15–30 mm, flat to weakly canaliculate, spreading or suberect, bright green, upper surface with distinct depressed longitudinal grooves, lower surface suffused with dull magenta; leaf bases clasping, 10–20 mm long, subterranean, light green to white; primary seedling leaf terete, erect. *Inflorescence* corymbose, few- to many-flowered, sterile apex short; scape erect, 10–20 mm long, white, slender, subterranean; pedicels erect, 4–17 mm long, white; lower bracts ovate, becoming lanceolate above, 0.5–1.0 × 0.5–1.0 mm. *Perianth* actinomorphic, tubular, erect to suberect, strongly almond-scented; tube cylindrical, 20–27 mm long, slender, white; tepals becoming strongly rolled back when mature, light to bright mauve or white, tepal margins of mauve forms translucent white, outer tepals oblong, 8–10 × 2–3 mm; inner tepals linear-oblong, 8–9 × 2 mm. *Stamens* well exserted beyond recurved tepals; filaments mauve or white, 5–9 mm long, inner filament series longer than outer series, widely spreading. *Ovary* ellipsoid, 3–4 × 1.0–1.5 mm, bright green; style included to well exserted, more or less straight, 24–36 mm long, white. *Capsule* ellipsoid, 6–7 × 4–5 mm, bright green. *Seeds* globose, 1.3–1.4 × 1.1–1.2 mm, matte, black; primary sculpturing rugose; strophiole rudimentary, 0.1 mm long, ridged. *Chromosome number* unknown. Figures 278–280.

FLOWERING PERIOD. April to June, with a peak in May.

HISTORY. First described by N. J. Jacquin (1797a) as *Polyanthes pygmaea*, the species was transferred to *Polyxena* as *P. pygmaea* by Kunth (1843), a diagnosis that was upheld by Baker (1897b). The species was later placed in synonymy under *Polyxena ensifolia* by Jessop (1976) but reinstated as *Polyxena pygmaea* by Müller-Doblies & Müller-Doblies (1997). Jacquin's type material was lost between 1939 and 1945, and Müller-Doblies & Müller-Doblies (1997) mistakenly designated Jacquin's plate t. 380 as the 'iconotype' (it should have been designated a neotype). They also designated an epitype as a better match for Jacquin's protologue and Schönland's (1910) concept of *P. pygmaea*. Van der Merwe

(2002) upheld the view of Müller-Doblies & Müller-Doblies (1997), but unfortunately the species was relegated to synonymy under *L. ensifolia* when *Polyxena* was transferred to *Lachenalia* (Manning *et al.*, 2004). The species is morphologically distinct from *L. ensifolia* and is reinstated here under the new combination *L. pygmaea* (Jacq.) G. D. Duncan.

DISTINGUISHING CHARACTERS AND AFFINITIES. *L. pygmaea* is readily identified by its very slender and very long perianth tube (20–27 mm) and by its mauve or white outer and inner tepals that become strongly rolled back at the peak flowering stage. The species has well-exserted filaments (5–9 mm long) and a very long style (24–36 mm long), and the flowers are strongly almond-scented. The seeds are globose and have rugose primary sculpturing. Morphologically, the species appears most closely allied to *L. ensifolia*, which differs in its included or shortly exserted filaments (1–4 mm long), shorter style (11–20 mm long), usually much shorter perianth tube (12–20 mm long), spicy sweet-scented perianths and ovoid, glossy seeds.

DISTRIBUTION, HABITAT AND CONSERVATION STATUS. *L. pygmaea* has a disjunct distribution, occurring in Namaqualand and in the southern Cape (Map 128). In Namaqualand, it occurs on the Kamiesberg to the east and south-east of Kamieskroon, and south of Garies, and in the southern Cape, it is found in the De Hoop Nature Reserve and on the north-eastern slopes of the Potberg east of Bredasdorp, at Stilbaai and on the northern foothills of the Anysberg. The plants form large colonies, sometimes comprising thousands of individuals, in a variety of seasonally moist habitats. They traverse three vegetation types: Kamiesberg Granite Fynbos, North Swartberg Sandstone Fynbos and De Hoop Limestone Fynbos (Mucina & Rutherford, 2006). On the Kamiesberg, they grow in shallow gravel on granite rock sheets and flats, whereas on the Anysberg, they favour clay soils on north-facing slopes amongst karroid scrub. In De Hoop Nature Reserve, they are found in deep sand amongst

Figure 279 (left). Cultivated, mauve form of *Lachenalia pygmaea* from the Kamiesberg. Image: Graham Duncan.

Figure 280 (right). White form of *Lachenalia pygmaea* from limestone flats at De Hoop Nature Reserve. Image: Graham Duncan.

dense scrub as well as on calcareous outcrops, and at Stilbaai, on black loam flats overlaying limestone. Tepal colour is almost always bright mauve in the Namaqualand populations but white or light mauve in the southern Cape populations; both colour forms frequently occur within the same population in the latter region. *L. pygmaea* is not threatened.

NOTES. The heavily almond-scented flowers of *L. pygmaea* are pollinated by honey bees (*Apis mellifera*) and are also visited by the Painted Lady butterfly (*Vanessa cardui*). Before capsule dehiscence in spring, the scape elongates rapidly, forcing the infructescence upwards and then sidewards into a horizontal position. This results in some of the seeds being shed from the capsules a short distance away from the mother plant. The exceedingly narrow base of the scape eventually breaks, and the infructescence is blown away over the surface, liberating the remaining seeds.

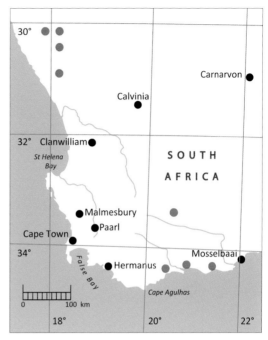

Map 128. Known distribution of *Lachenalia pygmaea*.

128. LACHENALIA LONGITUBA

Lachenalia longituba (A. M. van der Merwe) J. C. Manning & Goldblatt, *Edinburgh Journal of Botany* 60 (3): 565 (2004). *Polyxena longituba* A. M. van der Merwe, *South African Journal of Botany* 67: 44–46 (2001).
TYPE: South Africa, Northern Cape, 5 km north of Komsberg Pass summit, *J. C. Manning* 2165 (NBG!, holo.).

ETYMOLOGY. *longituba*: long perianth tubes.
DESCRIPTION. *Dwarf geophyte*, 20–45 mm high. *Bulb* globose, 10–15 mm in diameter, neck distinct, 5–10 mm long, offset-forming, white; tunic multilayered, outer layers dark brown, spongy; inner cataphyll translucent white, aerial portion with green longitudinal veins, loosely surrounding leaf bases, apex acute. *Leaves* 1–2, narrowly lanceolate, 40–90 × 5–13 mm, suberect to spreading, slightly to deeply canaliculate; leaf margins ciliolate; leaf bases clasping, 30–40 mm long, subterranean, white; primary seedling leaf terete, erect. *Inflorescence* corymbose, few-flowered, sterile apex short; scape erect, subterranean, 5–15 mm long, white; pedicels erect, 7–20 mm long, subterranean, white; lower bracts ovate, becoming lanceolate above, 0.5–1.0 × 0.5–2.0 mm, subterranean, white. *Perianth* actinomorphic, tubular, spicy sweet-scented, closing in cold weather; tube cylindrical, 15–27 mm long, widening gradually towards apex, lower two thirds subterranean, upper portion produced just above ground level, white; tepals lanceolate, light to deep mauve, keels light to dark purple on upper surface, dull blue on lower surface, slightly to strongly recurved; outer tepals 10–12 × 4–5 mm, apical gibbosity minute, white; inner tepals 9–12 × 3–4 mm. *Stamens* included, narrowly spreading; filaments white, 4–5 mm long; pollen yellow, ageing to cream. *Ovary* ellipsoid, 6–8 × 2 mm, light

green; style included, straight, 14–17 mm long, white. *Capsule* ovoid, 8 × 4 mm. *Seed* ovoid, 1.3–1.4 × 1.3–1.4 mm, glossy, black; strophiole rudimentary, 0.1 mm long, ridged. *Chromosome number* 2n = 28 (Hamatani *et al.*, 2007). Figures 32, 281.

FLOWERING PERIOD. April to July, with a peak in mid-April.

HISTORY. Before its formal publication, *L. longituba* had been grown in the United Kingdom for many years, mistakenly under the name *Polyxena odorata*. The provenance of these cultivated plants was the Komsberg, were plants were collected in the late 1960s by the former Kirstenbosch horticulturist, Harry Hall. *L. longituba* is cold tolerant and thrives in the United Kingdom, where it has become a common species in specialist bulb collections. *L. longituba* was first described as *Polyxena longituba* (van der Merwe & Marais, 2001) as part of a biosystematic study of the minor genera of the Hyacinthaceae. The taxon was rediscovered in the wild in 2000, when it became apparent that it represented an undescribed species.

DISTINGUISHING CHARACTERS AND AFFINITIES. *L. ensifolia* is the closest ally to *L. longituba* (Duncan *et al.*, 2005a), but differs in its usually shorter, aerial perianth tube (12–20 mm), shorter, narrower tepals (5–6 × 2–3 mm), and two broadly lanceolate leaves. Superficially, *L. longituba* resembles *L. corymbosa* in its similar recurved, light to deep mauve tepals, but *L. corymbosa* has a much shorter perianth tube (6 mm long) and 2–6 linear, slightly canaliculate leaves.

DISTRIBUTION, HABITAT AND CONSERVATION STATUS. *L. longituba* is restricted to a few populations on the Roggeveld Plateau south-east of Sutherland (Map 129). It occurs in Roggeveld Karoo vegetation (Mucina & Rutherford, 2006), in stony clay in seasonally wet, boggy sites that bake rock hard in summer, sometimes growing in association with *Romulea komsbergensis* (Iridaceae). The plants occur

Figure 281. *Lachenalia longituba* flowering on the Roggeveld Plateau near Sutherland in early autumn. Image: Graham Duncan.

in large colonies on flats or slightly sloping ground, in full sun. Occurring in an area where winter temperatures frequently drop to -9–15°C for short periods, the plants are fully hardy. *L. ensifolia* subsp. *maughanii* is sometimes found in slightly drier terrain adjacent to *L. longituba* populations. A conservation status of Rare is recommended for *L. longituba*.

NOTES. *L. longituba* is pollinated mainly by honey bees (*Apis mellifera*) and is also visited by the African monarch butterfly (*Danaus chrysippus aegyptius*). It has an almost entirely subterranean perianth tube, with this species and *L. pygmaea* having the longest perianth tubes in the genus, reaching up to 27 mm. Like those of *L. ensifolia* and several other geoflorous lachenalias, the subterranean scapes elongate considerably just before capsule dehiscence in mid-winter, resulting in the infructescence falling to the ground a short distance away from the plant, during which process most of the seeds are dispersed. The scape subsequently detaches and the remaining seeds are dispersed as it is moved around by wind. In cultivation, *L. longituba* sometimes produces a second flush of blooms in early summer, just before the dormant period (Duncan, 2003a).

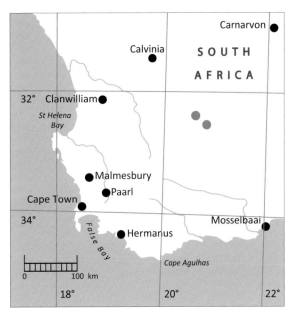

Map 129. Known distribution of *Lachenalia longituba*.

129. LACHENALIA ENSIFOLIA

Lachenalia ensifolia (Thunb.) J. C. Manning & Goldblatt, *Edinburgh Journal of Botany* 60 (3): 565 (2004). *Mauhlia ensifolia* Thunb., *Prodromus Plantarum Capensium* 1: t. 1 (1794); *Agapanthus ensifolius* (Thunb.) Willd., *Species Plantarum* 2: 48 (1799); *Massonia ensifolia* (Thunb.) Ker Gawl., *Curtis's Botanical Magazine* 16: t. 554 (1802); *Polyxena ensifolia* (Thunb.) Schönland, *Transactions of the Royal Society of South Africa* 1: 443 (1910). (See under subspecies below for further synonymy.)

TYPE: South Africa, Eastern Cape, 'inter Sondags et Visch River' [between Sundays and Fish Rivers], *C. P. Thunberg s.n.* (UPS!, holo., microfiche).

ETYMOLOGY. *ensifolia*: sword-shaped leaves.

DESCRIPTION. *Dwarf geophyte*, 10–50 mm high. *Bulb* subglobose, 15–25 mm in diameter, neck distinct, 6–10 mm long, offset-forming; tunic multilayered, outer layers dark brown, spongy; inner cataphyll translucent white, upper half with distinct longitudinal grooves, adhering to leaf bases, apex obtuse. *Leaves* 2, broadly lanceolate to ovate, 20–100 × 7–35 mm, spreading to suberect, glaucous to bright green, upper surface flat with depressed longitudinal grooves, lower surface plain or tinged with dark purple; margins ciliolate; leaf bases clasping, 10–40 mm long, aerial portion dark purple to glaucous or bright green, subterranean portion white; primary seedling leaf erect, terete. *Inflorescence* corymbose, few- to many-flowered, sterile apex short; scape erect, 5–10 mm

129. LACHENALIA ENSIFOLIA

Figure 282. *Lachenalia ensifolia* subsp. *ensifolia* on limestone flats near Bredasdorp. Image: Graham Duncan.

long at anthesis, subterranean, elongating in fruit, white; rachis white to light green; pedicels erect to suberect, 1–10 mm long, white; lower bracts ovate, becoming lanceolate above, 2–5 × 0.8–3.0 mm, translucent white. *Perianth* actinomorphic, tubular, white, rarely mauve, spicy sweet-scented; tube cylindrical, 12–20 mm long; outer tepals oblong or lanceolate, 5–6 × 2–3 mm, suberect to spreading, apices slightly recurved, apical gibbosities white or mauve; inner tepals oblong to lanceolate, 5–6 × 2 mm, apices slightly recurved. *Stamens* included to shortly exserted, straight or narrowly spreading; filaments 1–4 mm long, inner filament series included to shortly exserted beyond mouth of tube, white, rarely mauve; pollen yellow, ageing to black. *Ovary* ellipsoid, 3–5 × 1–2 mm, bright green; style included, straight, 11–20 mm long, white, rarely mauve. *Capsule* ovoid, 6–7 × 3–6 mm. *Seed* ovoid, 1.3–1.4 × 1.1–1.2 mm, glossy, black; strophiole rudimentary, 0.1 mm long, ridged. *Chromosome number* see subspecies below. Figures 282–284.

FLOWERING PERIOD. Late March to June, with a peak in April.

HISTORY. The taxon now known as *L. ensifolia* subsp. *ensifolia* has a complicated history. It was described by C. P. Thunberg (1794) in his *Prodromus Plantarum Capensium* as *Mauhlia ensifolia*, from specimens he collected at an unrecorded locality between the Sundays and Fish Rivers in the Eastern Cape. It was subsequently placed in *Agapanthus* (Willdenow, 1799), *Massonia* (Ker Gawler, 1802), *Polyxena* (Schönland, 1910) and *Lachenalia* (Manning *et al.*, 2004). Two described taxa have been found to be synonymous with it, *Massonia odorata* (Hooker, 1871), subsequently placed in *Polyxena* as *P. odorata* (Baker, 1897a) and *Massonia uniflora* (Baker, 1871), subsequently placed in *Polyxena* as *P. uniflora* (Durand & Schinz, 1895).

L. ensifolia subsp. *maughanii* was originally described as *Polyxena maughanii* by W. F. Barker (1931d), from material collected by Drs Herbert Maughan Brown and Louisa Bolus at Nieuwoudtville in 1930 and 1931, respectively. In her biosystematic study of the minor genera of the Hyacinthaceae (now part of the Asparagaceae), van der Merwe (2002) proposed that its status should be reassessed as *Polyxena ensifolia* var. *maughanii*. My view is that the taxon is most appropriately recognised at subspecific level.

DISTINGUISHING CHARACTERS AND AFFINITIES. *L. ensifolia* is allied to *L. longituba* (Duncan *et al.*, 2005a), the latter differing in its narrowly lanceolate, deeply canaliculate leaves, larger flowers with almost fully subterranean, usually much longer perianth tubes (up to 27 mm long) and larger tepals (outer 10–12 × 4–5 mm, inner 9–12 × 3–4 mm).

DISTRIBUTION AND HABITAT. *L. ensifolia* is widely distributed from central Namaqualand to Grahamstown, occurring in numerous vegetation types (Map 130). For further details see under subspecies below.

NOTES. The spicy sweet-scented flowers of *L. ensifolia* are pollinated primarily by honey bees (*Apis mellifera*) and are also visited by the Painted Lady butterfly (*Vanessa cardui*). Just before capsule dehiscence in spring, the subterranean scape elongates rapidly, pushing the infructescence upwards and then into a horizontal position, resulting in some of the seeds being dispersed a short distance from the mother plant. The scape detaches and is blown away, dispersing the remaining seeds.

Figure 283 (left). Cultivated specimens of *Lachenalia ensifolia* subsp. *ensifolia* from Van Rhyn's Pass. Image: Graham Duncan/SANBI.

Figure 284 (right). Cultivated specimens of *Lachenalia ensifolia* subsp. *maughanii* from Nieuwoudtville. Image: Graham Duncan/SANBI.

129. LACHENALIA ENSIFOLIA

Key to the subspecies

1a Filaments 3–4 mm long; inner (longer) stamen series exserted beyond mouth of tube; tube 15–20 mm long; tepals oblong; central Namaqualand to Eastern Cape . . . . .**a.** subsp. **ensifolia**
1b Filaments 1–2 mm long; inner (longer) stamen series included within tube; tube 12–15 mm long; tepals lanceolate; Bokkeveld Plateau to Roggeveld . . . .**b.** subsp. **maughanii**

a. subsp. ensifolia

SYNONYMY: *Massonia odorata* Hook.f., *Curtis's Botanical Magazine* 27: t. 5891 (1871); *Polyxena odorata* (Hook.f.) W. A. Nicholson, *Dictionary of Gardening* 3: 196 (1886), comb. inval.; *Polyxena odorata* (Hook.f.) Baker, in Thiselton-Dyer (ed.), *Flora Capensis* 6: 420 (1897). Type: South Africa, Northern Cape, Colesberg, *D. Arnot s.n.* (K!, holo.).
Massonia uniflora Sol. ex Baker, *Journal of the Linnean Society* (Botany) 11: 393 (1871); *Polyxena uniflora* (Sol. ex Baker) Benth. & Hook. ex Dur. & Schinz, *Conspectus Florae Africae* 5: 367 (1895). Type: South Africa, Cape, precise locality unknown, *F. Masson s.n.* (BM!, holo.).

DESCRIPTION. *Plant* 10–50 mm high. *Bulb* 15–22 mm in diameter. *Leaves* 20–100 × 7–30 mm; leaf bases 10–20 mm long. *Scape* 8–10 mm long; pedicels 1–10 mm long; bracts 2–4 × 0.8–1.0 mm. *Perianth tube* 15–20 mm long, white or mauve; outer tepals oblong, 5–6 × 2–3 mm, white or mauve; inner tepals oblong, 5–6 × 2 mm, white or mauve. *Filaments* straight or narrowly spreading, 3–4 mm long, inner (longer) stamen series exserted beyond mouth of tube. *Ovary* 3–4 × 1–2 mm; style 15–20 mm long. *Capsule* 6–7 × 5–6 mm. *Seed* 1.3–1.4 × 1.1–1.2 mm. *Chromosome number* $2n = 24, 26$ (Johnson & Brandham, 1997); $2n = 26$ (Hamatani *et al.*, 2007). Figures 282, 283.

FLOWERING PERIOD. March to June, with a peak in mid-May.

DISTRIBUTION, HABITAT AND CONSERVATION STATUS. *L. ensifolia* subsp. *ensifolia* has the widest distribution of the geoflorous *Lachenalia* taxa, extending from just south of Concordia in central Namaqualand, south-east to Nieuwoudtville on the Bokkeveld Plateau, to Calvinia and Sutherland in the western Karoo, west to the Cederberg, south-east to Bredasdorp, east to Grahamstown and inland to Victoria West, Richmond and Middelburg in the Great Karoo (Map 130). It occurs in 22 vegetation types: Namaqualand Blomveld, Namaqualand Klipkoppe Shrubland, Nieuwoudtville-, Montagu-, Matjiesfontein-, Central Rûens-, Uniondale-, Mossel Bay- and Humansdorp Shale Renosterveld, Eastern Upper-, Roggeveld-, Gamka-, Eastern Little-, Hantam- and Tanqua Karoo, Prince Albert Succulent Karoo, Cederberg Sandstone-, De Hoop Limestone- and Loerie Conglomerate Fynbos, Sundays Thicket, Albany Broken Veld and Bedford Dry Grassland (Mucina & Rutherford, 2006). In the northern and central parts of its range, *L. ensifolia* subsp. *ensifolia* is encountered mainly on granite (Namaqualand) and clay (Western Karoo) flats. In the southern parts, around Bredasdorp and eastwards to Grahamstown, it occurs on sandy flats and limestone ridges, and slightly inland in shale-derived loam of lower hill slopes. The plants usually occur among stones and low bushes in open aspects, in large colonies. *L. ensifolia* subsp. *ensifolia* is not threatened.

b. subsp. maughanii (W. F. Barker) G. D. Duncan, comb. et stat. nov. *Polyxena maughanii* W. F. Barker, *The Flowering Plants of South Africa* 11: t. 420 (1931); *Lachenalia maughanii* (W. F. Barker) J. C. Manning & Goldblatt, *Edinburgh Journal of Botany* 60 (3): 565 (2004).

TYPE: South Africa, Northern Cape, Nieuwoudtville, *L. Bolus s.n.* sub. BOL 19613 (BOL!, holo.; K!, iso.).

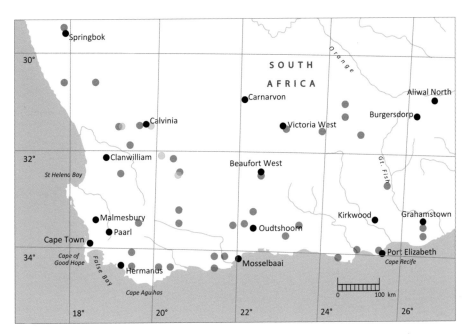

Map 130. Known distribution of *Lachenalia ensifolia*: ● = subsp. *ensifolia*; ◐ = subsp. *maughanii*.

ETYMOLOGY. *maughanii*: after Dr Herbert Maughan-Brown (1883–1940), South African physician and plant collector.

DESCRIPTION. *Plant* 15–30 mm high. *Bulb* 20–25 mm in diameter. *Leaves* 40–60 × 20–35 mm, bright green; leaf bases clasping, 30–40 mm long. *Scape* 5–8 mm long; pedicels 7–10 mm long; bracts 2–5 × 1–3 mm. *Perianth* white; tube 12–15 mm long; outer tepals lanceolate, 5–6 × 3 mm; inner tepals lanceolate, 6 × 2 mm. *Filaments* straight, 1–2 mm long, inner stamen series included within tube. *Ovary* 5 × 2 mm; style 11–12 mm long, slightly inflated. *Capsule* 6 × 3 mm. *Seed* 1.3–1.4 × 1.3–1.4 mm. *Chromosome number* unknown. Figure 284.

FLOWERING PERIOD. May to June.

DISTRIBUTION, HABITAT AND CONSERVATION STATUS. *L. ensifolia* subsp. *maughanii* is restricted to the Bokkeveld Plateau, Hantamsberg and Roggeveld, growing in colonies on shale flats and between crevices of rocky outcrops, in doleritic clay, in Nieuwoudtville-Roggeveld Dolerite Renosterveld vegetation (Map 130). The taxon is sympatric with subsp. *ensifolia* in parts of its range but always occurs as discrete populations. *L. ensifolia* subsp. *maughanii* is not threatened, but is known from only a few populations and a conservation status of Rare is recommended.

130. LACHENALIA CALCICOLA

Lachenalia calcicola (U. Müll.-Doblies & D. Müll.-Doblies) G. D. Duncan, comb. nov. *Polyxena calcicola* U. Müll.-Doblies & D. Müll.-Doblies, *Feddes Repertorium* 108: 85 (1997).

TYPE: South Africa, Western Cape, De Hoop Nature Reserve, calcareous rocks, *U. & D. Müller-Doblies* 84026a (PRE!, holo.; B, BOL!, BTU, G, GRA!, K, M, MO, NBG!, P, PRE!, S, UPS, Z, iso.).

ETYMOLOGY. *calcicola*: growing on limestone outcrops.

130. LACHENALIA CALCICOLA

DESCRIPTION. *Dwarf geophyte*, 15–30 mm high. *Bulb* globose, 10–15 mm in diameter, tunic multilayered, outer layers dark brown, spongy; inner cataphyll translucent white, adhering to leaf bases, apex obtuse. *Leaves* 2, lanceolate, 40–60 × 10–20 mm, spreading, light to dark green, slightly canaliculate, depressed longitudinal grooves prominent; margins coriaceous, crenulate, flat or rarely slightly to strongly undulate; leaf bases clasping, 10–20 mm long, subterranean, white; primary seedling leaf terete, erect. *Inflorescence* corymbose, few-flowered, sterile apex short; scape 3–5 mm long at anthesis, subterranean, elongating up to 75 mm long in fruit, light green; rachis light green; pedicels erect to suberect, 2–5 mm long, white; lower bracts ovate, becoming lanceolate above, minute, 1.0 × 0.5–1.0 mm, white. *Perianth* actinomorphic, tubular, uniformly white, unscented; tube cylindrical, 6–8 mm long; tepals lanceolate, narrowly spreading; outer tepals 3–5 × 2 mm; inner tepals 3–5 × 1 mm. *Stamens* included, straight; filaments white, 1–2 mm long. *Ovary* ellipsoid, 3–4 × 2–3 mm, light green; style included, straight, 3–5 mm long, white. *Capsule* broadly ellipsoid, 5–8 × 3–5 mm. *Seed* globose, 1.1–1.2 × 1.1 mm, glossy, black; strophiole rudimentary, 0.1 mm long, ridged. *Chromosome number* unknown. Figure 285.

FLOWERING PERIOD. May.

HISTORY. *L. calcicola* was collected initially by Dr Ute Müller-Doblies and Prof. Dietrich Müller-Doblies on the Arniston Flats in the southern Cape in August 1980, and slightly further east at Melkkamer Farm within the Denel missile testing range adjacent to the De Hoop Nature Reserve, in August 1984. The plants were collected in a fruiting state and cultivated in Germany. Flowering specimens from these cultivated plants were added to the fruiting, wild-collected type material of the Melkkamer collection, and the species was published in *Feddes Repertorium*; its authors stated its ploidy level to be tetraploid but did not provide its chromosome number (Müller-Doblies & Müller-Doblies, 1997). I re-collected the species in flower at Melkkamer Farm in May, 2010.

DISTINGUISHING CHARACTERS AND AFFINITIES. *L. calcicola* has not been phylogenetically analysed. Superficially, it resembles *L. ensifolia*, which has similar corymbose inflorescences but which differs in its longer perianth tube (12–20 mm long) and tepals (5–6 mm long), larger globose seed (1.3–1.4 × 1.1–1.2 mm) and broadly lanceolate to ovate leaves.

Figure 285. *Lachenalia calcicola* on limestone flats near De Hoop Nature Reserve. Image: Graham Duncan.

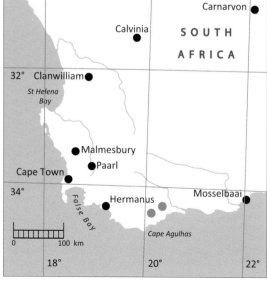

Map 131. Known distribution of *Lachenalia calcicola*.

DISTRIBUTION, HABITAT AND CONSERVATION STATUS. *L. calcicola* is recorded from the southern part of the Bredasdorp grid (3420) and is known from two localities, one from the Arniston Flats north-west of Waenhuiskrans, the other from Melkkamer Farm adjacent to De Hoop Nature Reserve (Map 131). It occurs in De Hoop Limestone Fynbos vegetation (Mucina & Rutherford, 2006) and is confined to calcareous outcrops on sandy flats. At Melkkamer Farm, it grows in association with a white form of *L. pygmaea*. *L. calcicola* is only known from two locations and a conservation status of Rare is recommended.

NOTES. The brush-like inflorescence is adapted to pollination by honey bees (*Apis mellifera*). In common with several other *Lachenalia* species that have geoflorous inflorescences, the scape of *L. calcicola* elongates markedly in late winter and early spring, causing the infructescence to be pushed into a vertical position. When the infructescence falls onto the ground, seed is dispersed a short distance away from the mother plant. The base of the scape then breaks and the infructescense is blown away, liberating the remaining seeds.

131. LACHENALIA PAUCIFOLIA

Lachenalia paucifolia (W. F. Barker) J. C. Manning & Goldblatt, *Edinburgh Journal of Botany* 60 (3): 565 (2004). *Hyacinthus paucifolius* W. F. Barker, *Journal of South African Botany* 7: 198–200 (1941); *Periboea paucifolia* (W. F. Barker) U. Müll.-Doblies & D. Müll.-Doblies, *Feddes Repertorium* 108: 84 (1997); *Polyxena paucifolia* (W. F. Barker) A. M. van der Merwe & J. C. Manning, *Cape Plants*: 714 (2000).

TYPE: South Africa, Western Cape, Vredenburg, *J. W. Mathews s.n.*, sub. NBG 689/30 (NBG!, holo.; PRE!, K!, iso.).

SYNONYMY. *Periboea oliveri* U. Müll.-Doblies & D. Müll.-Doblies, *Feddes Repertorium* 108: 84 (1997). Type: South Africa, Western Cape, open ground near hotel at Paternoster, *V. F. Anderson s.n.* sub *BTU* 6390 (PRE!, holo.; B, BTU, K, M, MO, NBG!, Z, iso.).

ETYMOLOGY. *paucifolia*: few-leafed.

DESCRIPTION. *Dwarf geophyte*, 30–60 mm high. *Bulb* globose, 10–15 mm in diameter, offset-forming; tunic multilayered, outer layers dark brown, spongy; inner cataphyll translucent white, adhering to leaf bases, apex obtuse. *Leaves* 1–2, narrowly lanceolate, 30–60 × 2–3 mm, spreading, deeply canaliculate, leathery, dark green, lower surface plain or tinged with brownish-maroon; leaf bases clasping, 5–10 mm long, subterranean, white; primary seedling leaf flat, prostrate. *Inflorescence* subcorymbose, few-flowered, sterile apex short; scape erect, 7–15 mm long, light green, mottled with dull brown; rachis slender, light green, mottled with brown; pedicels erect, 1–2 mm long, white; lower bracts ovate, becoming lanceolate above, 0.5–1.0 × 0.5–1.0 mm, white. *Perianth* actinomorphic, tubular, honey-scented, closing in cold weather; tube cylindrical, 6–7 mm long, translucent white; tepals widely spreading, slightly recurved, glossy; outer tepals oblong, 7–10 × 3 mm, bright lilac; keels and apical gibbosities deep lilac; inner tepals linear-oblong, 8–10 × 3 mm, bright lilac. *Stamens* included, straight, erect; filaments white, 1 mm long; pollen yellow. *Ovary* ellipsoid, 2.0 × 1.5 mm, dark green; style included, straight, 3 mm long, white. *Capsule* obovoid, 4 × 4 mm. *Seed* globose, 1.1–1.2 × 1.3 mm, matte, black; strophiole rudimentary, 0.1 mm long, ridged. *Chromosome number* $2n = 26$ (Hamatani *et al.*, 2007). Figure 286.

FLOWERING PERIOD. April to June, with a peak in May.

HISTORY. The first collection of *L. paucifolia* was made in 1930 by J. W. Mathews, first Curator of Kirstenbosch Botanical Garden, at Vredenburg on the Cape west coast. Originally described as *Hyacinthus paucifolius* by W. F. Barker (1941) in the *Journal of South African Botany*, the species was transferred to *Periboea* as *P. paucifolia* (Müller-Doblies & Müller-Doblies, 1997), then to *Polyxena* as *P.*

131. LACHENALIA PAUCIFOLIA

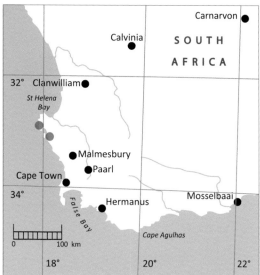

Map 132. Known distribution of *Lachenalia paucifolia*.

Figure 286. *Lachenalia paucifolia* in shallow, humus-rich soil on a granite outcrop near Langebaan. Image: Graham Duncan.

paucifolia (van der Merwe & Manning, 2000), and subsequently to *Lachenalia* as *L. paucifolia* (Manning et al., 2004). *Periboea oliveri* U. & D. Müll.-Doblies was published in 1997 with the type locality stated as 'Cape, open ground near to the hotel at Paternoster', but the plants at this locality do not significantly differ morphologically from *L. paucifolia*.

DISTINGUISHING CHARACTERS AND AFFINITIES. Phylogenetic evidence based on morphology (Duncan et al., 2005a) suggests that *L. paucifolia* is allied to *L. corymbosa*, which differs in usually having more numerous (up to four or more) linear leaves, and in having strongly recurved, mauve tepals with darker keels, shorter perianth tubes (3-6 mm long), longer filaments (4–7 mm long), and ovoid seeds with rugose primary sculpturing.

DISTRIBUTION, HABITAT AND CONSERVATION STATUS. *L. paucifolia* is restricted to the immediate surrounds of the coastal towns of Langebaan and Paternoster on the Cape west coast (Map 132). It occurs in Saldanha Granite Strandveld and Saldanha Limestone Strandveld vegetation (Mucina & Rutherford, 2006) and is locally common, forming colonies in humus-rich, shallow depressions of granite outcrops and on limestone flats at the base of granite hills, in full sun. On a granite outcrop at Langebaan, it grows in seasonally inundated rock pans, in association with the dwarf amaryllids *Strumaria chaplinii* and *S. tenella* (Amaryllidaceae), which flower simultaneously, and close to a population of *L. quadricolor*, which favours slightly drier habitat and flowers in July. *L. paucifolia* is threatened by housing and industrial development and has a conservation status of Endangered (Helme & Raimondo, 2009f).

NOTES. The widely campanulate, actinomorphic flowers of *L. paucifolia* are sweetly scented and reminiscent of honey or marzipan. They are pollinated by honey bees (*Apis mellifera*). Prior to capsule dehiscence, the scape elongates considerably, causing the infructescence to fall to the ground in a horizontal position and to liberate most of the seeds nearby. The scape eventually breaks off at its base and the whole infructescence is then carried away by wind, dispersing the remaining seeds.

132. LACHENALIA CORYMBOSA

Lachenalia corymbosa (L.) J. C. Manning & Goldblatt, *Edinburgh Journal of Botany* 60 (3): 565 (2004). *Hyacinthus corymbosus* L., *Mantissa plantarum Altera*: 223 (1771); *Massonia corymbosa* (L.) Ker Gawl., *Curtis's Botanical Magazine* 25: t. 991 (1807); *Scilla corymbosa* (L.) Ker Gawl., *Curtis's Botanical Magazine* 36: sub. t. 1468 (1812); *Periboea corymbosa* (L.) Kunth, *Enumeratio Plantarum* 4: 293 (1843); *Baeoterpe corymbosa* (L.) Salisb., *The Genera of Plants*: 18 (1866), nom. inval. *Polyxena corymbosa* (L.) Jessop, *Journal of South African Botany* 42: 429 (1976).

TYPE: South Africa, Cape, collector and precise locality unknown, *Herb. Linn.* 438.9 (LINN!, neotype designated by Stearn (1990)).

SYNONYMY: *Scilla brevifolia* Ker Gawl., *Curtis's Botanical Magazine* 36: t. 1468 (1812), non *Hyacinthus brevifolius* Thunb., *Prodromus plantarum Capensium* 63 (1794) [= *Dipcadi brevifolium* (Thunb.) Fourc.]. *Scilla brachyphylla* Roem. & Schultes, *Systema Vegetabilium* 7 (1): 573 (1829), nom. nov. for *Scilla brevifolia* Ker Gawl. *Periboea gawleri* Kunth, *Enumeratio plantarum* 4: 293 (1843), nom. nov. for *Scilla brevifolia* Ker Gawl.; *Hyacinthus gawleri* (Kunth) Baker, in Thiselton-Dyer (ed.), *Flora Capensis* 6: 472 (1897). Type: South Africa, Cape, collector and precise locality unknown, figure in *Curtis's Botanical Magazine* 36: t. 1468 (1812) (lectotype, designated here).

ETYMOLOGY. *corymbosa*: descriptive of the corymbose racemes in certain forms of this species.

DESCRIPTION. *Dwarf geophyte*, 30–100 mm high. *Bulb* globose, 15–20 mm in diameter, neck distinct, 10–15 mm long, offset-forming; tunic multilayered, outer layers dark brown, spongy; inner cataphyll translucent white, adhering to leaf bases, subterranean, apex obtuse. *Leaves* 2–3(–6), linear, opposite or rosulate, 1.5–2.0 × 2–4 mm long at anthesis, elongating to 12–20 mm long in fruit, suberect to spreading, slightly canaliculate, dark green, leathery; upper surface unmarked, lower surface tinged with dull purple; leaf bases loosely surrounding scape, 10–20 mm long, subterranean, white; primary seedling leaf flat, prostrate. *Inflorescence* a corymbose, subcorymbose or ordinary raceme, few- to many-flowered, sterile apex short; scape erect, 10–55 mm long, lengthening in fruit, light green or brownish-mauve to magenta; rachis light green, brownish-mauve or magenta, narrowing markedly toward apex; pedicels erect to suberect, 3–10 mm long, light green, light to dark mauve or magenta; bracts ovate, 1.0 × 1–2 mm, white. *Perianth* actinomorphic, tubular, erect to suberect, lightly to strongly honey-scented, closing in cold weather; tube cylindrical, 3–6 mm long, white; tepals oblong, 7–9 × 3–4 mm, widely spreading, apices strongly recurved, translucent white in lower half, shading to mauve in upper half, keels dark mauve to purple, outer tepal apical gibbosities dark mauve to purple. *Stamens* included, straight; filaments white, 4–7 mm long. *Ovary* obovoid, 2–3 × 2 mm, light green; style included, straight, 8–9 mm long, white. *Capsule* obovoid, 4–5 × 4–6 mm. *Seed* ovoid, 1.1–1.2 × 0.8–0.9 mm, matte, black, primary sculpturing rugose; strophiole rudimentary, 0.2 mm long, ridged. *Chromosome number* unknown. Figures 287, 288.

FLOWERING PERIOD. Early March to June, with a peak in April.

HISTORY. Carl Linnaeus (1771) originally described this taxon as *Hyacinthus corymbosus* from specimens gathered by his pupil, the Danish physician and naturalist J. G. Koenig (1728–1785). The collection was made on Lion's Head above Cape Town in April 1768, but its whereabouts has not been unequivocally established despite exhaustive searches by Jessop (1976) and myself. The main collection of Koenig specimens is housed in the University of Copenhagen Herbarium (C) and contains three specimens of *L. corymbosa* made by him at the Cape, but with only 'Cap. B. Spei' written onto each sheet by the previous owners: one belonged to Vahl, another to Schumacher and the third to Rottbøll; those of Vahl and Schumacher both have racemose inflorescences but the Rottbøll

132. LACHENALIA CORYMBOSA

Figure 287. Cultivated specimens of a corymbose form of *Lachenalia corymbosa* from Green Point, northern Cape Peninsula. Image: Graham Duncan.

specimen is corymbose and might possibly have been the one used by Linnaeus. According to Savage (1945), writing in the *Catalogue of the Linnean Herbarium*, the specimen in this herbarium, sometimes regarded as the type, was not annotated by Linnaeus nor identified as originating from Koenig, and cannot therefore be accepted as the holotype (Jessop 1976); this specimen has subsequently been designated as the neotype in *Annales Musei Goulandris* (Stearn, 1990).

Since its original publication, the species has been variously placed in *Massonia* (Ker Gawler, 1807b), *Scilla* (Ker Gawler, 1812b), *Periboea* (Kunth, 1843), *Polyxena* (Jessop, 1976) and *Lachenalia* (Manning *et al.*, 2004). Salisbury (1866) transferred it to *Baeoterpe* as *B. corymbosa* (L.) Salisb. but this name is invalid as no type was designated. The history of *L. corymbosa* in cultivation dates back to at least 1807 when it was illustrated on t. 991 (as *Massonia corymbosa*) of *Curtis's Botanical Magazine*, from a plant cultivated by the nurseryman Richard Williams of Turnham Green, London (Curtis, 1807). A fairly recent plate published in *Kew Magazine* (as *Polyxena corymbosa*) was painted from specimens introduced to Kew by W. Marais, which had originally been obtained from Kirstenbosch (Mathew, 1988).

DISTINGUISHING CHARACTERS AND AFFINITIES. *L. corymbosa* is variable in inflorescence type and leaf length. Plants occurring in the northern Cape Peninsula usually have corymbose racemes, although plants with ordinary racemes (such as *Pappe s.n.*, sub. SAM 23242) have rarely been recorded. Beyond the Peninsula, inflorescence types include ordinary, subcorymbose and corymbose racemes, sometimes within the same population, such as that in the Harmony Flats Geometric Tortoise Reserve at Somerset West (*Boucher* 5189, in NBG). In her biosystematic study of the minor genera of the Hyacinthaceae, van der Merwe (2002) proposed that plants at Greenpoint, Bantry Bay and Camps Bay in the northern Cape Peninsula conform to the species originally described as *Hyacinthus corymbosus* (now *L. corymbosa*), having corymbose racemes, whereas those with ordinary racemes and a slightly different DNA profile, originally described as *Scilla brevifolia* Ker Gawl. (1812b), should be published under the new combination *Polyxena brevifolia* (Ker Gawl.) A. M. van der Merwe. However, plants with corymbose racemes are not restricted to the Cape Peninsula, making this taxonomic suggestion unworkable.

The mature leaves of plants outside the Peninsula, such as those at Somerset West on the Cape Flats, are somewhat longer (up to 20 mm compared to 12 mm), but their shape is consistently linear with a slightly canaliculate upper surface. Plants from Somerset West have ordinary racemes, slightly longer perianth tubes and consistently flower up to a month earlier than those on the Peninsula, but the floral morphology of plants throughout the distribution range of this species is almost identical. The slight difference in molecular sequence data between the Peninsula plants and those further afield is insufficient justification for recognition of a separate taxon, considering that only one gene (the *trnL-F* chloroplast region) was studied (van der Merwe, 2002).

L. corymbosa is allied to *L. paucifolia* (Duncan *et al.*, 2005a), the latter differing in its longer perianth tube (6–7 mm long), slightly recurved, uniformly bright lilac tepals, and much shorter filaments (1 mm long). *L. paucifolia* differs in its solitary or paired, narrowly lanceolate, deeply canaliculate leaves, and it has globose seeds with a slightly shorter (0.1 mm long) strophiole.

DISTRIBUTION, HABITAT AND CONSERVATION STATUS. *L. corymbosa* extends from the Atlantic seaboard in the northern Cape Peninsula east to Stellenbosch and north to Clanwilliam in the Olifants River Valley (Map 133). The species occurs in Swartland Granite Renosterveld and Swartland Shale- and Peninsula Shale Renosterveld vegetation (Mucina & Rutherford, 2006). Judging from the large number of early collections, it must have been a very common species in the vicinity of Cape Town; its former habitat there is now almost entirely transformed into industrial, recreational and housing sites.

Figure 288. A form of *Lachenalia corymbosa* with ordinary racemes at Kraaifontein. Image: Graham Duncan.

Similarly, *L. corymbosa* used to be frequent on the Cape Flats around Strand, Stellenbosch and Gordon's Bay, but is rarely encountered there now mainly because of the large-scale destruction of its fertile habitat for agriculture and vast housing developments. The species remains plentiful in the Elandsberg Private Nature Reserve near Hermon. The plants occur in colonies in seasonally waterlogged, stony clay or sandy loam soils, usually on open flats or occasionally between rock cracks on granite hillsides. *L. corymbosa* has a conservation status of Vulnerable (Raimondo & Turner, 2009).

NOTES. The widely campanulate, honey-scented flowers of *L. corymbosa* are pollinated mainly by honey-bees (*Apis mellifera*) and are also visited by the beewolf (*Philanthus triangulum*), the drone-fly (*Eristalis tenax*) (Figure 68) and the Painted Lady butterfly (*Vanessa cardui*) (Figure 70). A form of this species with ordinary racemes from Somerset West on the Cape Flats is the second-earliest member of the genus to flower after *L. pearsonii*. The short scape elongates just prior to capsule dehiscence in July, resulting in the scape falling into a horizontal position and liberating the ripe seeds a short distance from the mother plant. The infructescence detaches and is carried away by wind, dispersing the remaining seeds. The seeds ripen in midwinter at a time of plentiful rainfall and germinate rapidly during the current season.

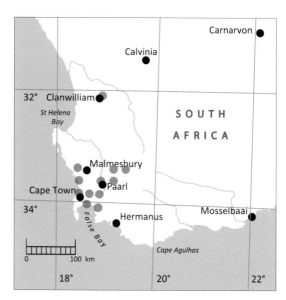

Map 133. Known distribution of *Lachenalia corymbosa*.

133. LACHENALIA ARGILLICOLA

Lachenalia argillicola G. D. Duncan, sp. nov.

A L. corymbosa, *bulbo ovoideo, foliis glaucis lanceolatis, ovario ellipsoidea et seminibus grandioribus politis differt.*

TYPE: South Africa, Western Cape, 3117 (Lepelfontein): Toringberg south of Kotzesrus, in seasonally inundated muddy clay pans (–BB), *N. A. Helme* 6082 (NBG!, holo.).

ETYMOLOGY. *argillicola*: growing in muddy clay.

DESCRIPTION. *Dwarf geophyte*, 30–70 mm high. *Bulb* ovoid, 15–20 mm in diameter, neck distinct, 5–10 mm long, offset-forming, deep-seated; tunic multilayered, outer layers dark brown, spongy; inner cataphyll translucent white, apex obtuse. *Leaves* 2, narrowly lanceolate, 25–85 × 5–13 mm, spreading to suberect, glaucous, moderately to strongly canaliculate, upper surface with depressed longitudinal grooves; leaf bases clasping, 20–60 mm long, subterranean, white; leaf margins smooth to ciliolate; primary seedling leaf terete, erect. *Inflorescence* subcorymbose to weakly racemose, few- to many-flowered, dense, sterile apex short; scape erect, 5–10 mm long at anthesis, elongating to 40 mm long in fruit, green to purplish-brown; rachis green; pedicels suberect to erect, 1–3 mm long, green to purplish-brown; lower bracts ovate, becoming lanceolate above, 1–3 × 1–2 mm, translucent white. *Perianth* actinomorphic, tubular, erect to suberect, lightly honey-scented; tube cylindrical, 6 mm long, white to light mauve; tepals lanceolate, widely spreading, apices recurved, white, outer tepals 7 × 3–4 mm, keels white to light lilac, broadening above, apical gibbosities dark mauve to purple; inner tepals 7 × 3–4 mm, keels white to light lilac. *Stamens* included, straight; filaments white to lilac, 3–4 mm long; pollen yellow at anthesis, ageing to black. *Ovary* ellipsoid, 3–4 × 2 mm; style included, straight, 8 mm long, white. *Capsule* ellipsoid, 8–9 × 5–6 mm. *Seed* ovoid, 3 × 2 mm, glossy, black; strophiole rudimentary, 0.1 mm long, ridged. *Chromosome number* unknown. Figure 289.

FLOWERING PERIOD. June.

HISTORY. *L. argillicola* was discovered in late June 2009 by Nick Helme, who was surveying the flora of the Toringberg south of Kotzesrus in the western Knersvlakte, in the extreme north-western corner of the Western Cape. Material collected from two populations in late flowering stage was pressed for incorporation into the Compton Herbarium. An additional collection of living specimens in fruit was made several months later, making it possible to study the species' capsule and seed morphology.

DISTINGUISHING CHARACTERS AND AFFINITIES. *L. argillicola* appears most closely allied to *L. corymbosa*, with which it shares similar tubular perianths with widely spreading tepals that have recurved apices. *L. corymbosa* differs mainly in its globose bulbs, linear leaves, longer filaments (4–7 mm long), obovoid ovaries and capsules, much smaller seeds (1.1–1.2 × 0.8–0.9 mm) that have a matte testa and rugose primary sculpturing, and slightly longer strophioles (0.2 mm long). Superficially, *L. argillicola* resembles *L. ensifolia*, but the latter differs in its subglobose bulbs, broadly lanceolate to ovate, non-canaliculate leaves, corymbose inflorescences, much longer perianth tubes (12–20 mm long) and much smaller seeds (1.3–1.4 × 1.1–1.2 mm). These two species perform differently under uniform conditions in cultivation: the leaves of *L. argillicola* become erect and conduplicate, and its scape elongates strongly at anthesis, whereas *L. ensifolia* more or less retains its wild growth habit.

Figure 289. *Lachenalia argillicola* in a seasonally inundated clay pan in the western Knersvlakte. Image: Nick Helme.

133. LACHENALIA ARGILLICOLA

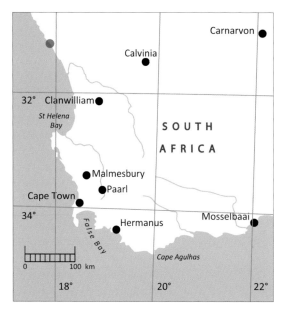

Map 134. Known distribution of *Lachenalia argillicola*.

DISTRIBUTION, HABITAT AND CONSERVATION STATUS. *L. argillicola* is known from only three populations on the Toringberg south of Kotzesrus in the western Knersvlakte, in the extreme north-western corner of the Western Cape (Map 134). It occurs in Namaqualand Heuweltjieveld vegetation (Mucina & Rutherford, 2006) in colonies in alkaline, seasonally inundated, granitic clay pans. *L. argillicola* is not under threat but in view of its highly restricted distribution, a conservation status of Rare is recommended.

NOTES. The tubular, lightly honey-scented perianths with widely spreading to recurved tepals indicate that honey bees (*Apis mellifera*) are probably the major pollinator of *L. argillicola*. The scape elongates considerably in early spring, causing it to fall into a horizontal position and release most of the seeds close to the mother plant. The infructescence eventually breaks off at its base and the remaining seeds are dispersed as it is blown about.

INSUFFICIENTLY KNOWN NAMES

Lachenalia cooperi Baker, *Journal of the Linnean Society* (Botany) 11: 409 (1871). Type not found, described from a cultivated plant in the garden of Mr T. Cooper, a possible hybrid.

Lachenalia glaucina Jacq. var. *pallida* Lindl. The combination appears to have first been used by W. F. Barker (1930b) for the taxon now known as *L. orchioides* (L.) Aiton subsp. *orchioides*, but no formal publication of the combination has been traced.

Lachenalia hyacinthoides (L.f.) Lam., *Encyclopédie Méthodique* 3: 373 (1792); *Phormium hyacinthoides* L.f., *Supplementum Plantarum* ed. 2: 204 (1781). Linnaeus fil. cited *Hyacinthus orchioides* L., which does not agree with the diagnosis given for *P. hyacinthoides*; no material or figure was cited.

Lachenalia patens Hort. ex Andrews, *The Botanist's Repository* 5: sub. t. 302 (1803). The name was mentioned in the text under *L. fragrans* Andrews (Andrews, 1803a). No description or figure was referred to and no material has been traced.

Lachenalia pulchella Kunth, *Enumeratio Plantarum* 4: 284 (1843). *Orchiastrum pulchellum* (Kunth) Lem., *L'Illustration Horticole* 2: 100 (1855). No figure was cited and the diagnosis does not clearly match any known species. No material has been traced.

Lachenalia pusilla Eckl. ex Kunth, *Enumeratio Plantarum* 4: 283 (1843). The name was cited under *Coelanthus complicatus* Willd. ex Roem. & Schult. (=*L. reflexa* Thunb.) without any reference, and no material has been traced. Nomenclaturally, the name is invalid, antedated by *Lachenalia pusilla* Jacq. (1797).

Lachenalia pustulata P. J. Bergius ex Kunth, *Enumeratio Plantarum* 4: 284 (1843). The name was cited under *Lachenalia pulchella* Kunth, and alluded to a specimen in B, but no material has been traced. Nomenclaturally, the name is invalid, antedated by *Lachenalia pustulata* Jacq. (1790).

Lachenalia tenella Schneev. ex Steud. The name was published by Steudel (1821–1824) in volume 1: 459 of his *Nomenclator Botanicus*, without any description or figure, and no material has been traced.

Lachenalia unicolor Eckl. ex Kunth, *Enumeratio Plantarum* 4: 284 (1843). The name was cited under *L. pulchella* Kunth, and alluded to a specimen in the Herbarium Luca Ghini at Pisa, but no material has been traced. Nomenclaturally, the name is invalid, antedated by *Lachenalia unicolor* Jacq. (1797).

EXCLUDED TAXA

Lachenalia graminifolia Sol. ex Baker = *Dipcadi hyacinthoides* (Spreng.) Baker

Lachenalia lanceaefolia Jacq. = *Ledebouria revoluta* (L.f.) Jessop

Lachenalia lanceaefolia Sims var. *maculata* Tratt. = *Ledebouria revoluta* (L.f.) Jessop

Lachenalia maculata Tratt. = *Ledebouria revoluta* (L.f.) Jessop

Lachenalia ramosa Lam. = *Phormium tenax* L.f.

Lachenalia serotina (L.) Willd. = *Dipcadi serotinum* (L.) Medik.

Lachenalia speciosa F. Dietr. = *Dipcadi glaucum* (Burch. ex Ker Gawl.) Baker

Lachenalia stolonifera Hort. = *Scilla hispanica* Mill.

Lachenalia viridis (L.) Ait. = *Dipcadi viride* (L.) Moench

Polyxena angustifolia (L.f.) Baker = *Massonia echinata* L.f.

Polyxena bakeri T. Durand & Schinz = *Massonia echinata* L.f.

Polyxena burchellii (Baker) Baker = *Daubenya zeyheri* (Kunth) J. C. Manning & A. M. van der Merwe

Polyxena comata (Burch. ex Baker) Baker = *Daubenya comata* (Burch. ex Baker) J. C. Manning & A. M. van der Merwe

Polyxena haemanthoides Baker = *Daubenya marginata* (Willd. ex Kunth) J. C. Manning & A. M. van der Merwe

Polyxena marginata (Willd. ex Kunth) Baker = *Daubenya marginata* (Willd. ex Kunth) J. C. Manning & A. M. van der Merwe

Polyxena rugulosa (Lichtenst. ex Kunth) Baker = *Daubenya marginata* (Willd. ex Kunth) J. C. Manning & A. M. van der Merwe

REFERENCES

Aiton, W. (1789). *Lachenalia. Hortus Kewensis* 1: 460–463. George Nicol, London.
Aiton, W. T. (1811). *Lachenalia. Hortus Kewensis* 2 (2nd edn): 284–289. Longman, Hurst, Rers, Orme & Brown, London.
Andrews, H. C. (1797). *Massonia violacea. The Botanist's Repository* 1: t. 46. T. Bensley, London.
Andrews, H. C. (1802). *Lachenalia purpureo-caerulea. The Botanist's Repository* 4: t. 251. T. Bensley, London.
Andrews, H. C. (1803a). *Lachenalia fragrans. The Botanist's Repository* 5: t. 302. T. Bensley, London.
Andrews, H. C. (1803b). *Lachenalia rosea. The Botanist's Repository* 5: t. 296. T. Bensley, London.
Andrews, H. C. (1807a). *Lachenalia flava. The Botanist's Repository* 7: t. 456. T. Bensley, London.
Andrews, H. C. (1807b). *Lachenalia sessiliflora. The Botanist's Repository* 7: t. 460. T. Bensley, London.
Anonymous (1872). Lachenalias. *The Gardener's Chronicle and Agricultural Gazette*: 290–291.
APG III (2009). An update of the Angiosperm Phylogeny Group classification for the orders and families of flowering plants. *Botanical Journal of the Linnean Society* 161: 105–121.
Ascherson, P. & Graebner, P. (1905). *Lachenalia. Synopsis der Mitteleuropäischen Flora*: 279–280. Wilhelm Engelmann, Leipzig.
Axelrod, D. I. (1972). Edaphic aridity as a means of Angiosperm evolution. *The American Naturalist* 106: 311–320.
Axelrod, D. I. & Raven, P. H. (1978). Late Cretaceous and Tertiary vegetation history of Africa. In: M. J. A. Werger (ed.), *Biogeography and ecology of southern Africa*. W. Junk Publishers, The Hague.
Baker, J. G. (1871). A revision of the genera and species of herbaceous capsular gamophyllous Liliaceae. 17. Lachenalia. *Journal of the Linnean Society of Botany* 11: 349–410.
Baker, J. G. (1874a). *Lachenalia campanulata. Journal of Botany* 12: 6.
Baker, J. G. (1874b). *Lachenalia trichophylla. Journal of Botany* 12: 368.
Baker, J. G. (1878). *Lachenalia wrightii. Journal of Botany* (new series) 7: 322.
Baker, J. G. (1884). New lachenalias. *The Gardener's Chronicle* 21 (Series 2): 668.
Baker, J. G. (1886). New Cape Liliaceae. *Journal of Botany* (London) 24: 335–336.
Baker, J. G. (1892a). *Lachenalia polyphylla. Botanische Jahrbücher* 15 (3): 7.
Baker, J. G. (1892b). *Lachenalia bachmanni. Botanische Jahrbücher* 15 (3): 8.
Baker, J. G. (1897a). *Lachenalia* Jacq. *Flora Capensis*, ed. 6: 421–436. Reeve & Co., London.
Baker, J. G. (1897b). *Polyxena* Kunth. *Flora Capensis* 6: 418–421.
Baker, J. G. (1901a). *Lachenalia brevipes. Bulletin L'Herbier Boissier* 1 (Ser. 2): 856.
Baker, J. G. (1901b). *Lachenalia petiolata. Bulletin L'Herbier Boissier* 1 (Ser. 2): 856.
Baker, J. G. (1904). *Lachenalia schlechteri. Bulletin L'Herbier Boissier* 4 (Ser. 2): 999.
Barker, W. F. (1930a). Plants — new and noteworthy. *South African Gardening and Country Life* 20: 14.
Barker, W. F. (1930b). *Lachenalia. Journal of the Botanical Society of South Africa* 16: 10–13.
Barker, W. F. (1931a). *Lachenalia mathewsii. The Flowering Plants of South Africa* 11: t. 422.
Barker, W. F. (1931b). *Lachenalia ovatifolia. The Flowering Plants of South Africa* 11: t. 401.

Barker, W. F. (1931c). *Lachenalia undulata. The Flowering Plants of South Africa* 11: t. 431.
Barker, W. F. (1931d). *Polyxena maughanii. The Flowering Plants of South Africa* 11: t. 420.
Barker, W. F. (1933a). *Lachenalia comptonii. The Flowering Plants of South Africa* 13: t. 507.
Barker, W. F. (1933b). *Lachenalia elegans. The Flowering Plants of South Africa* 13: t. 508.
Barker, W. F. (1933c). *Lachenalia gillettii. The Flowering Plants of South Africa* 13: t. 506.
Barker, W. F. (1933d). *Lachenalia salteri. The Flowering Plants of South Africa* 13: t. 505.
Barker, W. F. (1941). *Hyacinthus paucifolius* Barker. *Journal of South African Botany* 7: 198–200.
Barker, W. F. (1949). *Plantae Novae Africanae. Journal of South African Botany* 15: 37–39.
Barker, W. F. (1950). *Lachenalia*. In: R. S. Adamson & T. M. Salter (eds), *Flora of the Cape Peninsula*, pp. 198–202. Juta & Co. Ltd., Cape Town.
Barker, W. F. (1966). The rediscovery of two South African plants and the renaming of another. *Botaniska Notiser* 119 (2): 201–208.
Barker, W. F. (1969). A new combination in *Lachenalia* with notes on the species. *Journal of South African Botany* 35 (5): 321–322.
Barker, W. F. (1972). A new species of *Lachenalia* from the south-western Cape. *Journal of South African Botany* 38 (3): 179–183.
Barker, W. F. (1978). Ten new species of *Lachenalia* (Liliaceae). *Journal of South African Botany* 44 (4): 391–418.
Barker, W. F. (1979a). Ten more new species of *Lachenalia* (Liliaceae). *Journal of South African Botany* 45 (2): 193–219.
Barker, W. F. (1979b). *Lachenalia viridiflora. The Flowering Plants of Africa* 45 t. 1794.
Barker, W. F. (1980a). *Lachenalia trichophylla. The Flowering Plants of Africa* 46: t. 1808.
Barker, W. F. (1980b). Victor Stanley Peers, conservationist and cultivator of South African plants. *Veld & Flora* 66 (1): 25–27.
Barker, W. F. (1983a). A List of the *Lachenalia* species included in Rudolf Schlechter's collections made in 1891–1898 on his collecting trips in southern Africa, with identifications added. *Journal of South African Botany* 49 (1): 45–55.
Barker, W. F. (1983b). Six more new species of *Lachenalia* (Liliaceae). *Journal of South African Botany* 49 (4): 423–444.
Barker, W. F. (1984). Three more new species of *Lachenalia* and one new variety of an early species (Liliaceae). *Journal of South African Botany* 50 (4): 535–547.
Barker, W. F. (1987). Five more new species of *Lachenalia* (Liliaceae-Hyacinthoideae); four from the Cape Province and one from southern South West Africa/Namibia. *South African Journal of Botany* 53 (2): 166–172.
Barker, W. F. (1988). Historical background. In: G. D. Duncan (ed.), *The Lachenalia Handbook. Annals of Kirstenbosch Botanic Gardens* 17. National Botanical Gardens, Cape Town.
Barker, W. F. (1989). New taxa and nomenclatural changes in *Lachenalia* (Liliaceae-Hyacinthoideae) from the Cape Province. *South African Journal of Botany* 55 (6): 630–646.
Barnard, K. H. (1947). A description of the Codex Witsenii in the South African Museum. *Journal of South African Botany* 13: 1–51.
Barnhart, J. H. (1965). *Biographical Notes upon Botanists* 2: 332. G. K. Hall & Co., Boston.
Bentham, G. & Hooker, J. D. (1883). *Lachenalia pusilla. Genera Plantarum* 3. Reeve & Co., London.
Boyd, L. (1932). Monocotyledonous seedlings. Morphological studies in the post-seminal development of the embryo. *Lachenalia. Transactions of the Proceedings of the Botanical Society of Edinburgh* 31: 47.

Bremer, K. (1988). The limits of amino acid sequence data in angiosperm phylogenetic reconstruction. *Evolution* 42: 795–803.
Bremer, K. (1994). Branch support and tree stability. *Cladistics* 10: 295–304.
Breyne, J. (1689). *Prodromus Fasciculi Rariorum Plantarum Secundus*. Danzig.
Breyne, J. & Breyne, J. P. (1739). *Prodromi Fasciculi Rariorum Plantarum*. Danzig.
Buc'hoz, P-J. (1773–1778). *Histoire Universelle Règne Végétal* 2: t. 1. Brunet, Paris.
Burmann, N. L. (1768). *Prodromus Florae Capensis*. Johannes Schreuder, Amsterdam.
Buxbaum, J. C. (1729). *Plantarum minus cognitarum Centuria* III. Petropoly.
Cavanilles, A. J. (1802). *Hyacinthus bifolius*. Anales de Ciencias Naturales 5 (14): t. 41, fig. 1.
Chase, M. W., Reveal, J. L., & Fay, M. F. (2009). A subfamilial classification for the expanded asparagalean families Amaryllidaceae, Asparagaceae and Xanthorrhoeaceae. *Botanical Journal of the Linnean Society* 161 (2): 132–136.
Cirillo, D. (1788). *Plantarum Rariorum Regni Neapolitani* 1: 35, t. 12. Neapoli.
Cook, H. H. (1931). Propagation of lachenalias by leaf cuttings. *Journal of the Botanical Society of South Africa* 17: 12.
Cowling, R. M., Gibbs Russell, G. E., Hoffmann, M. T. & Hilton-Taylor, C. (1989). Patterns of plant species diversity in southern Africa. In: B. J. Huntley (ed.), *Biotic Diversity in Southern Africa: Concepts and Conservation*. Oxford University Press, Cape Town.
Cowling, R. M. & Pierce, S. M. (1999). *Namaqualand: a Succulent Desert*. Fernwood Press, Vlaeberg.
Crosby, T. S. (1978). Hybridisation in the genus *Lachenalia*. *Veld & Flora* 64 (3): 87–90.
Crosby, T. S. (1986). The genus *Lachenalia*. *The Plantsman* 8 (3): 128–166.
Curtis, W. (1789). *Lachenalia tricolor*. *Curtis's Botanical Magazine* 3: t. 82.
Curtis, W. (1807). *Massonia corymbosa*. *Curtis's Botanical Magazine* 25: t. 991.
David, J. (2009). *Lachenalia* in cultivation — history and hybridisation. *Amaryllids* 2009 (1): 11–16.
Dehnhardt, F. (1839). *Lachenalia pyramidalis*. Rivista Napolitana 1 (1): 162. Agnello, Naples.
De Wet, J. M. J. (1957). Chromosome numbers in the *Scilleae*. *Cytologia* 22: 145–159.
Diels, L. (1909). *Lachenalia dasybotrya* and *L. splendida*. Botanische Jahrbücher für Systematik Pflanzengeschichte und Pflanzengeographie 44: 116.
Dietrich, F. G. (1818). *Lachenalia reclinata*. Nachtrag der Lexicon der Gartnerei und Botanik: 291–293. Bierter Band, Berlin.
Dinter, M. K. (1920). *Lachenalia klinghardtiana*. Feddes Repertorium 16: 341–342.
Dinter, M. K. (1932). *Lachenalia buchubergensis*. Feddes Repertorium 30: 84.
Dold, A. P. & Phillipson, P. B. (1998). A revision of *Lachenalia* in the Eastern Cape, South Africa. *Bothalia* 28 (2): 141–149.
Duncan, G. D. (1978). Grow more lachenalias! *Bulletin of the Indigenous Bulb Growers Association of South Africa* 28: 2–4.
Duncan, G. D. (1986). The re-discovery of *Lachenalia mathewsii* W. Barker. *Veld & Flora* 72 (2): 40–41.
Duncan, G. D. (1987). *Lachenalia macgregoriorum*. *The Flowering Plants of Africa* 49: t. 1951.
Duncan, G. D. (1988a). *The Lachenalia Handbook*. Annals of Kirstenbosch Botanic Gardens 17. National Botanical Institute, Cape Town.
Duncan, G. D. (1988b). *Lachenalia arbuthnotiae*. *The Flowering Plants of Africa* 50: t. 1961.
Duncan, G. D. (1992a). The genus *Lachenalia*: its distribution, conservation status and taxonomy. *Acta Horticulturae* 325: 843–845.
Duncan, G. D. (1992b). Endangered monocots re-discovered in the Boland. *Veld & Flora* 78 (3): 72–73.

Duncan, G. D. (1993). *Lachenalia thomasiae*. *The Flowering Plants of Africa* 52 (2): t. 2061.
Duncan, G. D. (1994). The genus *Lachenalia*, and the discovery of a beautiful new species from the Western Cape (translation into Japanese). *Shin-Kaki* 163: 32–35.
Duncan, G. D. (1996). Four new species and one new subspecies of *Lachenalia* (Hyacinthaceae) from arid areas of South Africa. *Bothalia* 26 (1): 1–9.
Duncan, G. D. (1997). Five new species of *Lachenalia* (Hyacinthaceae) from arid areas of South Africa. *Bothalia* 27 (1): 7–15.
Duncan, G. D. (1998a). Five new species of *Lachenalia* (Hyacinthaceae) from arid areas of Namibia and South Africa. *Bothalia* 28 (2): 131–139.
Duncan, G. D. (1998b). Notes on the genus *Lachenalia*. *Herbertia* 53: 40–48.
Duncan, G. D. (1999a). *Lachenalia violacea*. *Curtis's Botanical Magazine* 16 (4): 252–255.
Duncan, G. D. (1999b). Notes on some rare and newly published species of *Lachenalia* from South Africa and Namibia. *Herbertia* 54: 171–179.
Duncan, G. D. (2001). *Lachenalia elegans* var. *flava*. *Curtis's Botanical Magazine* 18 (1): 18–22.
Duncan, G. D. (2003a). Polyxenas. *Veld & Flora* 89 (1): 22–26.
Duncan, G. D. (2003b). *Lachenalia valeriae*. *Flowering Plants of Africa* 58: 30–36.
Duncan, G. D. (2003c). *Lachenalia peersii*. *Curtis's Botanical Magazine* 20 (4): 202–207.
Duncan, G. D. (2003d). *Lachenalia salteri*. *Curtis's Botanical Magazine* 20 (4): 208–212.
Duncan, G. D. (2003e). *Lachenalia valeriae*. *Bulbs* 5 (1): 3–4.
Duncan, G. D. (2003f). Endangered geophytes of the Cape Floral Kingdom. *Curtis's Botanical Magazine* 20 (4): 245–250.
Duncan, G. D. (2005). Character variation and a cladistic analysis of the genus *Lachenalia* Jacq.f. ex Murray (Hyacinthaceae). M.Sc. thesis, University of KwaZulu-Natal.
Duncan, G. D. (2006). Synopsis of a morphological study of *Lachenalia*. *Bulletin of the Indigenous Bulb Association of South Africa* 55: 52–53.
Duncan, G. D. (2007). *Lachenalia sargeantii*. http://www.plantzafrica.com.
Duncan, G. D. (2008a). *Lachenalia sargeantii* revisited. *Bulbs* 7 (2): 26–27.
Duncan, G. D. (2008b). Reminiscences of Margaret Thomas. *Bulletin of the Indigenous Bulb Association of South Africa* 56: 27–29.
Duncan, G. D. (2009). *Lachenalia congesta*. *Curtis's Botanical Magazine* 26 (3): 210–220.
Duncan, G. D. (2010). *Grow Bulbs*. Kirstenbosch Gardening Series. South African National Biodiversity Institute, Cape Town.
Duncan, G. D. & Anderson, V. F. (1997). *Lachenalia rosea*. *Flowering Plants of Africa* 55: t. 2126.
Duncan, G. D. & Anderson, V. F. (1999). *Lachenalia convallarioides*. *Flowering Plants of Africa* 56: 24–29.
Duncan, G. D. & Anderson, V. F. (2003). *Lachenalia aloides* var. *vanzyliae*. *Flowering Plants of Africa* 58: 38–42.
Duncan, G. D. & Edwards, T. J. (2002). Notes on African Plants. Hyacinthaceae: Massonieae. A new species of *Lachenalia* from Namaqualand, South Africa. *Bothalia* 32 (2): 190–192.
Duncan, G. D. & Edwards, T. J. (2005). *Lachenalia sargeantii*. *Curtis's Botanical Magazine* 22 (3): 176–184.
Duncan, G. D. & Edwards, T. J. (2006). Three new species of *Lachenalia* (Hyacinthaceae: Massonieae) from Western and Northern Cape, South Africa. *Bothalia* 36 (2): 147–155.
Duncan, G. D. & Edwards, T. J. (2007). Notes on African Plants. Hyacinthaceae. A new pyrophytic *Lachenalia* species (Massonieae) from Western Cape, South Africa. *Bothalia* 37 (1): 31–34.
Duncan, G. D. & Linder Smith, C. (1999a). *Lachenalia duncanii*. *Flowering Plants of Africa* 56: 14–17.

Duncan, G. D. & Linder Smith, C. (1999b). *Lachenalia nervosa*. *Flowering Plants of Africa* 56: 18–23.

Duncan, G. D., Edwards, T. J. & Mitchell, A. (2005a). Character variation and a cladistic analysis of the genus *Lachenalia* Jacq.f. ex Murray (Hyacinthaceae). *Acta Horticulturae* 673:113–120.

Duncan, G. D., Edwards, T. J. & Anderson, F. (2005b). *Lachenalia verticillata*. *Flowering Plants of Africa* 59: 8–12.

Duncan, G. D., McMaster, C. & McMaster, R. (2005c). Out of the ashes. *Veld & Flora* 91 (2): 66–69.

Duncan, G. D. & Raimondo, D. (2009a). *Lachenalia alba*. In: D. Raimondo, L. von Staden, W. Foden, J. E. Victor, N. A. Helme, R. C. Turner, D. A. Kamundi & P. A. Manyama (eds), *Red List of South African Plants 2009. Strelitzia 25*. South African National Biodiversity Institute, Pretoria.

Duncan, G. D. & Raimondo, D. (2009b). *Lachenalia martiniae*. In: D. Raimondo, L. von Staden, W. Foden, J. E. Victor, N. A. Helme, R. C. Turner, D. A. Kamundi & P. A. Manyama (eds), *Red List of South African Plants 2009. Strelitzia 25*. South African National Biodiversity Institute, Pretoria.

Duncan, G. D. & Raimondo, D. (2009c). *Lachenalia neilii*. In: D. Raimondo, L. von Staden, W. Foden, J. E. Victor, N. A. Helme, R. C. Turner, D. A. Kamundi & P. A. Manyama (eds), *Red List of South African Plants 2009. Strelitzia 25*. South African National Biodiversity Institute, Pretoria.

Duncan, G. D. & Raimondo, D. (2009d). *Lachenalia peersii*. In: D. Raimondo, L. von Staden, W. Foden, J. E. Victor, N. A. Helme, R. C. Turner, D. A. Kamundi & P. A. Manyama (eds), *Red List of South African Plants 2009. Strelitzia 25*. South African National Biodiversity Institute, Pretoria.

Duncan, G. D., Raimondo, D. & Ebrahim, I. (2009). *Lachenalia mathewsii*. In: D. Raimondo, L. von Staden, W. Foden, J. E. Victor, N. A. Helme, R. C. Turner, D. A. Kamundi & P. A. Manyama (eds), *Red List of South African Plants 2009. Strelitzia 25*. South African National Biodiversity Institute, Pretoria.

Duncan, G. D. & Visagie, M. (2001). *Lachenalia zebrina*. *Flowering Plants of Africa* 57: 34–37.

Duncan, G. D. & Visagie, M. (2009). *Veltheimia bracteata*. *Flowering Plants of Africa* 61: 24–33.

Du Plessis, N. M. & Duncan, G. D. (1989). *Lachenalia*. *Bulbous Plants of Southern Africa*: 78–82. Tafelberg, Cape Town.

Durand, T. A. & Schinz, H. (1895). *Conspectus Florae Africae* 5: 367. Friedlander & Sohn, Berlin.

Du Toit, E. S., Robbertse, P. J. & Niederwieser, J. G. (2001). An evaluation of bulb growth and structure of *Lachenalia* cv. Ronina bulbs. *South African Journal of Botany* 67 (4): 667–670.

Edwards, P. (1964). Some manuscripts relating to South African Botany in the William Sherard Collection in the Bodleian Library at Oxford. *Journal of South African Botany* 30: 103–105.

Edwards, D. & Leistner, O. A. (1971). A degree reference system for citing biological records in southern Africa. *Mitteilungen der Botanischen Staatssammlung München* 10: 501–509.

Engler, H. G. A. (1899). *Lachenalia* Jacq. *Notizblatt des Koniglichen Botanischen Gartens und Museums zu Berlin* 2: 321.

Esau, K. (1977). *Anatomy of Seed Plants*. John Wiley and Sons, New York.

Esler, K. J., Rundel, P. W. & Vorster, P. (1999). Biogeography of prostrate-leaved geophytes in semi-arid South Africa: hypotheses on functionality. *Plant Ecology* 142: 105–120.

Faegri, K. & van der Pijl, L. (1979). *The Principles of Pollination Ecology*. Pergamon Press, Oxford.

Fernandes, A. & Neves, J. B. (1962). Sur la caryologie de quelques monocotyledones africaines. Compt. Rend. de la IV-e Reunioin Pleniere de l'Association pour l'Etude Taxinomique de la Flore d'Afrique Tropicale, Lisboa: 439–463.

Fourcade, H. G. (1934). Contributions to the flora of the Knysna and neighbouring divisions: *Lachenalia haarlemensis* and *L. subspicata*. *Transactions of the Royal Society of South Africa* 21: 79.

Fox, D. L. (1979). *Biochromy: Natural Coloration of Living Things*. University of California Press, Berkeley.

Garside, S. (1942). Baron Jacquin and the Schönbrunn Gardens. *South African Journal of Botany* 8: 201–224.

Glover, R. (1915). *Scilla pearsonii*. In: F. Bolus, L. Bolus & R. Glover (eds), *Flowering Plants and Ferns collected on the Great Karasburg by the Percy Sladen Memorial Expedition, 1912–1913, The Annals of the Bolus Herbarium* 1: 105–106.

Gouws, J. B. (1964). Cytological studies in the genus *Lachenalia* Jacq. *Annals of the University College of the Western Cape* 2: 1–7.

Grey, C. H. (1938). *Hardy Bulbs* 3: 344. Williams & Norgate, London.

Gunn, M. & du Plessis, E. (1978). *The Flora Capensis of Jakob and Johann Philipp Breyne*. The Brenthurst Press, Johannesburg.

Gunn, M. & Codd, L. E. (1981). *Botanical Exploration of Southern Africa*. A. A. Balkema, Rotterdam.

Hamatani, S., Hashimoto, K. & Kondo, K. (1998). A comparison of somatic chromosomes at metaphase in *Lachenalia* (Liliaceae). *Chromosome Science* 8: 55–61.

Hamatani, S., Ishida, G., Hashimoto, K. & Kondo, K. (2004). A chromosome study in ten species of *Lachenalia* (Liliaceae). *Chromosome Science* 2: 21–25.

Hamatani, S., Kondo, K., Kodaira, E. & Ogawa, H. (2007). Chromosome morphology of 12 species and one variety of *Lachenalia* and five species of closely related, allied genera (Liliaceae). *Chromosome Botany* 2: 79–86.

Hamatani, S., Masuda, Y., Kondo, K., Kodaira, E. & Ogawa, H., (2008). Molecular phylogenetic relationships among *Lachenalia*, *Massonia* and *Polyxena* (Liliaceae) on the basis of the internal transcribed spacer (ITS) region. *Chromosome Botany* 3: 65–72.

Hancke, F. L. & Liebenberg, H. (1990). B-chromosomes in some *Lachenalia* species and hybrids. *South African Journal of Botany* 56 (6): 659–664.

Hancke, F. L. & Liebenberg, H. (1998). Meiotic studies of interspecific *Lachenalia* hybrids and their parents. *South African Journal of Botany* 64: 250–255.

Harrower, A, & Raimondo, D. (2009). *Lachenalia leipoldtii*. In: D. Raimondo, L. von Staden, W. Foden, J. E. Victor, N. A. Helme, R. C. Turner, D. A. Kamundi & P. A. Manyama. (eds), *Red List of South African Plants 2009. Strelitzia 25*. South African National Biodiversity Institute, Pretoria.

Helme, N. A. (2009). *Lachenalia leomontana*. In: D. Raimondo, L. von Staden, W. Foden, J. E. Victor, N. A. Helme, R. C. Turner, D. A. Kamundi & P. A. Manyama (eds), *Red List of South African Plants 2009. Strelitzia 25*. South African National Biodiversity Institute, Pretoria.

Helme, N. A. & Raimondo, D. (2009a). *Lachenalia bachmannii*. In: D. Raimondo, L. von Staden, W. Foden, J. E. Victor, N. A. Helme, R. C. Turner, D. A. Kamundi & P. A. Manyama (eds), *Red List of South African Plants 2009. Strelitzia 25*. South African National Biodiversity Institute, Pretoria.

Helme, N. A. & Raimondo, D. (2009b). *Lachenalia capensis*. In: D. Raimondo, L. von Staden, W. Foden, J. E. Victor, N. A. Helme, R. C. Turner, D. A. Kamundi & P. A. Manyama (eds), *Red List of South African Plants 2009. Strelitzia 25*. South African National Biodiversity Institute, Pretoria.

Helme, N. A. & Raimondo, D. (2009c). *Lachenalia contaminata*. In: D. Raimondo, L. von Staden, W. Foden, J. E. Victor, N. A. Helme, R. C. Turner, D. A. Kamundi & P. A. Manyama (eds), *Red List of South African Plants 2009. Strelitzia 25*. South African National Biodiversity Institute, Pretoria.

Helme, N. A. & Raimondo, D. (2009d). *Lachenalia longibracteata*. In: D. Raimondo, L. von Staden, W. Foden, J. E. Victor, N. A. Helme, R. C. Turner, D. A. Kamundi & P. A. Manyama (eds), *Red List of South African Plants 2009. Strelitzia 25*. South African National Biodiversity Institute, Pretoria.

Helme, N. A. & Raimondo, D. (2009e). *Lachenalia nervosa*. In: D. Raimondo, L. von Staden, W. Foden, J. E. Victor, N. A. Helme, R. C. Turner, D. A. Kamundi & P. A. Manyama (eds), *Red List of South African Plants 2009. Strelitzia 25*. South African National Biodiversity Institute, Pretoria.

Helme, N. A. & Raimondo, D. (2009f). *Lachenalia paucifolia*. In: D. Raimondo, L. von Staden, W. Foden, J. E. Victor, N. A. Helme, R. C. Turner, D. A. Kamundi & P. A. Manyama (eds), *Red List of South African Plants 2009. Strelitzia 25*. South African National Biodiversity Institute, Pretoria.

Helme, N. A. & Raimondo, D. (2009g). *Lachenalia reflexa*. In: D. Raimondo, L. von Staden, W. Foden, J. E. Victor, N. A. Helme, R. C. Turner, D. A. Kamundi & P. A. Manyama (eds), *Red List of South African Plants 2009. Strelitzia 25*. South African National Biodiversity Institute, Pretoria.

Helme, N. A. & Raimondo, D. (2009h). *Lachenalia salteri*. In: D. Raimondo, L. von Staden, W. Foden, J. E. Victor, N. A. Helme, R. C. Turner, D. A. Kamundi & P. A. Manyama (eds), *Red List of South African Plants 2009. Strelitzia 25*. South African National Biodiversity Institute, Pretoria.

Helme, N. A., Raimondo, D. & Duncan, G. D. (2009). *Lachenalia margaretiae*. In: D. Raimondo, L. von Staden, W. Foden, J. E. Victor, N. A. Helme, R. C. Turner, D. A. Kamundi & P. A. Manyama (eds), *Red List of South African Plants 2009. Strelitzia 25*. South African National Biodiversity Institute, Pretoria.

Hooker, J. D. (1871). *Massonia odorata. Curtis's Botanical Magazine* 27: t. 5891.

Hooker, J. D. (1872). *Lachenalia tricolor* var. *aurea. Curtis's Botanical Magazine* 118: t. 5992.

Hutchinson, J. (1934). *The Families of Flowering Plants*, Vol. 2: Monocotyledons. Clarendon Press, Oxford.

Hutchinson, J. (1936). *Lachenalia mutabilis. Curtis's Botanical Magazine* 159: t. 9433.

Ingram, J. (1966). Notes on the Cultivated Liliaceae. 4. *Lachenalia. Baileya* 14 (3): 123–132.

Jacob, J. (1919). Freesias and lachenalias. *Journal of the Royal Horticultural Society* 45 (1): 29–38.

Jacquin, J. F. (1787). *Tria Genera Plantarum Nova. Lachenalia tricolor. Nova Acta Helvetica* 1: 34–41.

Jacquin, N. J. (1772). *Hortus Botanicus Vindobonensis* 2: t. 178. C. F. Wappler, Vienna.

Jacquin, N. J. (1781). *Lachenalia tricolor. Icones Plantarum Rariorum* 1: t. 61. C. F. Wappler, Vienna.

Jacquin, N. J. (1788). *Collectanea* 2. C. F. Wappler, Vienna.

Jacquin, N. J. (1790). *Collectanea* 3. C. F. Wappler, Vienna.

Jacquin, N. J. (1791). *Collectanea* 4. C. F. Wappler, Vienna.

Jacquin, N. J. (1795a). *Lachenalia pusilla. Icones Plantarum Rariorum* 2: 16, t. 385. C. F. Wappler, Vienna.

Jacquin, N. J. (1795b). *Lachenalia quadricolor. Icones Plantarum Rariorum* 2: 16, t. 396. C. F. Wappler, Vienna.

Jacquin, N. J. (1797a). *Collectanea* 5. C. F. Wappler, Vienna.

Jacquin, N. J. (1797b). *Lachenalia fragrans. Plantarum rariorum horti caesarei Schönbrunnensis* 1: 43, t. 83. C. F. Wappler, Vienna.

Jeppe, B. J. & Duncan, G. D. (1989). *Lachenalia. Spring and Winter Flowering Bulbs of the Cape*. Oxford University Press, Cape Town.

Jessop, J. P. (1964). Itinerary of Rudolf Schlechter's collecting trips in southern Africa. *Journal of South African Botany* 30 (3): 129–146.

Jessop, J. P. (1965). A volume of early water colours in the library of the Botanical Research Institute, Pretoria. *Journal of the South African Biological Society* 6: 38–52.

Jessop, J. P. (1975). Studies in the bulbous Liliaceae in South Africa: 5. Seed surface characters and generic groupings. *Journal of South African Botany* 41: 67–85.

Jessop, J. P. (1976). Studies in the bulbous Liliaceae in South Africa: 6. The taxonomy of *Massonia* and allied genera. *Journal of South African Botany* 42: 401–437.

Johnson, M. A. T. & Brandham, P. E. (1997). New chromosome numbers in petaloid monocotyledons and other miscellaneous angiosperms. *Kew Bulletin* 52: 121–138.

Johnson, S. D., Pauw, A. & Midgley, J. (2001). Rodent pollination in the African Lily *Massonia depressa* (Hyacinthaceae). *American Journal of Botany* 88 (10): 1768–1773.

Jones, R. N. & Rees, N. (1982). *B-Chromosomes*. Academic Press, London.

Journet, D. (2003). Growing and enjoying *Lachenalia*. *Bulbs* 5 (1): 24–33.

Jürgens, N. (1997). Floristic biodiversity and history of African arid regions. *Biodiversity and Conservation* 6: 495–514.

Kay, Q. O. N., Daoud, H. S. & Sirton, C. H. (1981). Pigment distribution, light reflection and cell structure in petals. *Botanical Journal of the Linnean Society* 83: 57–84.

Ker Gawler, J. B. (1802). *Massonia ensifolia*. *Curtis's Botanical Magazine* 16: t. 554.

Ker Gawler, J. B. (1807a). *Lachenalia tricolor* var. *luteola*. *Curtis's Botanical Magazine* 26: t. 1020.

Ker Gawler, J. B. (1807b). *Massonia corymbosa*. *Curtis's Botanical Magazine* 25: t. 991.

Ker Gawler, J. B. (1811). *Lachenalia lucida*. *Curtis's Botanical Magazine* 33: t. 1372.

Ker Gawler, J. B. (1812a). *Lachenalia nervosa*. *Curtis's Botanical Magazine* 36: t. 1497.

Ker Gawler, J. B. (1812b). *Scilla corymbosa*. *Curtis's Botanical Magazine* 36: t. 1468.

Ker Gawler, J. B. (1813). *Lachenalia racemosa*. *Curtis's Botanical Magazine* 37: t. 1517.

Ker Gawler, J. B. (1814). *Lachenalia bifolia*. *Curtis's Botanical Magazine* 40: t. 1611.

Kerkham, A. S. (1992a). The Claudius manuscript in the South African Public Library (Part 1). *Quarterly Bulletin of the South African Library* 46 (4): 146–153.

Kerkham, A. S. (1992b). The Claudius manuscript in the South African Public Library (Part 2). *Quarterly Bulletin of the South African Library* 47 (1): 28–36.

Kleynhans, R. (2006). *Lachenalia*. In: N. O. Anderson (ed.), *Flower Breeding and Genetics*, pp. 491–516. Springer, The Netherlands.

Kleynhans, R., Niederwieser, J. G. & Hancke, F. L. (2002). Development and commercialisation of a new flower crop. *Acta Horticulturae* 570: 81–86.

Kleynhans, R. & Spies, J. J. (1999). Chromosome number and morphological variation in *Lachenalia bulbifera* (Hyacinthaceae). *South African Journal of Botany* 65: 357–360.

Kleynhans, R. & Spies, J. J. (2000). Evaluation of genetic variation in *Lachenalia bulbifera* (Hyacinthaceae) using RAPDs. *Euphytica* 115: 141–147.

Kleynhans, R., Spies, J. J. & Spies, P. (2009). Cross-ability in the genus *Lachenalia*. *Acta Horticulturae* 813: 385–392.

Kunth, C. S. (1843). *Enumeratio Plantarum*. J. G. Cotta, Stuttgart.

Lamarck, J.-B. (1792). *Lachenalia*. *Encyclopédie Méthodique* 3 (2): 372–373. Charles-Joseph Panckouke, Paris.

Lamarck, J.-B. (1813). *Lachenalia*. *Encyclopédie Méthodique* 3 (1): 228–232. H. Agasse, Paris.

Langlois, A., Mulholland, D. A., Duncan, G. D., Crouch, N. R. & Edwards, T. J. (2005). A novel 3-benzylchromone from the South African *Lachenalia rubida* (Hyacinthaceae). *Biochemical Systematics and Ecology* 33 (9): 961–966.

Lemaire, C. A. (1855). Miscellanées. Du Genre *Lachenalia*. *L'Illustration Horticole* 2: 99–100.

Lemaire, C. A. (1856). Miscellanées. Du Genre *Lachenalia*. *L'Illustration Horticole* 3: 33–36.

Levin, D. A. (1973). The role of trichomes in defense. *Quarterly Review of Biology* 48 (1): 3–15.

Linder, H. P. (2006). Investigating the evolution of floras: problems and progress – an introduction. *Diversity and Distributions* 12 (1): 3–5.

Lindley, J. (1830). *Lachenalia pallida*. *Edwards's Botanical Register* 16: t. 1350.

Lindley, L. (1837). *Lachenalia pallida* var. *coerulescens*. *Edwards's Botanical Register* 23: t. 1945.

Lindley, L. (1856). New plants. 176. *Lachenalia aurea*. *The Gardener's Chronicle* (First Series): 404.

Linnaeus, C. (1753). *Species Plantarum*. Impensis Laurentii Salvii, Stockholm.
Linnaeus, C. (1762). *Species Plantarum*, 2nd edition. Impensis Laurentii Salvii, Stockholm.
Linnaeus, C. (1771). *Mantissa plantarum altera*. Impensis Laurentii Salvii, Stockholm.
Linnaeus, C. (1781). *Phormium aloides. Supplementum Plantarum et Specierum Plantarum* (second edition): 205. Impensis Orphanotrophei, Brunsvigae.
Linnaeus, C. (1799). *Species Plantarum*. Impensis G. C. Nauk, Berlin.
Loddiges, C. (1826a). *Lachenalia bicolor. Botanical Cabinet* 11: t. 1129.
Loddiges, C. (1826b). *Lachenalia fragrans. Botanical Cabinet* 11: t. 1140.
Loddiges, C. (1826c). *Lachenalia mutabilis. Botanical Cabinet* 11: t. 1076.
Loots, S. (2005). *Red data book of Namibian plants*. Southern African Botanical Diversity Network Report no. 38. SABONET, Pretoria and Windhoek.
Lubbinge, J. (1980). *Lachenalia* breeding. 1. Introduction. *Acta Horticulturae* 109: 289–295.
Macnae, M. M. & Davidson, L. E. (1969). The volume '*Icones Plantarum et Animalum*' in the Africana Museum, Johannesburg; and its relationship to the *Codex Witsenii* quoted by Jan Burman in his '*Decades Rariorum Africanum Plantarum*'. *Journal of South African Botany* 35 (2): 65–81.
Manning, J., Goldblatt, P. & Fay, M. (2004). A revised generic synopsis of Hyacinthaceae in sub-Saharan Africa, based on molecular evidence, including new combinations and the new tribe *Pseudoprosperae*. *Edinburgh Journal of Botany* 60 (3): 533–568.
Mason, H. & du Plessis, E. (1972). *Western Cape Sandveld Flowers*: 32–33. Struik, Cape Town.
Mathew, B. (1988). *Polyxena corymbosa. Kew Magazine* 5 (1): 5–7.
McMaster, R. & McMaster, C. (2006). *Lachenalia sargeantii* revisited. *Veld & Flora* 92 (1): 15.
Miller, P. (1768). *Muscari orchioides. The Gardeners Dictionary*, 8th edition. Rivington, London.
Moffett, A. H. (1936). The cytology of *Lachenalia. Cytologia* 7: 490–498.
Mogford, D. J. (1978). Centromeric heterochromatin in *Lachenalia tricolor* (L.) Thunb. *Journal of South African Botany* 44 (2): 111–117.
Moore, F. W. (1891). Lachenalias. *Journal of the Royal Horticultural Society* 13: 216–231.
Moore, F. W. (1905). *Lachenalia* hybrids. *The Gardener's Chronicle* (3rd series) 37: 210–211.
Morley, B. D. (1978). Lydia Schackleton's paintings in the National Botanic Gardens, Glasnevin. *Glasra* 2: 25–36.
Mucina, L. & Rutherford, M. C. (2006). *The Vegetation of South Africa, Lesotho and Swaziland. Strelitzia 19*. South African National Biodiversity Institute, Pretoria.
Müller-Doblies, U., Nordenstam, B. & Müller-Doblies, D. (1987). A second species in *Lachenalia* subgen. *Brachyscypha* (Hyacinthaceae): *Lachenalia barkeriana* sp. nov. from southern Little Namaqualand. *South African Journal of Botany* 53 (6): 481–488.
Müller-Doblies, U. & Müller-Doblies, D. (1997). A partial revision of the tribe *Massonieae* (Hyacinthaceae). *Feddes Repertorium* 108: 49–96.
Murray, J. A. (1784). *Lachenalia tricolor. Linnaeus Systema vegetabilium* 14: 314. Dieterich, Gottingen.
Nel, D. D. (1983). Rapid propagation of *Lachenalia* hybrids *in vitro*. *South African Journal of Botany* 2 (3): 245–246.
Nelson, E. C. (2000). *A Heritage of Beauty: 125*. Irish Garden Plant Society, Dublin.
Niederwieser, J. G., Anandajayasekeram, P., Coetzee, M., Martella, D., Pieterse, B. and Marasa, C. (1998). Research impact assessment as a management tool: *Lachenalia* research at ARC-Roodeplaat as a case study. *Journal of the South African Society for Horticultural Science* 8: 80–84.
Niederwieser, J. G. & Ndou, A. M. (2002). Review on adventitious bud formation in *Lachenalia*. *Acta Horticulturae* 570: 135–140.

Nixon, K. C. & Wheeler, Q. D. (1990). An amplification of the phylogenetic species concept. *Cladistics* 6: 211–223.

Nordenstam, B. (1982). Chromosome numbers of South African plants 2. *Journal of South African Botany* 48 (2): 273–275.

Oberlander, K. C., Dreyer, L. L., Bellstedt, D. U. & Reeves, G. (2004). Systematic relationships in southern African *Oxalis* L. (Oxalidaceae): congruence between palynological and plastid trnL-F evidence. *Taxon* 53 (4): 977–985.

Ornduff, R. & Watters, P. J. (1978). Chromosome numbers in *Lachenalia*. *Journal of South African Botany* 44 (4): 387–390.

Persoon, C. H. (1805). *Synopsis Plantarum* 1: 377. Tubingae, Cottam/Cramerum.

Petiver, J. (1709). *Gazophylacii Naturae & Artis seu Herbarium Capense, Decade* 9: t. 87 (8). Critoplorus Bateman, London.

Pfosser, M. & Speta, F. (1999). Phylogenetics of Hyacinthaceae based on plastid DNA sequences. *Annals of the Missouri Botanical Garden* 86: 852–875.

Pfosser, M., Wetschnig, W., Ungar, S. & Prenner, G. (2003). Phylogenetic relationships among genera of Massonieae (Hyacinthaceae) inferred from plastid DNA and seed morphology. *Journal of Plant Research* 116: 115–132.

Phillips, E. P. (1923). *Lachenalia roodeae*. *The Flowering Plants of South Africa* 3: t. 91.

Phillips, E. P. (1931). *Lachenalia longibracteata*. *The Flowering Plants of South Africa* 11: t. 405.

Plukenet, L. (1692). *Leonardi Plukenetii Phytographia*. London.

Prain, D. (1921). Obituary notices of fellows deceased. John Gilbert Baker, 1834–1920. *Proceedings of the Royal Society* B., 92 (642): 24–30.

Proctor, M. & Yeo, P. (1972). *The Pollination of Flowers*. Traplinger, New York.

Raimondo, D. (2009). *Lachenalia muirii*. In: D. Raimondo, L. von Staden, W. Foden, J. E. Victor, N. A. Helme, R. C. Turner, D. A. Kamundi & P. A. Manyama (eds), *Red List of South African Plants 2009. Strelitzia 25*. South African National Biodiversity Institute, Pretoria.

Raimondo, D. & Duncan, G. D. (2009a). *Lachenalia dehoopensis*. In: D. Raimondo, L. von Staden, W. Foden, J. E. Victor, N. A. Helme, R. C. Turner, D. A. Kamundi & P. A. Manyama (eds), *Red List of South African Plants 2009. Strelitzia 25*. South African National Biodiversity Institute, Pretoria.

Raimondo, D. & Duncan, G. D. (2009b). *Lachenalia maximiliani*. In: D. Raimondo, L. von Staden, W. Foden, J. E. Victor, N. A. Helme, R. C. Turner, D. A. Kamundi & P. A. Manyama (eds), *Red List of South African Plants 2009. Strelitzia 25*. South African National Biodiversity Institute, Pretoria.

Raimondo, D. & Duncan, G. D. (2009c). *Lachenalia orthopetala*. In: D. Raimondo, L. von Staden, W. Foden, J. E. Victor, N. A. Helme, R. C. Turner, D. A. Kamundi & P. A. Manyama (eds), *Red List of South African Plants 2009. Strelitzia 25*. South African National Biodiversity Institute, Pretoria.

Raimondo, D. & Duncan, G. D. (2009d). *Lachenalia physocaulos*. In: D. Raimondo, L. von Staden, W. Foden, J. E. Victor, N. A. Helme, R. C. Turner, D. A. Kamundi & P. A. Manyama (eds), *Red List of South African Plants 2009. Strelitzia 25*. South African National Biodiversity Institute, Pretoria.

Raimondo, D. & Duncan, G. D. (2009e). *Lachenalia purpureo-caerulea*. In: D. Raimondo, L. von Staden, W. Foden, J. E. Victor, N. A. Helme, R. C. Turner, D. A. Kamundi & P. A. Manyama (eds), *Red List of South African Plants 2009. Strelitzia 25*. South African National Biodiversity Institute, Pretoria.

Raimondo, D. & Duncan, G. D. (2009f). *Lachenalia sargeantii*. In: D. Raimondo, L. von Staden, W. Foden, J. E. Victor, N. A. Helme, R. C. Turner, D. A. Kamundi & P. A. Manyama (eds), *Red List of South African Plants 2009. Strelitzia 25*. South African National Biodiversity Institute, Pretoria.

Raimondo, D. & Ebrahim, I. (2009). *Lachenalia polyphylla*. In: D. Raimondo, L. von Staden, W. Foden, J. E. Victor, N. A. Helme, R. C. Turner, D. A. Kamundi & P. A. Manyama (eds), *Red List of South African Plants 2009. Strelitzia 25.* South African National Biodiversity Institute, Pretoria.

Raimondo, D. & Helme, N. A. (2009a). *Lachenalia mediana* var. *mediana*. In: D. Raimondo, L. von Staden, W. Foden, J. E. Victor, N. A. Helme, R. C. Turner, D. A. Kamundi & P. A. Manyama (eds), *Red List of South African Plants 2009. Strelitzia 25.* South African National Biodiversity Institute, Pretoria.

Raimondo, D. & Helme, N. A. (2009b). *Lachenalia mediana* var. *rogersii*. In: D. Raimondo, L. von Staden, W. Foden, J. E. Victor, N. A. Helme, R. C. Turner, D. A. Kamundi & P. A. Manyama (eds), *Red List of South African Plants 2009. Strelitzia 25.* South African National Biodiversity Institute, Pretoria.

Raimondo, D. & Turner, R. C. (2009). *Lachenalia corymbosa*. In: D. Raimondo, L. von Staden, W. Foden, J. E. Victor, N. A. Helme, R. C. Turner, D. A. Kamundi & P. A. Manyama (eds), *Red List of South African Plants 2009. Strelitzia 25.* South African National Biodiversity Institute, Pretoria.

Raimondo, D., Koopman, R., Victor, J. E. & Duncan, G. D. (2009). *Lachenalia liliiflora*. In: D. Raimondo, L. von Staden, W. Foden, J. E. Victor, N. A. Helme, R. C. Turner, D. A. Kamundi & P. A. Manyama (eds), *Red List of South African Plants 2009. Strelitzia 25.* South African National Biodiversity Institute, Pretoria.

Raimondo, D., Victor, J. E. & Duncan, G. D. (2009). *Lachenalia arbuthnotiae*. In: D. Raimondo, L. von Staden, W. Foden, J. E. Victor, N. A. Helme, R. C. Turner, D. A. Kamundi & P. A. Manyama (eds), *Red List of South African Plants 2009. Strelitzia 25.* South African National Biodiversity Institute, Pretoria.

Redouté, P. J. (1802). *Lachenalia tricolor. Les Liliacées* 1: t. 2. Didot Jeune, Paris.

Redouté, P. J. (1809). *Lachenalia luteola. Les Liliacées* 5: t. 297. Didot Jeune, Paris.

Reid, C. (1985). *Lachenalia aloides. The Flowering Plants of Africa* 48: t. 1910.

Reid, C. (1993). *Lachenalia*. In: T. H. Arnold & B. C. dse Wet (eds), *Plants of Southern Africa: Names and Distribution. Memoirs of the Botanical Survey of South Africa No. 62.* National Botanical Institute, Pretoria.

Reuthe, G. (1889). Die Lachenalien. *Gartenflora* 38: 155–158.

Riley, H. P. (1962). Chromosome studies in some South African monocotyledons. *Canadian Journal of Genetics and Cytology* 4: 40–55.

Roodbol, F. & Niederwieser, J. G. (1998). Initiation, growth and development of bulbs of *Lachenalia* 'Romelia' (Hyacinthaceae). *Journal of the South African Society for Horticultural Science* 8 (1): 18–20.

Rourke, J. P. (1995). Obituaries: Winsome (Buddy) Barker (1907–1994). *Bothalia* 25 (2): 255–259.

Salisbury, R. A. (1866). *The Genera of Plants*. John van Voorst, London.

Sâto, D. (1942). Karyotype alteration and phylogeny in Liliaceae and allied families. *Japanese Journal of Botany* 12: 57–161.

Savage, S. (1945). *Catalogue of the Linnean Herbarium*. Linnean Society, London.

Schlechter, R. (1924). Drei neue Gattungen der Liliaceen aus Südafrika. *Notizblatt des Konigl. Botanischen Gartens und Museums zu Berlin* 9: 145–151.

Schönland, S. (1910). Some flowering plants from the neighbourhood of Port Elizabeth. *Transactions of the Royal Society of South Africa* 1: 143–145.

Schubert, B. G. (1945). Publication of Jacquin's *Icones Plantarum Rariorum. Contributions from the Gray Herbarium of Harvard University*: 3–23.

Schultes, J. A. & Schultes, J. H. (1829). *Lachenalia orthopetala*. *Systema Vegetabilium* 7 (1): 601. J. G. Cottae, Stuttgart.

Schultes, J. A. & Schultes, J. H. (1830). *Coelanthus complicatus*. *Systema vegetabilium* 7 (2): xlvi. J. G. Cottae, Stuttgart.

Sims, J. (1815). *Lachenalia quadricolor* var. *lutea*. *Curtis's Botanical Magazine* 61: t. 1704.

Sölch, A. & Roessler, H. (1970). *Lachenalia* Jacq. *Prodromus einer Flora von Südwestafrika* 147: 52–54. Cramer, Lehre.

Sork, V. L. (1993). Evoutionary ecology of mast-seeding in temperate and tropical oaks (*Quercus* spp.). *Vegetatio* 107/108: 133–147.

Speta, F. (1998). Hyacinthaceae. In: K. Kubitzki (ed.), *The Families and Genera of Vascular Plants*, vol. 3. Springer-Verlag, Berlin.

Spies, P. (2004). *Phylogenetic Relationships of the Genus Lachenalia With Other Related Liliaceous Taxa*. M.Sc. thesis, University of the Free State.

Spies, J. J., du Preez, J. L., Minnaar, A. & Kleynhans, R. (2000). Chromosome studies on African Plants. 13. *Lachenalia mutabilis*, *L. pustulata* and *L. unicolor*. *Bothalia* 30 (1): 106–110.

Spies, J. J., Van Rooyen, P. & Kleynhans, R. (2002). The subgeneric delimitation of *Lachenalia* (Hyacinthaceae). *Acta Horticulturae* 570: 225–232.

Spies, J. J., Spies, P., Reinecke, S. M., Kleynhans, R., Duncan, G. D. & Edwards, T. J. (2008). *Lachenalia*. In: K. Marhold (ed.), IAPT/IOPB chromosome data 5, *Taxon* 57 (2): 554–555.

Spies, J. J., Spies, P., Minnaar, A., Reinecke, S. M., du Preez, J. L. & Kleynhaus, R. (2009). Hyacinthaceae. In: K. Marhold (ed.), IAPT/IOPB chromosome data 8. *Taxon* 58 (4): 1288–1289.

Sprenger, C. (1891). *Lachenalia Regeliana* Spr. – *L. reflexa* × *L. aurea*. *Gartenflora* 40: 356–359.

Stapf, O. (1923). *Lachenalia convallariodora*. *Curtis's Botanical Magazine* 168: t. 8955.

Stearn, W. T. (1990). The Linnean species of *Hyacinthus* (Liliaceae: Hyacinthaceae). *Annales Musei Goulandris* 8: 210, 214.

Stebbins, G. L. (1952). Aridity as a stimulus to plant evolution. *The American Naturalist* 826: 33–44.

Stebbins, G. L. (1974). *Flowering Plants: Evolution Above the Species Level*. Edward Arnold, London.

Steudel, E. (1821–1824). *Lachenalia tenella*. *Nomenclator Botanicus* 1: 459. J. G. Cottae, Stuttgart.

Sweet, R. (1832). *Lachenalia mutabilis*. *The British Flower Garden* 5 (series 2): t. 129. James Ridgway, London.

Swofford, D. L. (1999). *PAUP**. *Phylogenetic Analysis Using Parsimony (*and Other Methods)*. Version 4. Sinauer Associates, Sunderland, Massachusetts.

Thunberg, C. P. (1782). *Nova Genera Plantarum*, part 2. Edman, Uppsala.

Thunberg, C. P. (1784). *Nova Genera Plantarum*: 94–98. Edman, Uppsala.

Thunberg, C. P. (1794). *Prodromus Plantarum Capensium* 1: 64. Edman, Uppsala.

Tillich, H-J. (2000). In: K. L. Wilson & D. A. Morrison (eds), *Monocots: Systematics and Evolution*. CSIRO Publishing, Collingwood, Victoria.

Trattinnick, L. (1813). *Lachenalia pustulata* var. *densiflora*. *Archiv der Gewachskunde* 1: t. 94.

Trattinnick, L. (1814a). *Lachenalia botryoides*. *Archiv der Gewachskunde* 2: t. 140.

Trattinnick, L. (1814b). *Lachenalia latifolia*. *Archiv der Gewachskunde* 2: t. 142.

Trattinnick, L. (1814c). *Lachenalia luteola* var. *pallida*. *Archiv der Gewachskunde* 2: t. 150.

Van der Merwe, A. M. (2002). *A Biosystematic Study of the Seven Minor Genera of the Hyacinthaceae*. Ph.D dissertation, University of Stellenbosch.

Van der Merwe, A. M. & Manning, J. (2000). *Polyxena paucifolia*. *Cape Plants*, 714. *Strelitzia* 9. National Botanical Institute, Cape Town.

Van der Merwe, A. M. & Marais, E. M. (2001). A new species of *Polyxena* (Hyacinthaceae, tribe Massonieae) from Komsberg, Northern Cape Province. *South African Journal of Botany* 67: 44–46.

Van Rooyen, P., Spies, J. J. & Kleynhans, R. (2002). The species delimitation of *Lachenalia unifolia* and *L. hirta*. *Acta Horticulturae* 570: 395–401.

Van Wyk, A. E. & Smith, G. F. (2001). *Regions of Floristic Endemism*. Umdaus Press, Pretoria.

Victor, J. E. (2009). *Lachenalia concordiana*. In: D. Raimondo, L. von Staden, W. Foden, J. E. Victor, N. A. Helme, R. C. Turner, D. A. Kamundi & P. A. Manyama (eds), *Red List of South African Plants 2009. Strelitzia 25*. South African National Biodiversity Institute, Pretoria.

Victor, J. E. & Dold, A. P. (2009). *Lachenalia convallarioides*. In: D. Raimondo, L. von Staden, W. Foden, J. E. Victor, N. A. Helme, R. C. Turner, D. A. Kamundi & P. A. Manyama (eds), *Red List of South African Plants 2009. Strelitzia 25*. South African National Biodiversity Institute, Pretoria.

Victor, J. E. & Duncan, G. D. (2009a). *Lachenalia barkeriana*. In: D. Raimondo, L. von Staden, W. Foden, J. E. Victor, N. A. Helme, R. C. Turner, D. A. Kamundi & P. A. Manyama (eds), *Red List of South African Plants 2009. Strelitzia 25*. South African National Biodiversity Institute, Pretoria.

Victor, J. E. & Duncan, G. D. (2009b). *Lachenalia buchubergensis*. In: D. Raimondo, L. von Staden, W. Foden, J. E. Victor, N. A. Helme, R. C. Turner, D. A. Kamundi & P. A. Manyama (eds), *Red List of South African Plants 2009. Strelitzia 25*. South African National Biodiversity Institute, Pretoria.

Victor, J. E. & Duncan, G. D. (2009c). *Lachenalia doleritica*. In: D. Raimondo, L. von Staden, W. Foden, J. E. Victor, N. A. Helme, R. C. Turner, D. A. Kamundi & P. A. Manyama (eds), *Red List of South African Plants 2009. Strelitzia 25*. South African National Biodiversity Institute, Pretoria.

Victor, J. E. & Duncan, G. D. (2009d). *Lachenalia duncanii*. In: D. Raimondo, L. von Staden, W. Foden, J. E. Victor, N. A. Helme, R. C. Turner, D. A. Kamundi & P. A. Manyama (eds), *Red List of South African Plants 2009. Strelitzia 25*. South African National Biodiversity Institute, Pretoria.

Victor, J. E. & Duncan, G. D. (2009e). *Lachenalia kliprandednsis*. In: D. Raimondo, L. von Staden, W. Foden, J. E. Victor, N. A. Helme, R. C. Turner, D. A. Kamundi & P. A. Manyama (eds), *Red List of South African Plants 2009. Strelitzia 25*. South African National Biodiversity Institute, Pretoria.

Victor, J. E. & Duncan, G. D. (2009f). *Lachenalia macgregoriorum*. In: D. Raimondo, L. von Staden, W. Foden, J. E. Victor, N. A. Helme, R. C. Turner, D. A. Kamundi & P. A. Manyama (eds), *Red List of South African Plants 2009. Strelitzia 25*. South African National Biodiversity Institute, Pretoria.

Victor, J. E. & Duncan, G. D. (2009g). *Lachenalia minima*. In: D. Raimondo, L. von Staden, W. Foden, J. E. Victor, N. A. Helme, R. C. Turner, D. A. Kamundi & P. A. Manyama (eds), *Red List of South African Plants 2009. Strelitzia 25*. South African National Biodiversity Institute, Pretoria.

Victor, J. E. & Duncan, G. D. (2009h). *Lachenalia moniliformis*. In: D. Raimondo, L. von Staden, W. Foden, J. E. Victor, N. A. Helme, R. C. Turner, D. A. Kamundi & P. A. Manyama (eds), *Red List of South African Plants 2009. Strelitzia 25*. South African National Biodiversity Institute, Pretoria.

Victor, J. E. & Duncan, G. D. (2009i). *Lachenalia nordenstamii*. In: D. Raimondo, L. von Staden, W. Foden, J. E. Victor, N. A. Helme, R. C. Turner, D. A. Kamundi & P. A. Manyama (eds), *Red List of South African Plants 2009. Strelitzia 25*. South African National Biodiversity Institute, Pretoria.

Victor, J. E. & Duncan, G. D. (2009j). *Lachenalia polypodantha*. In: D. Raimondo, L. von Staden, W. Foden, J. E. Victor, N. A. Helme, R. C. Turner, D. A. Kamundi & P. A. Manyama (eds), *Red List of South African Plants 2009. Strelitzia 25*. South African National Biodiversity Institute, Pretoria.

Victor, J. E. & Duncan, G. D. (2009k). *Lachenalia schelpei*. In: D. Raimondo, L. von Staden, W. Foden, J. E. Victor, N. A. Helme, R. C. Turner, D. A. Kamundi & P. A. Manyama (eds), *Red List of South African Plants 2009. Strelitzia 25*. South African National Biodiversity Institute, Pretoria.

Victor, J. E. & Duncan, G. D. (2009l). *Lachenalia verticillata*. In: D. Raimondo, L. von Staden, W. Foden, J. E. Victor, N. A. Helme, R. C. Turner, D. A. Kamundi & P. A. Manyama (eds), *Red List of South African Plants 2009. Strelitzia 25*. South African National Biodiversity Institute, Pretoria.

Victor, J. E. & Duncan, G. D. (2009m). *Lachenalia viridiflora*. In: D. Raimondo, L. von Staden, W. Foden, J. E. Victor, N. A. Helme, R. C. Turner, D. A. Kamundi & P. A. Manyama (eds), *Red List of South African Plants 2009. Strelitzia 25*. South African National Biodiversity Institute, Pretoria.

Victor, J. E., Duncan, G. D. & Raimondo, D. (2009a). *Lachenalia angelica*. In: D. Raimondo, L. von Staden, W. Foden, J. E. Victor, N. A. Helme, R. C. Turner, D. A. Kamundi & P. A. Manyama (eds), *Red List of South African Plants 2009. Strelitzia 25*. South African National Biodiversity Institute, Pretoria.

Victor, J. E., Duncan, G. D. & Raimondo, D.(2009b). *Lachenalia lactosa*. In: D. Raimondo, L. von Staden, W. Foden, J. E. Victor, N. A. Helme, R. C. Turner, D. A. Kamundi & P. A. Manyama (eds), *Red List of South African Plants 2009. Strelitzia 25*. South African National Biodiversity Institute, Pretoria.

Victor, J. E. & Raimondo, D. (2009). *Lachenalia thomasiae*. In: D. Raimondo, L. von Staden, W. Foden, J. E. Victor, N. A. Helme, R. C. Turner, D. A. Kamundi & P. A. Manyama (eds), *Red List of South African Plants 2009. Strelitzia 25*. South African National Biodiversity Institute, Pretoria.

Vlok, J. H., Duncan, G. D. & Raimondo, D. (2009). *Lachenalia ameliae*.In: D. Raimondo, L. von Staden, W. Foden, J. E. Victor, N. A. Helme, R. C. Turner, D. A. Kamundi & P. A. Manyama (eds), *Red List of South African Plants 2009. Strelitzia 25*. South African National Biodiversity Institute, Pretoria.

Vlok, J. H. & Raimondo, D. (2009). *Lachenalia haarlemensis*. In: D. Raimondo, L. von Staden, W. Foden, J. E. Victor, N. A. Helme, R. C. Turner, D. A. Kamundi & P. A. Manyama (eds), *Red List of South African Plants 2009. Strelitzia 25*. South African National Biodiversity Institute, Pretoria.

Vogel, S. (1954). *Blutenbiologische Typen als Elemente der Sippengliederung dargestellt anhand der Flora Sudafrikas*. Gustav Fischer Verlag, Jena.

Vogelpoel, L. (1986). Flower colour — an appreciation. *South African Orchid Journal* 17 (3): 109–114.

Vogelpoel, L. (1995). Some observations on pigments of plastid origin. *South African Orchid Journal* 26 (2): 55–59.

Warner, B. & Rourke, J. P. (1990). Riebeek Kasteel — in the footsteps of de la Caille and Thunberg. *Sagittarius* 5 (1): 12–17.

Waterhouse, G. (1932). *Simon van der Stel's Journal of his Expedition to Namaqualand, 1685–6*. Hodges, Figgis and Co., Dublin.

Watson, W. (1904). *Lachenalia convallarioides*. The Garden 65: 264.

Watt, J. M. & Breyer-Brandwijk, M. G. (1962). *The Medicinal and Poisonous Plants of Southern and Eastern Africa* (2nd edition). E. & S. Livingstone Ltd., Edinburgh.

Wetschnig, W., Pfosser, M. & Prenner, G. (2002). Zur Samenmorphologie der Massonieae Baker 1871 (Hyacinthaceae) im Lichte phylogenetisch interpretierter molekularer Befunde. *Stapfia* 80: 349–379.

Wijnands, D. O., Wilson, M. L. & Toussaint van Hove, T. (eds), (1996). *Jan Commelin's Monograph on Cape Flora*. Published by the editors, Cape Town.

Willdenow, C. L. (1799). *Agapanthus ensifolius*. Species plantarum 2 (1): 48. G. C. Nauck, Berlin.

Wilson, M. L., van Hove-Exalto, Th. Toussaint & van Rijssen, W. J. J. (2002). *Codex Witsenii*. Iziko Museums, Cape Town.

Zakharyeva, O. I. & Makushenko, L. M. (1969). Chromosome numbers of monocotyledons belonging to the families Liliaceae, Iridaceae, Amaryllidaceae, Araccae. *Botanicheskii Zhurnal* 54: 1213–1227.

GLOSSARY

abaxial: of the lower surface
acropetalous: flowers opening in a longitudinal sequence from the base of the inflorescence upwards
actinomorphic: radially symmetrical
acute: sharp, ending in a point
adaxial: of the upper surface
adventitious: with respect to roots, not arising from primary root tissue
aerenchyma: non-pigmented tissue
allele fixation: when a form of a gene (allele) is carried by all members of a population, resulting in all individuals being homozygous for that allele and breeding true from seed
allopatric: occurring in areas geographically isolated from one another and thus unable to crossbreed
amphistomatic: stomata present on upper and lower leaf surfaces
anemochory: seed dispersal by wind
anther: the pollen-bearing part of the stamen at the top of the filament
anthesis: period during which the flowers are fully open and functional
anthocyanin: a water-soluble pigment present in the vacuoles of the subepidermal cell layer
apex: the tip or distal end
apiculate: terminated by a short, sharp, flexible point
arcuate: curved
attenuate: showing a long and gradual taper
autapomorphy: a derived character that differs from the ancestral condition
axile: ovules borne on the central axis

basifixed: attached at the base
bifacial: upper and lower surfaces with two distinct faces
binary characters: traits that have two character states
biseriate: arranged in two rows or whorls at different levels
bract: a much-reduced leaf subtending the base of a pedicel or flower
bulbil: an aerial bulblet, such as those that develop along the margins of *L. bifolia* leaves
bulblet: a subterranean offset or daughter bulb, usually formed on the perimeter of the basal plate, or rarely at the tip of a stolon

campanulate: bell-shaped
canaliculate: channelled
capitate: a compact cluster or head
capsule: a type of dried fruit
carotenoid: a yellow, orange or red pigment found in the mesophyll
cartilaginous: tough and hard
cataphyll: a rudimentary leaf preceding the true leaf or leaves
cernuous: slightly drooping
chalaza: the basal end of the seed, opposite the micropyle
chlorenchyma: tissue containing chloroplasts
chloroplast: specialised organelles that contain the green pigment chlorophyll
chromoplast: specialised organelles usually containing yellow or orange carotenoids pigments
ciliate: having fine hairs on the margins
ciliolate: having minute hairs on the margins
cladistics: a method of classifying living organisms on the basis of the relationships between phylogenetic branching patterns from a common ancestor
clavate: club-shaped
colliculate: covered with small rounded elevations
conduplicate: folded together lengthwise
contractile: (with reference to thick roots) pulling the bulb deeply into the soil
coriaceous: leathery
corymb: a more or less flat-topped raceme in which the pedicels arise from different points but all reach to more or less the same level
crisped: irregularly wavy and curled
cupule: a cup

declinate: bent or curved downwards or forwards
dehisce: the spontaneous opening of the seed capsule along structural lines

dichotomy: a mode of branching by constant forking
distal: situated farthest away
dorsifixed: attached to the back

edaphic: relating to soil
epicuticular: with reference to the outermost waxy layer of the leaves
epidermis: the outermost layer of cells covering the leaves
epithet: a name
eutrophication: water bodies with excessive runoff of nutrients from agriculture or industrial sites
exine: the outer layer of the pollen grain wall
exserted: protruding beyond the perianth, with reference to the stamens

flabellate: fan-shaped
fynbos: a derivative of the Dutch 'fijnbosch', referring to the community of small shrubs, evergreen and herbaceous plants occurring on nutrient-poor soils in the south-western and southern Cape, with a Mediterranean climate

gamophylly: fusion of leaf bases within the bulbs (as in *L. barkeriana* and *L. pusilla*)
gamotepalous: with fused tepals
genotype: the genetic make-up, as distinguished from the outward appearance (phenotype)
geoflorous: producing inflorescences at ground level
gibbosity: swelling
glabrous: smooth
glaucous: greyish- or bluish-green

heterochromatic: characterised by different colours
heuristic: relating to a usually speculative formulation that serves as a guide in the investigation or solution of a problem
hilum: the scar that indicates the point of attachment of a seed
homoisoflavonoid: a phytochemical compound
homoplasy: organs within different species that resemble each other and have the same functions, but which do not have a common ancestral origin or development
hydrochory: seed dispersal by water
hypogeal: remaining below ground, with respect to the bulb or the cotyledon during seed germination
hysteranthous: producing leaves after the flowers

included: not protruding beyond the perianth, with reference to the stamens

inflorescence: a group of flowers on the rachis
infructescence: the ripe fruiting stage of the inflorescence
ingroup: the group of immediate interest, i.e. *Lachenalia*
isodiametric: having equal diameters

keel: a ridge

lamina: a leaf blade or expanded portion
lanceolate: lance-shaped; much longer than broad
lignin: a cellulose-like substance that adds strength to cell walls.
linear: long and narrow, the sides parallel or nearly so

mast-flowering: massive flowering and fruiting at intermittent intervals
median: in the middle
mesophyll: the soft tissue inside a leaf (comprising green parenchyma), between the upper and lower epidermis
monophyletic: a group of organisms forming a clade, consisting of a last common ancestor and all of its descendents
monosulcate: pollen surface with a single groove
multistate characters: comprising three or more character states
myrmecochory: seed dispersal by ants

nectary: a glandular structure that secretes nectar
nototribic: (of pollen) deposited onto the back of the insect

obliquely: slanting, not perpendicular
oblong: obtuse at each end, with the sides almost parallel
offset: a small bulb arising from the base of a mature bulb
ontogeny: the origin and development of seedlings to maturity
outgroup: a taxon included in a study to polarise the characters of the ingroup
ovate: broadest at the lower end, resembling the longitudinal section of an egg

palisade: elongated cells of the mesophyll
parallelism: the development of similar characters (or states) separately in two or more lineages of common ancestry
parapatric: occupying the same geographic area but always separated by different substrates
paraphyletic: a group of taxa containing its last common ancestor but not containing all the descendants of that ancestor

parenchyma: tissue comprising thin-walled, unspecialised cells that forms the greater part of the leaves
parsimony: the principle that the simplest explanation that can explain the data is to be preferred
pedicel: the stalk of one flower
pendulous: hanging down
perianth: the floral envelope comprising the perianth tube, outer and inner tepals
phenotype: the outward appearance
phloem: the sugar-conducting tissue within plant stems
phylogeny: the evolutionary development and history of a species
placentation: the arrangement of ovules within an ovary
polychotomy: a section of a polychotomy in which the evolutionary relationships cannot be fully resolved to dichotomies
polymorphic: represented by two or more morphological forms
polyploidy: having a chromosome complement of more than two sets of the monoploid number
prostrate: lying absolutely flat on the ground
proteranthous: with the leaves produced and beginning to wither before flowering takes place
pustule: a blister-like protuberance

raceme: a simple, elongated inflorescence with pedicelled (stalked) flowers
rachis: the axis that bears the flowers
radially: perpendicular to the pedicel
raphe: the longitudinal ridge that forms below the hilum of a seed
raphides: needle-shaped crystals
recurved: bent or curved downward or backward
renosterveld: a derivative of the Dutch 'rhenosters bosch', referring to the shrubland community in the south-western and southern Cape that is dominated by members of the Asteraceae family, occurring on nutrient-rich, heavy soils
reticulate: netted
reversal: when an ancestor had a derived condition but the trait evolved back to the primitive condition
rugose: wrinkled

scape: the stalk of an inflorescence
sessile: without a stalk
soboliferous: producing bulbils at the tips of horizontal, subterranean stolons
spike: a simple, elongated inflorescence with sessile flowers

stellate: star-like
stomata: the pores in leaves through which transpiration takes place
strophiole: an appendage that occurs in the micropylar region of the seed
subspecies: a subdivision of a species consisting of interbreeding, usually geographically isolated populations
sympatric: occupying the same or overlapping geographic areas, without interbreeding
sympodial: bulbs consisting of two (inner and outer) consecutive units that are attached to a basal plate
synanthous: having leaves that are produced simultaneously with the flowers
synapomorphy: a derived trait that is shared by two or more taxa of shared ancestry
syncarpic: consisting of several united fruits

taxon: a general term applied to any taxonomic element, irrespective of its level of classification
tepals: the outer and inner units of the perianth
terete: circular in transverse section
testa: the outer coat of a seed
trichome: a hair or bristle
tunicate: the loose outer layers surrounding the bulb

undulate: wavy up and down
urceolate: urn-shaped

vacuole: a cavity in the cytoplasm of a cell
ventricose: having a swelling on one side
versatile: anthers swinging freely about the point of attachment
verticillate: arranged in whorls, or seemingly so
vesture: covering

xanthophyll: a light yellow plastid pigment
xerophyte: a plant of an arid habitat
xylem: the supporting and water-conducting tissue

zygomorphic: irregular, divisible into equal halves in one plane only

APPENDIX

LIST OF RUDOLF SCHLECHTER'S MANUSCRIPT NAMES, WITH IDENTIFICATIONS ADDED

Lachenalia Kunickiana Schltr. ms (*Schlechter* 10471 [Elim]) = *Lachenalia ensifolia* (Thunb.) J. C. Manning & Goldblatt subsp. *ensifolia* (B, BM, G, L, S, SAM, Z).

Lachenalia Loeseneriana Schltr. ms (*Schlechter* 8315 [Zout Rivier]) = *Lachenalia mutabilis* Lodd. ex Sweet (B, BM, BOL, GRA, K, PRE, S) and *Lachenalia framesii* W. F. Barker (G, L, PRE, Z).

Lachenalia montigena Schltr. ms (*Schlechter* 10816 [Pakhuisberg]) = *Lachenalia membranacea* (W. F. Barker) G. D. Duncan (BM, G, GRA, K, L, PRE, S, Z).

Lachenalia ophioglossoides Schltr. ms (*Schlechter* 7937 [Piqueniers Kloof]) = *Lachenalia mutabilis* Lodd. ex Sweet (BM, GRA, K, Z).

Lachenalia pachycaulos Schltr. ms (*Schlechter* 8260 [Klyn Fontein]) = *Lachenalia mutabilis* Lodd. ex Sweet (BM, G, GRA, K, PRE).

Lachenalia physopus Schltr. ms (*Schlechter* 10856 [Agtertuin]) = *Lachenalia violacea* Jacq. (B, BM, G, GRA, K, L, LD, S, Z).

Lachenalia picta Schltr. ms (*Schlechter* 10878 [Doornfontein]) = *Lachenalia schlechteri* Baker (B, BM, G, GRA, K, L, PRE, S, Z).

Lachenalia pulchella Schltr. ms (*Schlechter* 8968 [Mitchell's Pass]) = *Lachenalia pallida* Ait. (B, BM, G, GRA, K, L, PRE, S, Z).

Lachenalia satyrioides Schltr. ms (*Schlechter* 8053 [Lang Kloof NW of Clanwilliam]) = *Lachenalia undulata* Masson ex Baker (BM, G, GRA, K, L, PRE, S, Z).

Lachenalia Schwolkeana Schltr. ms (*Schlechter* 8771 [Koude Berg] = *Lachenalia membranacea* (W. F. Barker) G. D. Duncan (B, BM, G, GRA, K, L, PRE, S, Z).

Lachenalia sessiliflora Schltr. ms (*Schlechter* 10740 [Porterville]) = *Lachenalia lutea* G. D. Duncan (BM, G, GRA, K, PRE, S, Z).

Lachenalia sessiliflora Schltr. ms (*Schlechter* 11160, mixed sheet [Brakdam]) = *Lachenalia carnosa* Baker (BM, PRE) and *Lachenalia framesii* W. F. Barker (G, K, L, LD, S, Z).

GENERAL INDEX

Page numbers for illustrations are in **bold**.

Acocks, J. P. H. 234, 243, 410
Acta Helvetica 5
adventitious roots 58
aerenchymatous tissue 71
aerial scapes 70
aerodynamic capsules 75
Africana Museum 2
African honey bees 81
African Monarch butterfly 79
Aiton, William xiv, 7, 157, 198
Albany Museum 129
Albany Thicket 41
Aldborough Rectory 12
Alert List for Environmental Weeds 110
allotetraploids 87
Alpha Farm 308
Anderson, Fay 111, 112, 132, 154, 265, 273, 334, 386, 404, 414, 415
Andrews, H. C. 149, 267
anemochory 85
aneuploid series 86
aneuploidy 86
Annales Musei Goulandris 440
Annals of the Bolus Herbarium 366, 406
anthers 73, 74
anthesis 74
anthocyanins 61, 70, 73
ants 81, 85
aphids 29
Arbuthnot, Isobel A. 222
Archer, Fiona 249
Archiv der Gewachskunde 7, 203
artificial hybridisation 88
Attwell, Jill 275
Aus 362

Bachmann, Dr F. E. 341
Baileya 11
Bailey Hortorium 11
Baker, J. G. 7, 24, **24**, 298, 313, 317, 341, 348, 402
Banks Herbarium 139
Banks, Sir Joseph 332
Barber, M. E. 333
Barker, Winsome F. xiii, 9, **10**, 126, 140, 145, 147, 161, 166, 176, 186, 189, 202, 216, 222, 227, 231, 237, 243, 256, 258, 259, 263, 270, 271, 283, 297, 298, 300, 312, 313, 318, 323, 344, 369, 371, 380, 393, 395, 400, 407, 416, 421, 423
basal plate 58
Basel 6
basic chromosome numbers 86, 87
Batten, Dr Auriol 383
Baxter, William D. 270
Beaufort West 383
B-chromosomes 87
bee pollination 81
beewolfs 79, 442
Bellville 36
bird-pollination syndrome 83
blister beetles 79, 80
Bitterfontein 225
Bloutoring Station 419
Bodleian Library 3
Bokelmann, H. 212
Bolus, Dr Harry 354, 387 425
Bolus, Dr Louisa 9, 390
Bolus Herbarium 10, 227, 280, 410
Boos, Francis 4
Botanische Jahrbücher 235, 283, 317, 341

Botaniska Notiser 412
Bothalia 9, 128, 204, 215, 220, 233, 236, 238, 244, 248, 250, 255, 262, 275, 286, 292, 294, 322, 383, 391, 414
Botha, Prof. M. C. 135, 278, 295
Bot River 262
Bowie, James xv, 349
Bowker, Col. J. H. 239, 333
bracts 71
Bredasdorp 172
Breede Shale Renosterveld 41
breeding barriers 86, 87, 95
Brenthurst Library, The 2, 3
Brenthurst Press, The 2
Breyne, Jakob and Johann Philipp 114
British Museum (Natural History) 10, 12, 293, 332, 339, 413
Bruyns, Dr Peter 171, 367
Buc'hoz, Pierre-Joseph 156
Buchuberge 178
Buhr, H. 393, 400
bulb 58
  apex 59
  scales 59
  size 59
bulb mites 29
bulbils 57
bulblets 57
Bulletin de L'Herbier Boissier 168, 341
Burchell, W. J. 251
Burgers, C. J. 243
Burmann, N. L. 156, 162
butterfly species 79
Buxbaum, J. C. 2, 156, 162

Cabbage White butterfly 79
Calvinia 35, 36, 215, 233, 235, 244
Cam, Dr T. 12
Cape Agulhas 35
Cape honey bees 81, 82
Cape of Good Hope Nature Reserve 272
Cape porcupines 113, 330
Cape Town International Airport 224
capsule dehiscence 110, 191
capsules 75, 85
carotenoids 70
Catalogue of the Linnean Herbarium 139, 440
cataphylls 59, 60
caterpillars 29
Cederberg Sandstone Fynbos 41
centres of diversity 35
centres of endemism 41
chalazal region 79
Chaplin, P. 251
character states 92
   multistate characters 90
   qualitative characters 90
   quantitative characters 90
Charlie's Hoek 361
Chater, W. 205
Chelsea Physic Garden 12, 114
chlorophyll 70
chloroplasts 73
chromoplasts 73
chromosomal incompatibility 45, 87
chromosome numbers 86, 89
Cirillo, Domenico 157
Citrusdal 292
cladistic study 90
Clanwilliam 35, 36, 121, 128, 206
Claudius, Heinrich 1, 2
Codex Witsenii 1, 2
Codicis Comptoniani 2
Colesberg 36
Collectanea 4, 10, 114, 118, 136, 143, 160, 180, 188, 192, 201, 295, 328, 329, 331, 370, 426
Collett, Rhona 122
Colville, James 326
Commelin, Jan 1

Compton Herbarium 9, 10, 135, 152, 207, 279, 316, 365, 416, 443
Compton, Prof. R. H. 9, 245, 269, 303, 356, 396
Compton, Right Reverend Bishop Henry 2
conservation 33
   conservation status 33
   Custodians for Rare and Endangered Wildflowers 34
   Millenium Seed Bank 34
   South African National Biodiversity Institute (SANBI) 34
contractile roots 58
convergence 92
contact dermatitis 21
Courtenay-Latimer, Dr Marjorie 257
Crosby, Trevor 11, 12, 14
cross-pollination 79
cultivation 17
   aspect 17
   burning pyrophytic lachenalias 23
   feeding 23
   frost-hardy species 19
   fully hardy species 19
   growing lachenalias in containers 17
   growing medium 19
   hardiness 19
   lachenalias for the garden 25
   planting 20
   selection of best species for pot cultivation 25
   selection of most attractive xerophytic species 24
   selection of most strongly scented species 24
   selection of species for cut flowers 23
   selection of species for moist growing conditions 24
   species recommended for garden cultivation 26
   watering 21
Curtis's Botanical Magazine 6, 346, 389, 396, 401, 409, 434, 439

Curtis, William 6

damping-off fungi 30
Darwin, Charles 12, **13**
dehiscence 76
De Hoop Nature Reserve 242, 435
de Lachenal, Professor Werner 5, **6**
desert biome 41
Diels, F. L. E. 236, 284
Dinter, M. K. 179, 286, 289
diploids 86
dispersal strategies 79
   dispersal by ants 191
Dissertatio de Novis Generibus Plantarum 6, 141
Dold, A. P. 11, 335
dormancy 50
Drège, I. L. 129
Drège, J. F. 407
drone fly 79, 80
Dutch East India Company 1
Dwyka Tillite 38

East London Museum 257
Ecklon, C. F. 220, 313, 317, 336
ecology 41
   adaptive strategies 47
   altitudinal range 41
   aridity 47, 57
   biomes 41
   defence 49
   fire 50, 55, 175
   floristic regions 41
   habitat 41
   herbivore avoidance 49
   invertebrate predators 49, 50
   molerats 50
   natural hybrids 45
   precipitation 45
   sympatric species 42, 45, 87
   temperature 45
   vegetation types 41
   xeromorphic features 47
Edinburgh Journal of Botany 429, 431, 437, 439
Edith Stephens Wetland Park 224
Edwards, Prof. T. J. xiii
Ehlers, Dirk and Esna 113

elaiosome 85
Emperor Francis I 4
Emperor Joseph II 4
Encyclopédie Méthodique 7
endemism 41
Engler, H. G. A. 141
Enumeratio Plantarum 7, 109, 267, 426, 439
epidermis 62, 64
Erfdeel Farm 290
Esterhuysen, Elsie 313
evolution 86
Exbury Gardens xiv

Fagel, Baron Hendrick 1
Fauresmith 250
Feddes Repertorium 178, 288, 435
filaments 73, 74
 orientation of 74
fire-dependency 50, 55, 175
Flora Capensis 8, 239, 241, 280, 353, 439
floral symmetry 96
flowering 51, 55, 56, 87
 erratic 55
 phenology 51, 56, 87
Flowering Plants of Africa 9, 267, 335, 349, 386
flowers
 geoflorous 81
 narrowly campanulate 71
 oblong-campanulate 71
 structure of 79
 tubular 71
 urceolate 71
 widely campanulate 71
foetid scent 307
Fourcade, H. G. 241, 261
Frames, Percival Ross 410
fruits 75
fungal diseases 30, 31
fungal rotting 31
fusiform cells 64
fynbos 41

gamoep 248
gamophylly 59
Gartenflora 12
Garten-Zeitung 12

Gazophylacii Naturae & Artis seu Herbarium Capense 2
geoflorous species 59
Ghent 7
gibbosities 72
Giess, Johan Wilhelm 362
gifkop 186
Gilfillan, D. F. 251
Gillett, J. B. 152
Glasnevin 12, 13
Glover, R. 366
Gouda 168
Goudini 391
grassland biome 41
Groot Bo-Kouga 268
Grootbos Private Nature Reserve 275
Groot Karasberge 366
Guthrie, Louise 407
gynoecium 75

Haarlem 260
habit 57
Hall, Harry 112, 246, 430
Hansford, Gerard 149
Hantam Karoo 41
Hantamsberg 182
Harmony Flats Geometric Tortoise Reserve 440
Harrower, Adam 152, 392
Heady Maiden day-flying moth 79, 81
Hellskloof 377
Helme, Nick 209, 443
helmeted guineafowl 113
Herbert, William 348
Hermanus 281, 388, 390
Hex River 36
Hibbert, G. 405
Hiroshima University 11
Hitchcock, Anthony 152
Hofbaur, Joseph 5
Holgat River Canyon 416
homoisoflavonoids 164
homoplasy 91
Hondeklipbaai 350
honey bees 342, 442
Hooker, Joseph 8
Hooker, Sir William 8

Hopefield 341
Horto Dominae de Flines 114, 115
Hortus Kewensis 7, 113, 157, 195, 198, 338
hoverflies 79
hybrid swarms 88
hybridisation 12, 87, 88, 95
 history of 12

Icones Plantarum et Animalium 2
Icones Plantarum Rariorum 4, 5, 10, 114, 192, 203, 329, 331, 372, 426
Imperial Museum of Natural History 5
inflated scapes 70
inflorescence 68
infructescence 76, 84
Ingram, John 11
Isoetes Vlei 222

Jacob, Rev. Joseph 14
Jacquin Herbarium 5
Jacquin, J. F. 141
Jacquin, N. J. 10, 141, 181, 192, 198, 203, 295, 297, 328, 331, 339, 370
Joubert, Elbe 119, 172, 173
Journal of Botany 121, 226, 353, 370
Journal of Botany (new series) 324
Journal of South African Botany 2, 9, 110, 121, 125, 172, 175, 182, 184, 208, 213, 222, 252, 254, 257, 281, 290, 302, 304, 308, 309, 315, 358, 359, 362, 366, 368, 373, 377, 385, 388, 408, 419, 422, 424
Journal of the Botanical Society 10
Journal of the Linnean Society (Botany) 8, 311, 333, 336, 406, 434

Kamieskroon 299
Karoo Poort 368, 394
karyology 86
Ker Gawler, John Bellenden 348
Kew Herbarium 8, 10, 152, 251, 282, 313, 348
Kew Magazine 440
Kimberley 35
Kirstenbosch Botanical Garden 9, 58

Kirstenbosch Bulb Nursery xiii, xiv, 79, 125, 278
Kleinsee Namaqua Diamond Mine 417
Klinghardt Mountains 288, 289
Klipfontein 424
Klipkoppe Shrubland 41
Kliprand 229, 344
Koenig, J. G. 439
Komsberg Pass 429
Kotzesrus 442
Kouberg Farm 294
Kruger, Paul 418
Kunth, C. S. 7, 267

Lamarck, Jean-Baptiste 7
Lamberts Bay 217
Langhorne, Joanna 396
Lavranos, John 365, 416
leaf/leaves 60
    anatomy and micromorphology 62
    bases 61
    hysteranthous 50, 60
    juvenile macromorphology and ontogeny 62
    margin anatomy 66
    margin micromorphology 68
    morphology 60
    number classes 61
    orientation 60
    primordia 59
    proteranthous 55, 60
    shapes 62
    synanthous 50, 60
Leggatt, Hugo 349
Leiden 3, 4
Leipoldt, Dr C. Louis 292
Leipoldtville Sand Fynbos 41
Lemaire, C. A. 7
Lemoenpoort 315
Levyns, Dr M. R. 301
Lewis, Dr G. J. 291, 323, 410
L'Illustration Horticole 7, 204, 329
Linder Smith, Claire 141, 230, 296, 347
Linnaeus, Carl 3, 4, 114, 139, 439
Linnaeus, Carl junior 6
Linnaeus Systema Vegetabilium 6

Linnean Herbarium 139
Linnean Society 9
Loddiges, Conrad 381
longitudinal grooves 60
Lord Hill 367
Loubser, J. W. 391
Lourensford Farm 270
Lundgren, Dr J. 187, 226
Lutzeyer, Heiner 275

MacOwan Herbarium 123
MacOwan, Prof. Peter 354
maculation 61
Mader, P. A. 123
Malachite Sunbird 83
Malan, C. 421
Mantissa plantarum Altera 439
march flies 79, 80
Marloth, Dr Rudolf 215, 246, 303, 359, 384, 390, 420
Martin, Bina E. 126, 153, 308
Martley, Commander J. F. 270
Mary Gunn Library 3
Mason, Hilda 204
Masson, Francis 12, 114, 123, 135, 139, 144, 167, 227, 332, 339, 370, 372, 413
Matatiele 35
Mathews, J. W. 210, 212, 437
Maughan-Brown, Dr Herbert 435
Mauve, Amelia 421
McGregor Family 359
McMaster, Cameron 172
McMaster, Rhoda 172, 355
mealy bugs 30
median keels 72
Melkkamer Farm 426
mellitophilous brush mechanism 81
mellitophily 79, 81
Merxmüller, Hermann 365
mesophyll 64, 73
Messrs Lee and Kennedy 348
Middelpos 231, 410
Moedverloor 170
Moffett, A. A. 11
Molyneux Nature Reserve 414
monkey beetles 79, 80, 342
Montagu 322
Montagu Pass 280

Moore, F. W. 12, 13
morphological divergence 86
morphology 57
Mostert, Louis 263
Mostert's Ravine 121
moth 79
Mount Arthur Range 333
Muir, Dr J. 323, 405
Müller-Doblies, Dr Ute 436
Müller-Doblies, Prof. Dietrich 436
Murray, J. A. 6, 141
Murraysburg 36
myrmecochory 85

Nama Karoo biome 41
Namaqualand 1, 41
Nardousberge 175, 204
National Herbarium 391, 421
natural hybrids 87, 88, 95
Naturhistoriches Museum 6
Naturhistoriska Riksmuseet 133
nectar 81
Nelson, Reverend John G. 12
Nieuwoudtville 125, 236, 359, 361, 393, 397, 399, 433, 434
Niven, James 267
non-scented species 81
Nordenstam, Prof. Bertil 187, 226, 377
northern Cape Peninsula 36
Nortier, Hermione 177
Notizblatt des Koniglichen Botanischen Gartens und Museums zu Berlin 139, 157
Nova Acta Helvetica 6, 139, 141
Nova Genera Plantarum 144, 148, 320

octoploids 86
O'kiep 373, 387
Oliver, Dr Ted 112
Ongeluks Nek 35
Orange-breasted Sunbird 83
Oranjemund 35
Ornithogalum mosaic virus 15, 31
ornithophily 79, 83
outgroup 90
ovary 75
ovules 75

GENERAL INDEX

Paarden Island 213
Paine, Q. V. 212
Painted Lady butterfly 79, 81, 337, 342, 442
Pakhuis Mountains 36, 398
Pakhuis Pass 208
Palmiet Valley Farm 391
Pappe, C. W. L. 192
Parnell, Prof. John 313
Patensie 257
Paternoster 437
Pearson, Prof. H. H. W. 366
pedicels 70, 76
Peers, Victor Stanley 388, 390
Percy Sladen Memorial Expedition 366
Percy Sladen Memorial Trust 366
perianth 71, 72
   length 72
   shape 71
   tubes 72
Perry, P. L. 256, 316
Persoon, Dr C. H. 7
pests and diseases 29
Petiver, James 2
phenology 50, 55
phenotypic and genotypic variation 96
phenotypic plasticity 93, 96
phloem 66
pyrophytic species 50, 55, 175
Philadelphia 36
Phillips, Dr Edwin Percy 219, 285
Phillipson, P. B. 11
phylogeny 89, 90, 92
   data matrix 90
   informative characters 89
   informative gene regions 89
   ingroup 90
   macro-morphological characters 90
   molecular analyses 89
   morphological analyses 89
   natural classification 90
   outgroup 91
   parsimony analysis 90
   relationships 89, 91
   reproductive characters 92, 93
   strict consensus cladogram 91

   synapomorphies 91, 92
   terminal species pairs 91
   unweighted trees 91
   vegetative characters 92, 93
   weighted tree 91
phytogeography 35
   distribution 35
   allopatric speciation 37
   allopatry 37
   aridity 37
   centres of diversity 35
   endemism 41
   geographic speciation 37
   habitat heterogeneity 41
   parapatric species 38
   ploidy 37
   speciation 37
   topographical diversity 37
Phytographia 2
pigments 73
Piketberg 131, 153
Pillans, Neville 424
Plantarum minus cognitarum Centuria III 2, 162
Plantarum rariorum horti caesarei Schönbrunnensis 165, 348
Plantarum rariorum Regni Neapolitani 157
plastid pigments 73
ploidy levels 86, 87
Plukenett, Leonardi 2
Pole Evans, I. B. 251
pollen 74
pollen deposition 81
pollination biology 79
   bird pollination 83, 84
   rodent pollination 84
pollination syndromes 79, 83
polychotomy 91
polymorphism 61, 86, 96
polyploid complex 158
polyploidy 86, 87
Port Elizabeth 128
Prieska 36
primary sculpturing 78
primary seedling leaf orientation 62
Prince Albert District 384
Prodromi Fasciculi Rariorum Plantarum 3, 114, 339

Prodromus einer Flora von Südwestafrika 365, 372
Prodromus Florae Capensis 156
Prodromus Plantarum Capensium 6, 141, 320, 321, 432, 439
propagation 27
   bulblets, bulbils and division 28
   bulbils 28
   hand pollination 28
   leaf cuttings 28
   micropropagation 29
   seed 27
   stolons 28
pustules 61
Pym, C. A. 335
pyrophytes 50, 55, 175

racemes 70
   ordinary 68
   corymbose 68
   subcapitate 68
   subcorymbose 68
racemose inflorescences 68
rachis 70
   colour 70
   shape 70
Redouté, P. J. ii, 137
Regius, Karen 362
Rhodes University 129
Riebeek Kasteel 134
Rietfontein Farm 374
Rijksherbarium 3
Riversdale 343
rodent-pollinated flowers 84
Rogers, Rev. W. M. 192
Rondekop 238
Roodeplaat 13, 14, 144
roots 58
   adventitious 58
   contractile 58
   fibrous 58
root hairs 62
Rosh Pinah 364
Ross-Frames, Percival 227, 412
Rourke, Dr J. P. 135
Royal Horticultural Society 9, 12, 14
rust fungus 31

Salisbury, R. A. 7, 167
Salter, Paymaster Captain T. M. 177, 214, 226, 259, 272, 293, 387
Sargeant, Percival (Percy) A. 172
scales 47
scapes 70
  colouration 70
scented species 81
Schäfer, Dr Fritz 289
Schelpe, Prof. E. A. 183, 345
Schlafkuppe 286
Schlechter, Maximilian 254
Schlechter, Rudolf 9, **9**, 168, 177, 245, 252, 255, 282, 295, 374, 422
Scholl, George 4
Schönbrunn 4
Schönland, Selmar 129
Schultes, J. A. 332
Schultes, J. H. 332
Scott, Robert 172
secondary sculpturing 78
seed
  characters 77
  dispersal 84, 110
  shape 77
  germination 62
Selmar Schönland Herbarium 129
Shackleton, Lydia 13
Simons Bay 324
Simonstown 392
slugs 30
Smith, E. D. 381
Smitswinkel Bay 150
snails 30
solitary bees 82, 83, 182, 342
Solly, J. F. 220
Somerset East 354
South African Astronomical Observatory 121, 194, 201
South African Gardening and Country Life 356, 390, 403
South African Journal of Botany 9, 186, 225, 229, 242, 278, 299, 343, 344, 350, 364, 394, 397, 399
South African Museum 1
South African National Biodiversity Institute 3, 9

South African Public Library 1
southern Cape Peninsula 36
Southern Double-collared Sunbird 82
Sparrman, Anders 133
speciation 86, 95
Species Plantarum 3, 7
spicate inflorescences 68
spikes 70
Spring and Winter Flowering Bulbs of the Cape 171
Springbok 36, 252
Staatsbibliothek Preussischer Kulturbesitz 1
stable characters 95
stamens 73
  exsertion 73
Stayner, Frank J. 309, 410
Stearn, William 114
steenbuck 113
Steinkopf 374, 385, 422
stellate trichomes 64
stigmas 75
stolons 57
stomata 64
strophioles 78, 79, 85
  length 77
styles 75
subcapitate racemes 79
subterranean scapes 70
Succulent Karoo 41
Summerfield, Dr Alison 11
Summerfield, Gordon 171, 231, 359, 374
sunbird-pollinated species 83, 84
Supplementum Plantarum et Specierum Plantarum 6, 139
Sutherland 246, 408, 430
Swartland Granite Renosterveld 41
Swartland Shale Renosterveld 41
Swedish Museum of Natural History 377
Sweet, Robert 326, 381
Swellendam 278
sympatric species 42, 45, 87
Synnot, W. 326
Synopsis der Mitteleuropäischen Flora 157
Synopsis Plantarum 7

Systema Vegetabilium 109, 439

taxonomic treatment 95
  cryptic species 95
  keys 98
  sectional subdivision 96
  species arrangement 96
  species concepts 95
  subgeneric classification 96
tepals 73
  inner 73
  outer 72
  posture 71
  shape 71
terminal flowers 72
testa surfaces 78
tetraploids 86
The Botanist's Repository 146, 149, 203, 264, 403
The British Flower Garden 326, 379
The Flora Capensis of Jakob and Johann Philipp Breyne 2, 3, 114, 339
The Flowering Plants of Africa 206, 361
The Flowering Plants of South Africa 9, 210, 217, 272, 357, 393, 407, 413, 434
The Gardener's Chronicle 149, 401
The Gardener's Chronicle and Agricultural Gazette 149
The Genera of Plants 7, 167, 204, 439
The Lachenalia Handbook 11, 89, 280
The New Zealand Gardener 14
The Plantsman 11
Thomas, Margaret 206, 207, 208, 222, 269, 349, 359
Thomas, Vicki 276, 287, 389
Thunberg, C. P. 6, 109, 134, 188, 432
Transactions of the Royal Society of South Africa 128, 260
Trattinnick, Leopold 7, 203, 348
trichomes 62, 64
  simple 64
  stellate 64
Trinity College, Dublin 1, 313, 335

Trinity College Library 1
Tsarisberge 35
tubular flowers 71
Tulbagh 36, 317
Tulbagh Valley 35
tunics 59

uninformative characters 95
University of Basel 6
University of Berlin 284
University of Cape Town 10
University of Dublin 1
University of KwaZulu-Natal xiii
University of Leeds 11, 13, 14
University of the Free State 11
University of Zurich 169
Uppsala 10
Uppsala University 135
Uromyces lachenaliae 127
urticaria 21

van Berkel, Nicky 286
van der Merwe, Dr F. Z. 391
van der Stel, Simon 1
van Jaarsveld, Ernst 417
Vanrhynsdorp 283, 412
van Warmelo, Vivienne Judith 268

vascular bundles 66
Vegetable and Ornamental Plant Research Institute 13, 14
Vergelegen 220
Vienna 4, 5, 10, 372
viral disease 31
Visagie, Marieta 305, 306, 409
von Jacquin, Baron Joseph Franz 5, **5**
von Jacquin, Baron Nicolaus Joseph 4, **4**
Vredenburg 110, 210, 437

Wakehurst Place 35
Ward-Hilhorst, Ellaphie 211, 212, 222, 223, 360
Ware, T. S. 402
Waterhouse, Gilbert 1
wax platelets 64
Wench, George 198
Western Cape Sandveld Flowers 204
Whitehill 302, 304, 356
Whitehill Railway Station 302
Wildschutskraal 184
Willdenow, C. L. 109
William Sherard Collection 3
Williamson, Dr Graham 377

Williamson, T. 335
Williams, Richard 149, 162, 203, 440
Windhoek Herbarium 363
Wisura, Walter 416
Witputs-Süd 365
Witsen, Nicolaas 1
Witzenberg 336
Wolley-Dod, Anthony 115, 152, 185, 214
Worcester 35, 255, 309
Worcester Valley 392
Wright, Charles 324
Wupperthal 254, 420

xeromorphic features 47
xylem 66

yeast-scented species 81
Young, E. W. 280

Zaris Farm 363
Zeyher, Carl L. 220, 282, 293, 313, 317, 336
zygomorphic perianths 96

INDEX OF SCIENTIFIC NAMES

Accepted names are in **bold**, synonyms and invalid names in *italic*. Page numbers for principal references and illustrations are in **bold**.

Agapanthus ensifolius (Thunb.) Willd. 431
Aletris bifolia Burm.f. 155, 156, 157
Aletris linguaeformis Burm.f. 160, 162
Aloe brevifolia 26
    ferox 262
    framesii 417
Amata cerbera 81
Amegilla nivea 82, 182
Anthophora diversipes 82
Apis mellifera subsp. capensis 82
Aristea spiralis 403
Babiana secunda 332
    villosa 150, 168
Baeoterpe corymbosa (L.) Salisb. 439, 440
Brachyscypha Baker 97
Brachyscypha undulata (Thunb.) Baker 97, 188
Brunsvigia radula 352, 413
Centropodia glauca 376
Chloriza Salisb. 7, 97
Chondropetalum tectorum 267
Cinnyris chalybeus 82
Coelanthus Willd. ex J. A. Schultes & J. H. Schultes 96
Coelanthus complicatus Willd. ex Roem. & Schult. 7, 8, 96, 108, 109, 444
Colchicum latifolium 411
Convallaria majalis 333
Cynodon dactylon 260
Dorotheanthus bellidiformis 26
Drosanthemum hispidum 213
Empodium veratrifolium 113, 146
Erica cerinthoides 185
Eristalis tenax 80
Eucomis regia 349
Euphorbia mauritanica 308
Euryops speciosissimus 308
Family Asparagaceae (Hyacinthaceae s.l.) 95
Felicia dubia 26
Freesia andersoniae 252

Gazania krebsiana 317
Geissorhiza inflexa 150, 168
Gladiolus alatus 337
    equitans 352
    priorii 146
Haemanthus canaliculatus 275
Himas Salisb. 7, 96
Holothrix villosa 150
Hyacinthus 9
    bifolius Boutelou ex Cav. 426
    corymbosus L. 96, 439, 440
    gawleri (Kunth) Baker 439
    orchioides L. 3, 113, 114, 339
    paucifolius W. F. Barker 10, 437
Ixia vinacea 168
    curta 204
Jordaaniella dubia 26
Lachenalia J. Jacq. ex Murray **96**
    section **Angustae** 326
    section **Lachenalia** 108
    section **Latae** 346
    section **Oblongae** 191
    section **Urceolatae** 378
    subgenus *Brachyscypha* (Baker) Baker 97
    subgenus *Chloriza* (Salisb.) Baker 97
    subgenus *Coelanthus* (Willd. ex J. A. Schultes & J. H. Schultes) Baker 96
    subgenus **Lachenalia** 108
    subgenus *Orchiops* (Salisb.) Baker 97
    subgenus **Polyxena** 426
    alba W. F. Barker ex G. D. Duncan 33, 38, 51, 101, **236**, **237**
    albida Tratt. 338
    algoensis Schönland 24, 51, 75, 92, 93, 99, 116, **128**, **129**, **130**
    aloides (L.f.) Engl. 5, 6, 12, 23, 25, 26, 33, 51, 70, 73, 74, 81, 83, 84, 86, 88, 91, 92, 96, 107, 135, **139**, **140**, **141**, 143, 144, 150, 152, 153

var. *aurea* (Lindl.) Engl. 146
var. *quadricolor* (Jacq.) Engl. 12, 143, 144
var. *vanzyliae* W. F. Barker 131, 133
aloides (L.f.) Hort. ex Asch. & Graebn. 139
aloides (L.f.) Pers. 133, 139
ameliae W. F. Barker 19, 20, 51, 62, **63**, 64, 68, 92, **419**, **420**
angelica W. F. Barker 20, 24, 25, 26, 33, 41, **48**, 49, 51, 55, 60, 61, **63**, 64, 68, 70, 71, 75, **77**, 78, 105, **350**, **351**, 372, 375
anguinea Sweet 19, 20, 24, 38, **39**, 42, 49, 51, 57, 59, 60, 70, 75, **76**, 104, 286, 306, **326**, **327**, 413
angustifolia Jacq. 338, 339
arbuthnotiae W. F. Barker 23, 24, 25, 26, 33, 42, 49, 51, 61, **67**, 68, 81, 101, 220, 221, **222**, **223**, 224
argillicola G. D. Duncan 27, 51, 68, 71, 76, 84, 85, 107, 110, **442**, **443**
attenuata W. F. Barker ex G. D. Duncan 19, 51, 102, 303, **322**, **323**
aurea Lindl. 12, 146, 149
aureo-reflexa Baker 12
aurioliae G. D. Duncan 19, 31, 51, 57, 106, 299, 357, **383**, **384**
bachmannii Baker 19, 21, 24, 33, 42, **43**, 51, **80**, 104, 195, **341**, **342**
barkeriana U. Müll.-Doblies, B. Nord. & D. Müll.-Doblies 10, 21, 24, 27, 51, 59, 68, 70, 71, 74, 76, 79, 81, 84, **85**, 91, 92, 100, 110, **186**, 188, 190, 191
bicolor Lodd. 12, 196, 198, 298
bifolia (Burm. f.) W. F. Barker ex G. D. Duncan 13, 14, 15, 20, 23, 25, **26**, 28, 35, 37, **38**, **41**, 45, 51, 57, 58, **59**, 60, 61, **65**, 71, 72, 74, 75, **76**, 79, 83, 84, 86, 87, 90, 91, 92, 95, 96, 99, 141, **155**, **157**, **158**, 162, 163, 164
bifolia 'Boundii' 14
bifolia Ker Gawl. 2, 264, 267
bolusii W. F. Barker 25, 35, 41, 51, 62, 106, **424**, **425**
botryoides Tratt. 201, 203
bowieana Baker 346, 348, 349
bowkeri Baker 19, 36, **37**, 47, 51, 70, 71, 100, 103, **239**, **240**, 243, 333
brevipes Baker 341
bruynsii G. D. Duncan 51, 98, **170**, **171**
buchubergensis Dinter 21, 35, 49, 51, 64, 75, **77**, 78, 85, 91, 100, **178**, **179**, 307, 378

bulbifera (Cirillo) Engl. 2, 12, 86, 155, 157
calcicola (U. Müll.-Doblies & D. Müll.-Doblies) G. D. Duncan 27, 51, 68, 70, 71, 76, 84, 85, 107, 110, **435**, **436**
callista G. D. Duncan & T. J. Edwards 23, 25, 51, 70, 71, **72**, 83, 88, 92, 99, 134, 142, **153**, **154**, **155**
campanulata Baker 19, 25, 27, 35, 41, **43**, 45, 50, 51, 55, 61, **67**, 68, 75, 105, 313, 344, **353**, **354**, 355, 362
canaliculata G. D. Duncan 19, 37, 45, 51, 100, 182, 245, **246**, **247**, 304
capensis W. F. Barker 17, 24, 33, 42, 51, 64, 70, 73, 81, 100, **184**, 392, 402
carnosa Baker 28, 51, 59, 96, 106, 346, **406**, **407**
cernua G. D. Duncan 28, 51, 57, **69**, 106, 214, **391**, 392
comptonii W. F. Barker 19, 24, 25, 45, 49, 51, 55, 62, **63**, 64, 68, 71, **72**, **77**, 105, 168, 304, 346, **356**, 359, 369
concordiana Schltr. ex W. F. Barker 51, 100, **252**, **253**, 387
congesta W. F. Barker 17, 19, 24, 27, 42, 45, 51, 56, 59, 60, 61, 62, 70, 75, 92, 106, 345, **408**, **409**, 410
contaminata Aiton 3, 14, 15, 21, 24, 25, 26, 42, 45, 51, 57, 59, 74, 79, **81**, 90, 104, 114, 319, 337, **338**, **339**, 340, 342, 405
convallariodora Stapf 401, 402
convallarioides Baker 25, 33, 50, 51, 55, 74, 104, **333**, **334**, **335**, 337
var. *robusta* Baker 333, 335
cooperi Baker 444
corymbosa (L.) J. C. Manning & Goldblatt 17, **18**, 19, 21, 24, 25, 27, 33, 50, 51, 57, 60, 61, 66, 68, **69**, 71, 76, **77**, **80**, **81**, 84, 85, 96, 107, 110, 430, 438, **439**, **440**, **441**, 443
dasybotrya Diels 19, 33, 51, **78**, 79, 92, 101, **235**, 289
dehoopensis W. F. Barker 33, 51, 64, **65**, 102, 241, **242**, **243**
doleritica G. D. Duncan 19, 33, 38, **39**, 52, 71, **72**, 101, 183, **233**, **234**, 236
duncanii W. F. Barker 24, 33, 41, 42, 49, 52, **66**, **67**, 68, 101, 181, 182, 228, **229**, **230**
elegans W. F. Barker 24, 25, 31, 33, 52, **65**, 107, **393**, **394**, 399, 401, 405, 425, 426
var. *flava* W. F. Barker 393, 394, 396
var. *membranacea* W. F. Barker 393, 397, 398

THE GENUS LACHENALIA
INDEX OF SCIENTIFIC NAMES

var. *suaveolens* W. F. Barker 393, 399, 400
ensifolia (Thunb.) J. C. Manning & Goldblatt 19, 24, 25, 27, **66**, 68, 70, 71, 75, 76, **77**, 79, 81, 84, 85, 107, 110, 430, **431**, 435, 436, 443
   subsp. **ensifolia** 38, 52, **432**, **433**
   subsp. **maughanii** (W. F. Barker) G. D. Duncan 10, 52, 431, **433**, **434**
esterhuysenae W. F. Barker 311, 313
fistulosa Baker 17, 24, 25, 26, 33, 42, 52, 55, 72, 81, 107, 185, **401**, **402**
flava Andrews 12, 14, 23, 25, 26, 41, 45, **46**, 52, 55, **69**, 70, **83**, 84, 92, 99, 142, **146**, **147**, **148**, **149**, 152, 219, 396
fragrans Andrews 346, 348, 444
fragrans Jacq. 195, 199, 348
fragrans Lodd. 338
framesii W. F. Barker 24, 25, 38, **39**, 52, 59, 73, **77**, 78, 107, 227, 285, **412**, 417, 418
giessii W. F. Barker 24, 27, 35, 52, 59, 64, 68, 92, 105, **362**, **363**, 366, 367
gillettii W. F. Barker 196, 198, 199, 200
glauca (W. F. Barker) G. D. Duncan 23, 24, 52, 81, 103, 298, **299**, **300**, **301**
glaucina Jacq. 14, 115, 118, 405
   var. *parviflora* W. F. Barker 115, 121
   var. *pallida* Lindl. 444
glaucophylla W. F. Barker 19, 24, **48**, 49, 52, 60, 61, 72, **80**, 105, 357, **358**, 367
haarlemensis Fourc. 33, 52, 62, 70, 102, 259, **260**, **261**
hirta (Thunb.) Thunb. **1**, 2, 25, 49, 52, 61, 62, **63**, 68, **77**, 78, 102, 171, 316, **320**, **321**, 399, 425
   var. *exserta* W. F. Barker 320, 321
hyacinthina Hort. 338
hyacinthoides Jacq. 338, 339
hyacinthoides (L. *f.*) Lam. 444
inconspicua G. D. Duncan 19, 41, 52, 100, 226, **248**, **249**, 251, 295, 376
isopetala Jacq. 19, 24, 27, 45, **47**, 52, 55, 59, 60, 64, **74**, 77, **78**, 79, 82, 83, 100, **180**, **181**, 248, 308
judithiae G. D. Duncan 25, 52, 79, 104, **268**, 272
juncifolia Baker 19, 24, 26, 27, 45, 52, 57, 74, **77**, 79, 90, 96, 101, 260, 304, **311**, **312**, **313**, **314**, 317, 344, 369, 405
   var. *campanulata* W. F. Barker 313, 343, 344
karooica W. F. Barker ex G. D. Duncan 19, 20, 35, 41, 45, 52, 55, 70, 100, 103, 199, 242, 249, **250**, 385

karoopoortensis G. D. Duncan 52, 75, 106, **394**, **395**, 398, 401
klinghardtiana Dinter 20, 21, 28, 35, 49, 52, 61, 70, 102, 286, **288**, **289**, 378
kliprandensis W. F. Barker 17, 24, 25, 28, 52, 57, 59, 61, 62, 70, 104, **344**, **345**, 408, 410
krugeri G. D. Duncan 33, 52, 106, **417**, **418**, **419**
lactosa G. D. Duncan 33, 52, 70, 79, 103, **262**, 390
lanceaefolia Jacq. 4
latifolia Tratt. 7, 346, 348, 349
latimeriae W. F. Barker 42, 52, **78**, 79, 103, 256, **257**, **258**, **259**, 261
leipoldtii G. D. Duncan 33, 52, **65**, **67**, 68, 103, 291, **292**, **293**
leomontana W. F. Barker 17, 41, 52, **67**, 68, 102, **278**, **279**, 280, 281
lilacina Baker 118
liliiflora Jacq. **18**, 25, 33, 52, 55, 61, **80**, 88, 104, **328**, **329**, **330**
linguiformis Lam. 156, 157, 162
longibracteata E. Phillips 45, 52, 70, 71, 90, 101, 103, 150, **217**, **218**, 222
longituba (A. M. van der Merwe) J. C. Manning & Goldblatt 19, 24, **40**, 41, 45, 50, 52, 68, 70, 71, 75, **76**, 79, 81, 84, 85, 107, 110, **429**, **430**, 433
lucida Ker Gawl. 196, 198, 348
lutea G. D. Duncan 24, 25, 26, 52, **65**, **67**, 68, 101, 118, 219, **220**, **221**, 224, 270, 393
luteola Jacq. 7, 12, 14, 23, 25, 26, **49**, 52, 70, 73, **83**, 84, 92, 99, **136**, **137**, 138, 149
   var. *pallida* Tratt. 136, 137
lutzeyeri G. D. Duncan 23, 27, 38, 41, **40**, 50, 52, 55, 58, 59, 104, 267, **275**, **276**, **277**, 283
macgregoriorum W. F. Barker 33, 41, 49, 52, 55, 60, 66, 105, 354, **359**, **360**, **361**
macrophylla Lem. 146
magentea G. D. Duncan 52, 104, 313, **343**
margaretiae W. F. Barker 17, 25, 33, 41, 42, 52, **65**, 103, **208**, **209**
marginata W. F. Barker 24, 72, **125**
   subsp. **marginata** 52, **66**, 98, **126**, **127**
   subsp. **neglecta** Schltr. ex G. D. Duncan 33, 53, 100, **126**, **128**, 217
marlothii W. F. Barker ex G. D. Duncan 19, 24, 31, 53, 70, 101, 102, 126, **215**, **216**
martiniae W. F. Barker 33, 53, 57, 73, 92, 102, 299, **308**, **309**
martleyi G. D. Duncan 33, 53, 104, 269, **270**, **271**

massonii Baker 121, 123, 124
mathewsii W. F. Barker 18, 24, 25, 26, 33, 34, 38, 41, 42, 53, **65**, **74**, 87, 102, **210**, **211**, **212**, 255, 421
maughanii (W. F. Barker) J. C. Manning & Goldblatt 434
maximiliani Schltr. ex W. F. Barker 47, 53, 57, 102, **254**
mediana Jacq. 76, 97, 104, **192**, 207, 270, 272
    subsp. **mediana** 33, 53, **193**, **194**
    subsp. **rogersii** (Baker) G. D. Duncan 33, 42, 53, 192, **193**, 194, **195**, 342
        var. *mediana* 193
        var. *rogersii* (Baker) W. F. Barker 192, 195
membranacea (W. F. Barker) G. D. Duncan 25, 26, 53, 71, **72**, 83, 86, 87, 106, 396, **397**, 400, 401
minima W. F. Barker 24, 31, 33, 53, 59, 72, 101, **225**
moniliformis W. F. Barker 23, 25, 28, 31, 34, **40**, 41, 53, 57, 61, 62, **67**, 68, 104, **315**, **316**, 321
montana Schltr. ex W. F. Barker 22, 23, 38, 45, 50, 53, 55, 57, 58, 59, 68, 70, **77**, 78, 102, 172, 174, 175, 278, **281**, **282**, 283, 391
muirii W. F. Barker 403, 405
multifolia W. F. Barker 19, 24, 45, 53, 57, 61, 64, 66, 76, 105, 178, 319, 367, **368**, **369**
mutabilis Lodd. ex Sweet 14, 23, 25, 28, 53, 60, 70, 72, 74, 86, 87, 90, 95, 106, 118, 127, 128, 178, 208, 248, **379**, **380**, **381**
namaquensis Schltr. ex W. F. Barker 24, 25, 28, 50, 53, 57, 107, 212, 302, **422**, **423**
namibiensis W. F. Barker 20, 24, 25, 47, 53, 59, 68, **78**, 79, 92, 105, **364**, **365**, 367, 372
nardousbergensis G. D. Duncan 25, 53, 101, **204**, **205**, 206
neglecta Schltr. 126
neilii W. F. Barker ex G. D. Duncan 33, 38, 41, 53, 57, **74**, 101, 233, 237, **238**, **239**
nervosa Ker Gawl. 7, 24, 33, 49, 53, 55, 57, 60, 61, **67**, 68, 79, 81, 105, 231, 310, **346**, **347**, 349
nordenstamii W. F. Barker 53, 59, 64, 68, 74, 75, **77**, 85, 91, 92, 105, 179, 307, **377**
nutans G. D. Duncan 42, 49, 53, 102, **286**, **287**, 295, 327, 328, 378
obscura Schltr. ex G. D. Duncan 19, 53, 59, **67**, 68, 83, 101, 248, **244**
odoratissima Baker 196, 198
orchioides (L.) Aiton 2, 12, 14, 17, 24, 25, 26, 42, 55, 72, 86, 87, 90, 97, 99, **113**, 129, 192, 220, 241, 381

    subsp. **glaucina** (Jacq.) G. D. Duncan 2, **3**, 14, 33, 53, 114, **118**, **119**, **120**
    subsp. **orchioides** 2, **3**, 53, **65**, **69**, 83, 114, **115**, **117**, 185, 219, 221, 317, 403, 444
    subsp. **parviflora** (W. F. Barker) G. D. Duncan 33, 53, **120**, **121**
        var. *glaucina* (Jacq.) W. F. Barker 115, 118
orthopetala Jacq. 19, **20**, 21, 24, 25, 26, 33, 42, 53, 55, 57, 59, 61, 70, 71, **72**, **74**, 88, 104, 326, 329, 330, **331**
ovatifolia L. Guthrie 406, 407
pallida Aiton 14, 20, **21**, 23, 24, 25, 26, 41, 42, **43**, 45, **46**, 53, 59, 61, **78**, **82**, 83, 87, 90, 96, 104, 192, **195**, **197**, **198**, **199**, **200**, 201, 212, 213, 325, 330, 348, 399
pallida Lindl. 118
    var. *coerulescens* Lindl. 115, 118
patens Hort. ex Andrews 444
patentissima G. D. Duncan 41, 53, 83, 92, 99, 142, **150**, **151**, 392
patula Jacq. 20, 21, 24, 25, 27, **47**, 49, 53, 59, 60, 70, 71, 73, **77**, 78, 85, 92, 105, 365, **370**, **371**
paucifolia (W. F. Barker) J. C. Manning & Goldblatt 10, 24, 27, 33, 42, 53, 59, 68, 71, 74, 75, 76, **78**, 81, 84, 85, 107, 110, **437**, **438**, 441
pearsonii (R. Glover) W. F. Barker 14, 17, 21, 24, 41, 47, 50, 53, 57, 64, 68, 92, 105, 363, 365, **366**, 367, 442
peersii Marloth ex W. F. Barker 23, 24, 26, 27, 33, 45, 53, 55, 57, 64, 81, 106, 263, 264, 283, **388**, **389**, **390**
pendula Aiton 12, 141, 156, 157
perryae G. D. Duncan 53, 102, **255**, **256**, 311
petiolata Baker 188
physocaulos W. F. Barker 20, 33, 53, **65**, 70, 103, **290**, **291**, 293
piersonii 14
polyphylla Baker 24, 33, 42, 45, 53, 60, 68, 70, **77**, 78, 92, 104, **317**, **318**, **319**, 369
polypodantha Schltr. ex W. F. Barker 24, 57, 64, 68, 71, 105, 124, 352, **373**
    subsp. **eburnea** G. D. Duncan 54, 295, **374**, **376**
    subsp. **polypodantha** 54, 295, **374**, **375**, 376
pulchella Kunth 444, 445
punctata Jacq. 2, 3, 13, 14, 17, **18**, 19, 31, 37, 38, 41, **42**, 49, 50, 54, 58, 59, 60, 61, 75, 79, 81, 83, 84, 85, 91, 92, 99, 157, 158, 159, **160**, **161**, **162**, **163**
purpurea Jacq. 196, 198, 199

purpureo-caerulea Jacq. 23, 26, 33, 54, 55, 61, 101, 200, **201**, **202**, **203**, 206
pusilla Jacq. **20**, 24, 27, 54, 57, 59, 60, 61, **65**, 68, **69**, 70, 71, 74, 76, 79, 81, 84, **85**, 91, 97, 100, 110, 187, **188**, **189**, **190**, 349, 444
pusilla Eckl. ex Kunth 444
pustulata P. J. Bergius ex Kunth 445
pustulata F. Dietr. 196
pustulata Jacq. 195, 198, 199, 200, 251, 381, 445
　　var. *densiflora* Tratt. 196, 198
pygmaea (Jacq.) G. D. Duncan 4, 24, 25, 27, 54, 68, 70, 71, 76, 81, 84, 85, 96, 107, 110, 191, **426**, **427**, **428**, 431, 437
pyramidalis Dehnh. 196, 198
quadricolor Jacq. 12, 14, 20, 21, 23, 25, 26, 33, 42, **44**, 54, 73, 81, 83, 84, 92, 138, 142, **143**, **145**, 438
　　var. *lutea* Sims 146
racemosa Ker Gawl. 196, 198
reclinata F. Dietr. 196, 198
reflexa Thunb. 6, 7, 8, 12, 24, 27, 33, 42, 54, 71, 74, 75, 76, 79, 83, 84, 85, 86, 93, 96, 99, **108**, **109**, 112, 144, 444
rhodantha Baker 353, 354
roodeae E. Phillips 283
rosea Andrews 26, 35, 42, 54, 55, **67**, 68, 76, 103, 181, **264**, **265**, **266**, 269, 274, 278
rubida Jacq. 13, 160, 162
　　var. *punctata* (Jacq.) Baker 160, 162
　　var. *tigrina* (Jacq.) Baker 160, 162
salteri W. F. Barker 17, 20, 23, 24, 25, 26, 33, **34**, 38, 42, 50, 54, 55, **66**, **67**, 68, 73, 103, 267, **272**, **273**, **274**
sanguinolenta Willd. ex Kunth 264, 267
sargeantii W. F. Barker 23, 33, **34**, 38, **39**, 41, 50, 54, **55**, 57, 58, 59, **65**, 68, 78, **172**, **173**, **174**, 278, 283
schelpei W. F. Barker 19, 31, 33, 41, 54, **59**, 100, **182**, **183**, 233
schlechteri Baker 9, 54, 99, 167, **168**, **169**
sessiliflora Andrews 17, 27, 33, 35, 38, 50, 54, 55, 60, 72, 75, 106, 393, **403**, **404**, **405**
splendida Diels 14, 25, 26, 42, **45**, 54, 59, 100, 226, **283**, **284**, **285**
stayneri W. F. Barker 33, 41, 54, 59, 61, 64, **65**, 68, 101, **309**, **310**, 317, 349
suaveolens (W. F. Barker) G. D. Duncan 24, 45, 54, 106, 238, 369, 398, **399**, **400**
subspicata Fourc. 239, 241, 243

succulenta Masson ex Baker 370, 372
summerfieldii G. D. Duncan 19, 33, 41, 54, 101, **231**, **232**
tenella Schneev. ex Steud. 445
thomasiae W. F. Barker ex G. D. Duncan 17, 23, 25, 26, 33, 41, 42, 54, 103, 192, **206**, **207**, 210
thunbergii G. D. Duncan & T. J. Edwards 25, 26, 41, 54, 83, 92, 99, 133, **134**, 141, 142, 153
tigrina Jacq. 160
trichophylla Baker 24, 25, 41, 49, **50**, 54, 60, 61, **63**, 64, 68, 70, **77**, 96, 98, **121**, **122**, **123**, **124**, 352
tricolor J. Jacq. 5, 6, 7, 96, 139, 141
tricolor Thunb. 137, 139, 142, 144
　　var. *aurea* (Lindl.) Baker 146
　　var. *aurea* (Lindl.) Hook.f. 146, 149
　　var. *luteola* (Jacq.) Baker 136, 137, 149
　　var. *luteola* (Jacq.) Ker Gawl. 137, 146, 149
　　var. *quadricolor* (Jacq.) Baker 143, 144
undulata Masson ex Baker 31, 47, 54, 59, 71, 96, 101, 190, **226**, **227**, 231, 413
unicolor Eckl. ex Kunth 445
unicolor Jacq. 198, 199, 200, 280, 445
　　var. *fragrans* (Jacq.) Baker 330
　　var. *purpurea* (Jacq.) Baker 196
unifolia Jacq. 37, 45, 54, 61, 87, 99, 127, 150, **165**, **166**, **167**, 168, 170, 219, 325, 425
　　var. *pappei* Baker 264, 267
　　var. *rogersii* Baker, 192, 195
　　var. *schlechteri* (Baker) W. F. Barker 167, 168
　　var. *wrightii* (Baker) Baker 167, 324
ustulata Banks ex Schult. & Schult. f. 331, 332
valeriae G. D. Duncan 24, 38, 54, 61, 107, 187, 413, **414**, **415**, **416**
vanzyliae (W. F. Barker) G. D. Duncan & T. J. Edwards 17, 25, 26, 54, **59**, 62, **78**, 79, 83, 84, 88, 92, 93, 99, **131**, **132**, 135, 142, 155
variegata W. F. Barker 54, 102, **213**, **214**, 392
ventricosa Schltr. ex W. F. Barker 20, 28, 33, 54, 64, **78**, **79**, 100, **175**, **176**, **177**, 382
versicolor Baker 198
　　var. *fragrans* (Jacq.) Baker, 195
　　var. *purpurea* (Jacq.) Baker 196
　　var. *unicolor* (Jacq.) Baker, 196
verticillata W. F. Barker 24, 49, 54, 57, 61, 62, **66**, **67**, 68, **69**, 73, 106, 253, **385**, **386**, **387**
violacea Jacq. 19, 23, 24, 45, 54, 70, 90, 103, 178, **295**, **296**, **297**, **298**, 299, 301, 302, 308, 369
　　var. *glauca* W. F. Barker 298, 299, 301

viridiflora W. F. Barker 14, 17, 24, 25, **33**, 34, 41, 42, 45, 54, 57, 61, 73, 75, **78**, **82**, 83, 84, 93, 96, 99, 109, **110**, **111**, **112**, 133, 152
whitehillensis W. F. Barker 19, 24, 54, 70, 103, **302**, **303**, 323
wrightii Baker 45, 54, 102, 167, **324**, **325**
xerophila Schltr. ex G. D. Duncan 21, 31, 54, 57, 70, 79, 85, 103, 286, 287, **294**, 376
× boundii 14
× Cami 12, 13
× comesii Sprenger 12, 144
× nelsonii 12, **13**, 14, 137
× regeliana 12, 13
youngii Baker 33, 54, 73, 102, 209, 241, 277, 279, **280**
zebrina W. F. Barker 19, 24, 37, 42, 54, 57, 61, 64, 70, 71, 75, **76**, 85, 92, 102, **304**, **305**, **306**, 378
  forma *densiflora* W. F. Barker 304, 306
  forma *zebrina* W. F. Barker 306
zeyheri Baker 24, 25, 42, **43**, 54, 104, 335, **336**, **337**

Lapeirousia fabricii 208
Ledebouria revoluta 4
Lithops karasmontana 368
Lytta nitidula 80
Massonia corymbosa (L.) Ker Gawl. 439, 440
  depressa 84, 90, 187, 191, 257
  ensifolia (Thunb.) Ker Gawl. 431
  odorata Hook.f. 432, 434
  undulata Thunb. 97, 188
  uniflora Sol. ex Baker 434
  violacea Andrews 426
Mauhlia ensifolia Thunb. 431, 432
Micranthus alopecuroides 208
Mimetes cucullatus 174, 278
Monoestes Salisb. 7, 96
Moraea macronyx 337
  marlothii 233
  pallida 252
  ramosissima 393
  serpentina 413
  tricuspidata 403
  tripetala 224
Muscari orchioides (L.) Mill. 113
Nemesia strumosa 204
Nylandtia spinosa 337
Orchiastrum 7
  aitonii Lem. 117

  glaucinum (Jacq.) Lem. 118
  mutabile (Lodd. ex Sweet) Lem. 379
  pallidum (Aiton) Lem. 195
  pulchellum (Kunth) Lem. 444
  virenti-flavum Lem. 117
Orchiops Salisb. 7, 97
Order Asparagales 95
Ornithogalum thyrsoides 195
  xanthochlorum 328
Oxalis 93
  obtusa 257
Pachycnema crassipes 80
Pauridia longituba 113
Pelargonium capitatum 224
  fulgidum 113
  triste 224
Periboea Kunth 7, 96
  corymbosa (L.) Kunth 96, 439
  gawleri Kunth 439
  oliveri U. Müll.-Doblies & D. Müll.-Doblies 437, 438
  paucifolia (W. F. Barker) U. Müll.-Doblies & D. Müll.-Doblies 437
Phormium aloides L.f. 6, 7, 131, 139, 141, 142, 144, 148
  bulbiferum Cirillo 155, 157
  hirtum Thunb. 6, 320
  orchioides Thunb. 6, 109, 114, 339
    var α 109
  tenax 6
Pinus pinea 120, 185, 194, 403
Platyestes Salisb. 7, 96
Polyanthes pygmaea Jacq. 4, 96, 426, 427
Polygala myrtifolia 260
Polyxena Kunth 4, 7, 8, 11, 89, 91, 96, 188
  calcicola U. Müll.-Doblies & D. Müll.-Doblies 435
  corymbosa (L.) Jessop 439, 440
  ensifolia (Thunb.) Schönland 431
    var. *maughanii* 433
  longituba A. M. van der Merwe 429, 430
  maughanii W. F. Barker 10, 433, 434
  odorata (Hook.f.) Baker 430, 432, 434
  odorata (Hook.f.) W. A. Nicholson 434
  paucifolia (W. F. Barker) A. M. van der Merwe & J. C. Manning 437, 438
  pusilla (Jacq.) Schltr. 188
  pygmaea (Jacq.) Kunth 96, 426, 427
  uniflora (Sol. ex Baker) Benth. & Hook. ex Dur. & Schinz 432, 434

Protea cynaroides 174, 393
Pterygodium catholicum 118
Romulea komsbergensis 430
Rhus burchellii 252
Scilla brachyphylla Roem. & Schultes 439
    brevifolia Ker Gawl. 439, 441
    corymbosa (L.) Ker Gawl. 439
    pearsonii R. Glover 14, 366
Scillopsis 7
    anguinea (Sweet) Lem. 326, 327
    angustifolia (Jacq.) Lem. 338
    bifolia (Ker Gawl.) Lem. 264
    contaminata (Aiton) Lem. 338
    fragrans (Jacq.) Lem. 195
    hyacinthoides (Jacq.) Lem. 338
    isopetala (Jacq.) Lem. 180, 181
    liliiflora (Jacq.) Lem. 328, 329
    lucida (Ker Gawl.) Lem. 196
    mediana (Jacq.) Lem. 192
    nervosa (Ker Gawl.) Lem. 346
    orthopetala (Jacq.) Lem. 331, 332
    patula (Jacq.) Lem. 370, 372
    purpurea (Jacq.) Lem. 196
    purpureo-caerulea (Jacq.) Lem. 201
    pustulata (Jacq.) Lem. 195
    racemosa (Ker Gawl.) Lem. 196
    rosea (Andrews) Lem. 264
    unicolor (Jacq.) Lem. 196
    unifolia (Jacq.) Lem. 165, 167
    violacea (Jacq.) Lem. 298
Sparaxis bulbifera 224
    grandiflora 168
Steirodiscus tagetes 26
Strumaria chaplinii 438
    tenella 438
    truncata 413
Subfamily Scilloideae (Hyacinthoideae s.l.) 95
Tribe Hyacintheae (Massonieae s.l.) 95
Vanessa cardui 81
Watsonia borbonica 185
    hysterantha 146
    stenosiphon 40
    tabulans 393
Wurmbea stricta 342